Polar
Microbiology
Life in a Deep Freeze

Polar
Microbiology
Life in a Deep Freeze

Edited by
Robert V. Miller and Lyle G. Whyte

ASM
PRESS

Washington, DC

Library of Congress Cataloging-in-Publication Data

Polar microbiology: life in a deep freeze / edited by Robert V. Miller and Lyle G. Whyte.
   p. cm.
  Includes bibliographical references and index.
  ISBN-13: 978-1-55581-604-9 (hardcover: alk. paper)
  ISBN-10: 1-55581-604-5 (hardcover: alk. paper) 1. Extreme environments—Microbiology. 2. Microbial ecology—Polar regions.
  3. Microbiology—Research—Polar regions.   I. Miller,
  Robert V. (Robert Verne), 1945– II. Whyte, Lyle G.

QR100.9.P645 2011
577.5'8—dc23

                                              2011032860

eISBN: 978-1-55581-718-3

Address editorial correspondence to ASM Press, 1752 N St., N.W., Washington, DC 20036-2904, USA

Send orders to ASM Press, P.O. Box 605, Herndon, VA 20172, USA
Phone: 800-546-2416; 703-661-1593. Fax: 703-661-1501.
E-mail: books@asmusa.org
Online: http://estore.asm.org

CONTENTS

CONTRIBUTORS

Héctor L. Ayala–del–Río
Department of Biology, University of Puerto Rico at Humacao,
Humacao, PR 00791

Corien Bakermans
Division of Mathematics and Natural Sciences, Altoona College,
Pennsylvania State University, 3000 Ivyside Park,
Altoona, PA 16601

Peter W. Bergholz
Department of Food Science, Cornell University,
Stocking Hall, Room 412, Ithaca, NY 14853

Erik Broemsen
Department of Biological Sciences, Louisiana State University,
202 Life Sciences Building, Baton Rouge, LA 70803

S. Craig Cary
Department of Biological Sciences, University of Waikato, Hamilton 3240,
New Zealand, and College of Earth, Ocean and Environment,
University of Delaware, Lewes, DE 19958

Brent Christner
Department of Biological Sciences, Louisiana State University,
202 Life Sciences Building, Baton Rouge, LA 70803

Don A. Cowan
Institute for Microbial Biotechnology and Metagenomics,
Department of Biotechnology, University of the Western Cape,
Bellville 7535, South Africa

Markus Dieser
Department of Biological Sciences, Louisiana State University,
202 Life Sciences Building, Baton Rouge, LA 70803

Shawn Doyle
Department of Biological Sciences, Louisiana State University,
202 Life Sciences Building, Baton Rouge, LA 70803

Samantha Easton
Institute for Microbial Biotechnology and Metagenomics,
Department of Biotechnology, University of the Western Cape,
Bellville 7535, South Africa

Georges Feller
Laboratory of Biochemistry, Centre for Protein Engineering,
University of Liège, B-4000 Liège, Belgium

Charles W. Greer
National Research Council of Canada, Biotechnology Research Institute,
Montreal, QC H4P 2R2, Canada

Nina Gunde-Cimerman
Biology Department, Biotechnical Faculty, University of Ljubljana,
1000 Ljubljana, Slovenia

Anne Jungblut
Department of Botany, The Natural History Museum, Cromwell Road,
London SW7 5BD, United Kingdom

Karen Junge
Applied Physics Laboratory, University of Washington,
1013 NE 40th Street, Seattle, WA 98105

Bronwyn M. Kirby
Institute for Microbial Biotechnology and Metagenomics,
Department of Biotechnology, University of the Western Cape,
Bellville 7535, South Africa

Susanne Liebner
Department of Arctic and Marine Biology, University of Tromsø,
9037 Tromsø, Norway

Connie Lovejoy
Département de Biologie, Institut de biologie intégrative et des systèmes
(IBIS)/Québec-Océan, Université Laval, Québec City,
QC G1V 0A6, Canada

Rosa Margesin
Institute of Microbiology, University of Innsbruck, Technikerstrasse 25,
A-6020 Innsbruck, Austria

Ian R. McDonald
Department of Biological Sciences, University of Waikato,
Hamilton 3240, New Zealand

Christopher P. McKay
NASA Ames Research Center, Moffett Field, CA 94035

Robert V. Miller
Department of Microbiology and Molecular Genetics,
Oklahoma State University, 307 Life Sciences East, Stillwater, OK 74078

Nadia C. S. Mykytczuk
Department of Natural Resource Sciences, McGill University,
MacDonald Campus, 21,111 Lakeshore Road, Ste. Anne de Bellevue,
Quebec H9X 3V9, Canada

Thomas D. Niederberger
College of Earth, Ocean and Environment, University of Delaware,
Lewes, DE 19958

P. Buford Price
Physics Department, University of California, Berkeley, CA 94720

Debora F. Rodrigues
Department of Civil and Environmental Engineering, University of Houston,
N107 Engineering Building 1, Houston, TX 77204-4003

Mark Skidmore
Department of Earth Sciences, Montana State University, Bozeman, MT 59717

Silva Sonjak
Biology Department, Biotechnical Faculty, University of Ljubljana,
1000 Ljubljana, Slovenia

James M. Tiedje
Center for Microbial Ecology, 540 Plant and Soil Sciences Building,
Michigan State University, East Lansing, MI 48824-1325

I. Marla Tuffin
Institute for Microbial Biotechnology and Metagenomics,
Department of Biotechnology, University of the Western Cape,
Bellville 7535, South Africa

Matthew Urschel
Department of Microbiology, Montana State University, Bozeman, MT 59717

Tatiana A. Vishnivetskaya
Center for Environmental Biotechnology, The University of Tennessee,
676 Dabney-Buehler Hall, 1416 Circle Drive, Knoxville, TN 37996-1605,
and Biosciences Division, Oak Ridge National Laboratory,
Oak Ridge, TN 37831-6038

Helen A. Vrionis
Department of Natural Resource Sciences, McGill University,
Macdonald Campus, 21,111 Lakeshore Road, Ste.-Anne-de-Bellevue,
QC H9X 3V9, Canada

Dirk Wagner
Alfred Wegener Institute for Polar and Marine Research, Telegrafenberg A43,
14473 Potsdam, Germany

Karen Warner
Department of Microbiology and Molecular Genetics,
Oklahoma State University, 307 Life Sciences East, Stillwater, OK 74078

Lyle G. Whyte
Department of Natural Resource Sciences, McGill University,
Macdonald Campus, 21,111 Lakeshore Road, Ste.-Anne-de-Bellevue,
QC H9X 3V9, Canada

Roland Wilhelm
Department of Natural Resource Sciences, McGill University,
Ste.-Anne-de-Bellevue, QC H9X 3V9, Canada

Etienne Yergeau
National Research Council of Canada, Biotechnology Research Institute,
Montreal, QC H4P 2R2, Canada

Polona Zalar
Biology Department, Biotechnical Faculty, University of Ljubljana,
1000 Ljubljana, Slovenia

PREFACE

An Exciting Era in Polar Microbiology

Polar microbiology is a fast-growing field that can tell us much about the fundamentals of life on earth and the microbial contributions and consequences to such important global environmental issues as warming earth, ozone depletion, and elemental cycling. Polar microbiology has also recently received considerable attention because polar microbial communities are considered important analogues for astrobiology investigations looking for life on other very cold solar system bodies.

Over the last 20 to 30 years, microbiologists have had increasing access to the previously difficult to reach polar regions, which has resulted in tremendous progress in understanding the microbial ecology of these regions. This research has opened a door to a greater understanding of the physiology of the hardy microbial inhabitants of these extremely cold environments. During the International Polar Year of 2008, many international microbial investigations focusing on Arctic and Antarctic regions were conducted. They often lasted for a 3- to 4-year period. This book is a celebration of research undertaken in this exciting field of microbiology during the International Polar Year and over the 10 to 20 years that preceded it.

This is the ideal time to summarize the research carried out over the last decade that has increased our knowledge of the microbiology of the Arctic and Antarctic regions in our world. Much of the research initiated during the International Polar Year has been completed and the data compiled and analyzed. Now is the time to reflect on the major findings and conclusions that can be drawn from the International Polar Year's activities. This book was inspired in part by the presentations at the 2008 International Polar and Alpine Microbiology Conference in Banff, Canada, where many of the contributors to *Polar Microbiology* first discussed the project.

Polar Microbiology has been created as a book that addresses polar microbiology in a general fashion and is designed to inform a broad audience of microbiologists on the microbial ecology and physiology of this fascinating world of ice and snow. *Polar Microbiology* is targeted toward a general microbiology audience rather than to just the polar microbiology community because it is our hope that this book will become a useful reference and general polar microbiology textbook for scientists and students in all areas of biology and geomicrobiology.

To this end, we asked the authors to first give a general overview of their particular area of polar expertise, outlining the major advances and general themes and principles of the subject. The authors were also asked to include the most relevant highlights of recent findings and to provide future questions that should be explored in their field. They were encouraged to compare the Arctic and Antarctic environments wherever appropriate and to comment on the effects of climate change on these fragile ecosystems. The High Arctic in particular is experiencing the greatest increases in temperatures on the planet, with subsequent detrimental effects, including habitat destruction (Vincent et al., 2009).

The world's leading scientists in Arctic and Antarctic microbiology have written *Polar Microbiology*. The book is organized into four major thematic sections. In Part I, "Microbial Diversity of Polar Environments," we start with a survey of what is currently known (surprisingly, quite a bit!) and what is not known (paradoxically, still quite a bit!) about the microbial inhabitants of polar environments. Here, the diversity of the four major microbial groups—bacteria, archaea, viruses, and eukaryotes—found in polar environments is presented separately in four chapters.

Part II addresses the adaptations and physiology of cold-adapted microorganisms. General aspects of the theme are discussed in Chapter 5, while exciting new discoveries revealed by genomic, proteomic, and metagenomic analyses are described in Chapters 6 and 7. This section concludes with a chapter on how this information is being used to increasingly develop and utilize cold-adapted microorganisms for biotechnological applications.

In Part III, "Ecology and Biogeochemical Cycling of Polar Microbiology Communities," the significant ecological role and the importance of polar microbial communities in biogeochemical cycling is addressed through specific contributions focused on the major polar environments. These include (i) polar terrestrial systems, especially permafrost; (ii) polar marine systems; and (iii) cryosphere environments, including glaciers, ice shelves, and sea ice ecosystems.

Part IV of the book presents the challenges microorganisms face living in polar and subpolar environments and explores the low-temperature limits of microbial life. Growth, metabolism, and activity are addressed in Chapter 12. How climate change and ozone depletion are affecting polar communities is outlined in Chapter 13. Chapter 14 discusses how polar microbiology has become increasingly important to astrobiology and the search for microbial life on other worlds. Indeed, the primary targets for astrobiology investigations of other solar system bodies are Mars, in the short term, as well as Europa and Enceladus, in the mid- to longer term. Extremely cold temperatures characterize these targets, and as such, the best terrestrial analogues may be Earth's polar regions.

We envision *Polar Microbiology* as a summation for a general audience of the major aspects of our current knowledge of the amazing diversity, ecology, adaptations, and utility of microorganisms living and thriving in the coldest regions of our planet. We hope that you, the reader, both enjoy it and find it to be a useful resource. We offer our sincere thanks to the amazing authors who spent considerable time and effort preparing and revising their contributions to this endeavor; we are truly grateful for their insights expressed in such a readable way.

REFERENCE

Vincent, W. F., L. G. Whyte, C. Lovejoy, C. W. Greer, I. Laurion, C. A. Suttle, J. Corbeil, and D. R. Mueller. 2009. Arctic microbial ecosystems and impacts of extreme warming during the International Polar Year. *Polar Sci.* **3:**171–180.

ROBERT V. MILLER
LYLE G. WHYTE

MICROBIAL DIVERSITY IN POLAR ENVIRONMENTS

I

BACTERIAL DIVERSITY IN POLAR HABITATS

Bronwyn M. Kirby, Samantha Easton,
I. Marla Tuffin, and Don A. Cowan

I

INTRODUCTION

We live on a "cold" planet. Over 80% of the earth's ecosystems are considered psychrophilic, including many marine habitats, with an average temperature of 15°C or less. Distinct psychrophilic ecosystems exist, experiencing a unique set of climatic and environmental conditions, many of which seem inhospitable to life. Apart from low temperatures, many psychrophilic habitats withstand high levels of solar radiation, low water availability, strong winds, and high salinity. Antarctic terrestrial environments range from ice-free nunataks, to nutrient-rich coastal regions colonized by birds, to the arid and oligotrophic Dry Valleys deserts. Conversely, Arctic terrestrial ecosystems include Arctic tundra, permafrost, glaciers, and high-altitude alpine sites. Psychrophilic habitats may appear devoid of visible life; however, early classic microbiology studies (relying on standard culturing techniques and microscopy) demonstrated that these unique environmental niches were colonized by a wide range of bacterial and eukaryotic species. The development

of culture-independent, molecular methods has revolutionized the field and our understanding of molecular ecology. Through the use of these techniques, it is now apparent that the earlier culture-based studies were not a representative reflection of the dominant microorganisms in many psychrophilic habitats. Recent phylogenetic studies conducted on Antarctic and Arctic soils, sea ice, and cyanobacterial mats suggest that microbial diversity in psychrophilic habitats may be greater than ever appreciated, and that the majority of organisms identified in these studies remain uncultured (Table 1).

ANTARCTICA

Antarctica is the fifth-largest continent and has a total landmass of approximately 14 million km². Over 98% of the continent is covered by a thick ice sheet with an average thickness of 1.6 km. Antarctica is the coldest, driest continent and experiences severe environmental and climatic conditions—extremely low temperatures, low atmospheric humidity, low liquid water availability, and periods of high incidence of solar radiation coupled to long periods of complete darkness (Cowan and Ah Tow, 2004). The lack of liquid water severely restricts both macro- and microscopic life, and higher eukaryotes are confined to the more temperate northerly latitudes of the Antarctic Peninsula.

Bronwyn M. Kirby, Samantha Easton, I. Marla Tuffin, and Don A. Cowan, Institute for Microbial Biotechnology and Metagenomics, Department of Biotechnology, University of the Western Cape, Bellville 7535, South Africa.

Polar Microbiology: Life in a Deep Freeze
Edited by Robert V. Miller and Lyle G. Whyte © 2012 ASM Press, Washington, DC

TABLE 1 Summary of the findings from key bacterial diversity studies for Antarctic habitats as determined by culture-based and metagenomic techniques

Habitat, location, and sample date	Techniques used	Summary of findings	Reference
Sea ice			
First-year pack ice in Weddell Sea (1986) and Lazarev Sea (1994)	16S rRNA gene sequencing of 87 pure cultures isolated from pack ice and environmental samples. FISH for community composition analysis.	High incidence of clones and isolates with closely overlapping 16S rRNA gene sequences. The diversity for Antarctic sea ice was lower than for Arctic samples, with only 20 phylotypes. Most sequences were >98% identical to species already isolated from Antarctic/Arctic sea ice.	Brinkmeyer et al., 2003
Pack ice between Casey and Davis Bases (1996) using a SIPRE corer. Fast ice from McMurdo Sound (1997) and a single sample from the Canadian Arctic (Baffin Bay).	Phylogenetic analysis of 16S rRNA gene clone libraries (PCR with universal and archaeal primers). Clones were dereplicated by RFLP and 50–60 clones were analyzed per ice sample.	Algal biomass detected in McMurdo, Arctic, and some Southern Ocean pack ice samples (four of the six samples analyzed), and biodiversity was linked to increasing algal density. Algal-rich samples (including the single Arctic sample) had 11–18 phylotypes, while the two Antarctic samples with low algal levels had 6–7 phylotypes. Increasing algal density also appeared to increase the proportion of CFB clones from 0–7% to 20–27%.	Brown and Bowman, 2001
Vestfold Hills coast and Ellis Fjord, Eastern Antarctica	Culture-based analysis of sea ice and under-ice bacteria compared by 16S rRNA gene analysis	Bacterial diversity significantly higher in sea ice samples containing platelet and bottom-ice diatom assemblages. Of the 26 genera identified in the sea ice samples, 45% were psychrophilic. Fourteen genera were present only in sea ice samples, not in under-ice water samples.	Bowman et al., 1997
Palmer Station (1986) and McMurdo Station (1992)	Culture-based study. Characterization against the group database of 34 gas-vacuolated strains. 16S rRNA gene analysis on each strain.	Found that gas-vacuolate bacteria are phylogenetically diverse. Isolates belonged to the α-, γ-, and β-*Proteobacteria* and the *Flavobacterium-Cytophaga* group. No isolates were identical to any other known bacterial species.	Gosink and Staley, 1995
Fast ice from Armitage Point, McMurdo Sound (1987)	Bacterial counts of total and viable bacteria, and measurement of chlorophyll *a* concentrations	Bacterial numbers were greatest where the chlorophyll *a* concentration was highest, thereby indicating a correlation between bacterial numbers and the concentration of ice algae. Very low recovery of heterotrophic bacteria. High percentage of pigmented bacteria. Gas-vacuolated bacteria were distributed in both the surface and deeper layers.	Staley and Gosink, 1999

Location	Methods	Findings	Reference
Weddell Sea, Antarctica (1986)	Total bacterial counts and viable counts; culture-based analysis of temperature optima	Samples of older sea ice contained more bacterial diversity than fresh ice, and were thought to contribute more to the productivity of the Weddell Sea. Psychrophilic bacteria dominated in consolidated sea ice while facultative psychrophiles prevailed in young sea ice and the water column.	Helmke and Weyland, 1995
Lakes			
Lake Vostok	Culture-based study. Isolation on a variety of media. Phylogenetic analysis of the 16S rRNA gene.	Culture-based approaches identified strains closely related to *Methylobacterium*, *Paenibacillus*, and *Sphingomonas* species. Five phylogenetic lines were identified by 16S rRNA gene analysis: α- and β-*Proteobacteria*, CFB group, and low- and high-G+C Gram-positive groups.	Christner et al., 2001
Lake Bonney, McMurdo Dry Valleys	Metagenomics-based study. Phylogenetic analysis of the 16S rRNA gene.	Clone library was dominated by cyanobacteria, *Proteobacteria*, and members of the *Planctomycetales*. Additionally, the library also included members of the *Acidobacterium/Holophaga* division, the green nonsulfur division, and actinobacteria. Probes for the six most abundant clades identified dominant members of this community elsewhere in the region.	Gordon et al., 2000
Lake Fryxell, Taylor Valley	PCR amplification of the *pufM* gene and DGGE analysis	The photosynthesis-specific gene *pufM* was analyzed to detect the activity of purple phototrophic bacteria present at all depths in the water column. No *pufM* gene amplicons were identical to known purple sulfur bacteria, indicating the presence of novel purple bacteria in Lake Fryxell.	Karr et al., 2003
Lake Vida, Victoria Valley, McMurdo Dry Valleys	Culture-based study. Enrichment prior to isolation. Phylogenetic analysis of the 16S rRNA gene.	All isolates were related to the genera *Psychrobacter* or *Marinobacter* (γ-*Proteobacteria*). Isolates had slow growth rates, indicating that they were likely to be indigenous members of the community.	Mondino et al., 2009

(Continued)

TABLE 1 Summary of the findings from key bacterial diversity studies for Antarctic habitats as determined by culture-based and metagenomic techniques—*Continued*

Habitat, location, and sample date	Techniques used	Summary of findings	Reference
Lake Vida, Victoria Valley, McMurdo Dry Valleys	Direct bacterial counts by DAPI straining. PCR with prokaryote-, archaeal-, and eukaryote-specific primers. DGGE analysis.	Phylotypes found related to several *Marinobacter* species and other γ-*Proteobacteria*. Actinobacteria dominated (42%) at 4.8-m depth and γ-*Proteobacteria* dominated (52%) at 15.9-m depth. No identical phenotypes were found at both depths, suggesting deeper-ice ecosystems have been derived from the water column and those above have a terrestrial origin.	Mosier et al., 2006
Organic Lake, Vestfold Hills	Phylogenetic analysis of the 16S rRNA gene amplification	Phylogenetic analysis grouped Antarctic clones with members of the *Flavobacterium-Bacteroides* phylum and the genus *Cytophaga*.	Dobson et al., 1993
Microbial mats			
Lake Fryxell, Taylor Valley	Culture-based and metagenomic study. 16S rDNA gene amplification with bacterial and archaeal primers.	16S rDNA clone library represented 70 higher taxonomic groups and 133 different species. Little overlap between the clone library and isolates.	Brambilla et al., 2001
23 lakes and ponds in Larsemann Hills, Bolingen Islands, Vestfold Hills, Rauer Islands, and McMurdo Dry Valleys	Culture-based analysis. Screening for antimicrobials and cytotoxins.	Isolated 51 cyanobacterial strains. Of the 126 extracts screened, 17 showed antimicrobial activity and 25 had a cytotoxic effect.	Biondi et al., 2007
Shallow-water environments in McMurdo Sound	Taxonomic and pigment analysis	Mat morphologies highly diverse. The thick mucilaginous mats were dominated by members of the genus *Oscillatoria*. Although purple sulfur bacteria were identified, green sulfur bacteria were notably missing from the analysis.	Vincent et al., 2010
Lake Fryxell, Taylor Valley	Light microscopy. Analysis of the 16S gene (DGGE) and the ITS region.	Eleven phylogenetic lineages identified, including 3 that were exclusively Antarctic and 2 novel lineages. Morphological analysis identified 8 morphotypes compared to the 15 phylotypes identified by molecular analysis.	Taton et al., 2003
Lake Reid, Heart Lake, Lake Rauer, and Ace Lake, eastern Antarctica	16S rRNA gene clone libraries and ITS analysis. Microscopy.	A higher number of OTUs (28) than morphospecies (17) were detected in this study. Detected a high degree of bacterial endemism; 36 out of 53 OTUs were uniquely Antarctic.	Taton et al., 2006

Location	Methods	Results	Reference
Lakes Hoare, Fryxell, Ace, Druzhby, Grace, Highway, Pendant, Organic, Watts, and Reid	Culture-based study included fatty acid analysis. Metagenomic study included analysis of the 16S rRNA gene and the ITS region.	Fatty acid analysis grouped the isolates into 41 clusters and 31 single-strain branches. 16S rRNA gene clones were related to members of the α-, β-, and γ-Proteobacteria; high- and low-G+C gram-positives; and the CFB phyla.	Van Trappen et al., 2002
Lake Reid, Heart Lake, Lake Rauer, and Ace Lake, eastern Antarctica	Light microscopy. Analysis of the 16S gene (DGGE) and the ITS region.	Seventeen morphospecies belonging to 12 genera were identified by microscopic analysis, of which 9 were related to Oscillatoriales. Twenty-eight 16S rRNA gene OTUs belonging to Oscillatoriales, Nostocales, and Chroococcales were also identified. The molecular approach suggested that endemic Antarctic species were more abundant than previously estimated by morphological methods.	Taton et al., 2006
Fresh, Orange, and Salt ponds, McMurdo Ice Shelf	Light microscopy. Phylogenetic analysis of 16S rRNA gene clone libraries.	Identified 22 phylotypes belonging to Anabaena, Lyngbya, Nodularia, Nostoc, Oscillatoria, and Phormidium species groups.	Jungblut et al., 2005
Rod Bay and Evans Cove, Ross Sea	Metagenomic study. Construction of 16S rRNA gene and 16S rDNA gene libraries and reverse transcriptase PCR.	Combining RNA and DNA analysis provides a more complete overview of the community structure, and 93.6% of the phylotypes were found in RNA libraries and 70.5% were retrieved from total community DNA libraries. Clones from both libraries were related to γ-Proteobacteria and Bacteroidetes.	Gentile et al., 2006
Palmer Station	Construction of a fosmid library and small-subunit rRNA gene analysis	Six fosmid clones were analyzed. All contained small-subunit rRNA genes. The small-subunit rRNA genes spanned four phyla including Bacteroidetes, Gemmatinonadetes, Proteobacteria, and high-G+C gram-positive bacteria.	Grzymski et al., 2006

(Continued)

TABLE 1 Summary of the findings from key bacterial diversity studies for Antarctic habitats as determined by culture-based and metagenomic techniques—*Continued*

Habitat, location, and sample date	Techniques used	Summary of findings	Reference
Endolithic			
Cryptoendolithic communities from Beacon sandstone, McMurdo Dry Valleys. Collected in 1985–1986 summer season.	Phylogenetic analysis of 16S rRNA gene clone libraries	Over 1,100 clones were analyzed from cyanobacterium- and lichen-dominated communities. Found that no phylotypes were shared between the two community types. The dominant phylotypes detected from the lichen-dominated community were related to fungi (*Texosporium sancti-jacobi*), green algae (*Trebouxia jamesii*), as well as chloroplast sequences. The cyanobacterium-dominated communities were dominated by Antarctic cyanobacteria, α-Proteobacteria (predominantly *Blastomonas ursincola* species), and *Deinococcus* species. No archaeal phylotypes were detected in either library.	de La Torre et al., 2003
Dry Valleys			
Analyzed three soil biotopes: Miers Valley mineral soil, soil from underneath a crabeater seal carcass (Bratina Island), and a high-altitude site (Penance Pass)	Phylogenetic analysis of 16S rRNA gene clone libraries	The species diversity at all sites was relatively low. Eight broad phylotypic groups were detected: *Acidobacteria, Actinobacteria, Bacteroidetes,* chloroflexi, *Cyanobacteria, α-Proteobacteria, β-Proteobacteria,* and *Verrucomicrobia. Acidobacteria, Actinobacteria,* and *Bacteroidetes* were detected at all three sites. Cyanobacteria (*Oscillatoriales* and *Nostocales*) were restricted to high-altitude sites, while acidobacterial signals were highest in the nutrient-rich Bratina site.	Smith et al., 2006
Soil from underneath a seal carcass, Cape Evans, McMurdo Dry Valleys (January 1995)	Phylogenetic analysis of 16S rRNA gene clone libraries. Detection of ammonia-oxidizing bacteria by PCR with primers targeting an ammonia monooxygenase (*amoA* gene).	Over 50% of clones were affiliated with CFB phylum. Phylotypes related to α-, β-, γ-*Proteobacteria; Thermus-Deinococcus;* and high-G+C Gram-positive group were also detected. Ammonia-oxidizing genera detected included *Nitrosomonas* and *Nitrospira.* The presence of ammonia-oxidizing bacteria was confirmed by the presence of the *amoA* gene.	Shravage et al., 2007

Continental and Peninsula

Four sites in Luther Vale, Northern Victoria Land (two low- and two high-productivity sites)	Phylogenetic analysis of 16S rRNA gene clone libraries. Culture-based study included geochemical analysis of soil and morphological analysis of soil metazoan and invertebrate communities.	Analyzed 323 clones. Phylotypes related to *Deinococus* and *Bacteroidetes* species were unique to low-productivity soils, whereas *Cyanobacteria*-, *Verrucomicrobia*-, and β-*Proteobacteria*-related clones were only found in high-productivity soils. Low-productivity libraries were dominated by *Acidobacteria* (68%) and *Gemmatimonas* (55%) species. High-productivity sites were dominated (84%) by γ-*Proteobacteria* (genera *Rhizobiales*, *Sphingomonadales*, and *Xanthomonas*).	Niederberger et al., 2008
Soil samples from Scott Base (Ross Sea), Marble Point (continental), near Lake Vanda, Bull Pass (Wright Valley), and a penguin colony on Cape Hallett	Total bacterial counts and soil analysis. Construction of 16S rRNA gene clone libraries and RFLP analysis.	Phylotypes were related to *Acidobacteria*, *Actinobacteria*, *Bacteroidetes*, *Cyanobacteria*, *Deinococus-Thermus*, *Firmicutes*, *Proteobacteria*, and candidate TM7 species. The dominant phylotypes from the ornithogenic soil were most closely related to the endospore formers (*Oceanobacillus profundus*, *Clostridium acidiurici*, and *Sporosarcina aquimarina*), while soil sample clones were related to uncultured environmental clones. *Actinobacteria*, *Bacteroidetes*, *Deinococus-Thermus*, and γ-*Proteobacteria* were more prevalent in the drier soils.	Aislabie et al., 2008

In contrast, microorganisms colonize diverse environmental niches throughout the continent, and the distribution of bacteria is affected by the physical and chemical characteristics of the soil (pH, moisture, nutrient levels, and salinity), as well as site-specific factors (slope, altitude, drainage, and solar radiation) (Cameron et al., 1970). Antarctic soils are spatially complex ecosystems. Typically, microbial diversity is highest in the moist, coastal soils, while the mineral soils of the Dry Valleys are relatively barren (Smith et al., 2006). Additionally, there is a vertical distribution of microbes in soil, with the highest numbers occurring at the surface (O'Brien et al., 2004), while others have reported that the largest bacterial populations colonize the soil-ice cemented permafrost interface (Vishniac, 1993). Generally, microbial diversity also decreases with increasing altitude (Tindall, 2004).

Early cultivation-based microbial studies identified pigmented heterotrophic aerobes at the surface soils, while nonpigmented strains were found to dominate the subsurface layers (Aislabie et al., 2006b). Traditionally, Antarctic microbial populations were considered to be cosmopolitan—"everything is everywhere"—with *Arthrobacter*, *Bacillus*, *Micrococcus*, and *Pseudomonas* species frequently isolated from different habitats (Aislabie et al., 2006a; Cameron et al., 1971).

Antarctic Maritime and Peninsula Environments

The northerly latitudes of the Antarctic Peninsula tend to be warmer, and the most widespread terrestrial ecosystems in the Peninsula are fellfields. Most fellfield soils are dominated by cryptogamic vegetation (plants that reproduce by spores) (Cowan and Ah Tow, 2004). Fellfield soils are characteristically either moist with a high silt content or are drier, sand/gritty ash. While the total macro- and microscopic species diversity of Antarctic Peninsula soils is low, they are colonized by complex communities of algae, mosses, lichens, and liverworts. Fungal species play a critical role in these ecosystems as decomposers and may be essential for the transformation of soil nutrients and for

the development of soil structure (Wicklow and Söderström, 1997). Bacterial productivity in fellfield soils is positively correlated to increasing water content. Baker (1970) reported that soil bacterial numbers peaked after snowmelt on Signy Island, South Orkneys, while similar findings were reported for inland nanataks on Dronning Maud Land (Harris and Tibbles, 1997) and Mars Oasis, Alexander Island (Newsham et al., 2010). The Coal Nunatak, Fossil Bluff, and Ellsworth Mountains, situated in the southern Antarctic Peninsula, are exposed to high levels of UV radiation in summer, low levels of water availability, and low nutrient levels. These sites are dominated by pink-pigmented methylotrophic *Methylobacterium* species and *Sphingobacteriales* strains (Yergeau et al., 2007).

Conversely, the coastal regions of Antarctica experience smaller temperature fluctuations compared to the Peninsula. Maritime soils have a high water content and may be colonized by flowering plants and cryptogams (algae, lichens, and mosses) (Ludley and Robinson, 2008). These soils are also rich in marine-derived organic nutrients (typically penguin guano and macroalgae) that contribute to the development of rich heterotrophic communities at maritime sites.

Cyanobacteria are the dominant phototroph in Antarctic freshwater and terrestrial habitats, and predominantly colonize the upper 1 mm of fellfield soils (Davey and Clarke, 1991). Original biodiversity studies were based on microscopic observations, and it was felt that cyanobacterial diversity was limited (Taton et al., 2003). However, molecular, culture-independent studies have shown that these populations are diverse, and endemic species have been detected (Taton et al., 2003, 2006). Several cyanobacterial species associate with plant communities in these soils, including *Phormidium autumnale* and *Pseudoanabaena catenata* species. Smaller populations of *Achnanthes lapponica*, *Chlamydomonas chlorostellata*, *Cosmarium undulatum*, *Cylindrocystis brebissonii*, *Netrium* sp., *Nostoc* sp., and *Planktosphaerella terrestris* have also been detected by microscopy.

Antarctic Continental Environments

The Oasis Regions of continental Antarctica (including Schirmacher and Bunger Oasis) are only ice free during the Antarctic winter, and the levels of precipitation are greater than in the Dry Valleys (Walton, 1984). Early microbial studies on lake sediments from Schirmacher Oasis (Queen Maud Land, Eastern Antarctica) detected the presence of heterogeneous microbial populations (rods, cocci, and filamentous cells) with cell counts ranging from 2×10^3 to 1.2×10^5 per g of soil. The majority (60%) of the cells were Gram-positive rods; however, psychrotrophic, halotolerant, gram-negative rods were also detected (Shivaji et al., 1989). Conversely, other studies reported that soils from the same regions were dominated by Gram-negative rods, predominantly the genus *Pseudomonas* (Shivaji et al., 1989). Other Gram-negative bacilli isolated from continental soils include *Achromobacter* and *Alcaligenes* species (Vincent, 1988). Many of the cultured *Psychrobacter* species from Antarctic marine and terrestrial habitats are psychrotolerant or psychrophilic and/or also halotolerant, which may account for their ubiquitous distribution (Bowman et al., 1996).

Antarctic Ice-Free Regions

Less than 2% of the Antarctic terrestrial habitat is ice free, which includes the Dry Valleys situated along glacial tills, nunataks (mountain summits that protrude through the surrounding ice sheets), nutrient-rich ornithogenic soils (along the coast), thermal sites, and seasonally saturated lake sediments (Taton et al., 2006; Vishniac, 1993).

ORNITHOGENIC SOILS

Ornithogenic soils associated with penguin colonies are the most abundant source of organic carbon in terrestrial Antarctica. Although ornithogenic sites are mainly found in coastal regions, they also occur in ice-free regions such as the Vestfold Hills, Schirmacher Oasis, and geothermal sites. Direct microscopic counts of 2×10^{10} cells per g of soil have been reported, illustrating that these nutrient-rich sites support very large microbial populations (Bowman et al., 1996). During the breeding season 7,000 to 11,000 breeding pairs of Adélie penguins nest in rookeries on Admiralty Bay, King George Island. These sites are colonized by diverse microbial populations, and elevated levels of in situ respiration have been detected during the summer breeding season (Barrett et al., 2006), and may be attributed, in part, to the significantly higher soil temperature within colonized rookeries compared to the surrounding vicinity (Barrett et al., 2006; Aislabie et al., 2008). The penguins directly influence the biological and chemical nature of rookery soils, and microorganisms found in penguin guano are largely of marine origin, including organisms that associate with krill (Adélie penguins' main food source). Studies performed at Arctowski Station (situated at Admiralty Bay) showed that the total number of culturable bacteria increased to 3.83×10^{11} per g of soil during the breeding season, and that the number of copiotrophic bacteria (microorganisms found in nutrient-rich environments) was 2 orders of magnitude greater than the number of oligotrophic bacteria. The bacterial species isolated included *Agrobacterium radiobacter*, *Pasteurella* sp., *Pseudomonas fluorescens*, and *Sphingobacterium multivorum* (Zdanowski et al., 2005).

Numerous studies have shown that ornithogenic soils are dominated by *Psychrobacter* species, and the isolation of strains from these soils that are able to utilize uric acid (and its metabolite allantoin) as a sole carbon source could indicate the significance of this. Another study reported that the dominant ribotype in rookery soils from Cape Hallett was most similar to *Clostridium acidiurici*, a strain that can utilize uric acid and purines. Interestingly, while the ornithogenic soils of Cape Hallett have high levels of nitrates due to nitrification, no nitrifying bacteria were detected (Aislabie et al., 2008). Additionally, *Streptomyces* species have been isolated from Cape Bird, Ross Island, while the novel species *Arthrobacter gangotriensis* was characterized from a penguin rookery at an Indian station (Gupta et al., 2004).

GEOTHERMAL SITES

Geothermal sites provide a sheltered micro-environment (ice hummocks) for microbial colonization with constant temperatures and moisture (due to condensation of steam emissions). The most extensively studied geothermal site is "Cryptogam Ridge" located on the rim of Mount Melbourne crater. Investigation of the microbial diversity of this ridge showed that it was dominated by the moss *Campylopus pyriformis* (Broady et al., 1987). In addition, several other eukaryotic species (fungi and liverworts) and bacterial (actinobacteria) species were identified (Broady et al., 1987). In subsequent studies, novel thermophilic species, *Bacillus thermantarcticus* (formerly *Bacillus thermoantarcticus*) (Nicolaus et al., 1996) and *Bacillus fumarioli* (Logan et al., 2000), were isolated. Bargagli and coworkers conducted a study at a site 1.5 km from Cryptogam Ridge on the slopes of Mount Melbourne. They found that the site was dominated by a different moss species, *Pohlia nutans*, and molecular analysis revealed that this strain was related to a moss colonizing Mount Rittmann fumaroles (110 km north of Mount Melbourne). Several bacterial species were cultivated, including *Alicyclobacillus* and *Sulfobacillus* species (both thermophiles) and mesophilic *Paenibacillus* strains, as well as a psychrotrophic strain related to *Arthrobacter oxidans* (Bargagli et al., 2004). Other thermophilic species isolated from geothermal sites in Antarctica include *Alicyclobacillus pohliae* and *Brevibacillus levickii* from Mount Melbourne and *Alicyclobacillus acidocaldarius* subsp. *rittmannii*, *Aneurinibacillus terranovensis*, and *Anoxybacillus amylolyticus* from Mount Rittmann (Allan et al., 2005; Nicolaus et al., 1996; Poli et al., 2006; Logan et al., 2000).

Dry Valleys

The Dry Valleys of Eastern Antarctica are the largest ice-free region in Antarctica and account for 0.03% (4,800 km²) of the total land area. The Dry Valleys are considered to be the earth's coldest, driest desert, and experience extreme climatic conditions—low humidity, strong winds, and low precipitation that occurs only as snow, which is largely lost by sublimation (Doran et al., 2002a; Cowan and Ah Tow, 2004). The soils contain trace amounts of organic matter, have high salt concentrations, and lack bioavailable water (0.5 to 2% wt). Additionally, any organism inhabiting the Dry Valleys must be able to withstand wide temperature fluctuations and repeated freeze-thaw cycling (Cowan and Ah Tow, 2004). Studies conducted in the early 1970s reported that the microbial diversity and biomass levels of mineral soils were low, with direct microbial cell counts ranging from 10^2 to 10^4 per g of soil. However, recent studies have questioned these findings. Cowan et al. (2002) found that based on in situ ATP analysis microbial biomass levels were 3 to 4 orders of magnitude greater than previously reported. Similarly, total cell counts for surface mineral soils in the Dry Valleys ranged from 2.6×10^6 to 3.9×10^8 cells per g (Cowan et al., 2002), and molecular-based studies have confirmed that the bacterial diversity in these soils is far greater than anticipated (Smith et al., 2006; Babalola et al., 2009) (Color Plate 1).

Studies have found that organic nutrients in the Dry Valleys may be dispersed by wind from aquatic and high-productivity sites. Most microorganisms isolated from Dry Valleys mineral soils are aerobic, heterotrophic psychrotrophs, as psychrotrophic organisms are better adapted to survive the temperature fluctuations at the soil-air interface. Chromogenic bacteria are routinely isolated from the soil surface, while nonpigmented bacteria occur mainly below the soil surface (Cameron et al., 1970; Nienow and Friedmann, 1993; Cowan and Ah Tow, 2004). Several recent metagenomic studies have failed to detect archaea in Dry Valleys mineral soils (Smith et al., 2006; Pointing et al., 2009), and when archaea are present/detected they are limited to the globally ubiquitous group II low-temperature crenarchaeotes (Hogg et al., 2006).

Primary production in the Dry Valleys is limited as there are no vascular plant species and eukaryotes are restricted to areas with a higher water content (Nienow and Friedmann, 1993).

Until recently, it was proposed that Dry Valleys ecosystems were primarily driven by abiotic rather than biotic factors. However, it has become increasingly apparent that photosynthetic microorganisms may play an important role in these habitats. Cyanobacteria present in Dry Valleys mineral soils are considered to be the major primary producers (Cary et al., 2010) and contribute significantly to microbial diversity. Cyanobacteria increase both the stability of soil and the nutrient concentration by nitrogen fixation. The dominant cyanobacterial species identified in early microbial studies included *Calothrix*, *Nostoc*, and *Scytonema* species (Vincent, 1988). *Nostoc commune* species are ubiquitous in Victoria Land soil and have been detected as moss epiphytes (Adams et al., 2006). Large, visible populations of cyanobacteria occur in glacial streams and at lake edges, while mineral soils appear to lack any visible cyanobacterial (and algal) growth (Taton et al., 2006). A recent metagenomic study showed that the cyanobacterial ribotypes present in Dry Valleys mineral soils were most similar to clones from cyanobacterium–dominated cryptoendolithic communities in Beacon sandstone. The authors proposed that cyanobacterial ribotypes detected in mineral soils were in fact not actively growing but are dispersed by the wind from adjacent endolithic communities (Aislabie et al., 2006b) (Color Plate 1).

Actinobacterial populations appear to be cosmopolitan in Antarctic soils, and have been identified as the dominant phylotype in both culture-based and metagenomic studies. Cameron et al. (1972) reported that coryneforms were one of the most abundant bacteria isolated from Dry Valleys soils, of which the majority (85%) were strains of *Corynebacterium sepedonicum*. Subsequently, *Arthrobacter*, *Brevibacterium*, *Corynebacterium*, and *Micrococcus* species have been characterized from mineral soils (Aislabie et al., 2006a).

Several studies focusing on the diversity of actinobacteria present in Miers Valley mineral soils have shown that *Streptomyces* species account for over 80% of the culturable strains, although *Nocardia* and *Pseudonocardia* are also readily isolated (Cameron et al., 1972; Prabahar et al., 2004; Babalola et al., 2009). Similarly, *Arthrobacter*, *Friedmanniella*, and *Rubrobacter* species have been isolated from Wright Valley, and these genera appear to be more prevalent in drier soils (Aislabie et al., 2006b). Members of the class *Actinobacteria* were also among the heterotrophic microbial populations identified in early culture-based studies on McMurdo Dry Valleys soils, and included members of the genera *Arthrobacter*, *Corynebacterium*, *Micrococcus*, *Nocardia*, and *Streptomyces*. Other bacterial taxa isolated in these studies included *Achromobacter*, *Bacillus*, *Flavobacterium*, *Planococcus*, and *Pseudomonas* species (Cameron et al., 1972). Members of the *Thermus-Deinococcus* species group are the predominant ribotype in Wright Valley mineral soils. Several *Deinococcus* strains isolated from the Dry Valleys are resistant to radiation and are desiccation tolerant, which may account for their prevalence in drier soils (Carpenter et al., 2000; Aislabie et al., 2006b).

Lithic Communities

The undersides of porous translucent rocks offer a specialized microhabitat for microbial colonization and act as a climatic refuge for photosynthetic organisms, offering physical protection from UV radiation and scouring by wind. Additionally, rocks trap moisture and act as a water source (Cockell et al., 2003; Cockell and Stokes, 2004b). For microbial colonization to establish, the rock substrate must be sufficiently translucent to allow light to penetrate. In the Dry Valleys, fine-grained Beacon sandstone is most often colonized, although quartz rock and limestone may also be inhabited (Cowan and Ah Tow, 2004; Friedmann, 1993). Lithic communities are classified by the specific environmental niche they reside in, and hypoliths, chasmoliths, and cryptoendoliths are discussed below.

Hypoliths are photosynthetic microbial communities that colonize the undersides of translucent rocks and stones (Broady, 1981; Thomas, 2005). Hypolithons have been found in cold and hot deserts where temperatures under the stone can reach 65°C. Blackbody heating of the rock may warm the sublithic environment to temperatures 5 to 15°C above the ambient air temperature (Broady, 1981). Although the rock itself limits the amount of radiation that can penetrate, it has been shown that photosynthesis can occur at irradiance levels less than 0.1% of the total incident light (Thomas, 2005). The productivity of hypolithic communities may be equal to or greater than that of the aboveground biomass (Cockell and Stokes, 2004b). In polar deserts, hypoliths are often found in stone fields that are formed by paraglacial activity (freeze-thaw cycling of groundwater) that sorts rocks into patterns (Cockell and Stokes, 2004b). The most commonly observed types of hypolithon are green- and/or red-pigmented cyanobacterial layers that occur at the rock-soil interface and colonize the base and sides of rocks. Hypolithic communities are dominated by microthallate cyanobacteria, including *Leptolyngbya* and *Phormidium* species, as well as *Chroococcidiopsis*- and *Synechococcus*-like cells (Smith et al., 2000; Cockell and Stokes, 2004b; Pointing et al., 2009).

Chasmoendoliths inhabit cracks in weathering rocks, which are common in the ice-free areas where repeated freeze-thaw cycling causes fracturing of the rocks (Broady, 1981; Cowan and Ah Tow, 2004). Chasmoendolithic communities are colonized by endolithic lichens (fungal mycobionts with the algal phycobiont *Trebouxia*) associated with cyanobacteria, typically *Chroococcidiopsis* or *Gloeocapsa* species (Nienow and Friedmann, 1993).

Cryptoendolithic organisms reside in the interstices of crystalline rocks, and the extent of microbial colonization is dependent on the geological features of the rock substrate and its geographical location (Friedmann and Ocampo, 1976). Generally, in the Dry Valleys the "warmer" north-facing slopes are colonized by endolithic communities while the "colder" south-facing slopes are uncolonized (McKay and Friedmann, 1985). Cryptoendolithic communities consist of primary producers (algae or cyanobacteria) and consumers/degraders (fungi or heterotrophic bacteria). Generally, the biomass levels of cryptoendolithic communities are

low, and calculations suggest that a lichen-dominated endolithic community could actively metabolize for only 375 to 700 hours per year (Friedmann et al., 1987). The structure of cryptoendolithic communities is altered by different host substrates, and the activity of microorganisms can indirectly affect the local environment (Omelon et al., 2007). The physical properties of the crust may allow supercooling to occur (temperatures as low as −17°C have been recorded), and as such the endolithic communities may not freeze for most of the Antarctic summer. In winter, the temperatures within the layers would need to increase to 0°C for the community to thaw; therefore, once the first freezing event of winter has occurred, the endolithic communities are likely to remain frozen for the majority of the winter. *Chloroglea* species isolated from these crusts have been shown to produce exopolysaccharides that conserve moisture and may provide protection against desiccation, thereby allowing photosynthesis to occur for extended periods (McKay and Friedmann, 1985; Potts, 1994; Hughes and Lawley, 2003).

Two types of cryptoendolithic communities have been identified, namely lichen- and cyanobacterium-dominated assemblages. Cryptoendolithic communities have clear zonal stratification, and occasionally both community types will develop within the same rock. Lichen-dominated communities typically have an upper lichenized zone of mycorrhizal fungi associated with the green algal symbiont *Trebouxia*, while the lower zone is dominated by cyanobacteria (Siebert et al., 1996). Several fungal species inhabiting endolithic communities produce light-shielding pigments and can therefore reside closer to the rock surface, where the light intensity is greater. The algal species within the lichenized top layers may be protected by fungal pigments or are themselves less light sensitive, while light-susceptible green algae and cyanobacteria thrive in the lower layers. Excretions produced by these primary producers can support small heterotrophic communities. Heterotrophic organisms frequently identified from these communities include Gram-positive cocci

(Siebert et al., 1996; Nienow and Friedmann, 1993), while recent studies have found that the dominant phylotypes in a cyanobacterium-dominated cryptoendolithic community (in McMurdo Dry Valleys) were α-*Proteobacteria*. Several genera belonging to this phylum have been shown to play a role in aerobic anoxygenic photosynthesis (de La Torre et al., 2003).

Friedmann and coworkers reported that the species diversity of a lichen-dominated crypto-endolithic community inhabiting a sandstone boulder in Linnaeus Terrace, Asgard Range, was low and the community was dominated by Gram-positive cocci. The upper black layer was formed by lichen associated with green algae (*Trebouxia* and *Pseudotrebouxia*), unlichenized fungi, and black yeasts. The black layer was followed by a white zone of fungi and bacteria, while the third zone was colonized by green algae (*Hemichloris antarctica* and *Stichococcus* sp.) and/or cyanobacteria (*Chroococcidiopsis* sp.). The lower brown, iron oxide–rich zone contained heterotrophic bacteria (*Deinococcus* sp.). Some sandstone sampled lacked the upper lichen layer but instead contained a dark red cyanobacterium (*Gloeocapsa* sp.)-dominated layer (Friedmann et al., 1988). Endolithic communities can also develop in gypsum salt crust layers that form on the surface of ice-free rocks. *Chloroglea*, *Sphingomonas*, and *Verticillium* species were isolated from a sandstone community colonizing Two Step Cliffs, Alexander Island (Antarctic Peninsula). Dark-pigmented fungi were also detected in the crusts; however, they could not be cultured (McKay and Friedmann, 1985).

Quartz stone sublithic cyanobacterial communities are common throughout the Vestfold Hills. A comparison of the microbial diversity in sublithic communities to the surrounding soils showed that while the direct epifluorescent 4′,6-diamidino-2-phenylindole (DAPI) bacterial counts of the sublithic growth and underlying soil were similar (1.1×10^9 cells g^{-1} dry weight and 0.5×10^9 cells^{-1}, respectively), the viable counts for the sublithic samples (2.1×10^7 CFU g^{-1}) were on average 3 orders of magnitude higher (2.3×10^4 CFU

g^{-1}) (Smith et al., 2000). The dominant cyanobacterial species present were nonhalophilic and related to the *Phormidium* subgroup. The heterotrophic bacterial populations colonizing both the subliths and the underlying soil were related to psychrotolerant taxa typically found in Antarctic soil. Psychrophilic and halophilic bacteria, mostly α-*Proteobacteria* and members of the order *Cytophagales*, were abundant in the sublithic growth film (accounting for 50% of isolated individual taxa) but were absent from the underlying soils. Certain taxa were ubiquitous in all samples, and these included *Arthrobacter*, *Gelidibacter*, *Janibacter*, *Micrococcus*, *Pseudomonas*, *Psychrobacter*, *Rhodococcus*, and *Stenotrophomonas* species (Smith et al., 2000).

Cryoconites

Cryoconites are unique microhabitats found in polar regions that also act as a biological refuge for microorganisms. Cryoconite holes are formed when ionic impurities or particulates, which are insoluble in the crystal structure of ice, are excluded from the ice matrix. Particulates on the surface are heated by solar irradiation and melt into the underlying ice. This results in a network of micron-diameter veins forming in the ice (Christner et al., 2003; Price, 2006). These veins contain liquid water and nutrients/ions derived from particulates and melted glacial ice. The diameter of the veins is dependent on temperature, and narrowing occurs as the temperature decreases; at −50°C the diameter is ~1 μm, while at elevated temperatures (2°C) the diameter is ~10 μm. The concentration of ions increases as the veins become narrower, maintaining the liquid phase. Epifluorescence microscopy of lake and sea ice showed that most bacterial cells (eubacteria and archaea) were excluded from the ice phase into brine channels/veins, while larger eukarya (larger than 5 μm) were trapped in the solid phase (Junge et al., 2004a). Microbial cell counts within the cryoconite vary. Christner and coworkers (2006) reported values for Lake Vostok of 98 to 430 cells per ml in type I accretion ice and 77 to 89 cells per ml in type II accretion ice, while other authors have reported significantly lower levels (0.1 to 10 cells per ml) (Price, 2006). Cryoconite holes are seasonally active, and in the warmer summer months there are sufficient nutrients present for photosynthesis to occur (by cyanobacteria and algae), which supports complex communities including algae, bacteria, diatoms, fungi, and rotifers. Viable microorganisms cultured from Canada Glacier in McMurdo Dry Valleys belonged to eight taxa: *Acidobacterium, Actinobacteria, Cyanobacteria, Cytophagales, Gemmimonas, Planctomycetes, Proteobacteria,* and *Verrucomicrobia*. This study found that a greater number of bacteria were isolated at 4 and 15°C compared to 22°C, and obligately psychrophilic strains related to *Flavobacterium* spp. and *Cryobacterium psychrophilum* were also isolated. Sequence analysis of 16S rRNA gene clone libraries revealed that most ribotypes were related to species detected in adjacent lake ice and microbial mats, supporting the idea that cryoconite holes are seeded by particulates from surrounding environments. The dominant phylotypes detected were dominated by cyanobacteria related to *Chamaesiphon, Leptolyngbya,* and *Phormidium* species (Christner et al., 2003).

Lakes

Lakes, ponds, and streams are prevalent at many coastal and inland sites in polar regions, and encompass a range of different ecosystems from dilute meltwaters to hypersaline brines. Microbial mats, which are ubiquitous in polar lakes, are discussed separately (see "Microbial Mats," below).

Many Antarctic lake studies focus on the Antarctic Dry Valleys. These lakes are formed by glacial erosion and are unique in many ways. The constant cold temperatures, permanent ice cover, absence of wind mixing, and lack of higher organisms make the water columns in these lakes extremely stable, and the lake ecosystems are entirely microbial (Mondino et al., 2009). Although the major Dry Valleys lakes (Lake Hoare, Lake Fryxell, and Lake Bonney) are geographically close to one another, their limnology is dramatically different and therefore each has a unique geochemistry. Lake Hoare is a freshwater lake, while Lake

Bonney is hypersaline and supports a robust nitrogen cycle. Lake Fryxell is a weakly saline lake with an active sulfur cycle. Lakes Vanda, Vida, Joyce and Miers are permanently ice covered. The distribution of microorganisms in Antarctic lakes varies with lake type, the physical and chemical environment, and local food webs, including any input from terrestrial or mammalian sources. Within Lake Vida in the McMurdo Dry Valleys, several studies have shown that both heterotrophic and phototrophic processes occur, and culture-dependent analysis identified γ-*Proteobacteria* as the dominant community members (Mosier et al., 2006; Mondino et al., 2009). Strains related to the genera *Psychrobacter* and *Marinobacter*, which are routinely isolated from other marine and polar habitats, were isolated in pure cultures (Mondino et al., 2009). Phylogenetic studies similarly identified that γ-*Proteobacteria* (52%) dominated the deep-ice community while *Actinobacteria* (42%) dominated the surface waters (Mosier et al., 2006) (Color Plate 2). Members of the α- and β-*Proteobacteria*, including members of the phototrophic purple bacteria, were also identified and are routinely found in the other lake environments (Christner et al., 2001; Mosier et al., 2006; Mondino et al., 2009). In Lake Fryxell, the phototrophic purple bacteria form part of a novel community and display a high degree of diversity (Karr et al., 2003). Enrichment cultures for purple bacteria from this lake were successful in isolating only two morphotypes, both of which contained gas vesicles previously unknown in purple nonsulfur bacteria. It has been proposed that these buoyancy structures may be necessary for these organisms to position themselves at specific depths within the nearly freezing water column.

THE ARCTIC

Arctic terrestrial habitats cover more than 7 million km^2 and include parts of Alaska, Canada, Europe, Greenland, Iceland, and Russia. The Arctic Circle experiences long, dark winters broken by short summer periods (may be less than 6 weeks per annum) when the average temperature during the day is above 0°C. In addition to extremely low temperatures, several other factors place stress on organisms inhabiting the Arctic. Soils may be flooded during the spring thaw and then experience periods of drought during the late summer because of inconsistent rainfalls. Microorganisms colonizing tundra soils are further stressed by freeze-thaw cycles, while permafrost and persistent snow coverage place additional stress on soil communities (Callaghan et al., 2010). Arctic terrestrial habitats include polar deserts (in the High Arctic), permafrost areas, tundra, and alpine soils. Some Arctic soils are rich in minerals and fossil fuel reserves that can support large numbers of organisms, both macro- and microscopic (Callaghan et al., 2010), and the total bacterial cell counts in the active layer of an Arctic tundra birch hummock at the end of winter was 5×10^9 cells per g of soil (Buckeridge and Grogan, 2008).

Arctic Tundra Soils

Globally, tundra and boreal forest ecosystems cover 22% of the terrestrial landmass and are estimated to contain 27% of the earth's soil carbon. Tundra habitats are sensitive to temperature fluctuations, and a change in their productivity may ultimately have a substantial impact on global climate change (Schimel and Mikan, 2005). Northern Hemisphere tundra is divided into two subgroups, namely Arctic tundra and alpine tundra. As tree growth is restricted by low temperatures, both Arctic and alpine tundra are devoid of higher vascular plants and these areas are colonized by dwarf shrubs, lichens, and mosses. Arctic tundra is found north of the Taiga Belt, and this region experiences long, cold winters with short summers when the warmer temperatures allow the soils to thaw. Organic carbon accumulates in tundra soils as microbial decomposition is limited to the summer months and the large reservoirs of stored organic carbon favor the growth of methanotrophs and hydrocarbon-degrading microbes (discussed in "Methanogens" and "Hydrocarbon-Degrading Bacteria," below) (Wallenstein et al., 2007). In addition to meth-

anotrophs, spore-forming microorganisms have been described from tundra habitats, and members of the spore-forming genera *Bacillus*, *Gracilibacillus*, *Oceanobacillus*, *Ornithinibacillus*, and *Virgibacillus* have been isolated from a glacial moraine in Qaanaaq, Greenland (Yukimura et al., 2009).

A recent metagenomics-based study found that tussock and intertussock bacterial communities were dominated by *Acidobacteria* species, while *Proteobacteria* dominated in shrub soils. Tussocks are colonized by moss and graminoid vegetation, and these low-quality carbon sources may favor the growth of slow-growing *Acidobacteria* species. While there was generally no temporal shift in the microbial community, actinobacterial populations were found to be more abundant in August (prefreeze) compared to June (postthaw), and it is speculated that filamentous bacteria may be ruptured by freeze-thaw cycling. Since there was no detectable difference in the abundance of microorganisms pre- and postfreeze, it is probable that bacteria survive the freezing winter through resistance mechanisms (compared to "resilient" populations, which regrow after the winter thaw), and that vegetation type is the main factor determining microbial community structure (Wallenstein et al., 2007) (Color Plate 3).

METHANOGENS

Methanogenic archaea belong to the kingdom *Euryarchaeota*. They are ubiquitous in anoxic environments and play an important role in tundra wetlands and permafrost. During the Arctic summer the upper permafrost layer thaws and the soils become water saturated, which results in the anaerobic degradation of complex organic matter into acetate, H_2, CO_2, formate, and methanol by fermentative bacteria. These serve as substrates for methanogenic archaea, which produce the greeenhouse gas methane (Garcia et al., 2000). The estimated methane emissions from tundra and permafrost soils vary from 20 to 40 teragrams per year (Cao et al., 1996). A culture- and metagenomics-based study conducted in the Siberian Laptev Sea coast permafrost soils showed that methanogenesis occurred in a distinct vertical profile and that the metha-nogens found in the deep active-layer zone were better adapted to low temperatures. Denaturing gradient gel electrophoresis (DGGE) analysis re-vealed that the largest number of ribotypes were present in the 22- to 29-cm-depth samples and diversity decreased with increasing depth. The dominant ribotypes were found to be closely related to the methanogenic archaeal families *Methanomicrobiaceae* (12 ribotypes), *Methanosar-cinaceae* (11), and *Methanosaetaceae* (3) (Ganzert et al., 2007).

Type II methanotrophic bacteria (α-*Proteo-bacteria*) have also been isolated from Arctic tun-dra. Two novel genera, *Methylocella* (*Methylocella palustris*, *Methylocella silvestris*, and *Methylocella tundrae*) and *Methylocapsa* (*Methylocapsa acidiph-ila*), were isolated from spagnum peat bogs of boreal and tundra soils in Western Siberia and North Russia. The two genera are related to the genera *Methylosinus* (which has previously been observed in Arctic soils [Wartiainen et al., 2003]) and *Methylocystis*. Similarly, *Methylobacter psychrophilus*, *Methylobacter luteus*, and *Methylo-bacter tundripaludum* have been observed in Ca-nadian Arctic tundra soil (Omelchenko et al., 1996; Tourova et al., 1999; Pacheco-Oliver et al., 2002; Wartiainen et al., 2006).

HYDROCARBON-DEGRADING BACTERIA

Hydrocarbon fuel is the primary energy source in the Arctic region, and hydrocarbon pollu-tion is common. In polluted soils the active zone contains the highest concentration of hy-drocarbons and is the most biologically active. During the summer, the active zone thaws for 1 to 2 months, resulting in steep temperature gradients with near-freezing temperatures at the permafrost interface to 20°C at the sur-face (Mohn and Stewart, 2000). *Acetobacterium tundrae* is a novel acetogenic organism that was isolated from slightly acidic (pH of 6.1) tundra wetland soil of the Polar Ural (Simankova et al., 2000). Additionally, a lipolytic psychrotrophic *Acinetobacter* strain was identified from Siberian tundra soils (Suzuki et al., 2009). Complex communities of oil-degrading, cold-adapted

heterotrophs were detected in a study com-paring the diversity of hydrocarbon-degrading bacteria in pristine and contaminated soils from Tyrol, western Austria (600 to 2,900 m above sea level). Phylogenetic investigations revealed the presence of members of the class *Actino-bacteria*, as well as α-, β- and γ-*Proteobacteria* (Labbé et al., 2007). In a similar comparative study of oil-contaminated versus pristine soils from various Alpine sites in Tyrol, phylotypes related to *Acinetobacter* sp., *Mycobacterium* sp., *Pseudomonas putida*, and *Rhodococcus* sp. were identified through 16S rRNA gene sequence analysis (Margesin et al., 2003). In addition, phenol-degrading *Arthrobacter psychrophenolicus* and *Pseudomonas* and *Rhodococcus* spp. have been isolated from contaminated alpine soils (Margesin et al., 2003, 2004, 2005). Margesin and coworkers reported that the number of cold-adapted hydrocarbon degraders was 2 to 3 orders of magnitude greater than the mesophilic populations in hydrocarbon-contaminated al-pine sites (Margesin et al., 2003).

Alpine Sites

Alpine tundra is defined as the high-lying re-gions (between 1,800 and 2,500 m above sea level) that occur above mountainous forests, while the subalpine belt includes the forest-tundra region (Löve, 1970). Compared to Arc-tic tundra, the European Alpine region expe-riences higher maximum and lower minimum temperatures, increased levels of precipitation, and more humidity. Wider temperature fluc-tuations lead to frequent freeze-thaw cycles, and these regions also experience elevated solar radiation levels (Margesin et al., 2004). Microbial biomass peaks in late winter and then steadily declines as soil temperatures in-crease to 0°C, possibly as bacterial cells rupture due to freeze-thaw cycles and soluble organic nutrients become depleted by late winter (Nemergut et al., 2005; Jefferies et al., 2010). Additionally, snow insulates the underlying soil, resulting in warmer soil temperatures in late winter when the snow is deepest (Buckeridge and Grogan, 2008). A deep snowpack can maintain soil temperatures at or around 0°C, thereby maintaining the presence of liquid water (Nemergut et al., 2005).

Lapland is situated in the Arctic Circle and includes parts of Finland, Norway, Russia, and Sweden. In a study focusing on the bacterial populations of the Finnish Lapland, several oligotrophic, lichen-governed forest soils were sampled. The isolates belonged to six phylogenetic lineages: Actinobacteria (*Rhodococcus* and *Streptomyces*), Bacteroidetes (*Pedobacter*), α-Proteobacteria (*Sphingomonas*), β-Proteobacteria (*Burkholderia*, *Collimonas*, and *Duganella/Janthinobacterium*), γ-Proteobacteria (*Frateuria*, *Pseudomonas*, and *Yersinia*), and the low-G+C Gram-positive bacteria (*Bacillus* and *Paenibacillus*). The predominant bacterial group was the *Proteobacteria* (Männistö and Häggblom, 2006), with β-*Proteobacteria* being more prevalent during the spring thaws, whereas *Acidobacteria* were the most abundant during winter (Männistö et al., 2008). Members of the γ-*Proteobacteria* including *Aeromonas*, *Pseudomonas*, and *Xanthomonas* species were isolated from high alpine soils and glacial cryonites in the Eastern and Western Alps of Europe (Schinner et al., 1992). Psychrotolerant *Pedobacter* species related to *Pedobacter cryoconitis* and *Pedobacter himalayensis* have been isolated from soils and decaying lichen from forests in the Finnish Arctic (Männistö et al., 2008).

Permafrost
Permafrost refers to soil that is permanently (for periods longer than 2 years) at or below the freezing point of water (0°C). Globally, permafrost accounts for 26% of the terrestrial soil ecosystem, and nearly a quarter (24%) of the Northern Hemisphere landmass (Williams and Smith, 1989). While most permafrost is located at high latitudes, alpine permafrost is found at high-altitude sites at lower latitudes. Overlying the permafrost is a thin active layer that seasonally thaws during the warmer summer months. The thickness of the active layer varies with location and temperature, and can range in depth from 0.6 to 4 m. The active layer is thicker in regions that experience harsh winters and continuously have permafrost.

Permafrost active layers can support limited plant life during the summer months.

Studies have shown that Siberian permafrost harbors diverse populations of viable microorganisms including heterotrophs (anaerobic and aerobic), methanogens, and sulfur-reducing bacteria (Gilichinsky et al., 1989). Microscopic analysis revealed that permafrost is dominated by bacterial cells with altered ultrastructures (including thickened capsular layers and intracellular aggregates), as well as the presence of "dwarf cells" with a diameter of 0.1 to 0.4 μm (Vorobyova et al., 2001). Generally, both the number of viable microorganisms and bacterial diversity are inversely proportional to the age of the permafrost, and cell counts vary between sites. Bacterial cell counts of 5 to 130 cells per g (including mesophilic and thermophilic bacteria) were reported for permafrost from the western Arctic (at a depth of 1.8 m) (Boyd and Boyd, 1964). Similarly, the largest microbial population present in Siberian permafrost core samples was 108 cells per g, which included all viable fungi, yeast, algae, and unicellular and filamentous bacteria (Boyd and Boyd, 1964; Shi et al., 1997). The highest culturable bacterial cell count for Canadian permafrost samples was 6.9×10^3 per g of soil, while total bacterial counts determined by 5-(4,6-dichlorotriazinyl) aminofluorescein (DTAF) staining were 3.56×10^7 per g (Steven et al., 2007). However, several studies have failed to observe cells undergoing cellular division, suggesting permafrost may contain large populations of dormant bacteria (Friedmann, 1994; Vorobyova et al., 2001).

The majority of bacteria isolated from permafrost are aerobic and include a number of coryneforms, endospore formers, sulfate reducers, nitrifying and denitrifying bacteria, and cellulose degraders. Genera isolated from permafrost include *Acinetobacter*, *Aeromonas*, *Arthrobacter*, *Brevibacterium*, *Bacillus*, *Cytophaga*, *Deinococcus*, *Flavobacterium*, *Micrococcus*, *Nitrosospira*, *Nitrobacter*, *Promicromonospora*, *Pseudomonas*, and *Rhodococcus* (Zvyagintsev, 1992). In a more recent study investigating the bacterial diversity in permafrost from Eureka, Ellesmere Island, Canada, bacteria belonging to three phyla were isolated: *Firmicutes* (*Bacillus*, *Paenibacillus*, and *Sporosarcina*), *Actinobacteria* (*Arthrobacter, Micrococcus, Kocuria*, and *Rhodococcus*), and *Proteobacteria* (*Pseudomonas*) (Steven et al., 2007). The spore-forming *Firmicutes* were the most abundant and the majority (83%) were psychrotolerant. In contrast, the dominant bacterial phylotypes in metagenomic 16S rRNA gene clone libraries were related to *Actinobacteria* and *Proteobacteria*, and an archaeal library was dominated by phylotypes related to halophilic archaea (Steven et al., 2007). Similar findings were reported for permafrost cores collected from Kolyma-Indigirka lowland, northeast Siberia. Over 30% of the isolates were related to *Actinobacteria* (*Arthrobacter globiformis* and *Rhodococcus fascians*), 21% belonged to the β-*Proteobacteria* (of which four were closely related to *Alcaligenes faecalis*), 14% were assigned to the γ-*Proteobacteria* (*Escherichia coli* and *Serratia marcescens*), and the majority (34%) were related to either *Bacillus subtilis* or *Bacillus psychrophilus* (Shi et al., 1997).

Lakes

Most Arctic lake research has been concerned with the temporal distribution of bacterial groups (Lindström et al., 2005), seasonal variation in bacterioplankton numbers and species diversity (Hobbie et al., 1983; Crump et al., 2003), and the effects of climate change on this habitat. Diversity studies of Arctic Lake Toolik found that the lake and terrestrial bacterial diversity was similar, most notably for the β-*Proteobacteria* (Bahr et al., 1996). This reinforces the hypothesis that Arctic ecosystems contain land-based bacterial species as the Arctic Ocean is almost entirely surrounded by landmasses (Brinkmeyer et al., 2003), and the theory is further supported by the reisolation of β-*Proteobacteria* isolates that are adapted to life in Arctic lakes (Brinkmeyer et al., 2004). Through 16S rRNA gene analysis the phylogeny of Arctic lake bacteria appears to be dominated by α-and β-*Proteobacteria* (Bahr et al., 1996; Brinkmeyer et al., 2004), while culture-based studies have shown γ-*Proteobacteria* as the dominant members in lakes and runoff water (Boyd and Boyd, 1963).

The analysis of photosynthetic signature pigments has been applied in numerous polar lake studies to infer bacterial community structure. Using this technique, the surface layers of Lake A, Ellesmere Island were found to be dominated by picocyanobacteria, while green sulfur bacteria dominated the lower, anoxic layers (Antoniades et al., 2009). This sedimentary record of bacterial pigments in polar meromictic lakes could potentially be used to reconstruct past changes in ice cover and therefore shed light on the history of Arctic climate change.

MICROBIAL MATS

Both the Arctic and Antarctic polar regions document the presence of a variety of microbial mat communities growing on the surface of ice shelves in pools of meltwater. Variations in light, temperature, conductivity, and oxygen between such habitats may explain many of the differences seen in species composition and mat morphology. These communities develop undisturbed over long periods of time and present an opportunity to isolate novel bacteria. The upper layer of microbial mats is typically dominated by cyanobacteria and diatoms, and these oxygenic phototrophs are the primary producers in such environments. Anoxygenic phototrophs such as green and purple sulfur bacteria occupy the underlying layer(s). Cyanobacteria can produce mucilaginous organic secretions that bind sediment particles together, and the accumulation of such dense materials offers an extensive habitat for microorganisms to multiply where there is an abundance of food and protection (Vincent et al., 2004). The potential of mat communities to produce antibiotics or other toxic molecules as a survival strategy means there may be a plethora of novel bacteriocidal molecules awaiting discovery (Biondi et al., 2007).

Benthic cyanobacterial mats are a major feature of the Antarctic lakes, especially Dry Valleys lakes. The microbial mat bacterial diversity of 10 Dry Valleys lakes was assessed by culturing techniques (heterotrophic growth conditions and fatty acid analysis). The 746 isolates were divided into 41 delineated clusters, while 31 strains formed single branches. Analysis of the 16S rRNA gene identified strains belonging to the α-, β-, and γ-*Proteobacteria* subdivision, the high- and low-G+C Gram-positives, and the CFB (*Cytophaga-Flavobacterium-Bacteroides*) branch (Van Trappen et al., 2002). Sixteen of the fatty acid clusters showed less than 97% identity to their nearest phylogenetic neighbors, indicating that these clusters contained uncultured taxa. The majority of microbial mats' 16S rRNA sequences are not represented in sequence databanks or in culture collections, highlighting their potential as a source for novel, biotechnologically exploit-able organisms (Brambilla et al., 2001).

A recent study investigated the taxonomic distribution of bacteria in a wide variety of marine environments conducive to microbial mat development, and focused on the pigments produced by these communities. High levels of β-carotene and other pigments were found in all mat layers. These pigments have a dual role in quenching free radicals and shielding bacterial cells from UV light (Vincent et al., 1993). A high level of protection is necessary for microorganisms inhabiting microbial mats due to the stressful conditions they encounter—high UV irradiation and very low temperatures, combined with low rates of metabolic repair. One particular species, *Oscillatoria priestleyi*, was found to migrate downward into the mat in order to escape the damaging effects of solar radiation, which may account for the success of this species in Antarctic lakes.

Although the 16S rRNA gene has frequently been used to determine the phylogenetic relationship of cyanobacteria, 16S rRNA gene sequences often do not allow for the discrimination of bacterial strains at the subgeneric level (species, subspecies, and strain) (Fox et al., 1992). The less conserved internal transcribed spacer (ITS) region has proven to be useful for taxonomic studies of cyanobacterial species (Ernst et al., 2003) and complements 16S rRNA gene analysis. In combination, ITS and 16S rRNA gene analysis were employed to identify endemic and cosmopolitan cyanobacterial taxa in Antarctica, and identified 17 morphospecies and 28 16S rRNA gene opera-

tional taxonomic units (OTUs) belonging to the *Oscillatoriales*, *Nostocales*, and *Chroococcales* (Taton et al., 2003, 2006).

A more comprehensive phylogenetic analysis of Lake Fryxell microbial mats revealed that the phylotypes represented 70 higher taxonomic groups (<98% similarity) and 133 species (>98% similarity) (Brambilla et al., 2001). The organisms identified included *Clostridium* and *Bacteroides*, which possibly play a role in anaerobic fermentation within the mat communities. The presence of anaerobic, saccharolytic organisms that potentially could form CO_2, H_2, and C_1-C_4 acids and alcohols for the subsequent metabolism by other community members was detected. In addition, polysaccharide-producing strains, decomposers, proteolytic bacteria, and fastidious organisms were characterized (Brambilla et al., 2001), reflecting the elements of the food web occurring in these mats.

Microbial abundances in Arctic mat communities are reported to be similar between ice shelves, and have some identity with those observed in the Antarctic mats (Van Trappen et al., 2002). Microbial mats from Markham and Ward Hunt Ice Shelves showed species homogeneity in the vertical profile, which has not been seen previously in Antarctic mats, possibly due to differences in mat thickness. The stratified Antarctic mats from the McMurdo Ice Shelf were up to 8 cm thick in places, while the Arctic mats in this study were ~2 cm (Bottos et al., 2008). As with other polar habitats, the Arctic mat communities are more phylogenetically diverse compared to Antarctic mats (Brinkmeyer et al., 2003); however, the three dominant phyla (*Proteobacteria*, *Bacteroidetes*, and *Actinobacteria*) are similar in both (Brambilla et al., 2001; Van Trappen et al., 2002), substantiating the importance of these organisms in these extreme environments.

SEA ICE MICROBIAL COMMUNITIES

Antarctic and Arctic Sea Ice

Sea ice is an environment of extremes. Aside from low temperatures, the sea ice matrix varies widely in terms of salinity, pH, dissolved inorganic nutrients and gas, and light penetration. As sea ice freezes, the expelled salt forms brine channels of up to 150% salinity that can, upon melting, dilute the salt concentration to less than 10%. Sea ice microbial community (SIMCO) assemblages generally form in the lower 10 to 20 cm of the ice column at the sea-ice interface. SIMCOs are complex, highly productive communities, dominated by primary producers (mainly diatoms) (Brown and Bowman, 2001). Bacterial attachment to algal cells facilitates enrichment during the formation of SIMCOs (Grossmann, 1994), thereby increasing the diversity of the microbial communities. Heterotrophic bacteria and unicellular algae have been studied extensively, and the driving force behind such studies is the desire to understand how the communities thrive in extremes of light, temperature, and salinity. The high biological productivity of sea ice is closely associated with the changeable sea ice constituents, which include the ice surface, the sea ice matrix, under-ice, and the internal brine channels formed as sea ice freezes.

Microorganisms are continually incorporated into SIMCOs, and the microbial population is dependent on the diversity and bacterial numbers when it forms. As sea ice forms, there is a reduction in overall population size, while the numbers of psychrophilic and culturable bacteria increase as sea ice ages (Helmke and Weyland, 1995). SIMCO composition undergoes seasonal shifts, namely an increase in bacterial production and heterotrophy in winter as algal cells lyse, thereby providing a nutrient source for bacteria (Helmke and Weyland, 1995). Bacterial heterotrophs are commensals of diatoms and as such are tightly coupled to primary productivity, by supplying inorganic nutrients to sustain primary production (Bowman et al., 1997b). Free-living bacteria experience diffusive limitation in nutrient uptake, enzymatic reactions, and metabolite exchange in viscous fluids (such as low-temperature and high-salinity sea ice), and it has been suggested that sea ice bacteria can survive decreasing winter temperatures by associating with sediment or other particles in the seawater (Junge

et al., 2004b). Similarly, the production of extrapolysaccharide could provide evidence that bacteria such as the CFB (known for abundant slime production) possess specific mechanisms to survive at low temperatures.

The Southern Ocean encircles the Antarctic continent and is fed by melting glaciers and is therefore not extensively ice covered. Conversely, as the sea temperatures off the coast of the Antarctic Peninsula do not exceed 2°C, it remains frozen (−1.8°C) for most of the year. The coastal waters experience high levels of solar radiation, as well as fluctuating levels of photosynthetic biomass. Despite this, many Antarctic bacteria have adapted to these conditions and can achieve growth comparable to bacteria in temperate environments (Murray et al., 1998; Grzymski et al., 2006). Cultivation-based studies providing the first data on the phylotypic diversity of Antarctic sea ice were initiated with a study on gas-vacuolate bacteria (Gosink and Staley, 1995) and a report on the general phylotypic diversity of sea ice (Bowman et al., 1997a). Both studies found that bacterial populations in sea ice are highly diverse with a similar phylogenetic distribution. The gas-vacuolate bacteria identified belonged to a novel phylogenetic lineage, and were the first to be reported from the β-*Proteobacteria* and the *Flavobacterium-Cytophaga* groups (Gosink and Staley, 1995). Overall, Antarctic sea ice bacteria belong to four phylogenetic groups: γ-*Proteobacteria* and α-*Proteobacteria*, such as *Colwellia psychrotrophica* (Bowman et al., 1998a); the CFB phylum (Brinkmeyer et al., 2003); and the Gram-positive branch (Bowman et al., 1997a; Brown and Bowman, 2001). In the late 1990s Bowman and collaborators published the description of many novel genera and species isolated from sea ice, namely *Colwellia demingiae*, *Colwellia hornerae*, *Colwellia rossensis*, *C. psychrotrophica*, *Gelidibacter algens* (novel genus), *Glaciecola punicea*, *Glaciecola pallidula*, *Pseudoalteromonas prydzensis*, *Psychrobacter glacincola*, *Psychroflexus torques* (genus), *Psychroflexus gondwanense*, *Psychroserpens burtonensis* (genus), *Shewanella gelidimarina*, and *Shewanella frigidimarina* (Bowman et al., 1997a, 1997b, 1997c, 1998a, 1998b, 1998c).

The Arctic Ocean is the smallest of the world's five major oceanic regions and covers an area of about 14 million km². Partial coverage by sea ice occurs throughout the year, and the area of the ice cover decreases from 15 million km² in March to 7 million km² in September, and this seasonal melting and freezing causes the temperature and salinity of the ocean to vary (Cavalieri et al., 1997; Shiklomanov et al., 2000). The main phylogenetic study on culturable sea ice bacterial communities in the Arctic found that up to 62% of sea ice bacteria were culturable using conventional culturing methods. Several new species were identified, although there was no novelty at the higher taxonomic levels (Junge et al., 2002). Novel gas-vacuolate bacteria have also been isolated from Arctic sea ice (Gosink et al., 1993), particularly in Point Barrow, Alaska, where they comprised 0.2% of the culturable isolates, and cell densities of 186 vacuolate bacteria per ml were reported (Gosink et al., 1993). As mentioned previously, the gas vacuoles may confer a selection advantage to these organisms, providing buoyancy in the vertical water column (Gosink and Staley, 1995). In the Arctic Fram Strait region, phylogenetic studies showed that sea ice communities were dominated by β-*Proteobacteria* and *Actinobacteria*, including limnic phylotypes (Brinkmeyer et al., 2003). In the Chukchi Sea, 10 samples (ranging from clear ice to algal-band ice) resulted in the culturing of 44 isolates, many of which were shared between samples from different psychrophilic ecotypes (Junge et al., 2002).

Culture-independent metagenomic approaches have been employed for the analysis of SIMCOs and particularly for comparative purposes (Arctic versus Antarctic). Interestingly, the findings from culture-independent phylogenetic studies of sea ice agree with those from culture-based studies. Many of the phylotypes identified from sea ice are represented in sequence databases of previously cultivated bacteria from the same region. This is in contrast to studies of other polar habits, such as seawater, where a significant proportion of the detected phylotypes have no cultured repre-

sentatives (Junge et al., 2004a). The limited diversity in sea ice could be attributed to three main reasons (Junge et al., 2002). Firstly, the geological youth of the sea ice environment has not given enough time for highly diverse sea ice bacteria to evolve. Secondly, the extreme thermal and saline conditions encountered may have resulted in organisms uniquely adapted to these stresses to dominate in these ecosystems. Thirdly, the porous nature of sea ice provides many attachment sites that may select for specific bacterial types. Certainly the higher species diversity in a Canadian sea ice sample was proposed to be attributed to the higher algal numbers in the sample, facilitating bacterial association and thereby enrichment (Brinkmeyer et al., 2003). Culture-independent studies have also been successful in supporting the bipolar distribution hypothesis, which states that the presence of the same bacterial species at both poles indicates a cosmopolitan nature of bacterial distribution, despite the inherent difficulties of dispersal (Mock and Thomas, 2005). The detection of identical phylotypes from the *Actinobacteria* and β-*Proteobacteria* groups in both Arctic and Antarctic samples (Bowman et al., 1997a; Junge et al., 2002; Brinkmeyer et al., 2003) and the identification of an *S. frigidimarina* strain from Canadian sea ice with 100% sequence similarity to an isolate from Antarctica support the bipolar distribution theory (Staley and Gosink, 1999). Furthermore, the detection of limnic phylotypes in sea ice communities tentatively points toward a terrestrial influence on SIMCOs (Brinkmeyer et al., 2003).

Coastal Waters

The productivity of coastal waters is subject to seasonal increases (during the summer) due to the influx of melted sea ice seeding the water column with phytoplankton species (Brierley and Thomas, 2002), leading to periodical algal blooms during the Antarctic summer. Coastal waters' microbial communities have not been fully characterized in terms of their diversity or physiology, and any interpretation of phylogenetic data from clone libraries (RNA or DNA) can be taken out of context. A report based on the full sequencing of six fosmid library clones (Grzymski et al., 2006) identified some phyla commonly found in polar regions (*Gemmatimonadetes*, α- and γ-*Proteobacteria*, *Bacteroidetes*, and high-G+C Gram-positive bacteria) and some unique members of the community with no close relatives. The study also provided an in-depth analysis of the protein sequences, and potential metabolism and stress functions that may facilitate survival in coastal waters. The dominance of γ-*Proteobacteria*, along with *Bacteroidetes* and a small proportion of α-*Proteobacteria*, is typical for sea ice communities, and as sea ice flows directly into coastal water as it melts, it is logical that these communities will bear some similarity to that of sea ice (Gentile et al., 2006).

EFFECTS OF CLIMATE CHANGE ON POLAR REGIONS

While human impact on the Antarctic continent and the Arctic is limited, these regions are affected indirectly by humans. The depletion of the ozone layer and global warming are beginning to impact on these delicate ecosystems, sadly before their biological potential has been fully explored. Globally, the average air temperature of the earth's surface increased by 0.06°C per decade during the 20th century, and elevated temperatures were detected in the Antarctic Peninisula during the second half of the 20th century (Doran et al., 2002b). Polar ecosystems are highly susceptible to environmental disturbances, and current climate change models predict that warming will be amplified in the North and South Polar regions, especially in spring and summer months (Anisimov et al., 2007). In Arctic terrestrial habitats climate change is likely to result in a shift of ecotypes as more bare ground will be exposed, resulting in boreal forest advancing into areas that were previously covered in snow/ice. In alpine regions global warming may increase the rate of nitrogen deposition, thereby resulting in soil becoming saturated with nitrogen. Additionally, Arctic tundra contains large reservoirs of soil carbon and plays an important role in the global carbon cycle. Elevated temperatures will result in the soil thawing for extended periods, thereby resulting in increased microbial respiration, and the mineralization of soil carbon will inadvertently have a positive feedback on global warming (Nemergut et al., 2005).

As our knowledge of microbial ecology under extreme stress conditions advances, so the potential to use microorganisms as bioindicators of climate change becomes a possibility (Pointing et al., 2009). This has been highlighted in the comparison of hypolithic colonization in the Antarctic Dry Valleys (Pointing et al., 2009) to colonization frequencies in maritime polar (Cockell and Stokes, 2004a) and nonpolar deserts (Warren-Rhodes et al., 2006, 2007), the results of which suggest that landscape-scale patterns may be closely related to climatic variables (Pointing et al., 2009). However, in order to employ microorganisms in this way, it is imperative that the true microbial diversity and community structure in polar environments be sufficiently characterized. Without a complete knowledge of the microbiology, the ability to accurately assess the sensitivity of the communities to climate change (and other effects) will be limited.

CONCLUSIONS AND FUTURE WORK

Since the discovery of microorganisms that thrive in psychrophilic habitats, microbiologists have had a fascination with the earth's polar regions. Early microbial studies were limited to culture-based analysis, which focused predominantly on species distribution. While these traditional, culture-based studies were invaluable, the main disadvantage of such methods is that many slow-growing, coculture-dependent, or fastidious microorganisms may be missed (Smith et al., 2006). Culture-independent metagenomic methods have allowed researchers to take a whole-community-based approach to microbial ecology. Using metagenomic methods researchers can assess the diversity of culturable and uncultured organisms, including rare taxa. However, the reliability of a metagenomic approach has been called into question. Biases may be introduced during sample handling,

DNA extraction, and PCR (Moeseneder et al., 2005). Additionally, metagenomic studies are limited by the depth of sequence analysis as well as the inherent uncertainties associated with phylogenetic techniques. The identification of isolates and/or phylotypes using public databases (such as the Ribosomal Database Project) is limited to those sequences stored in the database; therefore, it is only an indication of the closest known species match, not an absolute measure of true species identity. This is especially relevant in polar regions, where fewer phylogenetic studies have been conducted in comparison to less extreme environments.

Technological advances in sequencing and annotation power over the last 15 years have facilitated the sustained discovery of novel gene sequences and their deposition in public databases. "Next-generation" sequencing technologies (Illumina, Roche-454, ABI SOLiD) (Bentley, 2006; Bentley et al., 2008; Pandev et al., 2008) now provide over 4 Gb sequence data per run, and improvements in computational capabilities have resulted in easy-to-use software such as CLC (www.clcbio.com) and MEGAN (Huson et al., 2007), which have simplified annotation and contig assembly. Despite these advances, metagenomic analyses are still influenced by the quality of genetic information available, and the deposition of additional sequences in the public databases can only serve to make such analyses more accurate. In order to fully assess the true microbial diversity in any polar ecosystem, a synergistic approach should be adopted. Traditional, culture-based studies can include isolation of strains on heterotrophic media, light microscopy, fluorescent in situ hybridization (FISH), oligonucleotide probes, and total and viable cell counts (DAPI staining). Metagenomic studies can be extended to include a comparison of rRNA and rDNA libraries, DGGE analysis, terminal restriction fragment length polymorphism (RFLP), and analysis of the ITS region.

Future research in polar microbiology should focus on the structure and function of microbial communities in these environments.

Psychrophiles play an invaluable role in the degradation of hydrocarbon pollutants, particularly in the Arctic tundra and boreal forests. However, studies investigating the functions of individual taxa in bioremediation processes are limited, and future studies should focus on identifying the key players in these habitats. Additionally, as global warming is likely to be more pronounced in polar regions, studies should focus on the sensitivity of endemic microorganisms to climate change. Future functional studies should not only focus on the role of individual taxa within an ecosystem but should ideally include whole-community analysis. Currently, functional studies are limited to quantitative analyses such as fixation and turnover rates, as well as the detection of the genes involved in these processes. Modern transcriptomic methods, when combined with high-throughput nucleotide sequencing, can greatly expand our understanding of these unique microhabitats.

REFERENCES

Adams, B. J., R. D. Bardgett, E. Ayres, D. H. Wall, J. Aislabie, S. Bamforth, R. Bargagli, C. Cary, P. Cavacini, L. Connell, P. Convey, J. W. Fell, F. Frati, I. D. Hogg, K. K. Newsham, A. O'Donnell, N. Russell, R. D. Seppelt, and M. I. Stevens. 2006. Diversity and distribution of Victoria Land biota. Soil Biol. Biochem. 38:3003–3018.
Aislabie, J. M., P. A. Broady, and D. J. Saul. 2006a. Culturable aerobic heterotrophic bacteria from high altitude, high latitude soil of La Gorce Mountains (86° 30' S, 147° W), Antarctica. Antarct. Sci. 18:313–321.
Aislabie, J. M., K. Chour, D. J. Saul, S. Miyauchi, J. Ayton, R. F. Paetzold, and M. R. Balks. 2006b. Dominant bacteria in soils of Marble Point and Wright Valley, Victoria Land, Antarctica. Soil Biol. Biochem. 38:3041–3056.
Aislabie, J. M., S. Jordan, and G. M. Barker. 2008. Relationship between soil classification and bacterial diversity in soils of the Ross Sea region, Antarctica. Geoderma 144:9–20.
Allan, R. N., L. Lebbe, J. Heyrman, P. De Vos, C. J. Buchanan, and N. A. Logan. 2005. Brevibacillus levickii sp. nov. and Aneurinibacillus terranovensis sp. nov., two novel thermoacidophiles isolated from geothermal soils of northern Victoria Land, Antarctica. Int. J. Syst. Evol. Microbiol. 55:1039–1050.

Anisimov, O. A., D. G. Vaughan, T. V. Callaghan, C. Furgal, H. Marchant, T. D. Prowse, H. Vilhjálmsson, and J. E. Walsh. 2007. Polar regions (Arctic and Antarctic), p. 653–685. *In* IPCC, *Climate Change 2007: Impacts, Adaptation and Vulnerability. Contribution of Working Group II to the Fourth Assessment Report of the Intergovernmental Panel on Climate Change.* Cambridge University Press, Cambridge, United Kingdom and New York, NY.

Antoniades, D., J. Veillette, M. J. Martineau, C. Belzile, J. Tomkins, R. Pienitz, S. Lamoureux, and W. F. Vincent. 2009. Bacterial dominance of phototrophic communities in a High Arctic lake and its implications for paleoclimate analysis. *Polar Sci.* **3:**147–161.

Babalola, O. O., B. M. Kirby, M. Le Roes-Hill, A. Cook, S. C. Cary, S. G. Burton, and D. A. Cowan. 2009. Phylogenetic analysis of actinobacterial populations associated with Antarctic Dry Valley mineral soils. *Environ. Microbiol.* **11:**566–576.

Bahr, M., J. E. Hobbie, and M. L. Sogin. 1996. Bacterial diversity in an arctic lake: a freshwater SAR11 cluster. *Aquat. Microb. Ecol.* **11:**271–277.

Baker, J. 1970. Yeasts, moulds and bacteria from an acid peat on Signy Island, p. 717–722. *In* M. Holdgate (ed.), *Antarctic Ecology,* 2nd ed. Academic Press, New York, NY.

Bargagli, R., M. L. Skotnicki, L. Marri, M. Pepi, A. Mackenzie, and C. Agnorelli. 2004. New record of moss and thermophilic bacteria species and physico-chemical properties of geothermal soils on the northwest slope of Mt. Melbourne (Antarctica). *Polar Biol.* **27:**423–431.

Barrett, J. E., R. A. Virginia, A. N. Parsons, and D. H. Wall. 2006. Soil carbon turnover in the McMurdo Dry Valleys, Antarctica. *Soil Biol. Biochem.* **38:**3065–3082.

Bentley, D. 2006. Whole-genome re-sequencing. *Curr. Opin. Genet. Dev.* **16:**545–552.

Bentley, D. R., S. Balasubramanian, H. P. Swerdlow, G. P. Smith, J. Milton, C. G. Brown, K. P. Hall, D. J. Evers, C. L. Barnes, and H. R. Bignell. 2008. Accurate whole human genome sequencing using reversible terminator chemistry. *Nature* **456:**53–59.

Biondi, N., M. R. Tredici, A. Taton, A. Wilmotte, D. A. Hodgson, D. Losi, and F. Marinelli. 2007. Cyanobacteria from benthic mats of Antarctic lakes as a source of new bioactivities. *J. Appl. Microbiol.* **105:**105–115.

Bottos, E. M., W. F. Vincent, C. W. Greer, and L. G. Whyte. 2008. Prokaryotic diversity of arctic ice shelf microbial mats. *Environ. Microbiol.* **10:**950–966.

Bowman, J. P., J. Cavanagh, J. J. Austin, and K. Sanderson. 1996. Novel *Psychrobacter* species from Antarctic ornithogenic soils. *Int. J. Syst. Bacteriol.* **46:**841–848.

Bowman, J. P., J. J. Gosink, S. A. McCammon, T. E. Lewis, D. S. Nichols, P. D. Nichols, J. H. Skerratt, J. T. Staley, and T. A. McMeekin. 1998a. *Colwellia demingiae* sp. nov., *Colwellia hornerae* sp. nov., *Colwellia rossensis* sp. nov., and *Colwellia psychrotropica* sp. nov.: psychrophilic Antarctic species with the ability to synthesize docosahexaenoic acid (22:6ω3). *Int. J. Syst. Bacteriol.* **48:**1171–1180.

Bowman, J. P., S. A. McCammon, J. L. Brown, and T. A. McMeekin. 1998b. *Glaciecola punicea* gen. nov., sp. nov., and *Glaciecola pallidula* gen. nov., sp. nov.: psychrophilic bacteria from Antarctic sea-ice habitats. *Int. J. Syst. Bacteriol.* **48:**1213–1222.

Bowman, J. P., S. A. McCammon, M. V. Brown, D. S. Nichols, and T. A. McMeekin. 1997a. Diversity and association of psychrophilic bacteria in Antarctic sea ice. *Appl. Environ. Microbiol.* **63:**3068–3078.

Bowman, J. P., S. A. McCammon, J. L. Brown, P. D. Nichols, and T. A. McMeekin. 1997b. *Psychroserpens burtonensis* gen. nov., sp. nov., and *Gelidibacter algens* gen. nov., sp. nov., psychrophilic bacteria isolated from Antarctic lacustrine and sea ice habitats. *Int. J. Syst. Bacteriol.* **14:**670–677.

Bowman, J. P., S. A. McCammon, T. Lewis, J. H. Skerratt, J. L. Brown, D. S. Nichols, and T. A. McMeekin. 1998c. *Psychroflexus torquis* gen. nov., sp. nov., a psychrophilic species from Antarctic sea ice, and reclassification of *Flavobacterium gondwanense* (Dobson et al. 1993) as *Psychroflexus gondwanense* gen. nov., comb. nov. *Microbiology* **144:**1601–1609.

Bowman, J. P., S. A. McCammon, D. S. Nichols, J. H. Skerratt, S. M. Rea, P. D. Nichols, and T. A. McMeekin. 1997c. *Shewanella gelidimarina* sp. nov. and *Shewanella frigidimarina* sp. nov., novel Antarctic species with the ability to produce eicosapentaenoic acid (20:5ω3) and grow anaerobically by dissimilatory Fe(III) reduction. *Int. J. Syst. Bacteriol.* **47:**1040–1047.

Boyd, W., and J. Boyd. 1963. A bacteriological study of an arctic coastal lake. *Ecology* **44:**705–710.

Boyd, W., and J. Boyd. 1964. The presence of bacteria in permafrost of the Alaskan arctic. *Can. J. Microbiol.* **10:**917–919.

Brambilla, E., H. Hippe, A. Hagelstein, B. J. Tindall, and E. Stackebrandt. 2001. 16S rDNA diversity of cultured and uncultured prokaryotes of a mat sample from Lake Fryxell, McMurdo Dry Valleys, Antarctica. *Extremophiles* **5:**23–33.

Brierley, A. S., and D. N. Thomas. 2002. Ecology of Southern Ocean pack ice. *Adv. Mar. Biol.* **43:**171–276.

Brinkmeyer, R., F. O. Glöckner, E. Helmke, and R. Amann. 2004. Predominance of β-Proteobacteria in summer melt pools on Arctic pack ice. *Limnol. Oceanogr.* **49**:1013–1021.

Brinkmeyer, R., K. Knittel, J. Jürgens, H. Weyland, R. Amann, and E. Helmke. 2003. Diversity and structure of bacterial communities in Arctic versus Antarctic pack ice. *Appl. Environ. Microbiol.* **69**:6610–6619.

Broady, P. 1981. The ecology of sublithic terrestrial algae at the Vestfold Hills, Antarctica. *Br. Phycol. J.* **16**:231–240.

Broady, P., D. Given, L. Greenfield, and K. Thompson. 1987. The biota and environment of fumaroles on Mount Melbourne, Northern Victoria Land. *Polar Biol.* **7**:97–113.

Brown, M. V., and J. P. Bowman. 2001. A molecular phylogenetic survey of sea-ice microbial communities (SIMCO). *FEMS Microbiol. Ecol.* **35**:267–275.

Buckeridge, K. M., and P. Grogan. 2008. Deepened snow alters soil microbial nutrient limitations in arctic birch hummock tundra. *Appl. Soil Ecol.* **39**:210–222.

Callaghan, T. V., S. Jonasson, H. Nichols, R. B. Heywood, and P. A. Wookey. 2010. Arctic terrestrial ecosystems and environmental change. *Philos. Trans. Phys. Sci. Eng.* **352**:259–276.

Cameron, R., J. King, and C. David. 1970. Microbiology, ecology and microclimatology of soil sites in Dry Valleys of Southern Victoria Land, Antarctica, p. 702–716. *In* M. Holdgate (ed.), *Antarctic Ecology*, 2nd ed. Academic Press, New York, NY.

Cameron, R., G. Lacy, F. Morelli, and J. Marsh. 1971. Farthest south soil microbial and ecological investigations. *Antarct. J.* **6**:105–106.

Cameron, R., F. Morelli, and R. Johnson. 1972. Bacterial species in soil and air of the Antarctic continent. *Antarct. J.* **7**:187–189.

Cao, M., S. Marshall, and K. Gregson. 1996. Global carbon exchange and methane emissions from natural wetlands: application of a process-based model. *J. Geophys. Res.* **101**:14399–14414.

Carpenter, E. J., S. Lin, and D. C. Capone. 2000. Bacterial activity in South Pole snow. *Appl. Environ. Microbiol.* **66**:4514–4517.

Cary, S. C., I. R. McDonald, J. E. Barrett, and D. A. Cowan. 2010. On the rocks: microbial ecology of Antarctic cold desert soils. *Nat. Rev. Microbiol.* **8**:129–138.

Cavalieri, D. J., P. Gloersen, C. L. Parkinson, J. C. Comiso, and H. J. Zwally. 1997. Observed hemispheric asymmetry in global sea ice changes. *Science* **278**:1104–1106.

Christner, B. C., B. H. Kvitko II, and J. N. Reeve. 2003. Molecular identification of Bacteria and Eukarya inhabiting an Antarctic cryoconite hole. *Extremophiles* **7**:177–183.

Christner, B. C., E. Mosley-Thompson, L. G. Thompson, and J. N. Reeve. 2001. Isolation of bacteria and 16S rDNAs from Lake Vostok accretion ice. *Environ. Microbiol.* **3**:570–577.

Christner, B., G. Royston-Bishop, C. F. Foreman, B. R. Arnold, M. Tranter, K. A. Welch, W. B. Lyons, A. I. Tsapin, M. Studinger, and J. C. Priscu. 2006. Limnological conditions in subglacial Lake Vostok, Antarctica. *Limnol. Oceanogr.* **51**:2485–2501.

Cockell, C., P. Rettberg, G. Horneck, K. Scherer, and M. D. Stokes. 2003. Measurements of microbial protection from ultraviolet radiation in polar terrestrial microhabitats. *Photochem. Photobiol.* **26**:62–69.

Cockell, C. S., and M. D. Stokes. 2004a. Hypolithic colonization of opaque rocks in the Arctic and Antarctic polar desert. *Arct. Antarct. Alp. Res.* **38**:335–342.

Cockell, C. S., and M. D. Stokes. 2004b. Ecology: widespread colonization by polar hypoliths. *Nature* **431**:414.

Cowan, D. A., and L. Ah Tow. 2004. Endangered Antarctic environments. *Annu. Rev. Microbiol.* **58**:649–690.

Cowan, D. A., N. J. Russell, A. Mamais, and D. M. Sheppard. 2002. Antarctic Dry Valley mineral soils contain unexpectedly high levels of microbial biomass. *Extremophiles* **6**:431–436.

Crump, B. C., G. W. Kling, M. Bahr, and J. E. Hobbie. 2003. Bacterioplankton community shifts in an Arctic lake correlate with seasonal changes in organic matter source. *Appl. Environ. Microbiol.* **69**:2253–2268.

Davey, M., and K. Clarke. 1991. The spatial distribution of microalgae in Antarctic fellfield soils. *Antarct. Sci.* **3**:257–263.

de La Torre, J. R., B. M. Goebel, E. I. Friedmann, and N. R. Pace. 2003. Microbial diversity of cryptoendolithic communities from the McMurdo Dry Valleys, Antarctica. *Appl. Environ. Microbiol.* **69**:3858–3867.

Dobson, S. J., R. R. Colwell, T. A. McMeekin, and P. D. Franzmann. 1993. Direct sequencing of the polymerase chain reaction-amplified 16S rRNA gene of *Flavobacterium gondwanense* sp. nov. and *Flavobacterium salegens* sp. nov., two new species from a hypersaline Antarctic lake. *Int. J. Syst. Bacteriol.* **43**:77–83.

Doran, P. T., C. P. McKay, G. D. Clow, G. L. Dana, A. G. Fountain, T. Nylen, and W. B. Lyons. 2002a. Valley floor climate observations from the McMurdo dry valleys, Antarctica, 1986–2000. *J. Geophys. Res.* **107**:4772.

Doran, P. T., J. C. Priscu, W. B. Lyons, J. E. Walsh, A. G. Fountain, D. M. McKnight, D. L. Moorhead, R. A. Virginia, D. H. Wall, G.

D. Clow, C. H. Fritsen, C. P. McKay, and A. N. Parsons. 2002b. Antarctic climate cooling and terrestrial ecosystem response. *Nature* **415:**517–520.

Ernst, A., S. Becker, U. I. Wollenzien, and C. Postius. 2003. Ecosystem-dependent adaptive radiations of picocyanobacteria inferred from 16S rRNA and ITS-1 sequence analysis. *Microbiology* **149:**217–228.

Fox, G. E., J. D. Wisotzkey, and P. Jurtshuk, Jr. 1992. How close is close: 16S rRNA sequence identity may not be sufficient to guarantee species identity. *Int. J. Syst. Bacteriol.* **42:**166–170.

Friedmann, E. I. (ed.). 1993. *Antarctic Microbiology*. Wiley-Liss, New York, NY.

Friedmann, E. I. 1994. Permafrost as microbial habitat, p. 21–26. *In* D. Gilichinsky (ed.), *Viable Microorganisms in Permafrost*. Russian Academy of Sciences, Pushchino, Russia.

Friedmann, E. I., M. Hua, and R. Ocampo-Friedmann. 1988. Cryptoendolithic lichen and cyanobacterial communities of the Ross Desert, Antarctica. *Polarforschung* **58:**251–259.

Friedmann, E. I., C. P. McKay, and J. A. Nienow. 1987. The cryptoendolithic microbial environment in Ross Desert of Antarctica: satellite-transmitted continuous nanoclimate data, 1984 to 1986. *Polar Biol.* **7:**273–287.

Friedmann, E. I., and R. Ocampo. 1976. Cyptoendolithic blue-green algae in the dry valleys: primary producers in the Antarctic desert ecosystem. *Science* **193:**1247–1249.

Ganzert, L., G. Jurgens, U. Münster, and D. Wagner. 2007. Methanogenic communities in permafrost-affected soils of the Laptev Sea coast, Siberian Arctic, characterized by 16S rRNA gene fingerprints. *FEMS Microbiol. Ecol.* **59:**476–488.

Garcia, J. L., B. K. Patel, and B. Ollivier. 2000. Taxonomic, phylogenetic and ecological diversity of methanogenic Archaea. *Anaerobe* **6:**205–226.

Gentile, G., L. Giuliano, G. D'Auria, F. Smedile, M. Azzaro, M. De Domenico, and M. M. Yakimov. 2006. Study of bacterial communities in Antarctic coastal waters by a combination of 16S rRNA and 16S rDNA sequencing. *Environ. Microbiol.* **8:**2150–2161.

Gilichinsky, D., G. Khlebnikova, D. Zvyagintsev, D. Fedorov-Davydov, and N. Kudryavtseva. 1989. Microbiology of sedimentary materials in the permafrost zone. *Int. Geol. Rev.* **31:**847–858.

Gordon, D. A., J. Priscu, and S. Giovannoni. 2000. Origin and phylogeny of microbes living in permanent Antarctic lake ice. *Microb. Ecol.* **39:**197–202.

Gosink, J. J., R. L. Irgens, and J. T. Staley. 1993. Vertical distribution of bacteria in arctic sea ice. *FEMS Microbiol. Lett.* **102:**85–90.

Gosink, J. J., and J. T. Staley. 1995. Biodiversity of gas vacuolate bacteria from Antarctic sea ice and water. *Appl. Environ. Microbiol.* **61:**3486–3489.

Grossmann, S. 1994. Bacterial activity in sea ice and open water of the Weddell Sea, Antarctica: a microautoradiographic study. *Microb. Ecol.* **28:**1–18.

Grzymski, J. J., B. J. Carter, E. F. DeLong, R. A. Feldman, A. Ghadiri, and A. E. Murray. 2006. Comparative genomics of DNA fragments from six Antarctic marine planktonic bacteria. *Appl. Environ. Microbiol.* **72:**1532–1541.

Gupta, P., G. S. Reddy, D. Delille, and S. Shivaji. 2004. *Arthrobacter gangotriensis* sp. nov. and *Arthrobacter kerguelensis* sp. nov. from Antarctica. *Int. J. Syst. Evol. Microbiol.* **54:**2375–2378.

Harris, J. M., and B. J. Tibbles. 1997. Factors affecting bacterial productivity in soils on isolated inland nunataks in continental Antarctica. *Microb. Ecol.* **33:**106–123.

Helmke, E., and H. Weyland. 1995. Bacteria in sea ice and underlying water of the eastern Weddell Sea in midwinter. *Mar. Ecol. Prog. Ser.* **117:**269–287.

Hobbie, J. E., T. L. Corliss, and B. J. Peterson. 1983. Seasonal patterns of bacterial abundance in an Arctic lake. *Arct. Alp. Res.* **15:**253–259.

Hogg, I. D., S. C. Cary, P. Convey, K. K. Newsham, A. G. Donnell, B. J. Adams, J. Aislabie, F. Frati, M. I. Stevens, and D. H. Wall. 2006. Biotic interactions in Antarctic terrestrial ecosystems: are they a factor? *Soil Biol. Biochem.* **38:**3035–3040.

Hughes, K. A., and B. Lawley. 2003. A novel Antarctic microbial endolithic community within gypsum crusts. *Environ. Microbiol.* **5:**555–565.

Huson, D. H., A. F. Auch, J. Qi, and S. C. Schuster. 2007. MEGAN analysis of metagenomic data. *Genome Res.* **17:**377–386.

Jefferies, R. L., N. A. Walker, K. A. Edwards, and J. Dainty. 2010. Is the decline of soil microbial biomass in late winter coupled to changes in the physical state of cold soils? *Soil Biol. Biochem.* **42:**129–135.

Jungblut, A. D., I. Hawes, D. Mountfort, B. Hitzfeld, D. R. Dietrich, B. P. Burns, and B. A. Neilan. 2005. Diversity within cyanobacterial mat communities in variable salinity meltwater ponds of McMurdo Ice Shelf, Antarctica. *Environ. Microbiol.* **7:**519–529.

Junge, K., H. Eicken, and J. W. Deming. 2004a. A microscopic approach to investigate bacteria under in situ conditions in Arctic lake ice: initial comparisons to sea ice, p. 381–388. *In* R. Norris and F. Stootman (ed.), *Life amongst the Stars*. Bioastronomy 2002; 213 edition. IAU Symposium Astronomical Society of the Pacific, San Francisco, CA.

Junge, K., H. Eicken, and J. W. Deming. 2004b. Bacterial activity at −2 to −20°C in Arctic wintertime sea ice. *Appl. Environ. Microbiol.* **70:** 550–557.

Junge, K., F. Imhoff, T. Staley, and J. W. Deming. 2002. Phylogenetic diversity of numerically important Arctic sea-ice bacteria cultured at subzero temperature. *Microb. Ecol.* **43:**315–328.

Karr, E. A., W. M. Sattley, D. O. Jung, M. T. Madigan, and L. A. Achenbach. 2003. Remarkable diversity of phototrophic purple bacteria in a permanently frozen Antarctic lake. *Appl. Environ. Microbiol.* **69:**4910–4914.

Labbé, D., R. Margesin, F. Schinner, L. G. Whyte, and C. W. Greer. 2007. Comparative phylogenetic analysis of microbial communities in pristine and hydrocarbon-contaminated Alpine soils. *FEMS Microbiol. Ecol.* **59:**466–475.

Lindström, E. S., M. P. Kamst-Van Agterveld, and G. Zwart. 2005. Distribution of typical freshwater bacterial groups is associated with pH, temperature, and lake water retention time. *Appl. Environ. Microbiol.* **71:**8201–8206.

Logan, N. A., L. Lebbe, B. Hoste, J. Goris, G. Forsyth, M. Heyndrickx, B. L. Murray, N. Syme, D. D. Wynn-Williams, and P. De Vos. 2000. Aerobic endospore-forming bacteria from geothermal environments in northern Victoria Land, Antarctica, and Candlemas Island, South Sandwich archipelago, with the proposal of *Bacillus fumarioli* sp. nov. *Int. J. Syst. Evol. Microbiol.* **50:**1741–1753.

Löve, D. 1970. Subarctic and subalpine: where and what? *Arct. Antarct. Alp. Res.* **2:**63–73.

Ludley, K. E., and C. H. Robinson. 2008. "Decomposer" Basidiomycota in Arctic and Antarctic ecosystems. *Soil Biol. Biochem.* **40:**11–29.

Männistö, M., and Häggblom, M. 2006. Characterization of psychrotolerant heterotrophic bacteria from Finnish Lapland 109. *Syst. Appl. Microbiol.* **29:**229–243.

Männistö, M., H. Kontio, M. Tiirola, and M. Häggblom. 2008. Seasonal variation in active bacterial communities of Fennoscandian tundra soil. 3rd Int. Conf. Polar Alp. Microbiol., Alberta, Canada, 11 to 15 May 2008.

Margesin, R., P. A. Fonteyne, and B. Redl. 2005. Low-temperature biodegradation of high amounts of phenol by *Rhodococcus* spp. and basidiomycetous yeasts. *Res. Microbiol.* **156:**68–75.

Margesin, R., D. Labbé, F. Schinner, C. W. Greer, and L. G. Whyte. 2003. Characterization of hydrocarbon-degrading microbial populations in contaminated and pristine Alpine soils. *Appl. Environ. Microbiol.* **69:**3085–3092.

Margesin, R., P. Schumann, C. Spröer, and A. M. Gounot. 2004. *Arthrobacter psychrophenolicus* sp.

nov., isolated from an alpine ice cave. *Int. J. Syst. Evol. Microbiol.* **54:**2067–2072.

McKay, C. P., and E. I. Friedmann. 1985. The cryptoendolithic microbial environment in the Antarctic cold desert: temperature variation in nature. *Polar Biol.* **4:**19-25.

Mock, T., and D. N. Thomas. 2005. Recent advances in sea-ice microbiology. *Environ. Microbiol.* **7:**605–619.

Moeseneder, M. M., J. M. Arrieta, and G. J. Herndl. 2005. A comparison of DNA- and RNA-based clone libraries from the same marine bacterioplankton community. *FEMS Microbiol. Ecol.* **51:**341–352.

Mohn, W. W., and G. R. Stewart. 2000. Limiting factors for hydrocarbon biodegradation at low temperature in Arctic soils. *Soil Biol. Biochem.* **32:**1161–1172.

Mondino, L. J., M. Asao, and M. T. Madigan. 2009. Cold-active halophilic bacteria from the ice-sealed Lake Vida, Antarctica. *Arch. Microbiol.* **191:**785–790.

Mosier, A. C., A. E. Murray, and C. H. Fritsen. 2006. Microbiota within the perennial ice cover of Lake Vida, Antarctica. *FEMS Microbiol. Ecol.* **59:**274–288.

Murray, A. E., C. M. Preston, R. Massana, L. T. Taylor, A. Blakis, K. Wu, and E. F. DeLong. 1998. Seasonal and spatial variability of bacterial and archaeal assemblages in the coastal waters near Anvers Island, Antarctica. *Appl. Environ. Microbiol.* **64:**2585–2595.

Nemergut, D. R., E. K. Costello, A. F. Meyer, M. Y. Pescador, M. N. Weintraub, and S. K. Schmidt. 2005. Structure and function of alpine and arctic soil microbial communities. *Res. Microbiol.* **156:**775–784.

Newsham, K. K., D. A. Pearce, and P. D. Bridge. 2010. Minimal influence of water and nutrient content on the bacterial community composition of a maritime Antarctic soil. *Microbiol. Res.* **165:**523–530.

Nicolaus, B., L. Lama, E. Esposito, M. Manca, G. Di Prisco, and A. Gambacorta. 1996. "*Bacillus thermoantarcticus*" sp. nov., from Mount Melbourne, Antarctica: a novel thermophilic species. *Polar Biol.* **16:**101–104.

Niederberger, T. D., I. R. McDonald, A. L. Hacker, R. M. Soo, J. E. Barrett, D. H. Wall, and S. C. Cary. 2008. Microbial community composition in soils of Northern Victoria Land, Antarctica. *Environ. Microbiol.* **10:** 1713–1724.

Nienow, J. A., and E. I. Friedmann. 1993. Terrestrial lithophytic (rock) communities, p. 343–412. *In* E. I. Friedmann (ed.), *Antarctic Microbiology.* Wiley-Liss, New York, NY.

O'Brien, A., R. Sharp, N. J. Russell, and S. Roller. 2004. Antarctic bacteria inhibit growth of food-borne microorganisms at low temperatures. *FEMS Microbiol. Ecol.* **48:**157–167.

Omelchenko, M., L. Vasilieva, G. Zavarzin, N. Saveliena, A. Lysenko, L. Mityushina, V. Khmelenina, and Y. Trotsenko. 1996. A novel psychrophilic methanotroph of the genus *Methylobacter*. *Microbiology* **65:**339–343.

Omelon, C. R., W. H. Pollard, and F. G. Ferris. 2007. Inorganic species distribution and microbial diversity within high Arctic cryptoendolithic habitats. *Microb. Ecol.* **54:**740–752.

Pacheco-Oliver, M., I. R. McDonald, D. Groleau, J. C. Murrell, and C. B. Miguez. 2002. Detection of methanotrophs with highly diverget *pmoA* genes from Arctic soils. *FEMS Microbiol. Lett.* **209:**313–319.

Pandev, V., R. C. Nutter, and E. Prediger. 2008. Applied Biosystems SOLiD™ System: ligation-based sequencing, p. 29–42. *In* M. Janitz (ed.), *Next Generation Genome Sequencing: Towards Personalized Medicine*. Wiley-VCH Verlag, Weinheim, Germany.

Pointing, S. B., Y. Chan, D. C. Lacap, M. C. Lau, J. A. Jurgens, and R. L. Farrell. 2009. Highly specialized microbial diversity in hyper-arid polar desert. *Proc. Natl. Acad. Sci. USA* **106:**19964–19969.

Poli, A., E. Esposito, L. Lama, P. Orlando, G. Nicolaus, F. de Appolonia, A. Gambacorta, and B. Nicolaus. 2006. *Anoxybacillus amylolyticus* sp. nov., a thermophilic amylase producing bacterium isolated from Mount Rittmann (Antarctica). *Syst. Appl. Microbiol.* **29:**300–307.

Potts, M. 1994. Desiccation tolerance of prokaryotes. *Microbiol. Rev.* **58:**755–805.

Prabahar, V., S. Dube, G. S. Reddy, and S. Shivaji. 2004. *Pseudonocardia antarctica* sp. nov. an Actinomycetes from McMurdo Dry Valleys, Antarctica. *Int. J. Syst. Evol. Microbiol.* **27:**66–71.

Price, P. B. 2006. Microbial life in glacial ice and implications for a cold origin of life. *FEMS Microbiol. Ecol.* **59:**217–231.

Schimel, J. P., and C. Mikan. 2005. Changing microbial substrate use in Arctic tundra soils through a freeze-thaw cycle. *Soil Biol. Biochem.* **37:**1411–1418.

Schinner, F., R. Margesin, and T. Pümpel. 1992. Extracellular protease-producing psychrotrophic bacteria from high alpine habitats. *Arct. Antarct. Alp. Res.* **24:**88–92.

Shi, T., R. H. Reeves, D. A. Gilichinsky, and E. I. Friedmann. 1997. Characterization of viable bacteria from Siberian permafrost by 16S rDNA sequencing. *Soil Sci.* **33:**169–179.

Shiklomanov, I. A., A. I. Shiklomanov, R. B. Lammers, B. J. Peterson, and C. J. Vörösmarty. 2000. The dynamics of river water inflow to the Arctic Ocean, p. 281–296. *In* E. L. Lewis, E. P. Jones, P. Lemke, T. D. Prowse, and P. Wadhams (ed.), *The Freshwater Budget of the Arctic Ocean*. Kluwer Academic Publishers, Dordrecht, The Netherlands.

Shivaji, S., N. S. Rao, L. Saisree, V. Sheth, G. S. Reddy, and P. M. Bhargava. 1989. Isolation and identification of *Pseudomonas* spp. from Schirmacher Oasis, Antarctica. *Appl. Environ. Microbiol.* **55:**767–770.

Shravage, B. V., K. M. Dayananda, M. S. Patole, and Y. S. Shouche. 2007. Molecular microbial diversity of a soil sample and detection of ammonia oxidizers from Cape Evans, McMurdo Dry Valley, Antarctica. *Microbiol. Res.* **162:**15–25.

Siebert, J., P. Hirsch, B. Hoffmann, C. G. Gliesche, K. Peissl, and M. Jendrach. 1996. Cryptoendolithic microorganisms from Antarctic sandstone of Linnaeus Terrace (Asgard Range): diversity, properties and interactions. *Biodivers. Conserv.* **5:**1337–1363.

Simankova, M. V., O. R. Kotsyurbenko, E. Stackebrandt, N. A. Kostrikina, A. M. Lysenko, G. A. Osipov, and A. N. Nozhevnikova. 2000. *Acetobacterium tundrae* sp. nov., a new psychrophilic acetogenic bacterium from tundra soil. *Arch. Microbiol.* **174:**440–447.

Smith, J. J., L. Ah Tow, W. Stafford, C. Cary, and D. A. Cowan. 2006. Bacterial diversity in three different Antarctic cold desert mineral soils. *Microb. Ecol.* **51:**413–421.

Smith, M. C., J. P. Bowman, F. J. Scott, and M. A. Line. 2000. Sublithic bacteria associated with Antarctic quartz stones. *Antarct. Sci.* **12:**177–184.

Staley, J. T., and J. J. Gosink. 1999. Poles apart: biodiversity and biogeography of sea ice bacteria. *Annu. Rev. Microbiol.* **53:**189–215.

Steven, B., G. Briggs, C. P. McKay, W. H. Pollard, C. W. Greer, and L. G. Whyte. 2007. Characterization of the microbial diversity in a permafrost sample from the Canadian high Arctic using culture-dependent and culture-independent methods. *FEMS Microbiol. Ecol.* **59:**513–523.

Suzuki, T., T. Nakayama, T. Kurihara, T. Nishino, and N. Esaki. 2009. Cold-active lipolytic activity of psychrotrophic *Acinetobacter* sp. strain no. 6. *J. Biosci. Bioeng.* **92:**144–148.

Taton, A., S. Grubisic, P. Balthasart, D. A. Hodgson, J. Laybourn-Parry, and A. Wilmotte. 2006. Biogeographical distribution and ecological ranges of benthic cyanobacteria in East Antarctic lakes. *FEMS Microbiol. Ecol.* **57:**272–289.

Taton, A., S. Grubisic, E. Brambilla, R. De Wit, and A. Wilmotte. 2003. Cyanobacterial diversity in natural and artificial microbial mats of Lake Fryxell (McMurdo Dry Valleys, Antarctica): a morphological and molecular approach. *Appl. Environ. Microbiol.* **69:**5157–5169.

Thomas, D. N. 2005. Photosynthetic microbes in freezing deserts. *Trends Microbiol.* **13:**87–88.

Tindall, B. J. 2004. Prokaryotic diversity in the Antarctic: the tip of the iceberg. *Microb. Ecol.* **47:**271–283.

Tourova, T. P., M. V. Omelchenko, K. V. Fegeding, and L. V. Vasilieva. 1999. The phylogenetic position of *Methylobacter psychrophilus* sp. nov. *Microbiology* **68:**437–444.

Van Trappen, S., J. Mergaert, S. Van Eygen, P. Dawyndt, M. Cnockaert, and J. Swings. 2002. Diversity of 746 heterotrophic bacteria isolated from microbial mats from ten Antarctic lakes. *Syst. Appl. Microbiol.* **25:**603–610.

Vincent, W. F. 1988. *Microbial Ecosystems of Antarctica.* Cambridge University Press, Cambridge, United Kingdom.

Vincent, W. F., M. T. Downes, R. W. Castenholz, and C. Howard-Williams. 1993. Community structure and pigment organisation of cyanobacterial-dominated microbial mats in Antarctica. *Eur. J. Phycol.* **28:**213–221.

Vincent, W. F., D. R. Mueller, and S. Bonilla. 2004. Ecosystems on ice: the microbial ecology of Markham Ice Shelf in the high Arctic. *Cryobiology* **48:**103–112.

Vishniac, H. 1993. The microbiology of Antarctic soils, p. 297–342. *In* E. I. Friedmann (ed.), *Antarctic Microbiology.* Wiley-Liss, New York, NY.

Vorobyova, E., N. Minkovsky, A. Mamukelashvili, D. Zvyagintsev, V. Soina, L. Polanskaya, and D. Gilichinsky. 2001. Microorganisms and biomarkers in permafrost, p. 527–541. *In* R. Paepe and V. Melnikov (ed.), *Permafrost Response on Economic Development, Environmental Security and Natural Resources.* Kluwer Academic Publishers, Dordrecht, The Netherlands.

Wallenstein, M. D., S. McMahon, and J. Schimel. 2007. Bacterial and fungal community structure in Arctic tundra tussock and shrub soils. *FEMS Microbiol. Ecol.* **59:**428–435.

Walton, D. W. H. 1984. The terrestrial environment, p. 1–60. *In* R. M. Laws (ed.), *Antarctic Ecology*, vol. 1. Academic Press, London, United Kingdom.

Warren-Rhodes, K. A., K. L. Rhodes, L. N. Boyle, S. B. Pointing, Y. Chen, S. Liu, P. Zhuo, and C. P. McKay. 2007. Cyanobacterial ecology across environmental gradients and spatial scales in China's hot and cold deserts. *FEMS Microbiol. Ecol.* **61:**470–482.

Warren-Rhodes, K. A., K. L. Rhodes, S. B. Pointing, S. A. Ewing, D. C. Lacap, B. Gómez-Silva, R. Amundson, E. I. Friedmann, and C. P. McKay. 2006. Hypolithic bacteria, dry limit of photosynthesis, and microbial ecology in the hyperarid Atacama Desert. *Microb. Ecol.* **52:**389–398.

Wartiainen, I., A. G. Hestens, I. R. McDonald, and M. M. Svenning. 2006. *Methylocystis rosea* sp. nov., a novel methanotrophic bacterium from Arctic wetland soil, Svalbard, Norway (786 N). *Int. J. Syst. Evol. Microbiol.* **56:**541–547.

Wartiainen, I., A. G. Hestens, and M. M. Svenning. 2003. Methanotrophic diversity in high arctic wetlands on the islands of Svalbard (Norway)—denaturing gel electrophoresis analysis of soil DNA and enrichment cultures. *Can. J. Microbiol.* **49:**602–612.

Wicklow, D. T., and B. Söderström (ed.). 1997. *Environmental and Microbial Relationships.* Springer, Berlin, Germany.

Williams, P. J., and M. W. Smith. 1989. *The Frozen Earth: Fundamentals of Geocryology.* Cambridge University Press, Cambridge, United Kingdom.

Yergeau, E., K. K. Newsham, D. A. Pearce, and G. A. Kowalchuk. 2007. Patterns of bacterial diversity across a range of Antarctic terrestrial habitats. *Environ. Microbiol.* **9:**2670–2682.

Yukimura, K., R. Nakai, S. Kohshima, J. Uetake, H. Kanda, and T. Naganuma. 2009. Spore-forming halophilic bacteria isolated from Arctic terrains: implications for long-range transportation of microorganisms. *Polar Sci.* **3:**163–169.

Zdanowski, M. K., M. J. Zmuda, and I. Zwolska. 2005. Bacterial role in the decomposition of marine-derived material (penguin guano) in the terrestrial maritime Antarctic. *Soil Biol. Biochem.* **37:**581–595.

Zvyagintsev, D. G. 1992. Microorganisms in permafrost, p. 229–232. *In Proceedings of the First International Conference on Cryopedology.* Pushchino, Russia.

ARCHAEA

Thomas D. Niederberger,
Ian R. McDonald, and S. Craig Cary

2

INTRODUCTION AND OVERVIEW

Modern comparative protein and genomic studies (reviewed by Gribaldo and Brochier-Armanet [2006]) have further validated the original assignment of *Archaea* as a third evolutionary domain separate from that of the *Bacteria* and *Eukarya* (Woese et al., 1990; Woese and Fox, 1997). *Archaea* were originally thought to only exist in extreme environments such as hot springs and deep-sea hydrothermal vents, as these were the only habitats to yield archaeal isolates via classical cultivation approaches. Modern molecular PCR-based methods, typically targeting the 16S rRNA gene, have now revolutionized the field of environmental microbiology and have allowed culture-independent surveys of natural in situ microbial communities. These new approaches have unearthed a wide diversity and ubiquitous presence of *Archaea* in nonex-treme environments such as soils, sediments, and oceans (Chaban et al., 2006).

The archaeal domain is split into two major phyla, the *Crenarchaeota* and *Euryarchaeota*. All cultured *Crenarchaeota* have been isolated from high-temperature environments and as such are (hyper)thermophilic, with the exception of a recently isolated low-temperature, ammonia-oxidizing archaeon (*Nitrosopumilus maritimus*) (Könneke et al., 2005). Although not widely accepted, the isolation of *N. maritimus* and the detection of similar uncultivated mesophilic *Crenarchaeota* have led to the recent proposal of a novel phylum, the *Thaumarchaeota*, encompassing *N. maritimus* and its nearest relatives (Brochier-Armanet et al., 2008). The cultured *Euryarchaeota* are more diverse than the *Crenarchaeota*, with members including methanogens (e.g., *Methanobacteriales*, *Methanococcales*), sulfate reducers (*Archaeoglobales*), halophiles (*Halobacterium*), and (hyper)thermophiles. In addition, two minor archaeal phyla have also been proposed, the *Korarchaeota* and the *Nanoarchaeota*. The *Korarchaeota* are a phylogenetically deep-branching group that have only been detected in hydrothermal environments. Members of the *Korarchaeota* have not yet been isolated in pure culture; however, the genome of a candidate member, "*Korarchaeum*

Thomas D. Niederberger, College of Earth, Ocean and Environment, University of Delaware, Lewes, DE 19958. *Ian R. McDonald*, Department of Biological Sciences, University of Waikato, Hamilton 3240, New Zealand. *S. Craig Cary*, Department of Biological Sciences, University of Waikato, Hamilton 3240, New Zealand, and College of Earth, Ocean and Environment, University of Delaware, Lewes, DE 19958.

Polar Microbiology: Life in a Deep Freeze
Edited by Robert V. Miller and Lyle G. Whyte © 2012 ASM Press, Washington, DC

cryptofilum," has recently been sequenced from a continuous coculture enrichment (Elkins et al., 2008). The phylogenetic position of the *Nanoarchaeota* phylum is not clear, and it contains a single cultured representative, *Nanoarchaeum equitans*, an extremely small, regular coccus (400-nm diameter) that lives attached to its archaeal host, *Ignicoccus hospitalis* (Huber et al., 2002). This relationship is the only known example of a parasitic/symbiotic partnership involving two *Archaea* (Forterre et al., 2009). The *Nanoarchaeota* have been shown through culture-independent analyses to inhabit both high-temperature terrestrial and marine biotopes and temperate hypersaline environments (Casanueva et al., 2008).

As *Archaea* are typically recalcitrant to cultivation, DNA-based, culture-independent studies, as mentioned above, have provided almost all our knowledge of archaeal diversity in natural environments. As a result, various 16S rRNA gene-based phylogenetic clades have been proposed to define uncultured archaeal 16S rRNA gene signatures detected in a variety of environments, as presented in Color Plate 4. A vast array of somewhat confusing acronyms exist; therefore, some of the commonly described lineages and lineages relevant to this chapter are briefly defined below; for a comprehensive review, see Chaban et al. (2006).

The crenarchaeotal Marine Group I (MGI) was originally detected in marine environments and has been broadened into five lineages: (i) 1.1a, dominated by sequences from marine plankton; (ii) 1.1b, dominated by sequences from soils and sediments; (iii) 1.1c, dominated by sequences from forest soils; (iv) 1.2, dominated by sequences from marine and lake environments; and (v) 1.3, from sediments, paleosoils, and anaerobic digestors (DeLong, 1998; Schleper et al., 2005). Group 1.3 has been further split into groups 1.3a and 1.3b (Ochsenreiter et al., 2003). Marine Groups II and III (MGII and MGIII) are euryarchaeotal lineages originally identified in lake and ocean environments, respectively, with a new lineage, tentatively titled Marine

Group IV (MGIV), branching off the haloarchaeal clade (López-García et al., 2001a; Chaban et al., 2006).

Euryarchaeota lineages associated with rice roots (Rice Clusters I through IV; RC-I to -IV) have also been proposed (Grosskopf et al., 1998) and include clades related to cultured methanogens (RC-I and -II), MGII (RC-III), nonthermophilic *Crenarchaeota* (RC-IV), and a euryarchaeotal clade (RC-V). The RC-V clade groups closely with the MGIII and the Lake Dagow Sediment (LDS) clade, originally identified in sediment from a eutrophic lake in northeastern Germany (Glissman et al., 2004). Uncultured *Archaea* inhabiting deep-ocean marine sediments and subsurface sites have also been grouped into various clades and comprehensively reviewed by Teske and Sorensen (2007). Lineages include the ubiquitous euryarchaeotal Marine Benthic Group B (MBG-B), which is equivalent to the Deep-Sea Archaeal Group (DSAG); the euryarchaeotal Marine Benthic Groups A and D (MBG-A and -D), which do not usually dominate deep-subsurface sediments (MBG-D being synonymous with MGIII); and the Miscellaneous Crenarchaeotic Group (MCG), which includes the Marine Benthic Group C subgroup (MBG-C), which have a wide habitat range and are typically detected in marine sediment environments. Uncultured *Euryarchaeota* implicated in anaerobic methane oxidation at deep-sea sediment sites have also been split into three independent phylogenetic lineages (ANME-1, -2, and -3), as reviewed by Chaban et al. (2006).

In spite of the extreme environmental conditions in the polar regions, through the application of culture-independent 16S rRNA gene-based surveys *Archaea* have been found to inhabit a wide range of polar environments. In this chapter we review the current literature describing archaeal presence and diversity in polar and subpolar habitats. *Archaea* isolated from polar habitats are listed in Table 1, and studies including the quantification of *Archaea* in polar habitats are listed in Table 2. Culture-independent 16S rRNA gene-based surveys

TABLE 1 *Archaea* isolated from polar environments

Isolate[a]	Location	Reference
Halorubrum lacusprofundi	Deep Lake, Antarctica	Franzmann et al., 1988
Methanococcoides burtonii	Ace Lake, Antarctica	Franzmann et al., 1992
Methanogenium frigidum	Ace Lake, Antarctica	Franzmann et al., 1997
"*Methanosarcina mazei*" strain MT★	Polar Urals, Russia	Simankova et al., 2003
"*Methanosarcina lacustris*" strain MS★	Subpolar pond, Russia	Simankova et al., 2003
"*Methanocorpusculum sp.*" strain MSP★	Subpolar pond, Russia	Simankova et al., 2003
"*Methanosarcina sp.*" strain FRX-1★	Lake Fryxell, Antarctica	Singh et al., 2005
"*Methanosarcina mazei*" strain JL01 VKM B-2370★	Arctic permafrost	Rivkina et al., 2007
"*Methanobacterium sp.*" strain M2 VKM B-2371★	Arctic permafrost	Rivkina et al., 2007
"*Methanobacterium sp.*" strain MK4 VKM B-2440★	Arctic permafrost	Rivkina et al., 2007
Methanobacterium arcticum	Arctic permafrost	Shcherbakova et al., 2010
Methanobacterium veterum	Siberian permafrost	Krivushin et al., 2010

[a]Isolates marked with a star are not validly named and characterized according to the *International Code of Nomenclature of Bacteria (1990 Revision). Bacteriological Code.* 1992. ASM Press.

of archaeal communities in natural polar environments are summarized in Table 3, with the locations of the associated study sites indicated in Color Plate 5. The oligonucleotides used to target *Archaea* in these PCR-based studies are also listed in Table 4. Distinct ecosystems are discussed individually for Antarctic and Arctic ecosystems, with final sections discussing comparative studies of archaeal communities between polar regions, the potential response and contribution of *Archaea* to future climate-change models, highlights of recent findings, and future research needs.

ANTARCTIC ENVIRONMENTS

Archaea in Terrestrial Environments

ARCHAEA IN SOIL HABITATS
Soils of the (sub)-Antarctic region encompass various biotopes ranging from arid, nutrient-limited mineral soils (e.g., McMurdo Dry Valleys), to ornithogenic soils (derived from the deposition of fecal matter from various species of birds), to rich soils located at higher latitude, e.g., the Antarctic Peninsula and sub-Antarctic islands. Few studies have described the microbial communities inhabiting Antarctic soils, and

they typically do not include the *Archaea*. Yergeau et al. (2007, 2009) recently surveyed microbial 16S rRNA gene signatures (including *Archaea*) and associated functional genes along an Antarctic latitudinal gradient using microarray-based analyses. The latitudinal gradient consisted of densely vegetated soils and bare fellfield soils ranging from the sub-Antarctic Falkland Islands (51°S) and Signy Island (South Orkney Islands, 60°S) to Anchorage Island (67°S) on the western Antarctic Peninsula and included frost-sorted polygon samples from Fossil Bluff (71°S) and Coal Nunatak (72°S) on the Antarctic Peninsula. Microbial diversity, based on the number of taxa detected, decreased with increasing latitude, with the Fossil Bluff and Coal Nunatak sites displaying the lowest diversity. Although the specific types of detected *Archaea* were not defined, both *Crenarchaeota* and *Euryarchaeota* were detected using PCR-based microarray analyses at the sub-Antarctic and Anchorage Island sites, but not from the frost-sorted polygon samples, i.e., Fossil Bluff and Coal Nunatak sites (Yergeau et al., 2009). Similarly, Pointing et al. (2009) did not detect *Archaea* in arid oligotrophic soils and associated samples (hypolith, chasmolith,

TABLE 2 Concentrations of *Archaea* in natural polar environments

Environment	Total cell counts[a]	Archaeal counts[a,b]	% Archaea of total counts	Method	Reference
Antarctic					
Marine waters	ND[c]	ND	21–34	rRNA hybridization	DeLong et al., 1994
	$1–3 \times 10^5$	ND	<25	rRNA hybridization	Massana et al., 1998
	ND	ND	<24	rRNA hybridization	Murray et al., 1998
	$1–2 \times 10^5$	$0.9–3 \times 10^4$	5–14	FISH	Murray et al., 1998
	$1–2 \times 10^8$	ND	7–21	rRNA hybridization	Murray et al., 1999
	$10^4–10^6$	$0.5–2 \times 10^4$	1–39	FISH	Church et al., 2003
	ND	$0.4–3 \times 10^4$	1–3	CARD-FISH	Topping et al., 2006
Marine sediment	ND	ND	0.4–1.2	rRNA hybridization	Bowman et al., 2003
	ND	ND	0.2	rRNA hybridization	Purdy et al., 2003
Lake sediment	ND	ND	34	rRNA hybridization	Purdy et al., 2003
Arctic					
Marine waters	$1–6 \times 10^5$	$0.3–20 \times 10^3$	0.1–13	FISH	Wells and Deming, 2003
	$0.5–9 \times 10^5$	$1–10 \times 10^5$	1–25	FISH	Wells et al., 2006
	$<10^9$	$0.1–2 \times 10^8$	8–30	FISH	Kirchman et al., 2007
	$0.2–1 \times 10^6$	ND	>26	CARD-FISH	Alonso-Sáez et al., 2008
Sea ice	ND	ND	<3	FISH	Junge et al., 2004
Marine sediment	$2–5 \times 10^9$	$<2 \times 10^8$	<6	FISH	Ravenschlag et al., 2001
Lake water	$0.7–3 \times 10^6$	$<3 \times 10^6$	ND	qPCR	Pouliot et al., 2009
Spring sediment	$4–6 \times 10^5$	$1–2 \times 10^4$	3–4	CARD-FISH	Niederberger et al., 2010
Permafrost soils	$0.1–2 \times 10^8$	$0.01–3 \times 10^8$	0.5–22	FISH	Kobabe et al., 2004
	ND	$0.9–7 \times 10^7$	ND	FISH	Morozova et al., 2007
	ND	$0.02–2 \times 10^4$	ND	qPCR	Yergeau et al., 2010
	$2–4 \times 10^7$	$0.4–1 \times 10^5$	ND	qPCR	Wilhelm et al., 2011

[a]Counts are represented as averages and include variations due to various sample types analyzed per study.
[b]Units: ml^{-1} for water samples; ml^{-1}/g^{-1} for sediment and soils samples; 16S rRNA gene copies for qPCR-based studies.
[c]ND, not determined.

and endolith) from McKelvey Valley in the McMurdo Dry Valleys.

The first wide-ranging PCR-based survey of archaeal 16S rRNA genes in terrestrial Antarctic sites has recently been reported by Ayton et al. (2010). Results from this study also indicate a patchy distribution and low concentration and diversity of *Archaea* in Antarctic soils. A total of 51 samples from 12 locations across the Ross Sea region including Ross Island (Scott Base, Cape Evans, Cape Bird), mainland coastal sites (Marble Point, Granite Harbour, Minna Bluff, Cape Hallett), and the McMurdo Dry Valleys (Beacon Valley, Victoria Valley, and Bull Pass, and soil near Lake Vanda and Mount Fleming in

Wright Valley) were screened for the presence of *Archaea* via PCR-based assays. Fluorescent in situ hybridization (FISH) analyses using both archaeal- and crenarchaeotal-specific probes were also trialed; however, cell densities were below detection levels (Ayton et al., 2010). Of the 51 samples, 18 proved positive for the presence of *Archaea*, mostly from the coastal sites (Scott Base, Marble Point, Granite Harbour, Minna Bluff) and a single Dry Valleys site (Victoria Valley). *Archaea* were dominated (>99%; 1,452 clones) by MGI 1.1b, with *Euryarchaeota* (<1%), related to MGIII, only detected in Granite Harbour soils. A low diversity (5 phylotypes from 85 clones) of archaeal 16S rRNA gene

TABLE 3 16S rRNA gene-based community surveys of *Archaea* in natural polar environments[a]

Pole	Habitat	Region	Sites[b]	Sample(s)	Phylum detected	Taxonomic groups detected	Reference	16S rRNA gene-based detection method: associated PCR primers (listed in Table 4)
Antarctic								
	Soils	Various sites (latitudinal gradient): sub-Antarctic Islands and western Antarctic Peninsula	Falkland Islands (1a), Signy Islands (1b), Anchorage Islands (1c), Fossil Bluff,★ Coal Nunatak★	Vegetated and fellfield	*Euryarchaeota* *Crenarchaeota*	ND[c] ND	Yergeau et al., 2009	Microarray: 4fa/1492R
		Ross Island	Scott Base (2a), Cape Evans,★ Cape Bird★	Mineral soil (ornithogenic for Cape Evans and Cape Bird)	*Crenarchaeota*	MGI 1.1b	Ayton et al., 2010	Clone library: ASF/ASR
		Mainland coast near Ross Island	Marble Point (2b), Minna Bluff, Granite Harbour (2c), Cape Hallett★	Mineral soil	*Crenarchaeota* *Euryarchaeota* (Granite Harbour)	MGI 1.1b MGIII (MBG-D)	As above	As above
		McMurdo Dry Valleys	Victoria Valley (2d), Mount Fleming,★ Beacon Valley,★ Wright Valley★ (near Lake Vanda and Bull Pass)	Mineral soil	*Crenarchaeota*	MGI 1.1b	As above	As above
		Mount Erebus	Tramway Ridge (3)	Hot (60–65°C) soil	*Crenarchaeota*	ND	Soo et al., 2009	Clone library: 21F/958R
	Freshwater to moderately saline lakes	Vestfold Hills	Clear Lake (4a), Pendant Lake (4b), Scale Lake (4c), Burton Lake (4d), Ace Lake (4e)	Sediment	*Euryarchaeota*	"Sediment *Archaea* group" MGIII (MBG-D) *Methanosarcinales*	Bowman et al., 2000a	Clone library: 530F/1492R

Region	Location	Lake (number)	Sample	Phylum	Group	Reference	Method
		Ace Lake (5)	Sediment and deep POM	*Euryarchaeota* *Crenarchaeota*	"Sediment *Archaea* group" MGIII (MBG-D) *Methanosarcinales* Unclassified sequences	Coolen et al., 2004#	DGGE band sequencing: 519F/915R
	Langhovde area	Lake Nurume-Ike (6)	Sediment	*Euryarchaeota*	MGIII (MBG-D) Unclassified sequences	Kurosawa et al., 2010#	Clone library: A21F/U1492R
	Taylor Valley, McMurdo Dry Valleys	Lake Fryxell (7)	Sediment and deep water	*Euryarchaeota* *Crenarchaeota*	*Methanosarcinales* *Methanomicrobiales* Unclassified sequences MBG-C	Karr et al., 2006	Clone library: A344F/A915R
		Lake Fryxell (8)	Mat from lake edge	*Euryarchaeota*	*Methanomicrobiales*	Brambilla et al., 2001	Clone library: arcF (10–30 nucleotides)/ 1,084–1,100 nucleotides
	Signy Island, South Orkney Islands	Lake Heywood (9)	Sediment	*Euryarchaeota*	*Methanosarcinales* *Methanomicrobiales*	Purdy et al., 2003	Clone library: 1AF/1100AR or seminested with 1AF/1404R
Hypersaline lakes and ponds	Vestfold Hills	Deep Lake (10a), Organic Lake (10b), Ekho Lake★	Sediment	*Euryarchaeota*	*Halobacteriales*	Bowman et al., 2000b	Clone library: 530F:1492R
	Taylor Valley, McMurdo Dry Valleys	Lake Bonney (11)	Water column	*Euryarchaeota*	*Halobacteriales*	Glatz et al., 2006	Clone library: A21F/A958R or seminested with A21F/1391R

(*Continued*)

TABLE 3 16S rRNA gene-based community surveys of *Archaea* in natural polar environments[a]—*Continued*

Pole	Habitat	Region	Sites[b]	Sample(s)	Phylum detected	Taxonomic groups detected	Reference	16S rRNA gene-based detection method: associated PCR primers (listed in Table 4)
Antarctic								
		Bratina Island, Ross Sea	Saline pond (12)	Sediment	*Crenarchaeota*	MGI	Sjöling and Cowan, 2003	Clone library: A21F/A958R
	Ocean	Antarctic Peninsula	Coastal region (13)	Seawater	*Crenarchaeota* *Euryarchaeota*	ND ND	Kalanetra et al., 2009	Clone library: 21F/958R
		Antarctic Peninsula	Gerlache Strait, Dallman Bay (14)	Seawater	*Crenarchaeota* *Euryarchaeota*	MGI MGII MGIII MGIV	Bano et al., 2004#	DGGE band sequencing; ARC344F/517R # and clone library: 21F/958R
		Antarctic Archipelago	Arthur Harbour (15)	Seawater	*Crenarchaeota* *Euryarchaeota*	MGI MGII	DeLong et al., 1994	Clone library: 21F/958R
		Antarctic Ocean	Near Scotia and Weddell Seas (16a) and Antarctic Peninsula (16b)	Seawater	*Crenarchaeota*	MGI	García-Martínez and Rodríguez-Valera, 2000	Clone library: F(7-26)/R(912-933)
		Antarctic Polar Front	Drake Passage (17)	Seawater (3,000-m depth)	*Euryarchaeota*	MGII MGIII MGIV	López-García et al., 2001b	Clone library: various (six total archaeal primer sets)
			As above (18)	As above	*Euryarchaeota*	MGII MGIII MGIV	López-García et al., 2001a	Clone library: various (as above) including MGIV specific (G-IV-2R and -1F)

Habitat	Location	Sample	Environment	Phylum	Group	Reference	Primers
Marine sediments	Weddell Sea	Bathypelagic zone, 2,165–3,406 m (19)	Sediment	*Crenarchaeota* *Euryarchaeota*	MGI Unknown	Gillan and Danis, 2007	Clone library: A21F:U1492R
	Mertz Glacier Polynya region	Mesopelagic, 761 m (20)	Sediment (0–21 cm)	*Crenarchaeota* *Euryarchaeota*	MGI Unknown Ace Lake clones Pendant Lake clones MGII MGIII	Bowman and McCuaig, 2003	Clone library: 519f/1492r
	Taynaya Bay	Burke Basin, 32 m (21)	Coastal sediment	*Euryarchaeota*	MGIII	Bowman et al., 2000a	Clone library: 530f/1492r
	Signy Island	Shallow Bay (22)	Coastal sediment	*Euryarchaeota*	*Methanogenium* *Methanolobus* *Methanococoides*	Purdy et al., 2003	Clone library: 1A (forward)/1100A (reverse) or seminested with 1A/1404R
Sponges	McMurdo Sound, Ross Sea	*Kirkpatrickia variolosa*, *Latrunculia apicalis*, *Mycale acerata*, *Homaxinella balfourensis*,★ and *Sphaerotylus antarcticus*★ (23)	Sponge tissue	*Crenarchaeota*	MGI	Webster et al., 2004	Clone library: A21F/A958R
Arctic							
Soils	Ellesmere Island, Canadian Arctic	Eureka, core Eur1 (24)	Permafrost: 9-m depth	*Crenarchaeota* *Euryarchaeota*	Unclassified sequences *Halobacteriales* Unclassified sequences	Steven et al., 2007	Clone library: 333Fa/A934R
		Eureka, core Eur3 (25)	Active-layer soil and permafrost table: 1-m depth	*Crenarchaeota*	Unclassified sequences	Steven et al., 2008	Clone library: 109F/934R

(Continued)

39

TABLE 3 16S rRNA gene-based community surveys of *Archaea* in natural polar environments[a]—*Continued*

Pole	Habitat	Region	Sites[b]	Sample(s)	Phylum detected	Taxonomic groups detected	Reference	16S rRNA gene-based detection method: associated PCR primers (listed in Table 4)
Arctic								
				Permafrost: 2-m depth	*Crenarchaeota* *Euryarchaeota*	Unclassified sequences *Halobacteriales* Unclassified sequences	As above	As above
				Ground ice: 7-m depth	*Crenarchaeota* *Euryarchaeota*	Unclassified sequences *Halobacteriales* *Methanobacteriales* Unclassified sequences	As above	As above
		Laptev Sea coast, Siberian Arctic	Lena River Delta, Samoylov Island (26a) and Lena-Anabar lowland near Cape Mamontovy Klyk (26b)	Active-layer soil: polygon center and floodplain	*Euryarchaeota*	*Methanosarcinales* *Methanomicrobiales* RC-III	Ganzert et al., 2007#	(Methanogen-specific) DGGE band sequencing: 357 +GC clamp/691R
		Spitsbergen, Norway	Various sites on west coast (27)	Peat, soils, and lake sediments	*Crenarchaeota* *Euryarchaeota*	MGI 1.3b *Methanosarcinales* Unclassified sequences	Høj et al., 2006#	DGGE band sequencing; PRARCH112F/ PRARCH1045r or PREA1100r nested with PARCH340/ PARCH519r
		Northwest Siberia (sub-Arctic)	Mire near Igarka (28)	Peat	*Crenarchaeota* *Euryarchaeota*	Unclassified sequences *Methanosarcinales* *Methanobacteriales*	Metje and Frenzel, 2007	Clone library: Ar109fa/Ar915r

	Location	Site	Sample	Phylum	Groups	Reference	Method
Wetlands	Kamchatka, Russia	Pushchino (29)	Permafrost soil	*Crenarchaeota*	MGI 1.1b	Ochsenreiter et al., 2003	Clone library: 20F/958R
	Axel Heiberg Island, Canadian Arctic	Near Expedition Fjord (30)	Active layer and permafrost: 1-m depth	*Crenarchaeota* *Euryarchaeota*	Unclassified sequences Unclassified sequences	Wilhelm et al., 2011#	Clone library: 109F/934R
	Spitsbergen, Norway	Solvatnet and Stuphallet (31)	Peat	*Crenarchaeota* *Euryarchaeota*	MGI 1.3b MGI RC-II *Methanomicrobiales* *Methanobacteriales* *Methanosarcinales*	Hoj et al., 2005#	DGGE band sequencing: RCH112F/ PREA1100R nested with PRARCH 340F+GC-clamp/ PARCH519R
Lakes	Northern coast of Ellesmere Island, Canadian High Arctic	Lake A (32)	Various: water column	*Euryarchaeota* *Crenarchaeota*	LDS (oxic depths) RC-V (anoxic) MGI (oxycline)	Pouliot et al., 2009	Clone library: A109F/A915R
	Western Canadian Arctic	Lake MacKenzie (stamukhi) (33)	Water	*Euryarchaeota*	LDS RC-V (anoxic)	Galand et al., 2008b	Clone library: A109F/A934R
Saline springs	Axel Heiberg Island, Canadian Arctic	Colour Peak spring (34a)	Source-pool sediment	*Crenarchaeota* *Euryarchaeota*	Unclassified sequences *Halobacteriales* Unclassified sequences	Perreault et al., 2007	Clone library: A344F/A934R
		Gypsum Hill spring (34b)	Source-pool sediment	*Crenarchaeota* *Euryarchaeota*	Unclassified sequences *Halobacteriales* *Methanosarcinales*	As above	As above
		Gypsum Hill spring (35)	Streamer biomass	*Crenarchaeota*	Unclassified sequences	Niederberger et al., 2009	Clone library: A344F/A934R

(Continued)

TABLE 3 16S rRNA gene-based community surveys of *Archaea* in natural polar environments[a]—*Continued*

Pole	Habitat	Region	Sites[b]	Sample(s)	Phylum detected	Taxonomic groups detected	Reference	16S rRNA gene-based detection method: associated PCR primers (listed in Table 4)
Arctic								
			Lost Hammer spring (36)	Source-pool sediment	*Euryarchaeota*	ANME-1a *Archaeoglobales Halobacteriales* Unclassified sequences	Niederberger et al., 2010#	Clone library: 109F/934R
	Ice shelf	Northern coast of Ellesmere Island, Canadian Arctic	Ward Hunt (37a) and Markham Ice Shelves (37b)	Microbial mat	*Euryarchaeota*	Unclassified sequences	Bottos et al., 2008	Clone library: 109F/915R
	Seafloor	Svalbard	Kongsfjorden (38)	Sediment core (15-cm depth)	*Crenarchaeota Euryarchaeota*	Unclassified MGI Unclassified	Tian et al., 2009	Clone library: 21F/958R
		Norwegian-Greenland Sea	Kolbeinsey (39a), Mohns (39b), Kinpovich ridges (39c)	Basalt and overlying bottom water	*Crenarchaeota*	MGI	Lysnes et al., 2004#	DGGE band sequencing: mPRA46f/ mPREA1100r nested with mPARCH340f/ 519r
	Ocean	Arctic Ocean	Central region (40)	Seawater (various locations and depths)	*Crenarchaeota Euryarchaeota*	ND ND	Kalanetra et al., 2009	Clone library: various
			Central region (41)	As above	*Crenarchaeota Euryarchaeota*	MGI MGII MGIII MGIV	Bano et al., 2004#	DGGE band sequencing: 344f/517r # and clone library: 21f/958r

Environment	Region	Location[b]	Sample type	Phylum	Groups	Reference[a]	Method
		Baffin Bay (42a), Canada Basin (42b), Chukchi and Beaufort Sea (42c), Franklin Bay (42d)	Seawater (various locations and depths)	*Crenarchaeota* *Euryarchaeota*	MGI MGII MGIII	Galand et al., 2009c#	Amplicon pyrosequencing: 958arcF/ 1048arcR-major and 1048arcR-minor
		Laptev Sea (43)	Seawater (various locations and depths)	*Crenarchaeota* *Euryarchaeota*	MGI MGII	Kellogg and Deming, 2009	Clone library: 21F/958R
		Beaufort Sea (44)	Coastal and open-ocean seawater	*Crenarchaeota* *Euryarchaeota*	MGI MGII RC-V *Methanomicrobiales*	Galand et al., 2008a	Clone library: 109F/934R
	As above	As above (45)	Coastal seawater	*Crenarchaeota* *Euryarchaeota*	MGI MGII RC-V *Methanomicrobiales*	Galand et al., 2006	Clone library: 109F/915R
	Northern waters	Between Ellsemere Island and Greenland (46)	Seawater (various locations)	*Crenarchaeota* *Euryarchaeota*	MGI MGII	Galand et al., 2009a	Clone library: 109f/915r
Sea ice	Western Arctic	Franklin Bay (47)	Sea ice	*Crenarchaeota* *Euryarchaeota*	MGI MGII	Collins et al., 2010	Clone library: Arch21F/ Arch 958R

[a]Sequences from these studies are included in the phylogenetic analysis shown in Color Plate 4. Sequences from DGGE-based studies and some recent studies (designated by # in the "Reference" column) are not included in Color Plate 4.

[b]Numbers in parentheses refer to locations indicated on map in Color Plate 5. *Archaea* were not detected at sites marked with a star.

[c]ND, not defined in the original study.

TABLE 4 Oligonucleotides utilized to target *Archaea* in 16S rRNA gene PCR–based culture-independent studies of polar environments

Oligonucleotide names[a]	Target	Sequence (5'-3')	Studies utilizing oligonucleotides in polar environments (as listed in Table 3)
PCR: forward primers			
1A	*Archaea*	TC YGK TTG ATC CYG SCR GAG	López-García et al., 2001a, 2001b; Purdy et al., 2003
4fa		TC CGG TTG ATC CTG CCR G	Yergeau et al., 2009
ARC-8F / 8F / ArcF		TC CGG TTG ATC CTG CC	Brambilla et al., 2001
Unnamed		TTC CGG TTG ATC CTG CCG GA	García-Martínez and Rodríguez-Valera, 2000; López-García et al., 2001a, 2001b
20F		TTC CGG TTG ATC CYG CCR G	Ochsenreiter et al., 2003
Arch21F / A21F / 21F / 21f		TTC CGG TTG ATC CYG CCG GA	DeLong et al., 1994; Sjöling and Cowan, 2003; Bano et al., 2004; Webster et al., 2004; Glatz et al., 2006; Gillan and Danis, 2007; Kalanetra et al., 2009; Kellogg and Deming, 2009; Soo et al., 2009; Tian et al., 2009; Collins et al., 2010; Kurosawa et al., 2010
23FLP	*Archaea*	YCT GGT YGA TYC TGC C	López-García et al., 2001a, 2001b
mPRA46f		YTA AGC CAT GYR AGT	Lysnes et al., 2004
109F / Arch109f / 109f / Ar109f	*Archaea*	ACK GCT CAG TAA CAC GT	Galand et al., 2006, 2008a, 2008b, 2009a; Metje and Frenzel, 2007; Bottos et al., 2008; Steven et al., 2008; Pouliot et al., 2009; Niederberger et al., 2010; Wilhelm et al., 2011
PRARCH112f / PRARCH112F		GCT CAI TAW CAC GTG G	Høj et al., 2005, 2006
333Fa	*Archaea*	TCC AGG CCC TAC GGG	Steven et al., 2007
ASF (nucleotides: 340–358)		CC AGG CCC TAC GGG GCG CA	Ayton et al., 2010
PARCH340f / PARCH340F		CCC TAC GGG GYG CAS CAG	Høj et al., 2005, 2006
mPARCH340f		TAY GGG GYG CAS CAG	Lysnes et al., 2004
A344F / ARCH 344F / ARC344f		AC GGG GTG CAG CAG GCG CGA	Bano et al., 2004; Karr et al., 2006; Perreault et al., 2007; Niederberger et al., 2009
357F	Methanogens	CCC TAC GGG GCG CAG CAG	Ganzert et al., 2007
519f	Prokaryotes	CAG CMG CCG CGG TAA TAC	Bowman and McCuaig, 2003

Name	Target	Sequence (5'→3')	References
Parch519f		CAG CCG CCG CGG TAA	Coolen et al., 2004
530f	Prokaryotes	GTG CCA GCM GCC GCG G	Bowman et al., 2000a, 2000b
G-IV-1F (nucleotides: 194–213)	MGIV	TAA TAA CAG GTA ATA CTC CT	López-García et al., 2001a
958arcF	Archaea	AAT TGG ANT CAA CGC CGG	Galand et al., 2009c
PCR: reverse primers			
517r	Prokaryotes	ATT ACC GCG GCT GCT GG	Bano et al., 2004
PARCH519r / PARCH519R		TT ACC GCG GCK GCT G	Lysnes et al., 2004; Høj et al., 2005, 2006
691R	Methanogens	GGA TTA CAR GAT TTC AC	Ganzert et al., 2007
G-IV-2R (nucleotides: 878–897)	MGIV	CCT ACG GCA CAG CAA AAG CA	López-García et al., 2001a
915R / Arch915r / ARC915r / 915r / Ar915r	Archaea	GTG CTC CCC CGC CAA TTC CT	García-Martínez and Rodríguez-Valera, 2000; Coolen et al., 2004; Galand et al., 2006, 2009a; Karr et al., 2006; Metje and Frenzel, 2007; Bottos et al., 2008; Pouliot et al., 2009
A934R / 934R		As for 915R	Perreault et al., 2007; Steven et al., 2007, 2008; Galand et al., 2008a, 2008b; Niederberger et al., 2009, 2010; Wilhelm et al., 2011
Arch958R / A958R / 958R / 958r	Archaea	YCC GGC GTT GAM TCC AAT T	DeLong et al., 1994; Ochsenreiter et al., 2003; Sjöling and Cowan, 2003; Bano et al., 2004; Webster et al., 2004; Glatz et al., 2006; Kalanetra et al., 2009; Kellogg and Deming, 2009; Soo et al., 2009; Tian et al., 2009; Collins et al., 2010
PRARCH1045r	Archaea	GG CCA TGC ACC WC CTC	Høj et al., 2006
1048arcR-major		CGR CGG CCA TGC ACC WC	Galand et al., 2009c
1048arcR-minor		CGR CRG CCA TGY ACC WC	Galand et al., 2009c
1100A	Archaea	T GGG TCT CGC TCG TTG	López-García et al., 2001a, 2001b
1100A		GGG TCT CGC TCG TTG	Purdy et al., 2003
Unnamed (nucleotides: 1084–1100)		GGG TCT CGC TCG TTA CC	Brambilla et al., 2001
PREA1100r / PREA1100R		Y GGG TCT CGC TCG TTR CC	Høj et al., 2005, 2006
mPREA1100r		B GGG TCT CGC TCG TTR CC	Lysnes et al., 2004

(Continued)

45

TABLE 4 Oligonucleotides utilized to target *Archaea* in 16S rRNA gene PCR-based culture-independent studies of polar environments—*Continued*

Oligonucleotide names[a]	Target	Sequence (5'-3')	Studies utilizing oligonucleotides in polar environments (as listed in Table 3)
PCR: reverse primers			
ASR (nucleotides: 1344–1362)	*Archaea*	GTG TGC AAG GAG CAG GGA C	Ayton et al., 2010
1391R	*Archaea*	GAC GGG CGG TGW G-R CA[b]	Glatz et al., 2006
1391R		GAC GGG CGG TGT GTR CA	López-García et al., 2001a, 2001b
1404R		CGG TGT GTG CAA GGR GC	Purdy et al., 2003
1492R / S (1492R)	Prokaryotes	GGC TAC CTT GTT ACG ACT T	López-García et al., 2001a, 2001b; Yergeau et al., 2009
1492R		TAC GGY TAC CTT GTT ACG ACT T	Bowman et al., 2000a, 2000b
1492r		TAC GGY TAC CTT GTT ACG AC	Bowman and McCuaig, 2003
U1492R		GGY TAC CTT GTT ACG ACT T	Kurosawa et al., 2010
U1492R		GGT TAC CTT GTT ACG ACT	Gillan and Danis, 2007

[a]Numbering associated with oligonucleotide names refers to the complementary nucleotide position on the target 16S rRNA gene.
[b]The mismatch, i.e., gap with the oligonucleotide sequence represented by the dash, may represent an error within the original publication (Glatz et al., 2006).

phylotypes has also recently been detected in geothermally heated soils on Tramway Ridge situated near the summit of Mount Erebus (Soo et al., 2009). These archaeal phylotypes were related to crenarchaeotal signatures previously detected in deep-subsurface environments.

ARCHAEA IN LAKE HABITATS

The majority of research describing archaeal diversity and ecology within Antarctic lake habitats has been undertaken on sediment samples collected from various meromictic lakes located in the Vestfold Hills region of East Antarctica (Bowman et al., 2000a, 2000b; Coolen et al., 2004). A meromictic lake is characterized as a lake containing stratified waters that remain perpetually unmixed with the remainder of the lake waters (Gibson, 1999; Bowman et al., 2000a), and it has been postulated that the Vestfold Hills lake system (68°S, 78°E) may contain the greatest concentration of meromictic lakes in Antarctica and possibly on Earth (Gibson, 1999).

Freshwater to Moderately Saline Lakes. Bowman et al. (2000a) described the microbial communities inhabiting anoxic sediments of five meromictic lakes in the Vestfold Hills: Clear, Pendant, Scale, Burton, and Ace Lakes. A total of 10 unique (≥98% sequence similarity) archaeal 16S rRNA gene phylotypes related to the *Euryarchaeota* were detected within sediments of all the lakes. A total of 7 phylotypes from Pendant, Burton, Clear, and Ace Lakes were phylogenetically related but were divergent from the *Euryarchaeota* phylum (45 to 91% sequence similarity) and were most closely related to uncultured clones from salt marsh and marine sediments and designated as the "Sediment *Archaea* group." The remaining phylotypes, including two phylotypes from Burton and Scale Lakes, were closely related to the MGIII (MBG-D) and a single phylotype from Scale Lake was related to *Methanosarcina barkeri*. Two strains of methanogenic *Archaea* (Table 1) have previously been isolated from the anoxic hypolimnion of Ace Lake, namely *Methanococcoides burtonii* (Franzmann et al., 1992) and *Methanogenium frigidum* (Franzmann et al.,

1997), and Coolen et al. (2004) have detected archaeal signatures in Ace Lake. The study conducted by Coolen et al. (2004) did not detect archaeal signatures via PCR in the oxygenated mixolimnion of Ace Lake; yet positive *Archaea* PCR assays were obtained from the anoxic, sulfidic bottom-water particulate organic matter (POM) and from a sediment core. Subsequent denaturing gradient gel electrophoresis (DGGE)-PCR profiling was undertaken and a total of 12 and 14 unique archaeal phylotypes were identified by DGGE band sequencing from the POM and sediment samples, respectively. The majority of the phylotypes were related to the *Euryarchaeota* phylum (10 and 12 phylotypes, respectively, for POM and sediment), with the remainder grouping within the *Crenarchaeota*. Notably, four of the POM phylotypes grouped most closely with the "Sediment *Archaea* group" as described above from sediments of other Vestfold Hills lake sediments (Bowman et al., 2000a), and the two *Crenarchaeota* phylotypes were only detected in the anoxic monimolimnion, i.e., absent from the anoxic chemocline and deepest waters. These were most closely related to sequences previously detected in Antarctic shelf sediments (Bowman and McCuaig, 2003). The sediment phylotypes were dominated (eight phylotypes) by sequences related to the MGIII and included two phylotypes that were only detected in the deepest section of the core, which were related to *Methanosarcina* spp., and a crenarchaeotal signature most closely related to low-temperature clones from terrestrial freshwater systems. The *Archaea* inhabiting anoxic sediment from another meromictic lake (Lake Nurume-Ike) located within the Langhovde area, approximately 1,500 km from the Vestfold Hills region, has been reported by Kurosawa et al. (2010). As was the case for the Vestfold Hills lakes, Lake Nurume-Ike sediment contained a low diversity of *Archaea* consisting of 3 phylotypes from a total of 205 archaeal 16S rRNA gene clones, of which 93% were related to the MBG-D.

Lake Fryxell is a permanently frozen freshwater lake located in Taylor Valley of the

McMurdo Dry Valleys. Karr et al. (2006) have investigated the diversity of archaea in the sediments and anoxic water column above sediments in Lake Fryxell, with Brambilla et al. (2001) describing archaeal diversity within a microbial mat collected from the moated region of the lake. At least four archaeal phylotypes were detected in the sediment and water samples analyzed by Karr et al. (2006), including two methanogenic signatures related to the genera *Methanosarcina* and *Methanoculleus* of the *Euryarchaeota* in the sediments, a potential methanotrophic marine-related euryarchaeotal signature in the anoxic water column, and a crenarchaeote phylotype related to the MBG-C in water below the oxycline. The presence of *Methanosarcina* species was further confirmed by the isolation of an archaeal isolate, *Methanosarcina* strain FRX-1 (Table 1) (Singh et al., 2005), from Lake Fryxell sediments. Brambilla et al. (2001) also report a low diversity of archaeal signatures in microbial mat collected from the wetted moat zone surrounding Lake Fryxell. Two dominant archaeal sequences were distinguished within a total of 72 screened archaeal 16S rRNA gene clones, with the majority (>90%) being distantly related to the genus *Methanoculleus*.

Lake Heywood is another freshwater lake, but is located on Signy Island, in the sub-Antarctic South Orkney Islands. Three dominant archaeal phylotypes were detected in sediment collected at the deepest (~7-m depth) section of the lake (Purdy et al., 2003). Of the 40 total clones, 73% were related to *Methanosaeta*, with the remaining 2 phylotypes (25 and 2%, respectively) related to *Methanogenium* spp.

Hypersaline Lakes. In contrast to the studies of low- to moderately saline meromictic lakes of the Vestfold Hills region described in the section above (Bowman et al., 2000a; Coolen et al., 2004), Bowman et al. (2000b) also investigated microbial communities, including *Archaea*, within sediments of three hypersaline lakes situated in the Vestfold Hills. Two of the studied lakes, Organic Lake (20% salinity) and Ekho Lake (15% salinity), are

characterized as meromictic; however, Deep Lake (32% salinity) is monomictic, i.e., mixing within the lake occurs once in the year during the summer thaw. In contrast to the meromictic lakes, the Deep Lake 16S rRNA gene clone library was dominated by archaeal signatures (universal archaeal and bacterial PCR primer-pairs were utilized). Of a total of 83 16S rRNA gene clones within the Deep Lake clone library, 76 were archaeal, consisting of 8 dominant phylotypes that were related members of the extremely halophilic *Halobacteriales* clade. Connected to this study, the first Antarctic archaeal strain, *Halobacterium lacusprofundi* (Table 1), was originally isolated from Deep Lake by Franzmann et al. (1988). A total of 6 archaeal 16S rRNA gene clones were represented in the Organic Lake clone library (76 total clones), with 3 detected archaeal phylotypes overlapping with the phylotypes detected in the Deep Lake sample. *Archaea* were not detected via PCR in the Ekho Lake sediment.

Lake Bonney is a permanently ice-covered lake situated in Taylor Valley of the McMurdo Dry Valleys, with its bottom waters having ~6 to 10 times the salinity of seawater (Aislabie and Bowman, 2010). Glatz et al. (2006) collected samples from depths of 13 and 16 m in the west lobe and 16, 19, and 25 m in the east lobe of Lake Bonney. Archaeal 16S rRNA genes were successfully PCR amplified from the two hypersaline depths, i.e., 16 and 25 m in the west and east lobes, respectively (Glatz et al., 2006). A single identical sequence related to halophilic *Euryarchaeota* (*Halobacterium*; 100% Ribosomal Database Project classifier [http://rdp.cme.msu.edu/]) sequences was present in both libraries and dominated the archaeal 16S rRNA clone library (>90% of 190 total clones), signifying low archaeal diversity. A low diversity of archaeal 16S rRNA gene signatures was also detected in saline maritime meltwater pond sediment (10- to 13-cm depth) from Bratina Island, Ross Sea, Antarctica (Sjöling and Cowan, 2003). A total of 7 archaeal phylotypes were detected from a total of 17 clones that were all related to the MGI.

ARCHAEA IN OTHER TERRESTRIAL ENVIRONMENTS

Attempts to detect *Archaea* inhabiting other Antarctic terrestrial environments utilizing 16S rRNA gene PCR-based analyses have previously failed, including sandstone crypto-endoliths of the McMurdo Dry Valleys (de la Torre et al., 2003), a cryconite hole in Canada Glacier of the McMurdo Dry Valleys (Christner et al., 2003), and accretion ice from Lake Vostok (Christner et al., 2001). However, it must be noted that the absence of a positive PCR amplification must be taken with some caution and may not be indicative of the absence of *Archaea* due to reported shortcomings of PCR from natural samples and/or PCR primer bias (Kanagawa, 2003).

Archaea in Marine Environments

ARCHAEA IN SEAWATER AND SEA ICE

In a pioneering study, DeLong et al. (1994) uncovered an unexpectedly high abundance of *Archaea* in surface waters of the Antarctic. By use of rRNA-specific oligonucleotide probes it was estimated that they constitute up to 34% of the total prokaryotic biomass in coastal pelagic surface waters near Arthur Harbour, East Antarctica, with an archaeal 16S rRNA gene clone library being dominated by MGI and including a minor MGII presence (DeLong et al., 1994). Subsequent studies have also proven a relatively high abundance of MGI in pelagic waters west of the Antarctic Peninsula (Church et al., 2003) and in both nearshore waters of Anvers Island and offshore in the Palmer Basin (Murray et al., 1998). Both studies also proved that MGI abundance varies in the nearshore waters during seasonal cycles, being more abundant during the winter. *Archaea* were also more abundant at deeper depths and varied seasonally between the winter and summer, whereas the MGII were persistently ~2% of the total picoplankton throughout the water column and between seasons (Church et al., 2003). *Crenarchaeota* were also proven to be absent from Antarctic surface waters west

of the Antarctic Peninsula during summer by Kalanetra et al. (2009).

Further rRNA hybridization-based studies have provided insight into both the circumpolar distribution of *Archaea* in water masses surrounding the Antarctic continent (Murray et al., 1999) and the vertical and temporal distribution of *Archaea* at a station located in the Gerlache Strait of the Antarctic Peninsula (Massana et al., 1998). The latter study proved that *Archaea* were more abundant at depth, with most of the archaeal signal attributable to the MGI (Massana et al., 1998). Archaeal assemblages have also shown via DGGE profiling to be more diverse at depth (Gerlache Strait and Dallman Bay), with samples at depth having a more complex DGGE profile (Bano et al., 2004). Archaeal 16S rRNA gene clone libraries from these sites included phylotypes related to the MGI, MGII, and MGIV (Bano et al., 2004).

Deep-sea planktonic communities at a depth of 3,000 m at the Antarctic Polar Front, represented by the Drake Passage, i.e., between the Antarctic Peninsula and South America, have also been investigated (López-García et al., 2001a, 2001b). Utilizing various archaeal-specific PCR primer sets, MGII were shown to be dominant, with minor phylotypes related to the MGIII and the halophilic *Archaea* (López-García et al., 2001b). In a related study, utilizing PCR primers specific to the previously detected halobacterium-related lineage, tentatively titled the MGIV, the MGIV were not detected in surface waters and it was shown that this clade is present in deep waters of the Antarctic area from the Southern Ocean to the South Atlantic and widely distributed in other deep oceanic regions (López-García et al., 2001a).

Archaeal presence in Antarctic pack ice has also been investigated (Brinkmeyer et al., 2003). FISH-based experimentation revealed that archaeal cells comprise less than 1% of the total prokaryotic cells and are below FISH detection levels. Likewise, *Archaea* concentrations in various Antarctic sea ice samples were below PCR-based detection limits (Brown and Bowman, 2001).

ARCHAEA IN SEAFLOOR ENVIRONMENTS

Studies describing the presence of *Archaea* in Antarctic marine sediments include a single study of bathypelagic sediments (1,000- to 3,500-m depth) of the Weddell Sea (Gillan and Danis, 2007), two studies of mesopelagic sediments (150- to 1,000-m depth) in the eastern part of the Southern Ocean (Bowman et al., 2003; Bowman and McCuaig, 2003), and studies of coastal sites (Bowman et al., 2000a; Powell et al., 2003; Purdy et al., 2003).

Gillan and Danis (2007) studied the archaeal communities of three sediment samples from the bathypelagic region (1,000 to 3,500 m) of the Weddell Sea. Archaeal 16S rRNA gene clone libraries were dominated (>99% of 146 clones) by *Crenarchaeota* of the MGI, with a single clone related to the *Euryarchaeota*, which was not affiliated with previously defined groups. The MGI also dominated archaeal clones (67 of 72, 15 of 19, and 5 of 18, respectively) from mesopelagic sediments (761-m depth) at core horizons of 0- to 0.4-cm, 1.5- to 2.5-cm, and 20- to 21-cm depths, respectively. *Euryarchaeota* were also detected but were minor components that were typically related to Ace and Pendant Lake (Vestfold Hills, eastern Antarctica) signatures and single sequences each within the MGII and MGIII (Bowman et al., 2000a; Bowman and McCuaig, 2003). In a related study in the same region and at similar depth (709 to 964 m), Bowman et al. (2003) describe the archaeal lipids and concentration of *Archaea* at this site, as presented in Table 2.

All *Archaea* detected in coastal marine basin sediment at a depth of 32 m in the Taynaya Bay fjord were related to the MGIII (Bowman et al., 2000a). Purdy et al. (2003) also failed to detect *Crenarchaeota* (using crenarchaeotal-specfic PCR primers) at Shallow Bay in the Signy Islands; however, euryarchaeotal-specific primers revealed the presence of signatures related to *Methanogenium cariaci*, a *Methanolobus* sp., and *M. burtonii*. *Archaea* have also been detected by PCR in hydrocarbon-impacted and pristine coastal

sediments in Brown Bay and O'Brien Bay, respectively, in the Casey Station region of the Windmill Islands, but the phylogeny of the detected archaeal sequences was not reported (Powell et al., 2003).

ARCHAEA IN OTHER MARINE ENVIRONMENTS

Archaea have also been detected inhabiting Antarctic marine sponges collected from McMurdo Sound in the Ross Sea. Webster et al. (2004) used 16S rRNA gene clone library-based methods to analyze archaeal communities from *Kirkpatrickia varialosa*, *Latrunculia apicalis*, *Mycale acerata*, *Homaxinella balfourensis*, and *Sphaerotylus antarcticus*. *Archaea* were not detected in *H. balfourensis* and *S. antarcticus*; however, a total of 150 archaeal 16S rRNA gene clones were screened from the remainder of the sponges. Four phylotypes were detected that were grouped within the MGI and were also distantly related to the first described crenarchaeotal endosymbiont, "*Candidatus* Crenarchaeum symbiosum," originally detected in a temperate sponge (Preston et al., 1996).

ARCTIC ENVIRONMENTS

Archaea in Terrestrial Environments

ARCHAEA IN SOIL HABITATS

As discussed in "*Archaea* in Polar Environments and Climate Change," p. 55, the majority of the literature describing *Archaea* in Arctic soils has been focused on members that may contribute to the production of the greenhouse gas methane (Wagner and Liebner, 2010). Biogenic methane is created by a group of *Archaea* within the *Euryarchaeota* called methanogenic *Archaea*, or methanogens (Wagner and Liebner, 2010). The content and production of biogenic methane has been well documented in Arctic permafrost environments (Rivkina et al., 2002, 2007; Wagner et al., 2005). Methanogens have been enriched from terrestrial Arctic samples, e.g., permafrost soils from the Lena Delta to the Kolyma Lowlands in Eastern Siberia (Rivkina et al., 2002,

2007; Gilichinsky et al., 2003; Wagner et al., 2005) and peat from both Spitsbergen, Norway (Høj et al., 2008) and northwest Siberia (Metje and Frenzel, 2007). Various culture-independent studies have also detected methanogens in natural Arctic samples, as discussed in more detail below. More recently, methanogenic strains have been successfully isolated from Arctic permafrost, as listed in Table 1 (Rivkina et al., 2007; Krivushin et al., 2010; Shcherbakova et al., 2010), with complementary studies proving that both methanogenic isolates and communities are well adapted to the inherently stressful permafrost environment (Morozova et al., 2007; Morozova and Wagner, 2007).

Ganzert et al. (2007) specifically targeted the methanogenic component of archaeal communities in Siberian permafrost using methanogen-specific PCR primers. The sequencing of DGGE bands revealed the presence of major methanogenic groups, including the *Methanosarcinaceae*, *Methanosaeta*, *Methanomicrobiales*, and RC-II. Through the use of universal archaeal PCR primers, methanogenic *Archaea* have also been shown to dominate mildly acidic peat from permafrost in the Siberian sub-Arctic (Metje and Frenzel, 2007). Specifically, archaeon 16S rRNA gene clones from initial enrichment communities were dominated by sequences within the *Methanosarcinales* and *Methanobacteriales*, with few uncultured and undefined crenarchaeotal phylotypes. Kobabe et al. (2004) also report methanogen dominance in permafrost-affected soil of the Lena Delta (Siberia), with almost the same concentration of archaeal cells detected via archaeal-specific FISH probes (see Table 2) also hybridizing with a methanogen-specific FISH probe. In contrast, Høj et al. (2008) uncovered diverse archaeal communities in 4-week-old slurries of weakly acidic peat from Spitsbergen incubated at 5 and 20°C, consisting not only of known methanogenic groups (*Methanobacteriales*, *Methanomicrobiales*, and *Methanosarcinales*) but also major groups related to the RC-V, LDS, and MGI 1.3b. Similarly, Høj et al. (2006) used DGGE band sequencing to investigate archaeal communities of Spitsbergen

soils; however, samples included soils exposed to differing natural water regimes, e.g., seasonally and permanently flooded soils. This study also revealed the presence of both *Crenarchaeota* and *Euryarchaeota*; yet methanogens (*Methanomicrobiales*, *Methanobacteriales*, and *Methanosarcinales*) were detected only in samples that were wet during most of the season and the MGI 1.3b were found to inhabit a wide range of wet soils. As part of a PCR-based survey of terrestrial and freshwater habitats, Ochsenreiter et al. (2003) have shown that the MGI 1.1b group occupies soils from all over the planet, including permafrost samples from Pushchino, Kamchatka, Russia.

Culture-independent analyses of archaeal communities of two High Arctic wetland sites from Spitsbergen (Høj et al., 2005) and the Canadian Arctic (Wilhelm et al., 2011) have also been described. A diverse archaeal community detected via DGGE band sequencing in the Spitsbergen wetland peat samples consisted of MGI, RC-II, *Methanomicrobiales*, *Methanobacteriales*, and *Methanosarcinales* (Høj et al., 2005). Wilhelm et al. (2011) also detected crenarchaeotal and euryarchaeotal 16S rRNA gene signatures in active layer and permafrost soil collected from a wetland on Axel Heiberg Island in the Canadian High Arctic; however, crenarchaeota dominated both horizons (71 and 95% for active layer and permafrost, respectively).

The microbial communities from multiple horizons from two permafrost cores collected near Eureka weather station in the Canadian High Arctic have been well studied (Steven et al., 2007, 2008; Yergeau et al., 2010). Steven et al. (2007) initially described the microbial community of a permafrost core (Eur1) at a depth of 9 m. The archaeal community consisted of 11 phylotypes (56 clones) related to both the *Crenarchaeota* (31%) and *Euryarchaeota* (61%), with the only classified phylotype being related to the halobacteria. Following this report, the same research group described the microbial communities of multiple horizons in a core (Eur3) collected near the Eur1 core (Steven et al., 2008). Only *Crenarchaeota* were detected in the active layer (surface) and permafrost table (1 m) horizons; however, *Euryarchaeota* dominated the permafrost sample at a depth of 2 m, consisting of unknown sequences and phylotypes related to *Halorubrum* genus. *Archaea* detected at the deepest horizon tested (ground ice at a depth of 7 m) were dominated by *Crenarchaeota*, with minor phylotypes related to unknown *Euryarchaeota* sequences and relatives of the *Halorubrum* and *Methanobrevibacter* genera. Recently, further in-depth metagenomic analyses have been conducted on the 2-m permafrost and active-layer horizon samples of the Eur3 core (Yergeau et al., 2010). Archaeal 16S rRNA genes were not detected in the metagenome from both horizons, although functional genes related to *Archaea* were detected, including methanogen-related genes, e.g., methylene-H_4 methanopterin reductase and formylmethanofuran dehydrogenase. The failure to detect archaeal 16S rRNA genes was proposed to be most likely due to the potential biases in the methodology employed, i.e., biases involved in whole-genome amplification (multiple displacement amplification). Quantitative PCR (qPCR) was also utilized to indicate that the *Archaea* were minor community members in both samples (Table 2).

ARCHAEA IN LAKE HABITATS

Pouliot et al. (2009) have investigated the archaeal community composition in the water column of a meromictic lake (Lake A) located on the northern side of Ellesmere Island in the Canadian High Arctic. Differences were revealed along the water column, i.e., the oxic layers (2- and 10-m samples) were dominated by a phylotype related to the LDS cluster of the *Euryarchaeota* and included minor phylotypes related to the RC-V and the MGI; the oxycline (12-m) consisted completely of MGI phylotypes; and the anoxic zone (29- and 32-m) was dominated by representatives of the RC-V and the LDS. Concentrations of archaeal 16S rRNA gene copy numbers were also determined at each depth in both Lake A and another meromictic lake, Lake C1, as provided in Table 2.

The archaeal communities of an Arctic stamukhi lake (Lake MacKenzie) in the western Canadian Arctic has also been described (Galand et al., 2008b). Stamukhi lakes are a result of freshwater being retained behind a barrier of ice (stamukhi) that forms on the outer limits of land-fast sea ice. Archaeal communities were similar between the lake and its riverine source, with 16S rRNA gene clones being related to the LDS and RC-V clades (Galand et al., 2008b).

ARCHAEA IN OTHER TERRESTRIAL ENVIRONMENTS

The microbial communities (including *Archaea*) inhabiting source-pool sediments of perennial cold saline springs situated in three different locations on Axel Heiberg Island in the Canadian High Arctic have been previously documented. Perreault and colleagues (2007) describe archaeal clone libraries from sediments in springs at both Gypsum Hill (6.9°C, 7.5% salinity) and Colour Peak (7.4°C, 15.5% salinity) in the Expedition Fjord region of Axel Heiberg Island. The Gypsum Hill 16S rRNA gene clone library was dominated (79% of 156 clones) by crenarchaeotal signatures related to phylotypes from diverse marine environments, with the remainder of the clones most closely related to the *Halobacteriales* and *Methanosarcinales* of the *Euryarchaeota*. The Colour Peak clone library consisted of 52% (164 total clones) *Crenarchaeota*, which were related to phylotypes previously detected in soil, deep-sea, and cold sulfidic spring environments. The remainder of the clones were related to both *Halobacteriales* and methanogenic signatures. Some of the Gypsum Hill crenarchaeotal signatures as detected in the Gypsum Hill source-pool sediment by Perreault et al. (2007) were also detected in the *Thiomicrospira* sp.-dominated sulfur-oxidizing streamers that form in the Gypsum Hill runoff streams during the winter months (Niederberger et al., 2009).

The archaeal community inhabiting a hypersaline (~24%) methane-emitting spring (Lost Hammer) located on Axel Heiberg Island has been reported by Niederberger et al. (2010). As presented in Table 2, catalyzed activated reporter deposition (CARD)-FISH analyses indicated that the *Archaea* comprised a small proportion of the total microbial cells detected in the spring sediment. Further CARD-FISH analyses indicated that almost all archaeal cells (2.2 to 3.8% of the total cell numbers) were related to the uncultured anaerobic methane-oxidizing ANME-1a clade of *Archaea*. The 16S rRNA gene clone library was also dominated by ANME-1a-related signatures (47% of 66 clones) and included phylotypes related to the *Archaeoglobales* and the *Halobacteriales* of the *Euryarchaeota*.

A single archaeal phylotype has been shown to dominate archaeal 16S rRNA gene clone libraries (80 total clones) constructed from DNA extracted from the microbial mats situated on both the Markham and Ward Hunt Ice Shelves that extend off the northern coast of Ellesmere Island in the Canadian High Arctic (Bottos et al., 2008). While the original study described the phylotype as only being distantly related (74% sequence identity) to its closest related signature in the NCBI database, more closely related (97% sequence identity) phylotypes have now been detected in the water column of meromictic lakes in the Canadian High Arctic (Pouliot et al., 2009).

Archaea in Marine Environments

ARCHAEA IN THE ARCTIC OCEAN AND SEA ICE

As presented in Table 1, FISH-based studies have shown that *Archaea* are typically minor components of pelagic microbial communities in Arctic water bodies (Wells and Deming, 2003; Wells et al., 2006; Kirchman et al., 2007; Alonso-Sáez et al., 2008). These studies have also shown that (i) *Crenarchaeota* are more abundant at depth and *Euryarchaeota* are at detection limits in the western Arctic Ocean (Kirchman et al., 2007); (ii) *Archaea* are more abundant at nepheloid layers (the layer of water near the sediment bottom that contains a high concentration of suspended sediment and organic matter) as compared to surface waters in the Canadian Arctic, including the Northwest Passage (Wells and Deming, 2003) and the Beaufort Shelf and Franklin Bay (Wells et al., 2006); and (iii) concentrations of *Archaea* decrease during the summer to almost undetectable levels in Franklin Bay surface waters (Alonso-Sáez et al., 2008).

A recently documented study has comprehensively (>~16,000 reads for each sample) identified the *Archaea* inhabiting the Arctic Ocean via modern tagged 16S rRNA gene pyrosequencing methodology (Galand et al., 2009c). In conclusion, it was shown that MGI were overall the most abundant archaeal group (27 to 63% of the tagged sequences) in the combined depths sampled in the Arctic Ocean, with MGIII being more abundant in the deep waters, MGII being more represented in coastal surface waters, and MGIV also detected in low abundance in the deeper waters. Earlier 16S rRNA gene-based studies of Arctic water bodies have shown similar trends, with MGI dominating the next most common group, the MGII, in marine waters (Galand et al., 2006, 2008a, 2009a; Kalanetra et al., 2009; Kellogg and Deming, 2009) and MGII dominating in coastal waters, most likely due to riverine influences (Galand et al., 2006, 2008a). Through DGGE-based profiling Bano et al. (2004) have also shown that archaeal assemblages are more diverse in deeper waters than those of the upper water column in the Arctic Ocean, with sequences from an associated 16S rRNA gene clone library related to the MGI, MGII, and MGIV. A large 16S rRNA gene-based pyrosequencing effort encompassing 740,353 sequences including both bacteria and *Archaea* has shown that the distribution of rare microbial phylotypes in the Arctic Ocean follows patterns similar to those of the more dominant phylotypes of the complete microbial community, indicating that rare members of the community do not follow a cosmopolitan distribution (Galand et al., 2009b).

Archaea in Arctic sea ice have been shown to be in low concentrations (Table 1), i.e., near the detection limits of FISH-based analyses (Brinkmeyer et al., 2003; Junge et al., 2004)

and below PCR detection limits (Brown and
Bowman, 2001). However, a recent molecular-
based survey of winter sea ice samples and as-
sociated under-ice water from the Franklin
Bay indicated a dominance of MGI phylotypes
(Collins et al., 2010).

ARCHAEA IN
SEAFLOOR ENVIRONMENTS
Primary studies of Arctic seafloor sediments indicated that *Archaea* comprise a minor component of the microbial inhabitants and are at the
limits of detection (Ravenschlag et al., 2001).
Specifically, Ravenschlag et al. (2001) reported
that 4.9% of the cells probed through a 15-
cm-deep core taken from Svalbard sediments
were *Archaea* (Table 1) and were at the limits
of FISH-based detection. Further studies of archaeal communities at Arctic seafloor habitats
are limited to 16S rRNA gene clone library
surveys in a shallow (20-m) site situated in the
Svalbard region (Tian et al., 2009) and basalt
at the Arctic spreading ridges situated in the
Norwegian-Greenland Sea (Lysnes et al., 2004).

Tian et al. (2009) investigated the archaeal
communities inhabiting Arctic seafloor sediment consisting of a 15-cm-deep core collected
from Kongsfjorden, Svalbard. *Crenarchaeota*
consisting of the MGI and unknown signatures
dominated (283 of 285 clones), with only 2 sequences related to the *Euryarchaeota*. The microbial inhabitants, including *Archaea*, of basalt
and the associated bottom seawater collected
at various sites along the Arctic spreading ridge
in the Norwegian-Greenland Sea have been
described by Lysnes et al. (2004). DGGE was
used to compare samples with all sequenced
DGGE bands related to the MGI. Results
from enrichments also showed the presence
of methanogens; however, these were not detected via molecular methods.

POLAR COMPARISONS OF
ARCHAEAL DIVERSITY
Few studies have undertaken direct comparisons of microbial communities between both
poles. To date, these few studies have indicated that *Archaea* in both Arctic and Antarctic

pack ice are near detection limits (Brown and
Bowman, 2001; Brinkmeyer et al., 2003) and
that oceanic archaeal assemblages are somewhat similar between both poles (Bano et al.,
2004; Kalanetra et al., 2009).

Comparisons between independent studies must be taken with some degree of caution due to the differences and biases inherent
in the culture-independent methodologies
utilized to survey archaeal diversity and distribution in polar habitats, i.e., DNA extraction methods, PCR biases and differing primer
sets, and classification of archaeal phylogenetic
divisions. Although differences in methodologies exist, generalities between Arctic and
Antarctic archaeal communities can be drawn.
For example, studies investigating microbial
communities of polar marine water bodies and
sediments indicate that these environments
are typically dominated by the MGI, with a
minor MGII, MGIII, and MGIV presence,
and lake sediments are dominated by euryarchaeotal phylotypes, with halobacteria present
in saline environments, i.e., springs and lakes.
Comparison of archaeal communities between
polar terrestrial soil environments is difficult
not only due to the contrasting soil biotopes in
these polar areas but also due to the fact that the
majority of Arctic studies have been focused
on methanogenic fractions and very few terrestrial Antarctic studies targeting *Archaea* exist
in the literature.

HIGHLIGHTS OF RECENT FINDINGS
The presence and distribution of *Archaea* in
polar marine and lake sites have continued
to be studied since the pioneering studies of
the mid-1990s (DeLong et al., 1994; DeLong,
1998). Likewise, *Archaea* inhabiting terrestrial
Arctic sites have received recent attention
due to climate-change predictions. However,
the presence and distribution of *Archaea* in
Antarctic terrestrial environments is not well
understood, with initial studies indicating an
absence or patchy distribution of *Archaea* in
these environments. For the first time, Ayton
et al. (2010) have undertaken a provisional
wide-ranging survey of *Archaea* in terrestrial

Antarctic sites and have shown that *Archaea* of the MGI 1.1b group are prevalent in a number of terrestrial sites in the Ross Sea region.

Recently, new high-throughput DNA pyrosequencing technologies have begun to offer a previously unsurpassed amount of genetic data from natural environments providing deep insight into the identity and genetic capabilities of these communities. These methodologies have recently been applied to the Arctic environment. The detailed functional potential of a High Arctic permafrost sample has been described by Yergeau et al. (2010) via the sequencing of a permafrost metagenome. A diverse number of key genes related to methane generation and oxidation and nitrogen cycling were identified in this study. Massively parallel pyrosequencing of thousands of 16S rRNA gene PCR amplicons from Arctic Ocean samples has also recently overcome the inherent complexity and great abundance of microbes of these in situ environments (Galand et al., 2009b, 2009c). These studies have identified both high- and low-abundance taxa (including *Archaea*) in various Arctic Ocean habitats and link community structure and distribution to biogeography. However, while these provocative high-throughput sequencing technologies appear to offer a glimpse into the minor member microbial complexity of these polar environments, these data should be viewed with some caution as the fidelity and accuracy of the methods have not been well tested. For example, recent studies are challenging the extent and even the existence of the very diverse rare microbial biosphere (Huse et al., 2007; Kunin et al., 2010).

ARCHAEA IN POLAR ENVIRONMENTS AND CLIMATE CHANGE

In recent decades the polar regions have exhibited different climate changes. Large-scale warming has been documented in the Arctic, whereas bulk warming trends in the Antarctic have not been observed. Rather, slight regional warming trends have been documented along the coast of East and West Antarctica, with the largest warming trend observed at the Antarctic Peninsula (Turner and Overland, 2009).

It has been estimated that approximately one-third of the global carbon pool is stored in the frozen northern latitudes. Due to the large-scale warming trends witnessed in the Arctic regions and with global warming predictions being most pronounced at these latitudes, the potential melting of permafrost and subsequent release of this large amount of carbon pose a serious problem. As temperatures increase and more organic matter becomes available due to thawing events, it is hypothesized that intensified greenhouse gas (i.e., carbon dioxide and methane) production by the in situ microbial communities will significantly affect ecosystem processes and global climate feedbacks due to increased carbon turnover. Therefore, there is considerable interest in identifying the native microbial communities at these sites and predicting how climate change will affect these communities and the global carbon budget in these ecosystems (Wagner and Liebner, 2010). As a result, most of the studies investigating Arctic microbial communities, as discussed earlier, have focused on the methane-producing archaeal group, the methanogens. These studies are only starting to provide insight into the in situ microbial communities at these vulnerable Arctic sites; therefore, more work is needed to understand and model the response of these communities to future climate-change predictions.

FUTURE WORK NEEDED

The future challenge of the study of *Archaea* in polar environments is to address the ecology, activity, and function of these microbial inhabitants. Manipulative experiments are difficult to conduct in situ due to the isolation of the environment, and the simulation of in situ environmental conditions in laboratory-based studies is also difficult to replicate. Therefore, the application of new metagenomic approaches has the potential to characterize these organisms in the absence of laboratory cultures/enrichments and will provide insight into potential functionality and physiology of the *Archaea*. Moreover, these methods will overcome the large problem of PCR primer bias that is common in PCR-based surveys of microbes in natural environments (Kanagawa, 2003) and will certainly reveal an as yet unknown archaeal component. Yet it must be recognized that these metagenomic approaches cannot specifically target the *Archaea* and will also include bacteria, which typically dominate the microbial communities at these polar sites, and also viruses and eukaryotes. As such, it can be postulated that state-of-the-art single-cell isolation and subsequent single-cell genomics (Ishii et al., 2010) may provide targeted methods to specifically target archaeal cells in future environmental microbiological studies in polar habitats. The information from these directed approaches can then be used in the development of media and isolation strategies.

RNA-targeted studies, targeting the transcribed component of the DNA, will exclude any exogenous DNA or dead and dormant cells that may be included in typical DNA-based studies. Therefore, coupled with next-generation pyrosequencing methodology, metatranscriptomics is the next logical step in understanding the "active" microbial inhabitants of natural polar populations and their influence on global chemical and nutrient cycling. These approaches are becoming more common in the literature as methods are quickly improving and beginning to provide high-quality RNA from low-biomass samples (van Vliet, 2010).

SUMMARY

The majority of microbes in any given environment are typically recalcitrant to laboratory cultivation, and as such only a handful of *Archaea* have been isolated from polar environments. Therefore, our current knowledge of archaeal presence and diversity in polar habitats has been almost completely dependent on culture-independent, DNA-based surveys. Few culture-independent studies have described *Archaea* in polar habitats, with polar marine waters being the most comprehensively studied. These approaches are beginning to provide insight into the diversity and distribution of *Archaea* in polar habitats and indicate that *Archaea* occupy a variety of polar habitats and represent a considerable fraction of the microbial biomass. However, as methodology and detection limits improve, it can be expected that the types of *Archaea* and the habitat ranges and niches occupied by *Archaea* in the polar regions will increase.

As few laboratory-cultivated archaeal strains are currently available, the specific metabolisms of the *Archaea* are unknown; thus, it may be too early to speculate on their potential role in global nutrient cycles and contribution to climate-change predictions. However, it can be expected that future studies investigating microbial diversity and function in the polar regions will include the application of modern state-of-the-art metagenomic and metatranscriptomic approaches that may be coupled to single-cell isolation techniques, thereby providing insights into the potential metabolisms of the *Archaea*. As discussed in chapter 7, preliminary metagenomic studies are beginning to indicate that the *Archaea* may play an important role in biogeochemical cycling, in particular nitrification and heterotrophy in polar environments (Schleper et al., 2005; Cavicchioli, 2006).

ACKNOWLEDGMENTS
This work was supported by grants from the National Science Foundation (OPP 0229836, 0739648, 0632250, 0944560) to S.C.C. and the Royal Society of New Zealand Marsden Fund to S.C.C. and I.R.M., and by Foundation for Research, Science and Technology of New Zealand through an IPY grant to S.C.C. Antarctica New Zealand and Raytheon Polar Services provided logistical support.

We especially thank Shulamit Gordon, Simon Trotter, and Rob Strachan of Antarctica New Zealand for field and logistical planning over the past 5 years and for their enthusiasm for what they cannot see.

REFERENCES
Aislabie, J., and J. P. Bowman. 2010. Archaeal diversity in Antarctic ecosystems, p. 31–60. *In* A. K. Bej, J. Aislabie, and R. M. Atlas (ed.), *Polar Microbiology: the Ecology, Biodiversity and Bioremediation Potential of Microorganisms in Extremely Cold Environments.* CRC Press, Boca Raton, FL.

Alonso-Sáez, L., O. Sánchez, J. M. Gasol, V. Balagué, and C. Pedrós-Alio. 2008. Winter-to-summer changes in the composition and single-cell activity of near-surface Arctic prokaryotes. *Environ. Microbiol.* **10:**2444–2454.

Ayton, J., J. Aislabie, G. M. Barker, D. Saul, and S. Turner. 2010. Crenarchaeota affiliated with group 1.1b are prevalent in coastal mineral soils of the Ross Sea region of Antarctica. *Environ. Microbiol.* **12:**689–703.

Bano, N., S. Ruffin, B. Ransom, and J. T. Hollibaugh. 2004. Phylogenetic composition of Arctic Ocean archaeal assemblages and comparison with Antarctic assemblages. *Appl. Environ. Microbiol.* **70:**781–789.

Bottos, E. M., W. F. Vincent, C. W. Greer, and L. G. Whyte. 2008. Prokaryotic diversity of arctic ice shelf microbial mats. *Environ. Microbiol.* **10:**950–966.

Bowman, J. P., S. M. Rea, S. A. McCammon, and T. A. McMeekin. 2000a. Diversity and community structure within anoxic sediment from marine salinity meromictic lakes and a coastal meromictic marine basin, Vestfold Hills, Eastern Antarctica. *Environ. Microbiol.* **2:**227–237.

Bowman, J. P., S. A. McCammon, S. M. Rea, and T. A. McMeekin. 2000b. The microbial composition of three limnologically disparate hypersaline Antarctic lakes. *FEMS Microbiol. Lett.* **183:**81–88.

Bowman, J. P., and R. D. McCuaig. 2003. Biodiversity, community structural shifts, and biogeography of prokaryotes within Antarctic continental shelf sediment. *Appl. Environ. Microbiol.* **69:**2463–2483.

Bowman, J. P., S. A. McCammon, J. A. Gibson, L. Robertson, and P. D. Nichols. 2003. Prokaryotic metabolic activity and community structure in Antarctic continental shelf sediments. *Appl. Environ. Microbiol.* **69:**2448–2462.

Brambilla, E., H. Hippe, A. Hagelstein, B. J. Tindall, and E. Stackebrandt. 2001. 16S rDNA diversity of cultured and uncultured prokaryotes of a mat sample from Lake Fryxell, McMurdo Dry Valleys, Antarctica. *Extremophiles* **5:**23–33.

Brinkmeyer, R., K. Knittel, J. Jürgens, H. Weyland, R. Amann, and E. Helmke. 2003. Diversity and structure of bacterial communities in Arctic versus Antarctic pack ice. *Appl. Environ. Microbiol.* **69:**6610–6619.

Brochier-Armanet, C., B. Boussau, S. Gribaldo, and P. Forterre. 2008. Mesophilic Crenarchaeota: proposal for a third archaeal phylum, the Thaumarchaeota. *Nat. Rev. Microbiol.* **6:**245–252.

Brown, M. V., and J. P. Bowman. 2001. A molecular phylogenetic survey of sea-ice microbial communities (SIMCO). *FEMS Microbiol. Ecol.* **35:**267–275.

Casanueva, A., N. Galada, G. C. Baker, W. D. Grant, S. Heaphy, B. Jones, M. Yanhe, A. Ventosa, J. Blamey, and D. A. Cowan. 2008. Nanoarchaeal 16S rRNA gene sequences are widely dispersed in hyperthermophilic and mesophilic halophilic environments. *Extremophiles* **12:**651–656.

Cavicchioli, R. 2006. Cold-adapted archaea. *Nat. Rev. Microbiol.* **4:**331–343.

Chaban, B., S. Y. Ng, and K. F. Jarrell. 2006. Archaeal habitats—from the extreme to the ordinary. *Can. J. Microbiol.* **52:**73–116.

Christner, B. C., E. Mosley-Thompson, L. G. Thompson, and J. N. Reeve. 2001. Isolation of bacteria and 16S rDNAs from Lake Vostok accretion ice. *Environ. Microbiol.* **3:**570–577.

Christner, B. C., B. H. Kvitko II, and J. N. Reeve. 2003. Molecular identification of Bacteria and Eukarya inhabiting an Antarctic cryoconite hole. *Extremophiles* **7:**177–183.

Church, M. J., E. F. DeLong, H. W. Ducklow, M. B. Karner, C. M. Preston, and D. M. Karl. 2003. Abundance and distribution of planktonic Archaea and Bacteria in the waters west of the Antarctic Peninsula. *Limnol. Oceanogr.* **48:**1893–1902.

Collins, R. E., G. Rocap, and J. W. Deming. 2010. Persistence of bacterial and archaeal communities in sea ice through an Arctic winter. *Environ. Microbiol.* **12:**1828–1841.

Coolen, M. J. L., E. C. Hopmans, W. I. C. Rijpstra, G. Muyzer, S. Schouten, J. K. Volkman, and J. S. Sinninghe Damsté. 2004. Evolution of the methane cycle in Ace Lake (Antarctica) during the Holocene: response of methanogens and methanotrophs to environmental change. *Org. Geochem.* **35:**1151–1167.

de La Torre, J. R., B. M. Goebel, E. I. Friedmann, and N. R. Pace. 2003. Microbial diversity of cryptoendolithic communities from the McMurdo Dry Valleys, Antarctica. *Appl. Environ. Microbiol.* **69:**3858–3867.

DeLong, E. F. 1998. Everything in moderation: archaea as 'non-extremophiles.' *Curr. Opin. Genet. Dev.* **8:**649–654.

DeLong, E. F., K. Y. Wu, B. B. Prézelin, and R. V. M. Jovine. 1994. High abundance of Archaea in Antarctic marine picoplankton. *Nature* **371:**695–697.

Elkins, J. G., M. Podar, D. E. Graham, K. S. Makarova, Y. Wolf, L. Randau, B. P. Hedlund, C. Brochier-Armanet, V. Kunin, I. Anderson, A. Lapidus, E. Goltsman, K. Barry, K. V. Koonin, P. Hugenholtz, N. Kyrpides, G. Wanner, P. Richardson, M. Keller, and K. O. Stetter. 2008. A korarchaeal genome reveals insights into the evolution of the Archaea. *Proc. Natl. Acad. Sci USA* **105**:8102–8107.

Forterre, P., S. Gribaldo, and C. Brochier-Armanet. 2009. Happy together: genomic insights into the unique *Nanoarchaeum/Ignicoccus* association. *J. Biol.* **8**:7.

Franzmann, P. D., Y. Liu, D. L. Balkwill, H. C. Aldrich, E. Conway de Macario, and D. R. Boone. 1997. *Methanogenium frigidum* sp. nov., a psychrophilic, H_2-using methanogen from Ace Lake, Antarctica. *Int. J. Syst. Bacteriol.* **47**:1068–1072.

Franzmann, P. D., N. Springer, W. Ludwig, E. Conway de Macario, and M. Rhode. 1992. A methanogenic archaeon from Ace Lake, Antarctica: *Methanococcoides burtonii* sp. nov. *Syst. Appl. Microbiol.* **15**:573–581.

Franzmann, P. D., E. Stackebrandt, K. Sanderson, J. K. Volkman, D. E. Cameron, P. L. Stevenson, T. A. McMeekin, and H. R. Burton. 1988. *Halobacterium lacusprofundi* sp. nov., a halophilic bacterium isolated from Deep Lake, Antarctica. *Syst. Appl. Microbiol.* **11**:20–27.

Galand, P. E., C. Lovejoy, A. K. Hamilton, R. G. Ingram, E. Pedneault, and E. C. Carmack. 2009a. Archaeal diversity and a gene for ammonia oxidation are coupled to oceanic circulation. *Environ. Microbiol.* **11**:971–980.

Galand, P. E., C. Lovejoy, and W. F. Vincent. 2006. Remarkably diverse and contrasting archaeal communities in a large arctic river and the coastal Arctic Ocean. *Aquat. Microb. Ecol.* **44**:115–126.

Galand, P. E., C. Lovejoy, J. Pouliot, and W. F. Vincent. 2008a. Heterogeneous archaeal communities in the particle-rich environment of an arctic shelf ecosystem. *J. Mar. Syst.* **74**:774–782.

Galand, P. E., E. O. Casamayor, D. L. Kirchman, and C. Lovejoy. 2009b. Ecology of the rare microbial biosphere of the Arctic Ocean. *Proc. Natl. Acad. Sci. USA* **106**:22427–22432.

Galand, P. E., C. Lovejoy, J. Pouliot, M. E. Garneau, and W. F. Vincent. 2008b. Microbial community diversity and heterotrophic production in a coastal Arctic ecosystem: a stamukhi lake and its source waters. *Limnol. Oceanogr.* **52**:813–823.

Galand, P. E., E. O. Casamayor, D. L. Kirchman, M. Potvin, and C. Lovejoy. 2009c. Unique archaeal assemblages in the Arctic Ocean unveiled by massively parallel tag sequencing. *ISME J.* **3**:860–869.

Ganzert, L., G. Jurgens, U. Münster, and D. Wagner. 2007. Methanogenic communities in permafrost-affected soils of the Laptev Sea coast, Siberian Arctic, characterized by 16S rRNA gene fingerprints. *FEMS Microbiol. Ecol.* **59**:476–488.

García-Martínez, J., and F. Rodríguez-Valera. 2000. Microdiversity of uncultured marine pro-

karyotes: the SAR11 cluster and the marine Archaea of Group I. *Mol. Ecol.* **9**:935–948.

Gibson, J. A. E. 1999. The meromictic lakes and stratified marine basins of the Vestfold Hills, East Antarctica. *Antarct. Sci.* **11**:175–192.

Gilichinsky, D., E. Rivkina, V. Shcherbakova, K. Laurinavichuis, and J. Tiedje. 2003. Supercooled water brines within permafrost—an unknown ecological niche for microorganisms: a model for astrobiology. *Astrobiology* **3**:331–341.

Gillan, D. C., and B. Danis. 2007. The archaebacterial communities in Antarctic bathypelagic sediments. *Deep Sea Res. Part 2 Top. Stud. Oceanogr.* **54**:1682–1690.

Glatz, R. E., P. W. Lepp, B. B. Ward, and C. A. Francis. 2006. Planktonic microbial community composition across steep physical/chemical gradients in permanently ice-covered Lake Bonney, Antarctica. *Geobiology* **4**:53–67.

Glissman, K., K. J. Chin, P. Casper, and R. Conrad. 2004. Methanogenic pathway and archaeal community structure in the sediment of eutrophic Lake Dagow: effect of temperature. *Microb. Ecol.* **48**:389–399.

Gribaldo, S., and C. Brochier-Armanet. 2006. The origin and evolution of Archaea: a state of the art. *Philos. Trans. R. Soc. Lond. B Biol. Sci.* **361**:1007–1022.

Grosskopf, R., S. Stubner, and W. Liesack. 1998. Novel euryarchaeotal lineages detected on rice roots and in the anoxic bulk soil of flooded rice microcosms. *Appl. Environ. Microbiol.* **64**:4983–4989.

Høj, L., R. A. Olsen, and V. L. Torsvik. 2005. Archaeal communities in High Arctic wetlands at Spitsbergen, Norway (78°N) as characterized by 16S rRNA gene fingerprinting. *FEMS Microbiol. Ecol.* **53**:89–101.

Høj, L., R. A. Olsen, and V. L. Torsvik. 2008. Effects of temperature on the diversity and community structure of known methanogenic groups and other archaea in high Arctic peat. *ISME J.* **2**:37–48.

Høj, L., M. Rusten, L. E. Haugen, R. A. Olsen, and V. L. Torsvik. 2006. Effects of water regime on archaeal community composition in Arctic soils. *Environ. Microbiol.* **8**:984–996.

Huber, H., M. J. Hohn, R. Rachel, T. Fuchs, V. C. Wimmer, and K. O. Stetter. 2002. A new phylum of Archaea represented by a nanosized hyperthermophilic symbiont. *Nature* **417**:63–67.

Huse, S. M., J. A. Huber, H. G. Morrison, M. L. Sogin, and D. M. Welch. 2007. Accuracy and quality of massively parallel DNA pyrosequencing. *Genome Biol.* **8**:R143.

Ishii, S., K. Tago, and K. Senoo. 2010. Single-cell analysis and isolation for microbiology and biotechnology: methods and applications. *Appl. Microbiol. Biotechnol.* **86:**1281–1292.

Junge, K., H. Eicken, and J. W. Deming. 2004. Bacterial activity at −2 to −20°C in Arctic wintertime sea ice. *Appl. Environ. Microbiol.* **70:**550–557.

Kalanetra, K. M., N. Bano, and J. T. Hollibaugh. 2009. Ammonia-oxidizing *Archaea* in the Arctic Ocean and Antarctic coastal waters. *Environ. Microbiol.* **11:**2434–2445.

Kanagawa, T. 2003. Bias and artifacts in multitemplate polymerase chain reactions (PCR). *J. Biosci. Bioeng.* **96:**317–323.

Karr, E. A., J. M. Ng, S. M. Belchik, W. M. Sattley, M. T. Madigan, and L. A. Achenbach. 2006. Biodiversity of methanogenic and other *Archaea* in the permanently frozen Lake Fryxell, Antarctica. *Appl. Environ. Microbiol.* **72:**1663–1666.

Kellogg, C. T. E., and J. W. Deming. 2009. Comparison of free-living, suspended particle, and aggregate-associated bacterial and archaeal communities in the Laptev Sea. *Aquat. Microb. Ecol.* **57:**1–18.

Kirchman, D. L., H. Elifantz, A. I. Dittel, R. R. Malmstrom, and M. T. Cottrell. 2007. Standing stocks and activity of Archaea and Bacteria in the western Arctic Ocean. *Limnol. Oceanogr.* **52:**495–507.

Kobabe, S., D. Wagner, and E. M. Pfeiffer. 2004. Characterisation of microbial community composition of a Siberian tundra soil by fluorescence in situ hybridisation. *FEMS Microbiol. Ecol.* **50:**13–23.

Könneke, M., A. E. Bernhard, J. R. de la Torre, C. B. Walker, J. B. Waterbury, and D. A. Stahl. 2005. Isolation of an autotrophic ammonia-oxidizing marine archaeon. *Nature* **437:**543–546.

Krivushin, K. V., V. A. Shcherbakova, L. E. Petrovskaya, and E. M. Rivkina. 2010. *Methanobacterium veterum* sp. nov., from ancient Siberian permafrost. *Int. J. Syst. Evol. Microbiol.* **60:**455–459.

Kunin, V., A. Engelbrektson, H. Ochman, and P. Hugenholtz. 2010. Wrinkles in the rare biosphere: pyrosequencing errors can lead to artificial inflation of diversity estimates. *Environ. Microbiol.* **12:**118–123.

Kurosawa, N., S. Sato, Y. Kawarabayasi, S. Imura, and T. Naganuma. 2010. Archaeal and bacterial community structures in the anoxic sediment of Antarctic meromictic lake Nurume-Ike. *Polar Sci.* **4:**421–429.

López-García, P., D. Moreira, A. López-López, and F. Rodríguez-Valera. 2001a. A novel halo-archaeal-related lineage is widely distributed in deep oceanic regions. *Environ. Microbiol.* **3:**72–78.

López-García, P., A. López-López, D. Moreira, and F. Rodríguez-Valera. 2001b. Diversity

of free-living prokaryotes from a deep-sea site at the Antarctic Polar Front. *FEMS Microbiol. Ecol.* **36:**193–202.

Lysnes, K., I. H. Thorseth, B. O. Steinsbu, L. Øvreås, T. Torsvik, and R. B. Pedersen. 2004. Microbial community diversity in seafloor basalt from the Arctic spreading ridges. *FEMS Microbiol. Ecol.* **50:**213–230.

Massana, R., L. T. Taylor, A. E. Murray, K. Y. Wu, W. H. Jeffrey, and E. F. DeLong. 1998. Vertical distribution and temporal variation of marine planktonic archaea in the Gerlache Strait, Antarctica, during early spring. *Limnol. Oceanogr.* **43:**607–617.

Metje, M., and P. Frenzel. 2007. Methanogenesis and methanogenic pathways in a peat from subarctic permafrost. *Environ. Microbiol.* **9:**954–964.

Morozova, D., and D. Wagner. 2007. Stress response of methanogenic archaea from Siberian permafrost compared with methanogens from nonpermafrost habitats. *FEMS Microbiol. Ecol.* **61:**16–25.

Morozova, D., D. Möhlmann, and D. Wagner. 2007. Survival of methanogenic archaea from Siberian permafrost under simulated Martian thermal conditions. *Orig. Life Evol. Biosph.* **37:**189–200.

Murray, A. E., C. M. Preston, R. Massana, L. T. Taylor, A. Blakis, K. Wu, and E. F. DeLong. 1998. Seasonal and spatial variability of bacterial and archaeal assemblages in the coastal waters near Anvers Island, Antarctica. *Appl. Environ. Microbiol.* **64:**2585–2595.

Murray, A. E., K. Y. Wu, C. L. Moyer, D. M. Karl, and E. F. DeLong. 1999. Evidence for circumpolar distribution of planktonic Archaea in the Southern Ocean. *Aquat. Microb. Ecol.* **18:**263–273.

Niederberger, T. D., N. N. Perreault, J. R. Lawrence, J. L. Nadeau, R. E. Mielke, C. W. Greer, D. T. Andersen, and L. G. Whyte. 2009. Novel sulfur-oxidizing streamers thriving in perennial cold saline springs of the Canadian high Arctic. *Environ. Microbiol.* **11:**616–629.

Niederberger, T. D., N. N. Perreault, S. Tille, B. S. Lollar, G. Lacrampe-Couloume, D. Andersen, C. W. Greer, W. Pollard, and L. G. Whyte. 2010. Microbial characterization of a subzero, hypersaline methane seep in the Canadian High Arctic. *ISME J.* **4:**1326–1339.

Ochsenreiter, T., D. Selezi, A. Quaiser, L. Bonch-Osmolovskaya, and C. Schleper. 2003. Diversity and abundance of *Crenarchaeota* in terrestrial habitats studied by 16S RNA surveys and real time PCR. *Environ. Microbiol.* **5:**787–797.

Perreault, N. N., D. T. Andersen, W. H. Pollard, C. W. Greer, and L. G. Whyte. 2007. Characterization of the prokaryotic diversity in cold saline perennial springs of the Canadian high Arctic. *Appl. Environ. Microbiol.* **73**:1532–1543.

Pointing, S. B., Y. Chan, D. C. Lacap, M. C. Lau, J. A. Jurgens, and R. L. Farrell. 2009. Highly specialized microbial diversity in hyperarid polar desert. *Proc. Natl. Acad. Sci. USA* **106**:19964–19969.

Pouliot, J., P. E. Galand, C. Lovejoy, and W. F. Vincent. 2009. Vertical structure of archaeal communities and the distribution of ammonia monooxygenase A gene variants in two meromictic High Arctic lakes. *Environ. Microbiol.* **11**:687–699.

Powell, S. M., J. P. Bowman, I. Snape, and J. S. Stark. 2003. Microbial community variation in pristine and polluted nearshore Antarctic sediments. *FEMS Microbiol. Ecol.* **45**:135–145.

Preston, C. M., K. Y. Wu, T. F. Molinski, and E. F. DeLong. 1996. A psychrophilic crenarchaeon inhabits a marine sponge: *Cenarchaeum symbiosum* gen. nov., sp. nov. *Proc. Natl. Acad. Sci. USA* **93**:6241–6246.

Purdy, K. J., D. B. Nedwell, and T. M. Embley. 2003. Analysis of the sulfate-reducing bacterial and methanogenic archaeal populations in contrasting Antarctic sediments. *Appl. Environ. Microbiol.* **69**:3181–3191.

Ravenschlag, K., K. Sahm, and R. Amann. 2001. Quantitative molecular analysis of the microbial community in marine Arctic sediments (Svalbard). *Appl. Environ. Microbiol.* **67**:387–395.

Rivkina, E., V. Shcherbakova, K. Laurinavichius, L. Petrovskaya, K. Krivushin, G. Kraev, S. Pecheritsina, and D. Gilichinsky. 2007. Biogeochemistry of methane and methanogenic archaea in permafrost. *FEMS Microbiol. Ecol.* **61**:1–15.

Rivkina, E. M., K. S. Laurinavichus, D. A. Gilichinsky, and V. A. Shcherbakova. 2002. Methane generation in permafrost sediments. *Dokl. Biol. Sci.* **383**:179–181.

Schleper, C., G. Jurgens, and M. Jonuscheit. 2005. Genomic studies of uncultivated archaea. *Nat. Rev. Microbiol.* **3**:479–488.

Shcherbakova, V., E. Rivkina, S. Pecheritsyna, K. Laurinavichius, N. Suzina, and D. Gilichinsky. 2010. *Methanobacterium arcticum* sp. nov., a methanogenic archaeon from Holocene Arctic permafrost. *Int. J. Syst. Evol. Microbiol.* **61**:144–147.

Simankova, M. V., O. R. Kotsyurbenko, T. Lueders, A. N. Nozhevnikova, B. Wagner, R. Conrad, and M. W. Friedrich. 2003. Isolation and characterization of new strains of methanogens from cold terrestrial habitats. *Syst. Appl. Microbiol.* **26**:312–318.

Singh, N., M. M. Kendall, Y. Liu, and D. R. Boone. 2005. Isolation and characterization of methylotrophic methanogens from anoxic marine sediments in Skan Bay, Alaska: description of *Methanococcoides alaskense* sp. nov., and emended description of *Methanosarcina baltica. Int. J. Syst. Evol. Microbiol.* **55**:2531–2538.

Sjöling, S., and D. A. Cowan. 2003. High 16S rDNA bacterial diversity in glacial meltwater lake sediment, Bratina Island, Antarctica. *Extremophiles* **7**:275–282.

Soo, R. M., S. A. Wood, J. J. Grzymski, I. R. McDonald, and S. C. Cary. 2009. Microbial biodiversity of thermophilic communities in hot mineral soils of Tramway Ridge, Mount Erebus, Antarctica. *Environ. Microbiol.* **11**:715–728.

Steven, B., G. Briggs, C. P. McKay, W. H. Pollard, C. W. Greer, and L. G. Whyte. 2007. Characterization of the microbial diversity in a permafrost sample from the Canadian high Arctic using culture-dependent and culture-independent methods. *FEMS Microbiol. Ecol.* **59**:513–523.

Steven, B., W. H. Pollard, C. W. Greer, and L. G. Whyte. 2008. Microbial diversity and activity through a permafrost/ground ice core profile from the Canadian high Arctic. *Environ. Microbiol.* **10**:3388–3403.

Teske, A., and K. B. Sorensen. 2007. Uncultured archaea in deep marine subsurface sediments: have we caught them all? *ISME J.* **2**:3–18.

Tian, F., Y. Yu, B. Chen, H. Li, Y. F. Yao, and X. K. Guo. 2009. Bacterial, archaeal and eukaryotic diversity in Arctic sediment as revealed by 16S rRNA and 18S rRNA gene clone libraries analysis. *Polar Biol.* **32**:93–103.

Topping, J. N., J. L. Heywood, P. Ward, and M. V. Zubkov. 2006. Bacterioplankton composition in the Scotia Sea, Antarctica, during the austral summer of 2003. *Aquat. Microb. Ecol.* **45**:229–235.

Turner, J., and J. Overland. 2009. Contrasting climate change in the two polar regions. *Polar Res.* **28**:146–164.

van Vliet, A. H. 2010. Next generation sequencing of microbial transcriptomes: challenges and opportunities. *FEMS Microbiol. Lett.* **302**:1–7.

Wagner, D., and S. Liebner. 2010. Methanogenesis in Arctic permafrost habitats, p. 655–663. *In* K. N. Timmis (ed.), *Handbook of Hydrocarbon and Lipid Microbiology.* Springer, Berlin, Germany.

Wagner, D., A. Lipski, A. Embacher, and A. Gattinger. 2005. Methane fluxes in permafrost habitats of the Lena Delta: effects of microbial community structure and organic matter quality. *Environ. Microbiol.* **7**:1582–1592.

Webster, N. S., A. P. Negri, M. M. Munro, and C. N. Battershill. 2004. Diverse microbial communities inhabit Antarctic sponges. *Environ. Microbiol.* **6**:288–300.

Wells, L. E., M. Cordray, S. Bowerman, L. A. Miller, W. F. Vincent, and J. W. Deming.

2006. Archaea in particle-rich waters of the Beaufort Shelf and Franklin Bay, Canadian Arctic: clues to an allochthonous origin? *Limnol. Oceanogr.* **51**:47–59.

Wells, L. E., and J. W. Deming. 2003. Abundance of Bacteria, the Cytophaga-Flavobacterium cluster and Archaea in cold oligotrophic waters and nepheloid layers of the Northwest Passage, Canadian Archipelago. *Aquat. Microb. Ecol.* **31**:19–31.

Wilhelm, R. C., T. D. Niederberger, C. Greer, and L. G. Whyte. 2011. Microbial diversity of active layer and permafrost in an acidic wetland from the Canadian High Arctic. *Can. J. Microbiol.* **57**:303–315.

Woese, C. R., and G. E. Fox. 1997. Phylogenetic structure of the prokaryotic domain: the primary kingdoms. *Proc. Natl. Acad. Sci. USA* **74**:5088–5090.

Woese, C. R., O. Kandler, and M. L. Wheelis. 1990. Towards a natural system of organisms: proposal for the domains Archaea, Bacteria, and Eucarya. *Proc. Natl. Acad. Sci. USA* **87**:4576–4579.

Yergeau, E., H. Hogues, L. G. Whyte, and C. W. Greer. 2010. The functional potential of high Arctic permafrost revealed by metagenomic sequencing, qPCR and microarray analyses. *ISME J.* **4**:1206–1214.

Yergeau, E., S. Kang, Z. He, J. Zhou, and G. A. Kowalchuk. 2007. Functional microarray analysis of nitrogen and carbon cycling genes across an Antarctic latitudinal transect. *ISME J.* **1**:163–179.

Yergeau, E., S. A. Schoondermark-Stolk, E. L. Brodie, S. Dejean, T. Z. DeSantis, O. Goncalves, Y. M. Piceno, G. L. Andersen, and G. A. Kowalchuk. 2009. Environmental microarray analyses of Antarctic soil microbial communities. *ISME J.* **3**:340–351.

BACTERIOPHAGES AT THE POLES

Robert V. Miller

3

Bacteriophages, viruses that infect bacteria, are the core of any ecosystem. Wommack and Colwell (2000) and Thingstad et al. (2008) have observed that bacteriophages regulate the size of bacterial populations in natural habitats. They increase bacterial diversity by stimulating lateral gene transfer through virus-mediated transduction (Miller, 1998a, 2004; Miller and Day, 2008) and have likely contributed to bacterial speciation (Miller, 1998a; Miller and Day, 2004b, 2008; Bull et al., 2006). Yet, how do bacteriophages survive in hostile environments such as the Arctic and Antarctic ecosystems in which their bacterial hosts usually exist under extreme starvation conditions (Koch, 1971, 1979; Morita, 1997)? In such environments the latency period for phage propagation is lengthened, the burst size is reduced (Kokjohn et al., 1991), and the rate of decay of virion infectivity is rapid (Miller, 2006). Still, high concentrations of phage-like particles have been observed in many environments, their numbers often exceeding 10^8 particles/ml in aquatic environments (Miller and Sayler, 1992; Wommack and Colwell, 2000; Miller, 2006; Miller and Day, 2008).

This chapter will explore the data that have been collected on the importance of bacteriophages to the ecology of the earth's polar regions. It will examine the numbers of phagelike particles that have been observed in these ecosystems and their importance in regulating bacterial numbers and the food chain of these extreme oligotrophic environments. The alternate lifestyles of phages (Miller and Day, 2008) will be described and the data (or lack thereof) on the relative importance of these lifestyles in the microbial communities of the polar regions will be discussed.

LIFE CHOICES

To understand the potential impact of bacteriophages on polar ecosystems, it is first necessary to understand the different life choices of phages and how environmental factors are known to affect them. There are three lifestyle choices that phages are known to make. These choices depend on the genetics of both phage and host and the physiological state of the host. They are commonly tagged as the productive, lysogenic, and pseudolysogenic lifestyles (Miller, 2006). For completeness we will discuss each of these life choices, but pseudolysogeny has not yet been studied in either the Artic or Antarctic.

Robert V. Miller, Department of Microbiology and Molecular Genetics, Oklahoma State University, 307 Life Sciences East, Stillwater, OK 74078.

Polar Microbiology: Life in a Deep Freeze
Edited by Robert V. Miller and Lyle G. Whyte © 2012 ASM Press, Washington, DC

The Productive Life Choice

As may be surmised from the name, the productive life cycle of bacteriophages leads to the production of progeny virions (Saye and Miller, 1989). All phages undergo productive life cycles at some point. The infection of the host is followed by the synthesis of new phage nucleic acid and capsid proteins that are assembled into new virions. Productive life cycles can be divided into two types: lytic and chronic (Miller and Day, 2008).

Lytic productive cycles terminate with the lysis of the host cell and the liberation of a burst of progeny phage particles. The complex double-stranded DNA (dsDNA) bacteriophages such as T4 and λ that infect *Escherichia coli* have lytic productive cycles (Saye and Miller, 1989), as illustrated in Fig. 1. Each specific species of phage's lytic life cycle can be described with several parameters that are affected by the environment. The latency period is the time between infection of the host and the burst of the infected cell to release the progeny virions. In oligotrophic environments, this period is typically longer than is observed with the same phage species in the laboratory setting, where the host has ready access to nutrients to supply energy and macromolecular building blocks (amino acids and nucleotides) for phage growth (Kokjohn et al., 1991). The burst size of progeny phage, which is a measure of the average number of progeny virions produced by an infected cell, is reduced in environments of low nutrient availability compared to the numbers observed in the laboratory (Kokjohn et al., 1991). This lifestyle is responsible for the vast majority of the phagelike particles observed in the environment.

In other phages, including many filamentous phages, the productive life cycle can be characterized as chronic. The chronic productive cycle does not terminate with the lysis of the host cell. Instead the infected cell continuously releases virions that bud through the cellular membrane of the host (Miller and Day, 2008). This slow production of phage particles continues until the host suffers sufficient damage to its cellular membrane and is no longer

viable. While potentially important to natural microbial ecosystems, chronic infection is difficult to identify in nature and has been little studied. It is assumed the vast majority, if not all, of the phagelike particles observed in environmental samples have been produced from lytic phages.

PHENOTYPIC CHANGES TO THE HOST CELL DURING THE LATENT PERIOD (LYTIC CONVERSION)

More often than was originally believed, bacteriophage genomes contain hostlike genes that alter the host's phenotype during the latent period (Miller and Day, 2008). The classic example of this phenomenon is the expression of diphtheria toxin. The gene for this virulence factor is not encoded in the bacterium's genome but is a part of the genome of corynebacteriophage β (Holmes and Barksdale, 1970; Barksdale and Ardon, 1974; Hyman and Abedon, 2008; Miller and Day, 2008). Other examples include cholera toxin (Hyman and Abedon, 2008) and *Shigella* toxin in *E. coli* O157:H7 (Canchaya et al., 2003). In each of these cases, the bacterium cannot make the toxin and, therefore, is not pathogenic unless infected or lysogenized by the appropriate bacteriophage (Miller and Day, 2008).

In addition to toxin production in several pathogenic bacteria, "lytic conversion" has been identified in a number of cyanophages. These phage genomes contain the genes (*psbA* and *psbD*) for polypeptides (D1 and D2) that make up part of the core of the photosystem II reaction center of their hosts (Mann et al., 2003; Bailey et al., 2004; Lindell et al., 2004, 2005; Clokie et al., 2006). These proteins are very labile and rapidly turn over in the cell. The phage genes are believed to be expressed during lytic growth to maintain active photocenters during lytic development of progeny virions (Mann et al., 2003; Bailey et al., 2004; Clokie et al., 2006; Lindell et al., 2005). In addition to *pbsA* and *pbsD*, cyanophages of *Prochlorococcus* contain analogues to *hli* (high-light-inducible protein, or HLIP), *petE* (plastocyanin), and *petF* (ferredoxin) (Lindell et al., 2004).

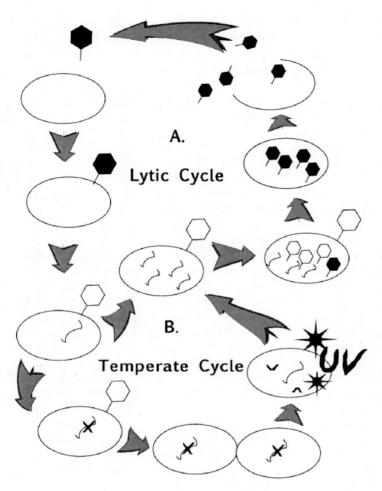

A.

Lytic Cycle

B.

Temperate Cycle

FIGURE 1 The lytic and lysogenic life cycles of bacteriophages. (A) The lytic life cycle, which leads to the production of progeny phages. (B) The temperate life cycle, in which a prophage is established in the host cell, producing a bacterium referred to as a lysogen. The prophage is not transcribed (indicated by the X's), but replicates with the host cell's genome and is partitioned into its daughter cells. Some prophages are integrated into the host genome, while other types are carried as plasmids. Note: The host genome is not illustrated in this figure. (Reprinted from *Microbial Evolution: Gene Establishment, Survival, and Exchange*, Miller and Day [2004a], with permission from ASM Press.)

The phenomenon of lytic conversion seen in cyanophages has the potential to facilitate the evolution of these genes and proteins among cyanobacteria through lateral gene transfer mediated by transduction (Miller, 2001, 2004; Replicon et al., 1995; Saye and Miller, 1989; Saye et al., 1987). Lindell et al. (2004) suggested that lateral gene transfer among cyanophages may play a role in ensuring the fitness of these organisms in environments such as the polar regions. Examination of uncultured environmental cyanophages and prophages of *Synechococcus* by Zeidner et al. (2005) revealed strong evidence that these genes are transduced among *Synechococcus* and *Prochlorococcus* populations. These gene products protect cells from photobleaching and are important to the pho-

tosynthetic process. By transfer of these genes through infection of their hosts, these phages have enriched the bacterial gene pool (Replicon et al., 1995) and have most likely played an important role in the evolution of their hosts (Zeidner et al., 2005; Miller, 2004).

Lysogeny

In addition to the productive life cycle, many phages can initiate a reductive infection in which the phage genome is retained in the host cell in a quiescent state and progeny virus particles are not produced. These reductive lifestyles are divided into lysogeny and pseudolysogeny (Miller, 2004; Miller and Day, 2008). In lysogeny, the phage genome (termed a prophage) establishes a long-term relationship in which prophage replication is coordinated with host-genome replication and progeny phage genomes are distributed to each of the daughter cells at division (Fig. 1). Depending on the specific virus species, the prophage may be integrated into the host's genome or maintained in the cytoplasm as a plasmid. At a later time, the prophage may become activated and enter a productive cycle leading to the production of progeny virions and the death of the host. This activation may occur either spontaneously or following exposure of the host to various types of environmental stressors. These stressors are usually those that affect DNA integrity such as UV radiation, mitomyocin C, or desiccation (Miller and Day, 2008).

Most prophages are kept from expressing their lytic functions by a repressor molecule that is encoded by the phage genome and expressed even during lysogeny (Ptashne, 2004; Miller and Day, 2008). The concentration of this repressor in the cytoplasm is autoregulated both positively and negatively. The stimulation of its degradation by exposure to the stressors mentioned above leads to the induction of the prophage and the initiation of lytic growth (Ptashne, 2004).

LYSOGENIC CONVERSION

The presence of prophage in the lysogenized cell may change not only the cell's genotype but also its phenotype (Miller and Day, 2008). The most common form of lysogenic conversion is the establishment of superinfection immunity (Ptashne, 2004). This process is initiated by the presence of the repressor of lytic expression. It renders the lysogenized bacterium immune to infection by virions of viruses of the same and sometimes closely related phages (Miller and Day, 2008). This is due to the presence of the soluble repressor of lytic expression in the cytoplasm of the lysogenized bacterium. These repressor molecules bind to the operator present in the phage/prophage genome, inhibiting their expression (Ptashne, 2004). The binding of repressor to the prophage promoter/operator is maintained by the balance in the concentration of bound and free repressor molecules

in the bacterial cytoplasm. When a superinfecting phage genome enters the cytoplasm, it is immediately bound by free repressor protein, inhibiting expression of its lytic functions and rendering the host immune to its infection.

Many phages contain other genes that may be expressed during reductive growth that alter the phenotype of the host. These can include alterations in cell envelope components. These alterations may render the host resistant to infection by certain bacteriophages while sensitizing it to others (Saye and Miller, 1989). Other alterations include acquisition of new metabolic systems (Cavenagh and Miller, 1986; Ptashne, 2004). As mentioned in the section above on lytic conversion, many cyanobacterium phages carry the genes for a portion of the photosynthetic complex that can be expressed during lysogeny as well as productive growth (Mann et al., 2003; Lindell et al., 2004, 2005). The acquisition of bacteriocin production is also sometimes associated with the presence of prophage in the host cell (Ivánovics et al., 1976), as is the synthesis of antibiotics (Martinez-Molina and Olivares, 1979), the acquisition of new restriction-modification systems (Rocha et al., 2001), and the production of pili (Karaolis et al., 1999). Lysogenic conversion as well as lytic conversion can lead to the expression of phage-encoded toxins (Mahenthiralingam, 2004) including cholera toxin (Hyman and Abedon, 2008) and diphtheria toxin (Barksdale and Ardon, 1974).

Most often lysogenic conversion increases fitness. Frye et al. (2005) suggested that increasing the flexibility and mobility of phenotypes leads to successful niche occupation. Bacterial community composition is influenced by introducing alternative phenotypes, sometimes leading to the dominance of the lysogens (Edlin et al., 1975). New ecological niches may be opened to lysogeny by enabling them to utilize new metabolites or express new virulence factors (Mahenthiralingam, 2004; Lindell et al., 2005; Hyman and Abedon, 2008). Thus lysogeny can be considered a positive symbiotic relationship between phage and bacterium

(Miller and Day, 2008). It provides safety to the phage genome from environmental insults that result in virion decay while providing the host with opportunities for occupancy of additional niches and increased fitness. Given the advantages of lysogeny, it should not be surprising that it is widespread in nature. Ackerman and DuBow (1987) estimated that between 21 and 60% of environmental bacteria harbor at least one prophage. Even this may be an underestimate. Ortmann et al. (2002), studying a fjord in British Columbia, found that 80% of heterotrophs were lysogens while at least 0.6% of cyanobacteria contained bacteriophage genomes that were inducible by mitomycin C. Similarly, Miller et al. (1992) found that 80% of *Pseudomonas aeruginosa* isolates from Fort Loudon Lake in Tennessee contained sequences that hybridized to *Pseudomonas* phage genomes. Sequencing of whole bacterial genomes has shown the presence of prophages and vestigial prophages in practically all bacteria. Many contain multiple prophages. For instance, *E. coli* O157:H7 strain Sakai contains parts of at least 18 prophages. This accounts for 16% of the genome and includes the *Shigella* toxin gene that makes O157:H7 such an important player in food contamination (Canchaya et al., 2003; Miller and Day, 2008).

Pseudolysogeny

Pseudolysogeny was first defined by Baess in 1971. In his review of the early literature on the subject, pseudolysogeny was identified as a phage-host interaction in which the phage, following infection of the host, elicits an unstable, nonproductive response (Baess, 1971; Miller and Ripp, 2002; Miller and Day, 2008). In this state neither lysogeny nor productive growth ensues. Rather the viral genome is maintained in the host cell in limbo for potentially extended periods of time as a preprophage (Miller and Ripp, 2002). Unlike in true lysogeny, the preprophage does not replicate and is inherited linearly as opposed to exponentially (Miller and Day, 2008). Miller and Ripp (2002) hypothesized that a pseudolysogenic preprophage is established in dramatically starved hosts that cannot provide the phage with the necessary energy and substrates required for either a productive or lysogenic response. Cellular starvation in environmental ecosystems is commonplace (Koch, 1971, 1979; Morita, 1997). This is certainly true of the polar regions and especially Antarctica. However, in all ecosystems there are time periods in which nutrients are available (Miller and Day, 2008). During these periods preprophages can be converted into true prophages or in some cases the productive life cycle of phage growth can be initiated (Ripp and Miller, 1997, 1998; Miller and Ripp, 1998, 2002; Miller and Day, 2008). To my knowledge, no studies on pseudolysogeny have yet been carried out in the polar regions.

NUTRIENT AVAILABILITY

Lifestyle decisions and their outcome for bacteriophages are to a great extent controlled by the availability of nutrients to the host (Ptashne, 2004). Ashelford et al. (1999, 2003) demonstrated that the frequency of lytic (those phages that are not known to establish lysogeny) and temperate (those that do establish lysogeny) phages varied seasonally in a sugar beet field. Early in the spring, when nutrients were plentifully produced by the sugar beets and were readily available to the bacterial host of these viruses, lytic phages predominated. In the fall, when nutrients were scarce, temperate phages predominated.

Effects on Productive Growth

Studies investigating the correlation between phage productivity and nutrient availability have been carried out under optimal laboratory conditions (Lenski, 1988) and under environmental conditions (Kokjohn et al., 1991; Proctor et al., 1993; Guixa-Boixereu et al., 1996; Schrader et al., 1997a, 1997b; Moebus, 1996; Middelboe, 2000). All of these studies have demonstrated a direct correlation between nutrient availability and phage productivity. Latency periods were lengthened and burst size reduced when nutrients were limited. For example, bacteriophage F116, which infects *P. aeruginosa*, has a latency period of 240 minutes in starved hosts, but there is only 100 to 110 minutes between infection and lysis in hosts grown in rich medium (Kokjohn et al., 1991). Likewise, burst sizes were reduced from 27 to 4 when hosts were starved. Schrader et al. (1997b) obtained similar results with a number of *P. aeruginosa* and *E. coli* phages, with burst sizes often reduced by 10-fold in starved versus fed cells.

Similar results were obtained with *P. aeruginosa* phage UT1 (Kokjohn et al., 1991), which was isolated from the eutrophic Fort Loudon Lake in Tennessee (Ogunseitan et al., 1990). This phage had a latency period of 70 to 80 minutes when infecting exponentially growing bacteria and 110 minutes in starved cells. Burst sizes were 65 in fed cells and 6 in starved cells (Kokjohn et al., 1991).

Effects on the Productive/Reductive Life Decision

Molecular studies of the *E. coli* phage λ have revealed a life-changing mechanism that determines the probability that a given host infection will lead to a productive or lysogenic infection. This decision is controlled by the metabolic state of the host. When ATP concentrations are high and cyclic AMP levels are low, lytic growth most often occurs. When the ratio of ATP to cyclic AMP is reversed, lysogeny is increased. Although in most cases the molecular mechanisms are not known, similar observations have been made with environmental phages (Farrah, 1987; Moebus, 1987; Williams et al., 1987; Miller and Ripp, 2002; Miller, 2006).

An additional complexity to understanding the lysogenic relationship between host and prophage was observed by Lin et al. as early as 1977. These authors found that the relative reproductive rate (fitness) of an *E. coli* lysogen was higher than that of a nonlysogen under similar environmental conditions. Lin et al. (1977) found that lysogenization enhanced growth rate of lysogens in aerobic, glucose-limited chemostats. However, rates were re-

tarded under anaerobic conditions. Several prophages were studied, including λ, P1, P2, and μ. All increased the growth rate of *E. coli*.

Inorganic Nutrients

Inorganic factors also influence phage growth. Wilson et al. (1996, 1998) determined that nitrate and phosphate limitation reduced burst sizes of the cyanophage S-PM2. Increasing the availability of inorganic nutrients has been shown to stimulate phage production in several environments (Hewson et al., 2001; Smith et al., 1996).

Phosphate is often found to have a greater effect on phage production than nitrate. This may be due to the higher ratio of nucleic acid to protein in viruses than in other organisms (Bratbak et al., 1993). Phosphate-stress genes have been identified in the genomes of the roseophage SIO1 (Rohwer et al., 2000), the broad-host-range, T4-like *Vibrio* spp. phage KVP40 (Miller et al., 2003), and two cyanophages (Sullivan et al., 2005). These cyanophages contained both the phosphate-induced genes *phoH* and *pstS* and the photosynthetic genes *psbA* and *hliP*. The likelihood that these genes are involved in the adaptation to phosphate stress is consistent with the observation that phosphate is often the limiting factor in marine environments (Sullivan et al., 2005).

PHAGES AT THE POLES

The ecology and molecular biology of bacteriophages have fascinated microbiologists and other scientists for decades. It was from the studies of *E. coli* and its phages T4 and λ that the foundations of molecular biology and our understanding of how the cell works were laid. Many studies on the molecular biology and ecology of phages have been carried out (Abedon, 2008). However, the polar regions have been neglected. Although a few studies to identify and characterize specific phages have been carried out on Arctic and Antarctic phages, the vast majority of studies have been catalogues of the number of viruslike particles (VLPs) that can be identified. Still, we can learn much about the ecology of these viruses and their importance to polar ecosystems from these studies.

Hendrix et al. (1999) carried out a study on the evolutionary relationships among bacteriophages throughout the world. Their studies indicated that genetic exchange among bacteriophages has gone on throughout evolutionary time, and their results imply that "very possibly all of the dsDNA tailed phages, share common ancestry" (Hendrix et al., 1999). As these are arguably the most abundant biological entities on earth, this is most exciting. The authors propose that the genomes of these complex phages are mosaics that through lateral gene transfer share in a common genetic pool. Thus, it may not be surprising that phages from polar regions share many characteristics with their more temperate relatives, yet adaptation to these special environments most likely has led to the development of unique characteristics. It will be exciting to explore the molecular genetics and biology of these viruses to discover how the most basic of organisms have learned to deal with the stresses to themselves and their hosts imposed by living in the extreme cold and diel variations of the polar regions.

It's a Numbers Game

Bacteriophagelike particles often outnumber their hosts in natural environments (Bergh et al., 1989; Miller and Sayler, 1992; Wommack and Colwell, 2000; Miller, 1998b, 2006; Miller and Day, 2008). Yet whenever the infective half-life of phage virions is examined, it is found to be of short duration (hours to a few days) (Miller and Sayler, 1992; Miller, 2001; Noble and Fuhrman, 1997). Thus any study that enumerates total VLPs must be tempered with the understanding that not all will be infective. Still, it is clear that the numbers of bacteriophages in most environments are sufficient to insure that contact between the host and phage is frequent and that infection is both expected and common (Miller and Sayler, 1992; Proctor et al., 1993).

While we often think of the poles of the earth as being very similar—cold—there are significant differences. Mammals are numerous in the Arctic regions but quite limited in the Antarctic. Humans have inhabited the North for an extended time, while exploration of the South has only occurred in the last century. Organic nutrient availability is greater in the Arctic than in the Antarctic. Given these differences, it seems advantageous to explore these two poles separately in this review before trying to reach general conclusions.

THE ARCTIC

Several studies of bacteriophages and their hosts have been carried out in Canadian Arctic waters and the North Atlantic. Wells and Deming (2006a) looked at the significance of viral predation on bacteria in the bottom waters of Franklin Bay in the Canadian Arctic. They used epifluorescence microscopy to estimate the abundance of VLPs and bacteria in subzero-temperature bottom waters during the winter months of February and March 2004. VLP concentrations were typically an order of magnitude greater than bacterial densities (1×10^6 to 4×10^6/ml and 1×10^5 to 3×10^5/ml of seawater, respectively). These authors found that viral-induced mortality of bacteria was substantial and was comparable to or exceeded the intrinsic bacterial growth rates, suggesting that bacteriophage infection was the most important regulator of bacterial populations in these oligotrophic waters. The concentrations of VLPs in these bottom waters were similar to concentrations seen at similar depths in summer months (Steward et al., 1996) but were higher than Arctic surface-water counts (Bergh et al., 1989).

In their studies of the Bering and Chukchi Seas during the summer months (August and September 1992), Steward et al. (1996) found that VLPs were abundant in seawater, ranging from 10^6 to 10^7/ml, while bacterial cell concentrations were around 5×10^5/ml. In most cases the VLP/bacterium ratios were around 5. These authors estimated that from 3 to 25% of the bacterial mortality (depending on station and depth) was due to viral lysis. Certainly, bacteriophage lysis was as important as or more so than grazing by flagellates in determining bacterial population size. Steward et al. (1996) reminds us that because viruses are species specific (Miller, 1998a; Ackermann and DuBow, 1987), viral mortality of specific bacterial components of these polar seas communities may be devastating. These conditions could certainly regulate species diversity, causing species succession or elimination (Fuhrman, 1992).

Payet and Suttle (2008) correlated phage and bacterial numbers spatially and seasonally with biotic and abiotic variables in the Beaufort Sea and Amundsen Gulf of the Canadian Arctic. They also found that VLPs ranged from 10^5 to 10^6/ml and were 1.5 to 2 times more abundant in the spring and summer months than in the winter. Bacteria and phytoplankton also increased four- to sixfold during these periods. These authors estimate that approximately 72% of the VLPs are bacteriophages, with the remainder infecting phytoplankton. Again their results point out the importance of bacteriophages and algal phages in controlling microbial numbers in Arctic waters.

Winter et al. (2004) carried out an interesting study in the North Sea between June 2001 and April 2002 where they looked at diel cycling of virus infection and production. They found that bacterial activity as measured by [^{14}C]leucine incorporation was several times higher during the day than at night, while the rates of phage infection were greatest at night, with as many as 55 to 64% of the bacterial cells appearing to be actively infected. Lysis and phage production were highest from noon to afternoon, when bacterial activity was highest. These data are consistent with the hypothesis that induction of productive growth occurs when host cells are most metabolically active and capable of producing the maximum number of virions.

Several papers have explored the viral activity in microbial populations in ice from polar regions. Säwström et al. (2002) examined cryoconite holes in an Arctic glacier. They discovered bacterial concentrations of 1×10^4 to 4.5×10^4/ml of water from these holes. Levels in bottom materials from the holes were slightly

higher. VLPs were two to three times more frequent, ranging from 4×10^4 to 1.3×10^5/ml of water and as high as 4×10^5/ml on the bottoms of the holes. VLP/bacterium ratios ranged from 0.2 to 8.1 in this study. There was some evidence that the VLPs as well as bacteria were a direct energy source for the heterotrophic nanoflagellates in the ecosystem.

Viral and bacterial concentrations in Arctic winter sea-ice brines were found to be higher than in seawater, increasing the probability of interaction between host and pathogen in these relatively closed ecosystems (Wells and Deming, 2006b). VLP levels were as high as 8×10^7/ml of brine, with the highest values obtained in the coldest ice horizons (−31 to −24°C). Bacterial counts reached 1×10^7 CFU/ml of brine but more commonly were from 2×10^6 to 4×10^6 CFU/ml. These data unambiguously show that viral production and bacterial growth can occur in temperatures as low as −12°C in sea-ice brines.

A number of studies have been carried out in Arctic and alpine freshwater environments including glacial and lake ecosystems (Säwström et al., 2007b; Hofer and Sommaruga, 2001; Maurice et al., 2010). Säwström et al. (2007b) found a range of 10^4 to 10^6 heterotrophic bacteria and 10^4 to 10^8 VLPs/ml of water in several freshwater Arctic habitats. The virus-to-bacterium ratio of 13 (average) and positive correlation between viral and bacterial concentrations were similar to those seen in temperate lakes. These authors were unable to detect lysogenic bacteria in these populations using mitomycin C induction of phage particle production as the criterion for lysogeny.

Maurice et al. (2010) identified lysogens in the populations of bacteria in three northern temperate lakes. They explored the seasonal variation in these populations, and found that the VLPs (10^7/ml) in these lake waters were consistently 10-fold more frequent than were bacteria (10^6/ml). Lower numbers were observed in the winter months than in the summer. The fraction of bacteria that were lysogenized by temperate phages was determined and observed in 23 of the 30 samples analyzed. Approximately 2% of the bacteria showed viral infection in all three lakes. There was limited seasonal variation in these populations, and the frequency of lysogenized bacteria varied from 5 to 73%. In the most eutrophic lake, lysogeny was observed at higher levels during the winter months and was minimal in the spring. The other two lakes were more oligotrophic and showed the lowest number of lysogens in the winter under ice. Their frequency increased dramatically in the spring. These results support the hypothesis that the metabolic level of the host determines the frequency of reductive infections.

An alpine lake was investigated for the seasonal dynamics of viruses by Hofer and Sommaruga (2001). They found that VLPs' abundance was the greatest under the ice that covered the lake in the winter (5×10^6/ml). A secondary peak of VLPs occurred in the autumn. Bacterial concentrations were somewhat lower than the VLP concentration, with the virus-to-bacterium ratio most frequently between 2 and 6. From 5 to 28% of the bacterial mortality was due to lysis by viruses. Interestingly, VLPs dropped to $<2 \times 10^4$/ml following the breakup of ice. The authors feel that this is likely due to the increased lethality of solar UV radiation that is encountered by these microbes when the ice cover is removed. Filamentous phages associated with filamentous bacteria were observed in this study. These microorganisms are unusual in freshwater aquatic systems.

The northern waters between Canada and Greenland were explored by Middelboe et al. (2002). They measured bacterial growth and phage lysis in subzero waters. Bacterial counts were approximately 10^6 CFU/ml at all of the sampling stations, while VLP concentrations were two- to fivefold higher. Viral mortality of bacteria ranged from 6 to 28%. They also isolated a *Pseudoalteromonas* sp. and characterized a bacteriophage that infected it. They detected this phage-host system throughout their sampling range and determined that phage production was significant even at 0°C. A latent period of 15 hours and a burst size of 18 ± 2 PFU/burst were observed under these conditions for this phage-host pair.

Three phage-host systems were isolated and characterized from Arctic sea ice collected northwest of Svalbard, Arctic, Norway (Borriss et al., 2003). Phages that infect *Flavobacterium hibernum*, *Shewanella frigidimarina*, and *Colwellia psychrerythraea* were studied. Both the bacteria and phages were adapted to growth in the cold. The hosts grew well at 0°C. The phages had growth maxima below 14°C and produced well-formed plaques at 0°C. Electron microscopy revealed that these phages belong to the *Siphoviridae* and *Myoviridae* families of tailed, dsDNA phages. They all were host specific, infecting only the species from which they were isolated.

A bacteriophage (ϕ11b) infecting a psychrophilic *Flavobacterium* sp. was isolated from Arctic sea ice (Borriss et al., 2007). Its genome is 36 kbp in length with a GC content of 30.6%. This is the lowest GC content of any known phage genome and may reflect the fact that the virus was isolated from a very cold habitat. Sequence and protein analysis suggest that this phage is related to mesophilic, non-marine siphoviruses and is most closely related to bacteriophage SPP1, which infects *Bacillus subtilis* (Alonso et al., 1997). This relationship between phages that infect very different hosts suggests that there has been a high level of genetic exchange among viruses during their evolution.

THE ANTARCTIC

VLPs have been reported in the Antarctic and sub-Antarctic oceans. In the Australian Southern Ocean, Evans et al. (2009) observed 7×10^6 VLPs/ml of seawater in the Polar Frontal Zone and as many as 2×10^7/ml in the sub-Antarctic zone. Viral production was estimated to range from 2×10^7 to 2×10^8/ml per day. From a low of 23% of the bacteria lysed by viruses in the Polar Frontal Zone, lysis rates reached as high as 40% in the eastern sub-Antarctic region. Perhaps not surprisingly, these data suggest that bacteriophages are an important shunt in carbon cycling in the Antarctic regions as well as in the Arctic.

Several studies have addressed seasonal variation in bacteriophage populations in Antarctic waters, in marine settings (Guixa-Boixereu et al., 2002; Marchant et al., 2000; Pearce et al., 2007), pack ice (Gowing et al., 2002, 2004), and Antarctic lakes (Laybourn-Parry et al., 2001; Madan et al., 2005). Marchant et al. (2000) studied viruses, bacteria, and cyanobacteria in the Southern Ocean during the spring (September to December) of 1996. They found that VLPs declined slightly in concentration from Tasmania (4×10^6/ml) to the Polar Front (1×10^6/ml). Interestingly, virus concentrations increased south of the Front to again reach concentrations of 4×10^6/ml. Phage-to-bacterium ratios varied from 3 to 15 in open water, with <8% of the bacteria being cyanobacteria. The ratios were considerably higher in sea ice (between 15 and 40). Guixa-Boixereu et al. (2002) studied Antarctic waters during the summer (December 1995 to February 1996). They also found relatively consistent concentrations of VLPs (7×10^6 to 2×10^7/ml) throughout their sampling area. Viral mortality of bacteria was always found to be greater than mortality due to predation by eukaryotic organisms. These data clearly demonstrate that bacteriophages are important regulators of bacterial population size and structure in the polar regions and indeed in all aquatic systems explored to date (Wommack and Colwell, 2000).

Virus activity in coastal waters of Antarctica was explored by Pearce et al. (2007). Metabolically active bacteria were estimated throughout the study and declined rapidly at the end of summer. The authors believe that this is due to increased levels of viral infection and microheterotrophic grazing. It coincided with the peak abundance of microheterotrophs and the highest concentrations of VLPs. The decline continued over the winter and only increased after the phytoplankton bloom during the following spring and summer.

Gowing et al. (2002, 2004) explored the viruses in pack ice in the Ross Sea in late autumn and summer. During the autumn these investigators observed many viruses that appeared to infect algae and protozoa. In the summer, VLPs likely infect bacteria at concentrations between 5×10^6 and 1×10^8/ml of melted sample. Bacteria were present in approximately 10-fold lower concentrations. Unlike many other studies, these authors conclude that neither bacteriovores nor bacteriophages are important regulators of bacterial numbers but that numbers are regulated by algae blooms and other factors that affect bacterial growth.

Laybourn-Parry et al. (2001) found that VLPs play a pivotal role in carbon cycling in Antarctic lakes. They investigated VLP abundances in nine freshwater and saline lakes in the Vestfold Hills of eastern Antarctica during December 1999. VLP concentrations ranged from 1×10^6 to 3×10^6/ml of lake water, with phage-to-bacterium ratios of between 23 and 50 observed. Ratios were higher in saline than in freshwater lakes. Madan et al. (2005) also studied the interaction between bacteria and phages in three saline Antarctic lakes during a complete annual cycle. They found concentrations of VLPs as high as 1×10^7/ml of lake water in the winter. During the summer months they increased to 1.2×10^8/ml. Virus-to-bacterium ratios ranged from 18 to as high as 126 depending on the time of year and specific lake. Madan and colleagues found that VLPs were more readily degraded in the summer months, probably due to solar UV radiation and higher temperatures. These data again indicate the importance of bacteriophages in the aquatic ecosystems of the Antarctic.

Lysogeny is an important aspect of microbial community structure in all environments. Several studies in Antarctic lakes have addressed the importance of lysogeny in this environment. Laybourn-Parry et al. (2007) investigated the patterns of lysogeny in two saline lakes. These investigators observed high concentrations of 10^8 VLPs/ml of lake water and virus-to-bacterium ratios as high as 70. Rates of lysogeny as judged by mitomycin C induction were 32% in Pendant Lake and 71% in Ace Lake, with the highest levels observed in winter and spring. These data are suggestive of an increase in lysogeny in environments where sources of energy and, therefore, biosynthetic capacity are limited.

Ten Antarctic lakes in the Vestfold Hills were studied by Säwström et al. (2008). Bacterial levels in most lakes were about 5×10^5/ml while VLPs averaged around 2×10^6/ml of lake water. Mitomycin C–inducible prophages were found in only 1 of the 10 lakes. This suggests that productive infection was common in these environments. The highest abundances of VLPs were found in the lakes with the highest concentrations of dissolved organic carbon and reactive phosphorus. These data suggest that viral abundance and lysogeny in these lakes were determined by the trophic status of the lake. It should be remembered that not all prophages are inducible by mitomycin C and that in environments of low carbon and energy availability pseudolysogeny is common (Ripp and Miller, 1997, 1998; Miller et al., 1992). It is likely that these investigators underestimated the occurrence of lysogeny in the Antarctic lakes under study.

Lake Druzhby and Crooked Lake are two large, highly oligotrophic freshwater lakes in the Vestfold Hills of eastern Antarctica. Seasonal variation in viral productivity and lysogeny in these lakes were investigated by Säwström et al. (2007a). The concentrations of VLPs in these lakes ranged from 1×10^5 to 1×10^6 particles/ml. Bacterial concentrations were observed at from 1×10^5 to 2×10^5 cells/ml, and phage-to-bacterium ratios were low (between 1 and 8). In this study, lysogeny was observed in 3 of 10 samples. Again, given the extreme oligotrophy of the lakes, this is likely to be an underestimate. When observed, between 18 and 73% of the bacteria were lysogenized. The data obtained in this study support the hypothesis that bacteriophages are of quantifiable significance in the carbon-flow cycle of Antarctic oligotrophic lakes.

Few studies have been carried out to investigate the occurrence of bacteriophages in polar soils. However, Williamson et al. (2007) did look at the frequency of lysogeny in temperate and extreme soil environments including some Antarctic soil samples. The fraction of cells that could be induced by mitomycin C to produce bacteriophage particles was lower in Antarctic soils (4 to 20%) than in temperate soil samples collected in Delaware (22 to 68%). Whether these data speak more to the frequency of lysogeny or the frequency of lysogens that are inducible was not addressed. It certainly reinforces the hypothesis that lysogeny is common among all soil bacteria from diverse environments.

Wilson et al. (2000) looked at VLPs in freshwater lakes on Signy Island, Antarctica. They found that these particles were ubiquitous, with frequencies ranging from 5×10^6 to 3×10^7 particles/ml, and that diversity was high. These viruses appeared to infect bacteria, cyanobacteria, and algae. An unusually large virus was isolated with a complex morphology. The viral head was 370×330 nm and its tail was 1.3 μm in length. While the host for this virus is unknown, algal phages of this size have been observed (Dodds and Cole, 1980) in coastal water in Arctic and sub-Arctic regions (Bratbak et al., 1992).

COMPARING THE POLES

Few researchers have investigated bacteriophages in both the Arctic and Antarctic in the same study. The numbers of visibly phage-infected bacteria in two ultra-oligotrophic eastern Antarctic lakes (Lake Druzhby and Crooked Lake) were compared with samples from the cryoconite holes on the Midre Løvenbreen glacier in the Arctic by Säwström et al. (2007c). Phage activity in each of these sites was quite comparable. They gave the lowest burst sizes of phages ever reported in the literature (4, 4, and 3, respectively). At the same time they recorded the highest frequencies of visibly infected bacterial cells (22.7, 34.2, and 11.3%, respectively). Although not confirmatory, these data suggest that lysogeny may be high in these ecosystems.

All of the data from the poles indicate that bacteriophage populations can be maintained in low-temperature, highly oligotrophic environments. This is true of freshwater and saline lakes and the oceans of the polar regions. While not an identified reservoir in all of the habitats studied, lysogeny has been observed frequently. It will be interesting to determine the role of pseudolysogeny (Miller and Day, 2008) as an alternate lifestyle for these bacteriophages.

Molecular Studies

Filée et al. (2005) explored the range of T4-like bacteriophages throughout the world. They examined samples from fjords and bays of British Columbia, the Gulf of Mexico, and the Arctic Ocean using DNA primers designed to amplify the T4 *g23* gene. Phylogenetic analysis of their data revealed five previously uncharacterized subgroups of T4-like bacteriophages in many environments, including the Arctic samples. Their data demonstrated the ubiquity and diversity of T4-like viruses worldwide and revealed that they have a much broader host range than previously imagined.

Phylogenetic diversity in cyanophages was explored by Chénard and Suttle (2008) and by Short and Suttle (2005). Chénard and Suttle examined isolates from the Arctic Ocean, Gulf of Mexico, and northeast Pacific Ocean as well as Lake Constance in Germany and several freshwater lakes in northern Ontario, Canada. Phages that infect *Synechococcus* and *Prochlorococcus* species were examined for the presence of the *psbA* gene. While the gene was found to be widespread among cyanophages, freshwater and marine cyanophages have distinct phylogenies. Likewise, phages that infect *Synechococcus* spp. have an evolutionary history distinct from those that infect *Prochlorococcus* spp. At some time, sequences from the same geographical origin were more closely related than those from other ecosystems, which clustered in different clades.

Short and Suttle explored the diversity of the conserved cyanophage structural gene *g20*. They found that this gene is distributed widely in samples from the Arctic, the Southern Ocean, the northeast and southeast Pacific Ocean, a catfish production pond, lakes in

Canada and Germany, and deep in the Chukchi Sea. They found nearly identical sequences of this gene to be widespread at all of these sites. These data indicate that closely related hosts and their viruses are distributed worldwide and that lateral gene transfer has most likely occurred among phage communities in vastly different habitats.

A metagenomic study of phages in four ocean regions including the Arctic Ocean, the Pacific off British Columbia, the Gulf of Mexico, and the Sargasso Sea revealed that most of the viral sequences were not similar to those currently in databases (Angly et al., 2006). Diversity was very high globally, with several hundred thousand species. Different phage assemblages predominated in the different regions. Cyanophages and a previously unknown group of single-stranded DNA (ssDNA) phages predominated in the Sargasso Sea, while prophagelike sequences suggesting a high level of lysogeny predominated in the Arctic. Most viral species were observed in all of the sampling regions. They varied according to which viral species predominated. No specific groups of phages were excluded from any sample (Angly et al., 2006).

Bacteriophage diversity in the North Sea was investigated by Wichels et al. (1998), who found high numbers of VLPs (10^4 to 10^7/ml). They investigated 22 different phage-host systems after collecting 85 different viruses near Helgoland, Germany. Morphology studies placed 11 of the phages in the *Myoviridae* and 7 in the *Siphoviridae*; 4 were podoviruses. DNA-DNA hybridization revealed that there were no sequences shared among the phages and that they belonged to 13 different species. All infected gram-negative, facultatively anaerobic, motile cocci belonging to the genus *Pseudoalteromonas*.

CONCLUSIONS

Each study of the polar regions that has been carried out has pointed to the great genetic diversity among bacteriophages worldwide. While they have adapted to cold climates, the bacteriophages of the poles are far from unique. They all are members of the major groups of bacteriophages. Still, it is daunting to think that while we have characterized only a small fraction of 1% of the bacteria in the environment, there are probably 10 to 100 times more species of bacteriophages to be investigated than bacteria.

The diversity of the viral community from an Antarctic lake was explored by López-Bueno et al. (2009). Their analysis revealed unexpected genetic richness and distribution among numerous viral families. This assemblage had a large proportion of sequences related to eukaryotic viruses. It also contained a number of ssDNA viruses that had not previously been identified in aquatic environments. Samples from the ice-covered lake collected in spring were dominated by ssDNA viruses, while ice-free summer waters contained dsDNA viruses. This and other studies reported here point to the vast richness of phage diversity and the many unknowns awaiting discovery as further study of phages at the poles is undertaken.

REFERENCES

Abedon, S. T. (ed.). 2008. *Bacteriophage Ecology: Population Growth, Evolution, and Impact of Bacterial Viruses*. Cambridge University Press, Cambridge, United Kingdom.

Ackerman, H. W., and M. S. DuBow. 1987. *Viruses of Prokaryotes: General Properties of Bacteriophages*. CRC Press, Boca Raton, FL.

Alonso, J. C., G. Luder, A. C. Stiege, S. Chai, R. Weise, and T. A. Trautner. 1997. The complete nucleotide sequence and functional organization of *Bacillus subtilis* bacteriophage SPP1. *Gene* **204**:201–212.

Angly, G. E., B. Felts, M. Breitbart, P. Salamon, R. A. Edwards, C. Carlson, A. M. Chan, M. Haynes, S. Kelley, H. Liu, J. M. Mahaffy, J. E. Mueller, J. Nulton, R. Olson, R. Parsons, S. Rayhawk, C. A. Suttle, and F. Rohwer. 2006. The marine viromes of four oceanic regions. *PLoS Biol.* **4**:2121–2131.

Ashelford, K. E., M. J. Day, and J. C. Fry. 2003. Elevated abundance of bacteriophage infecting bacteria in soil. *Appl. Environ. Microbiol.* **69**:285–289.

Ashelford, K. E., J. C. Fry, M. J. Bailey, A. R. Jeffries, and M. J. Day. 1999. Characterization of six bacteriophages of *Serratia liquefaciens* CP6 isolated from the sugar beet phytosphere. *Appl. Environ. Microbiol.* **65**:1959–1965.

Baess, I. 1971. Report on a pseudolysogenic mycobacterium and a review of the literature concerning pseudolysogeny. *Acta Pathol. Microbiol. Scand.* **79**:428–434.

Bailey, S., M. R. Clokie, A. Millard, and N. H. Mann. 2004. Cyanophage infection and photoinhibition in marine cyanobacteria. *Res. Microbiol.* **155**:720–725.

Barksdale, L., and S. B. Ardon. 1974. Persisting bacteriophage infections, lysogeny, and phage conversions. *Annu. Rev. Microbiol.* **28**:265–299.

Bergh, Ø., K. Y. Børsheim, G. Bratbak, and M. Heldal. 1989. High abundance of viruses found in aquatic environments. *Nature* **340**:467–468.

Borriss, M., E. Helmke, R. Hanschke, and T. Schweder. 2003. Isolation and characterization of marine psychrophilic phage-host systems from Arctic sea ice. *Extremophiles* **7**:377–384.

Borriss, M., T. Lombardot, F. O. Glöckner, D. Becher, D. Albrecht, and T. Schweder. 2007. Genome and proteome characterization of the psychrophilic *Flavobacterium* bacteriophage 11b. *Extremophiles* **11**:95–104.

Bratbak, G., J. K. Egge, and M. Heldal. 1993. Viral mortality of the marine alga *Emiliania huxleyi* (Haptophyceae) and termination of algal blooms. *Mar. Ecol. Prog. Ser.* **93**:39–48.

Bratbak, G., O. H. Huslund, M. Heldal, A. Naess, and T. Roeggen. 1992. Giant marine viruses? *Mar. Ecol. Prog. Ser.* **85**:201–202.

Bull, J. J., J. Millstein, J. Orcutt, and H. A. Wichman. 2006. Evolutionary feedback mediated through population density, illustrated with viruses in chemostats. *Am. Nat.* **167**:E39–E51.

Canchaya, C., C. Proux, G. Fournous, A. Bruttin, and H. Brüssow. 2003. Prophage genomics. *Microbiol. Mol. Biol. Rev.* **67**:238–276.

Cavenagh, M. M., and R. V. Miller. 1986. Specialized transduction of *Pseudomonas aeruginosa* PAO by bacteriophage D3. *J. Bacteriol.* **165**:448–452.

Chénard, C., and C. A. Suttle. 2008. Phylogenetic diversity of sequences of cyanophage photosynthetic gene *phbA* in marine and freshwaters. *Appl. Environ. Microbiol.* **74**:5317–5324.

Clokie, M. R., H. Shan, S. Bailey, Y. Jia, H. M. Krisch, S. West, and N. H. Mann. 2006. Transcription of "photosynthetic" T4-type phage during infection of a marine cyanobacterium. *Environ. Microbiol.* **8**:827–835.

Dodds, J. A. and A. Cole. 1980. Microscopy and biology of *Uronema gigas*, a filamentous eukaryotic green algae, and its associated tailed virus-like particle. *Virology* **100**:156–165.

Edlin, G., L. Lin, and R. Kudrna. 1975. λ-lysogens of *E. coli* reproduce more rapidly than nonlysogens. *Nature* **255**:735–737.

Evans, C., I. Pearce, and C. P. Brussaard. 2009. Viral-mediated lysis of microbes and carbon release in the sub-Antarctic and Polar Frontal zones of the Australian Southern Ocean. *Environ. Microbiol.* **11:**2924–2934.

Farrah, S. R. 1987. Ecology of phage in freshwater environments, p. 125–136. *In* S. M. Goyal, C. P. Gerba, and G. Bitton (ed.), *Phage Ecology.* John Wiley and Sons, New York, NY.

Filée, J., F. Tétart, C. A. Suttle, and H. M. Krisch. 2005. Marine T4-type bacteriophages, a ubiquitous component of the dark matter of the biosphere. *Proc. Natl. Acad. Sci. USA* **102:**12471–12476.

Frye, J. G., S. Porwollik, F. Blackmer, P. Cheng, and M. McClelland. 2005. Host gene expression changes and DNA amplification during temperate phage induction. *J. Bacteriol.* **187:**1485–1492.

Fuhrman, J. 1992. Bacterioplankton roles in cycling of organic matter: the microbial food web, p. 361–383. *In* P. G. Falkowski and A. D. Woodhead (ed.), *Primary Productivity and Biogeochemical Cycles in the Sea.* Plenum Press, New York, NY.

Gowing, M. M., D. L. Garrison, A. H. Gibson, J. M. Rupp, M. O. Jeffries, and C. H. Fritsen. 2004. Bacterial and viral abundance in Ross Sea summer pack ice communities. *Mar. Ecol. Prog. Ser.* **279:**3–12.

Gowing, M. M., B. E. Riggs, D. L. Garrison, A. H. Gibson, and M. O. Jeffries. 2002. Large viruses in Ross Sea late autumn pack ice habitats. *Mar. Ecol. Prog. Ser.* **241:**1–11.

Guixa-Boixereu, N., J. I. Calderón-Paz, J. Heldal, G. Bratbak, and C. Pedrnós-Alió. 1996. Viral lysis and bacterivory as prokaryotic loss factors along a salinity gradient. *Aquat. Microb. Ecol.* **11:**215–227.

Guixa-Boixereu, N., D. Vaqué, J. M. Gasol, J. Sánchez-Cámara, and C. Pedrós-Alió. 2002. Viral distribution and activity in Antarctic waters. *Deep Sea Res. Part 2 Top. Stud. Oceanogr.* **49:**827–845.

Hendrix, R. W., M. C. M. Smith, R. N. Burns, M. E. Ford, and G. F. Hatfull. 1999. Evolutionary relationships among diverse bacteriophages and prophages: all the world's a phage. *Proc. Natl. Acad. Sci. USA* **96:**2192–2197.

Hewson, I., J. M. O'Neill, J. A. Fuhrman, and W. C. Dennison. 2001. Virus-like particle distribution and abundance in sediments and overlying waters along eutrophication gradients in two subtropical estuaries. *Limnol. Oceanogr.* **47:**1734–1746.

Hofer, J. S., and R. Sommaruga. 2001. Seasonal dynamics of viruses in an alpine lake: importance of filamentous forms. *Aquat. Microb. Ecol.* **26:**1–11.

Holmes, R. K., and L. Barksdale. 1970. Comparative studies with *tox*+ and *tox*- corynebacteriophages. *J. Virol.* **5:**783–794.

Hyman, P., and S. T. Abedon. 2008. Phage ecology of bacterial pathogenesis, p. 353–385. *In* S. T. Abedon (ed.), *Bacteriophage Ecology: Population Growth, Evolution, and Impact of Bacterial Viruses.* Cambridge University Press, Cambridge, United Kingdom.

Ivánovics, G., V. Gaál, E. Nagy, B. Prágai, and M. Simon Jr. 1976. Studies on negacinogeny in *Bacillus cereus.* II. *Bacillus cereus* isolates characterized by prophage-controlled production of megacin A (phospholipase A). *Acta Microbiol. Acad. Sci. Hung.* **23:**283–291.

Karaolis, D. K. R., S. Somara, D. R. Maneval Jr., J. A. Johnson, and J. B. Kaper. 1999. A bacteriophage encoding a pathogenicity island, a type-IV pilus and a phage receptor in cholera bacteria. *Nature* **199:**375–379.

Koch, A. L. 1971. The adaptive responses of *Escherichia coli* to a feast and famine existence. *Adv. Microb. Physiol.* **6:**147–217.

Koch, A. L. 1979. Microbial growth in low concentrations of nutrients, p. 261–279. *In* M. Shilo (ed.), *Strategies of Microbial Life in Extreme Environments* (Berlin: Dahlem Konferenzen). Verlag Chemie, Weinheim, Germany.

Kokjohn, T. A., G. S. Sayler, and R. V. Miller. 1991. Attachment and replication of *Pseudomonas aeruginosa* bacteriophages under conditions simulating aquatic environments. *J. Gen. Microbiol.* **137:**661–666.

Laybourn-Parry, J., J. S. Hofer, and R. Sommaruga. 2001. Viruses in the plankton of freshwater and saline Antarctic lakes. *Freshw. Biol.* **46:**1279–1287.

Laybourn-Parry, J., W. A. Marshall, and N. J. Madan. 2007. Viral dynamics and patterns of lysogeny in saline Antarctic lakes. *Polar Biol.* **30:**351–358.

Lenski, R. E. 1988. Dynamics of interactions between bacteria and virulent bacteriophage. *Adv. Microb. Ecol.* **10:**99–108.

Lin, L., R. Bitner, and G. Edlin. 1977. Increased reproductive fitness of *Escherichia coli* lambda lysogens. *J. Virol.* **21:**554–559.

Lindell, D., J. D. Jaffe, Z. I. Johnson, G. M. Church, and S. W. Chisholm. 2005. Photosynthesis genes in marine viruses yield proteins during host infection. *Nature* **438:**86–89.

Lindell, D., M. B. Sullivan, Z. I. Johnson, A. C. Tolonen, F. Rohwer, and S. W. Chisholm. 2004. Transfer of photosynthesis genes to and from *Prochlorococcus* viruses. *Proc. Natl. Acad. Sci. USA* **101:**11013–11018.

López-Bueno, A., J. Tamames, D. Velázquez, A. Moya, A. Quesada, and A. Alcamí. 2009. High diversity of the viral community from an Antarctic lake. *Science* **326:**858–861.

Madan, N. J., W. A. Marshall, and J. Laybourn-Parry. 2005. Virus and microbial loop dynamics over an annual cycle in three contrasting Antarctic lakes. *Freshw. Biol.* **50:**1291–1300.

Mahenthiralingam, E. 2004. Gene associations in bacterial pathogenesis: pathogenicity islands and genomic deletions, p. 249–274. *In* R. V. Miller and M. J. Day (ed.), *Microbial Evolution: Gene Establishment, Survival, and Exchange.* ASM Press, Washington, DC.

Mann, N. H., A. Cook, A. Millard, S. Bailey, and M. Clokie. 2003. Marine ecosystems: bacterial photosynthesis genes in a virus. *Nature* **424:**741.

Marchant, J., A. Davidson, S. Wright, and J. Glazebrook. 2000. The distribution and abundance of viruses in the Southern Ocean during spring. *Antarct. Sci.* **12:**414–417.

Martinez-Molina, E., and J. Olivares. 1979. Antibiotic production by *Pseudomonas reptilivora* as a phage conversion. *Can. J. Microbiol.* **25:** 1108–1110.

Maurice, C. F., T. Bouvier, J. Comte, F. Guillemette, and P. A. del Giorgio. 2010. Seasonal variations of phage life strategies and bacterial physiological states in three northern temperate lakes. *Environ. Microbiol.* **12:**628–641.

Middelboe, M. 2000. Bacterial growth rate and marine virus-host dynamics. *Microb. Ecol.* **40:** 114–124.

Middelboe, M., T. G. Nielsen, and P. K. Bjørnsen. 2002. Viral and bacterial production in the North Water; in situ measurements, batch-culture experiments and characterization and distribution of a virus-host system. *Deep Sea Res. Part 2 Top. Stud. Oceanogr.* **49:**5063–5079.

Miller, E. S., J. F. Heidelberg, J. A. Eisen, W. C. Nelson, A. S. Durkin, A. Ciecko, T. V. Feldblyum, O. White, I. T. Paulsen, W. C. Nierman, J. Lee, B. Szczypinski, and C. M. Fraser. 2003. Complete genome sequence of the broad-host-range vibriophage KVP40: comparative genomics of a T4-related bacteriophage. *J. Bacteriol.* **185:**5220–5233.

Miller, R. V. 1998a. Bacterial gene swapping in nature. *Sci. Am.* **278:**66–71.

Miller, R. V. 1998b. Methods for enumeration and characterization of bacteriophages from environmental samples, p. 218–235. *In* R. Burlage (ed.), *Techniques in Microbial Ecology.* Oxford University Press, Oxford, United Kingdom.

Miller, R. V. 2001. Environmental bacteriophage-host interactions: factors contributing to natural transduction. *Antonie van Leeuwenhoek* **79:**141–147.

Miller, R. V. 2004. Bacteriophage-mediated transduction: an engine for change and evolution, p. 144–157. *In* R. V. Miller and M. J. Day (ed.), *Microbial Evolution: Gene Establishment, Survival, and Exchange.* ASM Press, Washington, DC.

Miller, R. V. 2006. Marine phages, p. 535–544. *In* R. Calendar (ed.), *The Bacteriophages,* 2nd ed. Oxford University Press, New York, NY.

Miller, R. V., and M. J. Day (ed.). 2004a. *Microbial Evolution: Gene Establishment, Survival, and Exchange.* ASM Press, Washington, DC.

Miller, R. V., and M. J. Day. 2004b. Horizontal gene transfer and the real world, p. 173–177. *In* R. V. Miller and M. J. Day (ed.), *Microbial Evolution: Gene Establishment, Survival, and Exchange.* ASM Press, Washington, DC.

Miller, R. V., and M. J. Day. 2008. Contribution of lysogeny, pseudolysogeny, and starvation to phage ecology, p. 114–143. *In* S. T. Abedon (ed.), *Bacteriophage Ecology: Population Growth, Evolution, and Impact of Bacterial Viruses.* Cambridge University Press, Cambridge, United Kingdom.

Miller, R. V., and S. Ripp. 1998. The importance of pseudolysogeny to *in situ* bacteriophage-host interactions, p. 179–191. *In* M. Syvanen and C. I. Kado (ed.), *Horizontal Gene Transfer.* Chapman & Hall, London, United Kingdom.

Miller, R. V., and S. A. Ripp. 2002. Pseudolysogeny: a bacteriophage strategy for increasing longevity *in situ*, p. 81–91. *In* M. Syvanen and C. I. Kado (ed.), *Horizontal Gene Transfer,* 2nd ed. Academic Press, San Diego, CA.

Miller, R. V., S. Ripp, J. Replicon, O. A. Ogunseitan, and T. A. Kokjohn. 1992 Virus-mediated gene transfer in freshwater environments, p. 50–62. *In* M. J. Gauthier (ed.), *Gene Transfers and Environment.* Springer, Berlin, Germany.

Miller, R. V., and G. S. Sayler. 1992. Bacteriophage-host interactions in aquatic systems, p. 176–193. *In* E. M. Wellington and J. D. van Elsas (ed.), *Genetic Interactions among Microorganisms in the Natural Environment.* Pergamon Press, Oxford, United Kingdom.

Moebus, K. 1987. Ecology of marine bacteriophages, p. 136–156. *In* S. M. Goyal, C. P. Gerba, and G. Bitton (ed.), *Phage Ecology.* John Wiley and Sons, New York, NY.

Moebus, K. 1996. Marine bacteriophage reproduction under nutrient-limited growth of host bacteria. I. Investigations with six phages. *Mar. Ecol. Prog. Ser.* **144:**1–12.

Morita, R. Y. (ed.). 1997. *Bacteria in Oligotrophic Environments: Starvation-Survival Lifestyle.* Chapman & Hall, New York, NY.

Noble, R. T., and J. A. Fuhrman. 1997. Virus decay and its causes in coastal waters. *Appl. Environ. Microbiol.* **63:**77–83.

Ogunseitan, O. A., G. S. Sayler, and R. V. Miller. 1990. Dynamic interactions of *Pseudomonas aeruginosa* and bacteriophages in lake water. *Microb. Ecol.* **19:**171–185.

Ortmann, A. C., J. E. Lawrence, and C. A. Suttle. 2002. Lysogeny and lytic viral production during a bloom of cyanobacterium *Synechococcus* spp. *Microb. Ecol.* **43:**225–231.

Payet, J. P., and C. A. Suttle. 2008. Physical and biological correlates of virus dynamics in the southern Beaufort Sea and Amundsen Gulf. *J. Mar. Syst.* **74:**933–945.

Pearce, I., A. T. Davidson, E. M. Bell, and S. Wright. 2007. Seasonal changes in the concentration and metabolic activity of bacteria and viruses at an Antarctic coastal site. *Aquat. Microb. Ecol.* **47:**11–23.

Proctor, L. M., A. Okubo, and J. A. Fuhrman. 1993. Calibrating estimates of phage-induced mortality in marine bacteria: ultrastructural studies of marine bacteriophage development from one-step growth experiments. *Microb. Ecol.* **25:**161–182.

Ptashne, M. 2004. *A Genetic Switch: Phage Lambda Revisited*, 3rd ed. Cold Spring Harbor Laboratory Press, Cold Spring Harbor, NY.

Replicon, J., A. Frankfater, and R. V. Miller. 1995. A continuous culture model to examine factors that affect transduction among *Pseudomonas aeruginosa* strains in freshwater environments. *Appl. Environ. Microbiol.* **61:**3359–3366.

Ripp, S., and R. V. Miller. 1997. The role of pseudolysogeny in bacteriophage-host interactions in a natural freshwater environment. *Microbiology* **143:**2065–2070.

Ripp, S., and R. V. Miller. 1998. Dynamics of pseudolysogenic response in slowly growing cells of *Pseudomonas aeruginosa*. *Microbiology* **144:**2225–2232.

Rocha, E. P. C., A. Danchin, and A. Viari. 2001. The evolutionary role of restriction and modification systems as revealed by comparative genome analysis. *Genome Res.* **11:**946–958.

Rohwer, F., A. Segall, G. Steward, V. Seguritan, M. Breitbart, F. Wolven, and F. Azam. 2000. The complete genomic sequence of marine phage Roseophage SIO1 shares homology with nonmarine phages. *Limnol. Oceanogr.* **45:**408–418.

Säwström, C., M. A. Anesio, W. Granéli, and J. Laybourn-Parry. 2007a. Seasonal viral loop dynamics in two large ultraoligotrophic Antarctic freshwater lakes. *Microb. Ecol.* **53:**1–11.

Säwström, C., J. Laybourn-Parry, W. Granéli, and A. M. Anesio. 2007b. Heterotrophic bacterial and viral dynamics in Arctic freshwaters: results from a field study and nutrient-temperature manipulation experiments. *Polar Biol.* **30:**1407–1415.

Säwström, C., W. Granéli, J. Laybourn-Parry, and A. M. Anesio. 2007c. High viral infection rates in Antarctic and Arctic bacterioplankton. *Environ. Microbiol.* **9:**250–255.

Säwström, C., P. Mumford, W. Marshall, A. Hodson, and J. Laybourn-Parry. 2002. The microbial communities and primary productivity of cryoconite holes in an Arctic glacier (Svalbard 79°N). *Polar Biol.* **25:**591–596.

Säwström, C., I. Pearce, A. T. Davidson, P. Rosén, and J. Laybourn-Parry. 2008. Influence of environmental conditions, bacterial activity and viability on the viral component in 10 Antarctic lakes. *FEMS Microbiol. Ecol.* **63:**12–22.

Saye, D. J., and R. V. Miller. 1989. The aquatic environment: consideration of horizontal gene transmission in a diversified habitat, p. 223–259. *In* S. B. Levy and R. V. Miller (ed.), *Gene Transfer in the Environment*. McGraw-Hill, New York, NY.

Saye, D. J., O. Ogunseitan, G. S. Sayler, and R. V. Miller. 1987. Potential for transduction of plasmids in a natural freshwater environment: effect of plasmid donor concentration and a natural microbial community on transduction in *Pseudomonas aeruginosa*. *Appl. Environ. Microbiol.* **53:**987–995.

Schrader, H. S., J. O. Schrader, J. J. Walker, N. B. Bruggeman, J. M. Vanderloop, J. J. Shaffer, and T. A. Kokjohn. 1997a. Effects of host starvation on bacteriophage dynamics, p. 368–385. *In* R. Y. Morita (ed.), *Bacteria in Oligotrophic Environments: Starvation-Survival Lifestyle*. Chapman & Hall, New York, NY.

Schrader, H. S., J. O. Schrader, J. J. Walker, T. A. Wolf, K. W. Nickerson, and T. A. Kokjohn. 1997b. Bacteriophage infection and multiplication occur in *Pseudomonas aeruginosa* starved for 5 years. *Can. J. Microbiol.* **43:**1157–1163.

Short, C. M., and C. A. Suttle. 2005. Nearly identical bacteriophage structural gene sequences are widely distributed in both marine and freshwater environments. *Appl. Environ. Microbiol.* **71:**480–486.

Smith, E., A. C. Wolters, H. Lee, J. T. Trevors, and J. D. van Elsas. 1996. Interactions between a genetically marked *Pseudomonas fluorescens* strain and bacteriophage φR2f in soil: effects of nutrients, alginate encapsulation, and the wheat rhizosphere. *Microb. Ecol.* **31:**125–140.

Steward, G. F., D. C. Smith, and F. Azam. 1996. Abundance and production of bacteria and viruses in the Bering and Chukchi Seas. *Mar. Ecol. Prog. Ser.* **131:**287–300.

Sullivan, M. B., M. L. Coleman, P. Weigele, R. Rohwer, and S. W. Chisholm. 2005. Three *Prochlorococcus* cyanophage genomes: signature features and ecological interpretations. *PLoS Biol.* **3:**e144.

Thingstad, T. F., G. Bratbak, and M. Heldal. 2008. Aquatic phage ecology, p. 251–280. *In* S. T. Abedon (ed.), *Bacteriophage Ecology: Population Growth, Evolution, and Impact of Bacterial Viruses*. Cambridge University Press, Cambridge, United Kingdom.

Wells, L. E., and J. W. Deming. 2006a. Significance of bacterivory and viral lysis in bottom waters of Franklin Bay, Canadian Arctic, during winter. *Aquat. Microb. Ecol.* **43:**209–221.

Wells, L. E., and J. W. Deming. 2006b. Modelled and measured dynamics of viruses in Arctic winter sea-ice brines. *Environ. Microbiol.* **8:**1115–1121.

Wichels, A., S. S. Biel, H. R. Gederblom, T. Brinkhoff, G. Muyzer, and C. Schütt. 1998. Bacteriophage diversity in the North Sea. *Appl. Environ. Microbiol.* **64:**4128–4133.

Williams, S., A. Mortimer, and L. Manchester. 1987. Ecology of soil bacteriophages, p. 136–156. *In* S. M. Goyal, C. P. Gerba, and G. Bitton (ed.), *Phage Ecology.* John Wiley and Sons, New York, NY.

Williamson, K. E., M. Radosevich, D. W. Smith, and K. E. Wommack. 2007. Incidence of lysogeny within temperate and extreme soil environments. *Environ. Microbiol.* **9:**2563–2574.

Wilson, W. H., N. G. Carr, and N. H. Mann. 1996. The effect of phosphate status on the kinetics of cyanophage infection in the oceanic cyanobacterium *Synechococcus* sp. WH7803. *J. Phycol.* **32:**506–516.

Wilson, W. H., D. Lane, D. A. Pearce, and J. C. Ellis-Evans. 2000. Transmission electron microscope analysis of virus-like particles in the freshwater lakes of Signy Island, Antarctica. *Polar Biol.* **23:**657–660.

Wilson, W. H., S. Turner, and N. H. Mann. 1998. Population dynamics of phytoplankton and viruses in a phosphate-limited mesocosm and their effect on DMSP and DMS production. *Estuar. Coast. Shelf Sci.* **46:**49–59.

Winter, C., G. J. Herndl, and M. G. Weinbauer. 2004. Diel cycles in viral infection of bacterioplankton in the North Sea. *Aquat. Microb. Ecol.* **35:**207–216.

Wommack, K. E., and R. R. Colwell. 2000. Virioplankton: viruses in aquatic ecosystems. *Microbiol. Mol. Biol. Rev.* **64:**69–114.

Zeidner, G., J. P. Bielawski, M. Shmoish, D. J. Scanlan, G. Sabehi, and O. Beja. 2005. Potential photosynthesis gene recombination between *Prochlorococcus* and *Synechococcus* via viral intermediates. *Environ. Microbiol.* **7:**1505–1513.

FUNGI IN POLAR ENVIRONMENTS

Polona Zalar, Silva Sonjak, and
Nina Gunde-Cimerman

4

INTRODUCTION

Cold environments represent reservoirs of rare and still unknown microbial species, and therefore also of novel biological processes and genes (Simon et al., 2009). Despite the difficulties that can be encountered in biodiversity studies of polar environments, research into fungi found in the extremely cold areas of the earth has a history that is more than a century long. Initial culture-based studies were concentrated on the soil, permafrost, vegetation, wood, aerosphere, and rock. More recent studies have discovered fungi also in the water-based environments, such as in seawater, freshwater, glacial meltwater, ice, and snow, as well as in animal dung and microbial mats. Studies of fungi have now covered geographically distant regions: Antarctica, the Arctic, the alpine regions of Europe, the Himalayas in Asia, the Andes in South America, and the Rocky Mountains and Great Smoky Mountains in the United States, among other places. The distribution of fungi ranges from circumpolar to endemic and cosmopolitan fungi. Although some fungi that are endemic to the extreme polar regions show psychrophilic behavior, the

majority are instead psychrotolerant and globally distributed, ranging from the Arctic to Antarctica, and including the temperate zones in between. It is important to note also that although their natural ecological niches are in the polar environments, such fungi can also grow in human proximity; they can inhabit freezers and cold-storage rooms, and refrigerated and even frozen food. Different species of the genera *Penicillium* and *Cladosporium* are particularly successful (Frisvad, 2008). These fungi can easily be transported via global water systems (Grabińska-Łoniewska et al., 2007) and air currents, which is also facilitated by the global warming phenomena.

The pioneering studies on extremophilic eukaryotes in polar environments concentrated on lichens, followed by numerous studies dealing with mainly "nonvisible" microfungi. Fungal representatives from all of the major phyla have been found and defined in the extremely cold environments, from the zoosporic *Chytridomycota* (Freeman et al., 2009) to the sporocarp-forming *Basidiomycota* (Ludley and Robinson, 2008). It appears that unlike in temperate zones, where filamentous fungi prevail, yeasts are particularly well adapted to the polar terrestrial and aquatic environments (Butinar et al., 2007). Among the filamentous fungi, the genera *Penicillium, Aspergillus,*

Polona Zalar, Silva Sonjak, and Nina Gunde-Cimerman, Biology Department, Biotechnical Faculty, University of Ljubljana, 1000 Ljubljana, Slovenia.

Polar Microbiology: Life in a Deep Freeze
Edited by Robert V. Miller and Lyle G. Whyte © 2012 ASM Press, Washington, DC

Cladosporium, and *Geomyces* have the broadest adaptive potential and the widest distributions across various polar niches. However, the fungi that have adapted the most to the most extreme polar conditions are the melanized, black, yeastlike fungi (Onofri et al., 1999); these can inhabit rocks and glacial ice with high salt concentrations (Butinar et al., 2011).

The methods used for fungal detection have been time appropriate, from the classical early microscope visualization to the more recent sophisticated DNA-based techniques (Fell et al., 2006), which have been complemented lately by metagenomic studies (Simon et al., 2009; Kennedy et al., 2010), although these have generally not been focused on fungi.

Disregarding the geographic distances, all of the polar and other extremely cold environments share the main stress factors that can affect fungal life. The main one of these is, of course, the low temperature, which is closely related to desiccation and low water availability due to the freezing, the relatively high concentrations of ions, and the generally low levels of nutrients in the liquid water, and sometimes high UV irradiation and hypoxia.

Most investigations on fungi in extremely cold environments have described their biodiversity. Due to their envisaged biotechnological potential, physiological and biochemical studies have mainly been oriented toward extracellular enzymatic activities, biosynthesis of secondary metabolites, and bioremediation at low temperatures (Onofri et al., 2000; Scorzetti et al., 2000; Margesin, 2009). Recently, fungal adaptations to cold at the molecular level have also received more attention, which has expanded the realm of extremophilic model organisms from prokaryotes to fungi, providing us with a better understanding of the complexity of these processes at "the edge of life" (Robinson, 2001; Ruisi et al., 2007; Xiao et al., 2010).

FUNGI IN VARIOUS EXTREMELY COLD ENVIRONMENTS

Most of our planet is permanently cold. Indeed, more than two-thirds of the earth is covered by seawater, with mostly deep oceans that have a constant low temperature of approximately 2°C. At the same time, the terrestrial and aquatic environments in the polar and alpine regions, the upper atmosphere, the plants and animals inhabiting cold regions, and the Antarctic rock together represent more than 80% of the biosphere (Russell et al., 1990; Cavicchioli et al., 2002). The important presence and role of fungi in ecological processes that occur in extremely cold environments is thus becoming increasingly evident (Fig. 1).

Vegetation

Fungi have been reported primarily in connection with the sub-Arctic vegetation and soil in the polar regions. Mainly basidiomycetous yeasts have been isolated from berries, flowers, vegetation of the littoral zone, soils, forest trees, grasses (Babjeva and Reshetova, 1998), and Antarctic moss (Tosi et al., 2002). Recently, fungi belonging to *Ascomycota* and *Basidiomycota* were discovered in abundance below the snow-covered tundra (Pennisi, 2003; Schadt et al., 2003). Their biomass can be up to 10 times higher than the biomass of prokaryotes (Schmidt and Bölter, 2002).

Soil and Permafrost

Permafrost in polar regions covers more than 25% of the land surface and significant parts of the coastal sea shelves (Wagner, 2008). Together with seasonally frozen soils, permafrost represents a large part (some 50%) of terrestrial Earth (Panikov, 2009).

Permafrost is defined as ground that is composed of soil, sediment, or rock, and includes ice and organic material, which remains at or below 0°C for at least two consecutive years. The permafrost can be hundreds to more than 1,000 m deep, and it can be divided into three temperature-depth layers: (i) the active layer, which is influenced by air temperature fluctuations, with a thickness of a few centimeters to 2 m and temperatures fluctuating from +15 to −35°C; (ii) the upper, perennially frozen permafrost sediments, with a thickness of 10 to

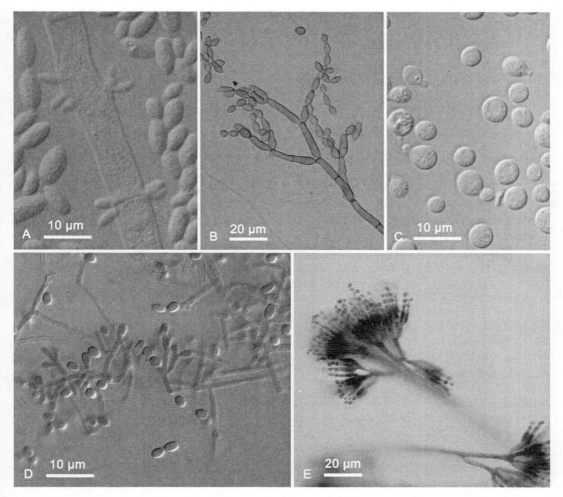

FIGURE 1 Micromorphological structures of the most common fungi in polar regions. (A) *Aureobasidium pullulans*; (B) *Cladosporium varians*; (C) a *Cryptococcus* sp.; (D) *Geomyces pannorum*; (E) a *Penicillium* sp.

20 m and temperatures from 0 to −15°C; and (iii) the deeper, stable permafrost sediments, with temperatures from −5 to −10°C (Wagner, 2008). The boundary between the active layer and the perennially frozen ground is called the permafrost table, which acts as a physical and chemical barrier. Thermal conditions influence the formation of different patterns of ground formations and various cryogenic structures, such as ice wedges, taliks, and cryopegs. Ice wedges are defined as cracks in the ground that are formed by a narrow or thin piece of ice that measures up to 3 to 4 m wide at ground level and extends downward into the ground, down to several meters. Taliks are layers of year-round unfrozen ground in permafrost areas, and cryopegs are layers of unfrozen ground that form perennially as part of the permafrost, while marine cryopegs are defined as lenses of brine (60 to 300 g of NaCl/liter). In Arctic regions, the permafrost is typically impregnated with ice, whereas in the Antarctic polar deserts and rocky areas where the interstitial water level is very low, the permafrost can be dry (Wagner, 2008). Different types of permafrost habitats are thus characterized by extremely

low temperatures, freeze-thaw cycles, and low water and nutrient availability (Morozova et al., 2007; Wagner, 2008). In the active layer and upper permafrost sediments, intensive physicochemical processes take place, whereas in the deeper permafrost sediments, where the conditions can remain stable for long periods of time, the microbial processes are limited (Wagner, 2008). Therefore, low gas permeability, the presence of liquid water, and a supply of mobile oxidizable compounds are the minimal criteria that can allow cell survival in permafrost (Panikov, 2009).

Aerobic and anaerobic microorganisms (bacteria, cyanobacteria, green algae, yeasts, and fungi) have been isolated from permafrost sediments that have remained in a frozen state for extended periods. Viable bacteria and even yeast have been isolated from up to three-million-year-old Siberian permafrost, although eukaryotes were preserved in considerably lower numbers than the prokaryotic cells (Golubev, 1998; Faizutdinova et al., 2005). In ancient permafrost deposits of up to 400,000 years old, zygomycetous, ascomycetous, and basidiomycetous fungi have been detected (Lydolph et al., 2005). Upon thawing, these microorganisms were able to resume their metabolic activities (Takano et al., 2004).

We now know that microbial communities in permafrost are very complex, and they are represented by dormant propagules of mesophilic cosmopolitan species that have been dispersed from the surrounding areas by wind and animals, and also by active indigenous culturable or nonculturable species of *Archaea*, *Bacteria*, and *Eukarya*, as indicated by their high numbers and psychrophilic nature (Ruisi et al., 2007; Panikov 2009). The total microbial biomass in permafrost is comparable to that of the communities of temperate soil ecosystems (Wagner et al., 2005; Wagner 2008). These microorganisms not only survive under permafrost conditions, but they can sustain active metabolism (Rivkina et al., 2004; Wagner, 2008). It appears that filamentous fungi and yeasts are more resistant to hostile permafrost environments and show more vigorous growth

in frozen habitats than bacteria (Steven et al., 2006). The levels of fungi present in permafrost can thus vary from 10 to almost 100,000 CFU g^{-1} material. Increased numbers of CFU may be detected in any portion of the sample, regardless of the depth or age of the sediments. The peaks of fungal populations in permafrost are thus microfocal, and importantly, paralleled with decreases in species number (Ozerskaya et al., 2009).

Fungal diversity in the Arctic and Antarctic permafrost has been studied intensively over the last decade, and it has been shown to have considerable taxonomic diversity, with significant numbers of new taxa (Ruisi et al., 2007). In some permafrost regions, yeasts represented an important, or even the major (up to 100%), part of all of the fungi isolated, and 20 to 25% of the total aerobic heterotrophs (Vorobyova et al., 1997; Steven et al., 2006). In the cold deserts in Antarctica, like in the McMurdo Dry Valleys and the Ross Desert in southern Victoria Land, which is considered one of the harshest environments known on Earth (Nienow and Friedmann, 1993), primarily xerophilic, basidiomycetous yeasts have been isolated (Vishniac and Onofri, 2003; Onofri et al., 2004; Takano, 2004). These were found at the highest frequencies in the youngest layers, which were less than 10,000 years old, although they have also been detected in three-million-year-old Pliocene samples (Dmitriev et al., 1997a, 1997b; Rivkina et al., 2000). The most frequently isolated yeasts from these permafrost sediments belong to the genera *Cryptococcus*, *Rhodotorula*, and *Saccharomyces*. Similarly, yeasts from the genera *Bullera*, *Candida*, *Debaryomyces*, *Mrakia*, *Pichia*, *Sporobolomyces*, *Tilletiopsis*, *Trichosporon*, and *Leucosporidium* have also been found in permafrost (Vishniac, 1993; Vorobyova et al., 1997; Xin and Zhou, 2007).

In both the active layer and the perennially frozen Arctic sediments, a large variety of filamentous fungi have been detected, belonging to *Ascomycota*, *Basidiomycota*, *Zygomycota*, *Chytridiomycota*, and *Glomeromycota* (Wallenstein et al., 2007). The most frequently occurring genera of the filamentous fungi are *Aspergillus*, *Cladosporium*, *Geomyces*, and *Penicillium*. This last showed the highest number of species. Additionally, the following ascomycetous genera have been isolated: *Acremonium*, *Alternaria*, *Acrodontium*, *Arthrinium*, *Aureobasidium*, *Bispora*, *Beauveria*, *Botrytis*, *Camposporium*, *Chaetomium*, *Chaetophoma*, *Chalara*, *Chrysosporium*, *Diatrype*, *Engyodontium*, *Epicoccum*, *Eurotium*, *Exophiala*, *Fusarium*, *Geotrichum*, *Gliocladium*, *Gymnascella*, *Helminthosporium*, *Hormoconis*, *Lecythophora*, *Malbranchea*, *Myceliophthora*, *Monodictys*, *Nectria*, *Paecilomyces*, *Papulaspora*, *Phialophora*, *Phoma*, *Racodium*, *Rhinocladiella*, *Scolecobasidium*, *Scopulariopsis*, *Stachybotrys*, *Sepedonium*, *Sphaeronaemella*, *Sporothrix*, *Sporotrichum*, *Stephanosporium*, *Thelebolus*, *Thysanophora*, *Trichoderma*, *Trichophyton*, *Tritirachium*, *Ulocladium*, *Verticillium*, *Wardomyces*, *Xylohypha*, and mycelia sterilia (Vishniac, 1993; Azmi and Seppelt, 1998; Ivanushkina et al., 2005; Kurek et al., 2007; Ozerskaya et al., 2008, 2009; Stakhov et al., 2008).

Fungi of the phylum *Zygomycota* have been isolated much less frequently. Most of these isolates belong to the genera *Mortierella*, *Mucor*, and *Rhizopus* (Vishniac, 1993; Kurek et al., 2007), while the culturable filamentous fungi from phylum *Basidiomycota* show much lower diversity. Recent molecular studies that allow the detection of nonculturable fungi have implied that the higher basidiomycetous fungi are much more abundant than has been assumed to date (Ozerskaya et al., 2009).

Most of the filamentous fungi and yeasts isolated from permafrost are cosmopolitan mesophilic species, which are characterized by the production of high numbers of conidia that are easily dispersed via the air, water, and animals (Samson et al., 2002). These fungi might represent recent contamination of the permafrost, and this thus makes the identification of the indigenous species more difficult (Lydolph et al., 2005; Ruisi et al., 2007). Indigenous species can be identified among some of these isolated fungi that are psychrophilic or psychrotolerant and can grow and reproduce in this extreme environment (Ruisi et al., 2007). For example, *Mucor flavus*, *Thelebolus microsporus*, and *Geomyces pannorum* are true psychrophiles; this last can grow at a rate of 0.05 mm per day when the temperature of the environment is −2°C (Hughes and Lawley, 2003).

Members of *Chytridiomycota* dominate the fungal biodiversity, and perhaps the decomposition processes, in plant-free, high-elevation soils from the highest mountain ranges on Earth, such as the Himalayas and the Rockies. In these types of soil, the *Chytridiomycota* ribosomal gene sequences of clone libraries can constitute over 70% of the total. Very few chytrids have been cultured (Freeman et al., 2009).

Fungi in Rock

The polar regions are rich in stony/rocky habitats. Rock represents a dwelling place as well as a substrate for endolithic microorganisms, including fungi. The conditions on the surface of the rock are the harshest ones seen, and they can only be sustained by lichens (Friedmann, 1982; Nienow and Friedmann, 1993; de los Ríos et al., 2003; Selbmann et al., 2005), assigned as epiliths. The interiors of the rock protect life from environmental extremes to some extent, which allows the growth of different kinds of endolithic microorganisms (Friedmann and Koriem, 1989). These are further divided into several categories, which depend on their microniche: chasmoendoliths inhabit rock cracks, fissures, and vesicles (de los Ríos et al., 2003, 2005; McLoughlin et al., 2007), while cryptoendoliths are found in structural cavities within porous and sandstone rock, and also other sedimentary rock. The most extreme microorganisms in rock are euendoliths, which actively burrow into the rock and create microtubular cavities (Golubic et al., 1981). The presence of microbes in rock is important, due to their involvement in geological and geochemical processes (Gadd, 2007). As lithotrophs, the microbes actively use the rock to obtain their energy, while cryptoendoliths and chasmoliths, in the process of microboring, biochemically dissolve carbonate grains by biosynthesizing organic acids (oxalic and gluconic acids) (Burford et al., 2003). Microbes can adopt various modes of life at different stages of their life cycles, or even in response to changes in their external environment (de los Ríos et al., 2005). Disregarding the endolithic subspecialization, life starts in all cases in the form of subaerial biofilms in between the atmosphere and the lithosphere (Gorbushina and Broughton, 2009).

Studies on cryptoendolithic fungi from Antarctica started with Friedmann in 1982 (Friedmann, 1982) and were continued primarily by Onofri and coworkers (Onofri et al., 1999, 2000, 2004; Selbmann et al., 2005, 2008). In contrast, studies on Arctic rock fungi have only recently been initiated (Omelon et al., 2006, 2007; L. Selbmann, personal communication). Nonlichenized, rock-inhabiting fungi in polar regions were described as cryptoendoliths. Together with cyanobacteria and algae, they represent a microbial community within the porous sandstone rock. Microorganisms form under an abiotic crust on successive layers of the rock surface, and they can be colored black, white, green, and sometimes blue-green. In the black and white zones, the filamentous fungi form lichen associations with chlorophycean algae, while darkly pigmented nonlichenized fungi inhabit the black layer beneath the hyaline fungi (Onofri et al., 2000).

Darkly pigmented, cryptoendolithic fungi from the rock of continental Antarctica have been isolated and newly described as anamorphic species of genera endemic to Antarctica: Friedmanniomyces (Onofri et al., 1999) and Cryomyces (C. antarcticus, C. minteri) (Selbmann et al., 2005). Additionally, the genus Recurvomyces has been newly described in Antarctic sandstone, and it has also been found on rock in the Italian Alps (Selbmann et al., 2008). In the genus Elasticomyces, the species E. elasticus was described based on an isolate from rock from the Andes (Argentina). Nonidentified species of the genus Verticillium, in addition to a noncultured, black-pigmented fungus, have been found as part of a cryptoendolithic community within the translucent gypsum salt crust on Antarctic sandstone (Hughes and Lawley, 2003).

Recent studies performed on cryptoendoliths in sandstone outcrops of the Canadian High Arctic have also revealed the presence of black-pigmented fungi, as well as filamentous hyaline fungi (Omelon et al., 2006, 2007). Although these isolates have not been identified, it appears that the general microbial diversity in the Arctic rock is higher in comparison to similar habitats in Antarctica. The main reasons for this are probably the higher temperatures and moisture, and to a lesser extent the lower pH conditions and higher concentrations of iron, aluminium, and silicon in the overlying surface (Omelon et al., 2007). The total time that is available for ideal metabolic activity within the cryptoendolithic environment in the Canadian High Arctic has been estimated as 2,500 hours per year (Omelon et al., 2006), whereas for the Antarctic Dry Valleys it has been estimated at between 50 and 500 hours per year (Omelon et al., 2006).

Studies performed over the last few years have shown that crypthoendolithic fungi are much more widespread and common than previously thought. These inhabit rock wherever the conditions are prohibitive enough to prevent the settlement and growth of cosmopolitan, fast-growing, and competitive fungi. Due to their ability for long-distance dispersal, they can occupy either very cold or very hot and dry rock (Selbmann et al., 2005, 2008; Ruibal et al., 2009). To our current knowledge, only the genera Friedmanniomyces and Cryomyces are endemic to Antarctica, while other recently described genera and species of Elasticomyces, Recurvomyces, and Acidomyces have also been found in different geographically remote mountain regions, e.g., the Andes of Argentina and the Italian Alps.

Multiple gene phylogenetic analyses have placed melanized Antarctic rock fungi mainly into Capnodiales and Dothideomycetes. The composition of some lineages that are rich with rock-inhabiting fungi suggests that rock can be the primary substrate for ancient fungi, and it has been hypothesized that there is a link between rock-dwelling fungi and lichenization for Dothideomycetes lineages, since they comprise lichens as well as fungi isolated from rock (Ruibal et al., 2009).

Fungi from the genera *Aspergillus, Candida, Fusarium, Mucor, Paecilomyces, Penicillium, Rhizopus, Rhodotorula,* and *Sporobolomyces* have also been reported as common fungal species inhabiting rock of different chemistries in the cold Antarctic deserts (Burford et al., 2003).

Microbes living on and within the rock are exposed to different kinds of stress factors, which, as indicated above, can include high UV irradiation, low temperatures, desiccation, repeated freeze-thawing, and lack of nutrients. The general response of cryptoendolithic fungi to these conditions is a high level of simplification. They primarily grow as multicellular clumps and have the ability to shift to a simpler life cycle and meristematic growth, which also includes the loss of sexual, and in some fungi even asexual, reproduction. Some species can reproduce just by a simple disarticulation of preexisting hyphae, as with *Friedmanniomyces endolithicus* (Selbmann et al., 2005). Through such simplification, these fungi can conclude their life cycle in a shorter time, with less energy needed and hence at lower metabolic cost (Ruisi et al., 2007).

Analyses of clone libraries of rRNA genes amplified from environmental DNA from the McMurdo Dry Valleys, Antarctica, have led to the detection of the yeast species *Bullera* aff. *unica* and the filamentous fungus *G. pannorum* (de la Torre et al., 2003).

Microorganisms living in rock can colonize environments on Earth that resemble most environments on the planet Mars, and therefore they have been used as the closest eukaryotic models for exobiological speculation. Their astonishing viability after freezing and thawing, and also after UV exposure, and their tolerance to osmotic imbalances has shown their uncommon ability to survive under harsh external pressures (Selbmann et al., 2005). Therefore, studies of their ability to withstand simulated space and the conditions on Mars have been performed on pure cultures of *C. antarcticus* and *C. minteri* and on fragments of rock with cryptoendolithic fungal communities (Onofri et al., 2008). Direct space exposure on the International Space Station has already been performed (Onofri et al., 2008), and the results show the incredible viability of the fungi that have been exposed (S. Onofri, personal communication).

Seawater and Sea Ice

Sea ice represents one of the coldest habitats on Earth, with temperatures ranging from −1°C to as low as −50°C in winter. Its temperature is highly variable, both temporally and spatially, and it is for the most part seasonally transient, being constantly broken up and reshaped by the wind and ocean currents. When seasonal temperatures drop, frozen seawater forms a semisolid matrix that is permeated by a network of channels and pores, and filled with brine formed from expelled salts as the ice crystals freeze. The salinity of sea-ice brines, which remain liquid to −35°C, can reach 20% NaCl. Changes in the salinity and temperature are thus the dominant factors that influence the biological communities within sea ice (Brown and Bowman, 2001; Thomas and Dieckmann, 2002). In winter, most of the microbial biomass can be detected in brine pockets within the central mass of the ice, while in spring, a dense microbial community sometimes develops at the bottom of the ice (Gosink et al., 1993). Although autotrophic algal communities are relatively well described, sea ice-dwelling fungal communities have been little documented (Brown and Bowman, 2001; Junge et al., 2002; Gunde-Cimerman et al., 2005). From Svalbard (Arctic), conventional culture-based methods with sea ice samples have yielded up to 7,000 CFU per liter of fungi (Gunde-Cimerman et al., 2003). Most of the isolates were different (85%), with basidiomycetous yeasts, and mainly *Cryptococcus adeliensis, Rhodotorula mucilaginosa,* and *Metschnikowia bicuspidate* (*Ascomycota*). Of the filamentous fungi species, the isolates included *Aspergillus, Aureobasidium, Cladosporium, Eurotium, Mucor, Penicillium, Phoma,* and *Trichoderma.* When DNA was extracted from several Antarctic and one

Arctic sea ice sample, some of the eukaryotic clones detected were described as fungal (Thomas and Dieckmann, 2002). When sea ice melts, the entrapped fungi are released into the oceans; however, few studies have reported on the occurrence of psychrotolerant and psychrophilic fungi in polar aquatic habitats. These fungi were associated with marine macroalgae from Antarctica (Loque et al., 2010). Mainly basidiomycetous yeasts have been isolated from polar offshore seawaters (Jones, 1976), while sequences belonging to *Eumycota* were detected in Antarctic Polar Front waters up to 3,000 m deep (López-García et al., 2001).

Polar Lakes

Most polar lakes are characterized by prolonged, sometimes perennial, ice cover, which is typically 3 to 6 m thick, and which contains a layer of sand and organic matter of Aeolian origin. This layer represents a dynamic equilibrium between the downward movement of sediments as a result of melting during the summer and the upward movement of ice from ablation at the surface and freezing at the bottom. During summer, when the solar radiation is continuous, there are liquid-water inclusions in this layer. The ice meltwater can support viable microbial communities that are associated with the sediment layer (Kriss et al., 1976; Priscu et al., 1998; Belzile et al., 2001; De Wit et al., 2003; Gaidos et al., 2004).

Primarily basidiomycetous yeasts and fungi have been isolated from freshwater samples, benthic microbial mats, and biofilms on pebbles beneath the ice of Antarctic lakes. Although aquatic hyphomycetes were absent from most samples, several soil hyphomycetes and aquatic phycomycetes have been retrieved regularly (Ellis-Evans, 1985; Baublis et al., 1991; De Wit et al., 2003). Fungi have also been found in the hypersaline Antarctic Lake Wanda (Kriss et al., 1976).

Antarctic Lake Vostok represents a special example of polar lakes. It comprises two types of ice; the upper 3,500-m layer of ice contains traces of nutrients of Aeolian origin, including various acids, sea salts, and mineral grains. The ice below 3,500 m comprises refrozen water from Lake Vostok that has been accreted to the bottom of the upper glacial ice. This layer is at least 420,000 years old, and it has a temperature of a few degrees below freezing. Nutrients in the accretion ice include salts and dissolved organic carbon and nitrogen. The total levels are high enough to support growth of heterotrophic microbial communities (Siegert et al., 2001). Microbiological analyses have revealed incorporation and respiration of radioactively labeled organic substrates. The accretion ice represents a habitat of interconnected liquid veins in which motile bacteria that probably originated from deep glacial ice have been detected by direct scanning (Christner et al., 2000; Siegert et al., 2001, 2003; Cavicchioli et al., 2002).

Fungi have been reported in the deep, ancient glacial layers of the core of Lake Vostok. Yeast species of the genera *Rhodotorula* and *Cryptococcus* have been identified by molecular methods, and fungal mycelia have been seen by direct epifluorescence microscopy (Poglazova et al., 2001; Abyzov et al., 2004). Individual fungal propagules have been detected in old Antarctic Lake Vostok ice-core sections that were 3,000 to 5,000 years old (D'Elia et al., 2009), and also in two deep glacial ice sections close to the bottom, which were 1,000,000 to 2,000,000 years old. These fungi were characterized by fluorescence microscopy, culturing, and analyses of their internal transcribed spacer sequences. Thirty-one unique fungal ribosomal DNA sequences were detected and compared to recent taxa. The sequences closest to *R. mucilaginosa* were the most frequent, with other basidiomycetes including isolates belonging to the genera *Cystofilobasidium* and *Cryptococcus*. The ascomycetes isolated were most closely related to *Aspergillus*, *Cladosporium*, *Penicillium*, and *Aureobasidium pullulans*. These data indicated that Lake Vostok contains a mixture of heterotrophic psychrotolerant fungal species (D'Elia et al., 2009).

Glaciers and Subglacial Environments

Glacial ice is formed from compacted snow at the poles, as well as at lower-latitude, high-altitude locations. Atmospheric samples and particulates prevailing at the time of snowfalls are preserved chronologically. Glacial ice was considered for a long time as an extremely stable and static environment; thus ice-core analyses have been used to document and date past climate changes, geological events, and human activities. Isolation of microorganisms from ice cores has demonstrated ancient microorganisms in the older layers, which originated from northern latitudes, and temperate as well as tropical regions (Abyzov, 1993; Ma et al., 1999, 2000). Low numbers of filamentous fungi were isolated from 10,000- to 13,000-year-old Greenland ice (Ma et al., 1999, 2000) and from Antarctic ice layers that are up to 38,600 years old (Abyzov, 1993). PCR amplification of fragments of the eukaryotic 18S rRNA gene that were extracted from 2,000-year-old and 4,000-year-old ice-core samples from north Greenland led to the identification of a diversity of fungi, plants, algae, and protists (Price, 2000). All of these findings of fungi in glacier ice cores have been interpreted as the result of coincidental Aeolian deposits of spores or mycelia into the ice during its geological history.

Recent investigations have shown that ice in glaciers is a much more dynamic habitat than previously thought, on the microscale as well as at the geomorphological level. Ice in temperate glaciers is permeated by a continuous network of aqueous veins that are formed due to sea salts deposited as aerosols, as they are essentially insoluble in ice crystals. These liquid veins can have high ionic strength (Rohde and Price, 2007). Due to the percolation of the salts from the top of the glacier to its bottom, these can accumulate to relatively high concentrations in the bottom parts of polythermal glaciers (Price, 2000). Due to quick seismic shifts (Ekstrom et al., 2003; Fahnestock, 2003) and the cryokarst phenomena, liquid water can also appear on the surface of glaciers and as caves, interglacial lakes, or moulins within the glaciers (Christner et al., 2000). If these

waters reach the glacier bed and mix with the groundwater and glacial basal meltwater, rock, and sediment, they create dynamic subglacial environments. Microbial communities in these special habitats are dominated by aerobic heterotrophic β-*Proteobacteria* (Skidmore et al., 2000; Foght et al., 2004), and based on the evidence obtained primarily from Arctic glaciers in Svalbard, they can also include basidiomycetous and ascomycetous yeasts, melanized fungi of the genera *Cladosporium* and *Aureobasidium*, and many different species of the genus *Penicillium* (Gunde-Cimerman et al., 2003, Sonjak et al., 2006, Butinar et al., 2007, 2011; Zalar et al., 2008).

The dominant species in these environments was *Penicillium crustosum*, which represented on average half of all of the isolated strains from the glaciers studied. The fungal counts detected in the subglacial samples were 2 orders of magnitude greater when compared with those from supraglacial samples, mainly due to yeast species (with up to 4×10^6 CFU liter^{-1}). In contrast to yeasts, which were primarily associated with the clear glacier ice, the highest counts for penicillia were obtained for debris-rich subglacial ice (Sonjak et al., 2006). Twenty-seven distinct species of ascomycetous and basidiomycetous yeasts were isolated, including four new species. According to the species diversity and the abundance, the majority of the species were assigned to the hymenomycetous yeasts (*Filobasidium*/*Cryptococcus albidus* taxa of the *Tremellales*). The stable core of the subglacial yeast communities was represented by *Cryptococcus liquefaciens*, *R. mucilaginosa*, *Debaryomyces hansenii*, and *Pichia guillermondii* (Butinar et al., 2011). Yeasts have also been isolated from ice and meltwater in Svalbard (Pathan et al., 2010; Butinar et al., 2011), from Italian Alpine glaciers (Turchetti et al., 2008; Branda et al., 2010; Thomas-Hall et al., 2010), and from Patagonian glaciers (de García et al., 2007).

Cryoconite Holes

Cryoconite holes occur in glaciers in the Arctic and Antarctic, as well as in alpine environments in the glacial ablation zone. They are formed when windblown particulates lodge on the surface in small depressions in the ice and cause vertical melting by absorbing more radiation than the surrounding ice. The result is a sediment inclusion, which is 90% immersed in water, with the remaining space filled with air. These holes range from less than 1 cm to 1 m in width. They seldom grow deeper than 60 cm, although they can coalesce into bigger holes or become interconnected by meltwater channels. As the solar radiation decreases, an ice cover can form at the water surface and grow downward. With faster freezing rates, solutes can become trapped in brine pockets with up to five times the salinity of seawater. These ponds can freeze only after reaching brine temperatures of −12°C. In relation to the melt regime of the glacier surface, cryoconite holes can last from several days to weeks, or for entire seasons. On some stable glaciers, they are thought to have remained for at least a hundred years, thus serving as biological refuges during periods of extreme cold (Mueller et al., 2001; Reeve et al., 2002; Vincent et al., 2004).

Although the vast majority of cryoconite holes can be classified as ultra-oligotrophic, they nevertheless provide a habitat for microbial colonization and growth. The dominant organisms are typically nitrogen-fixing, filamentous, mat-forming cyanobacteria. Light microscopy has also documented the presence of pollen grains, bacteria, algae, diatoms, filamentous fungi, yeasts, and occasional microinvertebrates. Psychrotrophs prevail, while psychrophiles have only rarely been detected (Mueller et al., 2001; Vincent et al., 2004). Mainly basidiomycetous yeasts were isolated as well (Margesin et al., 2002, 2007, Branda et al., 2010).

INDIGENOUS GROUPS OF FUNGI IN POLAR ENVIRONMENTS

The literature on Antarctic nonlichenized fungi has been compiled and well documented online by P. Bridge, B. Spooner, and P. Roberts (http://www.antarctica.ac.uk/bas_research/data/access/fungi/Speciespublic2.html#Use).

This list gathers information on fungi reported from the Antarctic region that has been published in the literature or deposited in major culture collections and herbaria. The list gives additional details of current nomenclature, synonyms, host/substrate, and location, along with information for the original citations. There have also been scattered reports of fungi from other cold regions, although some compilations do exist (Gilichinsky et al., 2007; Frisvad, 2008; Ludley and Robinson, 2008; Libkind et al., 2009; Ozerskaya et al., 2009; Branda et al., 2010). Public fungal databases, such as the CBS Fungal Biodiversity Center (http://www.cbs.knaw.nl/databases/) and the American Type Culture Collection (http://www.lgcstandards-atcc.org/), are also valuable sources for data on fungi from cold regions.

Ascomycetous and Basidiomycetous Yeasts

The first studies on Antarctic yeasts were performed in the early 1870s (http://www.antarctica.ac.uk/bas_research/data/access/fungi/Speciespublic2.html#Use). Yeasts were studied in soil and permafrost in particular (Gilichinsky et al., 2005; Fell et al., 2006; Vishniac, 2006; Margesin and Fell, 2008). Only recently, investigations have revealed that yeasts can be found at high CFU in Arctic polythermal glaciers. Biodiversity has been shown for large populations of yeast in subglacial ice and glacial meltwaters (Butinar et al., 2007) and in puddles in the vicinity of glaciers (Pathan et al., 2010). Yeast species have also been described from meltwaters of Patagonian glaciers (de García et al., 2007) and in high-altitude lakes (Libkind et al., 2009) and other glacial environments (Libkind et al., 2003). Yeasts from an Alpine glacier in Italy have also been documented (Turchetti et al., 2008; Branda et al., 2010; Thomas-Hall et al., 2010).

Detailed taxonomic analyses based on physiological profiles and relevant molecular markers have allowed the description of numerous new species originating from cold and polar environments (Montes et al., 1999; Scorzetti et al., 2000; Thomas-Hall and Watson, 2002;

Thomas-Hall et al., 2002, 2010; Vishniac, 2002; Libkind et al., 2005; Margesin et al., 2007; Xin and Zhou, 2007; Margesin and Fell, 2008; de García et al., 2010a, 2010b). According to the last compilation of data on yeasts, basidiomycetous yeasts prevail (Branda et al., 2010), and particularly different species of the genera *Bulleromyces*, *Cryptococcus*, *Cystofilobasidium*, *Dioszegia*, *Erythrobasidium*, *Filobasidium*, *Guehomyces*, *Leucosporidiella*, *Leucosporidium*, *Malassezia*, *Kondoa* (from the Arctics; N. Gunde-Cimerman, unpublished data), *Mastigobasidium*, *Mrakia*, *Mrakiella*, *Rhodosporidium*, *Rhodotorula*, *Sporidiobolus*, *Sporobolomyces*, *Trichosporon*, and *Udeniomyces*. Among these, *Cryptococcus* and *Rhodotorula*, which are both of polyphyletic origins that occur in several phylogenetic lineages of the *Agaricomycotina*, were the predominant genera, and these were also represented by the greatest numbers of species: 32 and 12, respectively.

The prevailing species in permafrost were those of the genera *Cryptococcus*, *Rhodotorula*, and *Saccharomyces*. Also, using standard culture techniques, yeast species from the genera *Bullera*, *Candida*, *Debaryomyces*, *Mrakia*, *Pichia*, *Sporobolomyces*, *Tilletiopsis*, *Trichosporon*, and *Leucosporidium* have also been found in permafrost (Vishniac, 1993; Vorobyova et al., 1997; Xin and Zhou, 2007). *Cryptococcus* species, *Debaryomyces*, and *Pichia* have been isolated from cryopegs (Gilichinsky et al., 2005).

The psychrophilic yeasts isolated from soils were from the genera *Bulleromyces*, *Candida*, *Cryptococcus*, *Dioszegia*, *Leucosporidium*, and *Mrakia* (Vishniac, 2006). Numerous mesophilic species have also been reported, from the genera *Bulleromyces*, *Candida*, *Clavispora*, *Cryptococcus*, *Debaryomyces*, *Guehomyces*, *Issatchenkia*, *Pseudozyma*, *Rhodosporidium*, *Rhodotorula*, *Sporidiobolus*, *Sporobolomyces*, *Stephanoascus*, *Torulaspora*, and *Trichosporon* (Thomas-Hall and Watson, 2002; Thomas-Hall et al., 2002; Vishniac 2002, 2006; Margesin et al., 2007; Xin and Zhou, 2007). Yeast species detected by molecular methods in extremely dry soils of the Antarctic Dry Valleys (0.2 to 1.3% moisture) were mainly represented by different species of the genus *Trichosporon*, and also

by different *Cryptococcus* species (Fell et al., 2006).

The prevailing species in the subglacial ice of the Arctic glaciers in Svalbard were *C. liquefaciens* and *R. mucilaginosa* (Butinar et al., 2007); in the Italian Alpine glaciers, *Cryptococcus gastricus* and *Rhodotorula psychrophenolica* (Branda et al., 2010); and in a Patagonian glacier, *Cryptococcus stepposus* (de García et al., 2007) and *R. mucilaginosa* (Libkind et al., 2003). Interestingly, yeasts with similar internal transcribed spacer sequences to *R. mucilaginosa* were also the most numerous in the very different environment of the Antarctic Lake Vostok accretion ice (D'Elia et al., 2009).

Ascomycetous yeasts have been found in glaciers of the Southern and Northern Hemispheres only occasionally. The dominant ascomycetous yeast species was *D. hansenii*, with different species of the genera *Candida*, *Metschnikowia*, and *Pichia* and the newly described species from Patagonian glaciers, *Wickerhamomyces patagonicus* (de García et al., 2010a).

Black Yeastlike Fungi

An important group of melanized yeastlike fungi are the rock-inhabiting cryptoendolithic, which are nonlichenized fungi that have been mainly described for continental Antarctica. The most important genera were *Friedmanniomyces* (*F. endolithicus*, *F. simplex*) (Onofri et al., 1999), *Cryomyces* (*C. antarcticus*, *C. minteri*) (Selbmann et al., 2005), and *Recurvomyces* (*R. mirabilis*) (Selbmann et al., 2008). Based on multigene phylogenetic analyses, Ruibal et al. (2009) reported the phylogenetic placement of rock Antarctic fungi in *Capnodiales* (particularly in the family *Teratosphaeriaceae s.l.*), as well as to some as yet uncharacterized groups with similarities to *Arthoniomycetes*, the sister class of *Dothideomycetes* (Ruibal et al., 2009). Previously undescribed rock isolates have also been found for the families *Davidiellaceae* and *Teratosphaeriaceae* (Ruibal et al., 2009).

Different black yeasts, like the fungal species of the genera *Exophiala*, *Cladophialophora*, and *Phialophora* (*Herpotrichiellaceae*, *Chaetothyriales*,

Chaetothyriomycetidae, Eurotiomycetes, Ascomycota), have been frequently isolated from different extremely cold environments, such as glaciers and substrates rich in organic deposits (ornithogenic soil and soil in the rhizosphere). Different species of the genus *Exophiala*, including the mesophilic human pathogens *E. spinifera* and *E. dermatitidis*, have been found in various Antarctic habitats, with the latter from an Italian glacier (Branda et al., 2010), and *E. jeanselmei* var. *heteromorpha* was found in permanently frozen volcanic deposits in Kamchatka (Ozerskaya et al., 2009). The related black yeast *Cladophialophora minutissima* was isolated from Arctic moss (CBS database), and species of the genus *Phialophora* (*Helotiales, Leotiomycetidae, Leotiomycetes, Ascomycota*)—*P. bubakii, P. danconii, P. fastigiata* and *P. malorum*—were isolated from Antarctic soil. The last two of these species were later reclassified to the genus *Cadophora*.

THE SPECIES
AUREOBASIDIUM PULLULANS
An important black yeastlike species that has been discovered repeatedly in diverse polar environments is the osmotolerant *A. pullulans* (*Dothioraceae, Dothideales, Dothideomycetidae, Dothideomycetes, Ascomycota*). This species has great biotechnological importance, due to its production of the extrapolysaccharide pullulan as well as numerous extracellular hydrolytic enzymes (Chi et al., 2009). Recently, *A. pullulans* has been described to comprise four varieties, of which three have been detected in polar environments. The most well known of these, *A. pullulans* var. *pullulans*, is a cosmopolitan taxon that has been found in a variety of osmotic substrates, such as fruit and plant surfaces and hypersaline waters (Zalar et al., 2008), and also at both poles, in glaciers (Zalar et al., 2008), in Greenland ice sheets (Starmer et al., 2005), in moss (Tosi et al., 2002), on mummified seals and ornithogenic soil (CBS database), and in cryopegs (Gilichinsky et al., 2005). *A. pullulans* var. *melanogenum* has been isolated mainly from oligotrophic watery habitats, such as the deep waters of the Pacific

Ocean, melted glacial water from the Italian Alps (Branda et al., 2010), and melted subglacial water of the Arctic Svalbard glaciers. The sequence data of five molecular markers have revealed an additional variety, named *A. pullulans* var. *subglaciale* (Zalar et al., 2008), which has so far been isolated only from the subglacial ice of Svalbard glaciers and the immediate surroundings. This variety probably represents an example of geographic isolation and initiation of speciation of this panglobal species (Gostinčar et al., 2010), indicating its high genetic variability. It appears likely that more varieties of *A. pullulans* will be discovered and described in polar environments in the future.

THE GENUS *CLADOSPORIUM*
The filamentous melanized genus *Cladosporium* (*Davidiellaceae, Capnodiales, Dothideomycetidae, Dothideomycetes, Ascomycota*) is among the most numerous and cosmopolitian genera of fungi on Earth. *Cladosporium* spp. have been isolated from almost all substrata, which is reflected in the total number of unrevised species names in this genus, which at present exceeds 776 (Dugan et al., 2004). Recent taxonomic studies supported by molecular data have shown that some species can be endemic in cold regions, as exemplified by *C. antarcticum*, which was newly described from an Antarctic lichen (Schubert et al., 2007). Others, such as *C. halotolerans* (found in salterns and Arctic subglacial ice) and *C. langeronii* (found in Arctic ice and Antarctic lake biomat), are examples of true cosmopolites (Zalar et al., 2007). A few isolates with the closest sequence similarities to the genus *Cladosporium* (D'Elia et al., 2009) that have not yet been identified to the species level have also been reported from accretion ice of Lake Vostok (D'Elia et. al, 2009). There have been many reports on *Cladosporium* spp. from polar and other cold environments, like soil and permafrost (Gilichinsky et al., 2005; Kurek et al., 2007; Ozerskaya et al., 2009). The fuller details on other *Cladosporium* species from polar environments are not included here, as many of the identifications have been based only on morphological characteristics

and are therefore probably not correct. Thus, based on methods of identification that need verification, and on our own unpublished observations, we can conclude that many more *Cladosporium* species will be described for polar environments in the future.

Filamentous Fungi

THE GENUS *GEOMYCES*

The genus *Geomyces* (*Incertae sedis*, *Leotiomycetes*, *Ascomycota*) is one of the most frequently encountered genera in cold regions. As the dominant genus in Arctic cryopegs (75% of the entire fungal community), it was represented by two species: *G. pannorum* var. *pannorum* and *G. vinaceus* (Gilichinsky et al., 2005). *G. pannorum* almost completely dominated ancient bluegrass seeds at 25°C, while at lower temperatures it was partially replaced by dark sterile mycelium with sclerotia (Stakhov et al., 2008). *G. pannorum* has also been reported as one of the three most numerously isolated species in High Arctic tundra soil (Kurek et al., 2007), and has been detected by molecular methods. Its widespread occurrence in polar environments is important in the light of the finding that a species closely related to *Geomyces* was identified as the causative agent of bat white-nose syndrome, a devastating disease of the bats that inhabit caves in northeastern United States (Blehert et al., 2009).

THE GENUS *PENICILLIUM*

At present, the anamorphic, ubiquitous, ascomycetous genus *Penicillium* (*Trichocomaceae*, *Eurotiales*, *Eurotiomycetidae*, *Eurotiomycetes*, *Ascomycota*) comprises 225 described species, which have mainly been isolated from food, soil, and air (Pitt et al., 2000). The entire genus shows tolerance to cold environments, as demonstrated by many species that can contaminate refrigerated food (Pitt and Hocking, 1999) and that inhabit high-mountain soils (Domsch et al., 1980). Many penicillia are xerotolerant (Gunde-Cimerman et al., 2005), and therefore they have been among the most frequently isolated filamentous fungi from

very dry or saline polar habitats, such as Arctic and Antarctic soils, permafrost, snow, sea ice, sea water (Vishniac, 1993; McRae et al., 1999; Gunde-Cimerman et al., 2003; Ivanushkina et al., 2005; Frisvad et al., 2006), and glacial ice cores up to 38,600 years old (Abyzov, 1993; Ma et al., 1999, 2000). An almost "tropical" high diversity of *Penicillium* spp. was isolated in our study of the coastal Arctic environment of the Svalbard archipelago (Gunde-Cimerman et al., 2003). Penicillia were the most frequently occurring filamentous fungi in all of our samples, including seawater, sea ice, snow/coastal ice in tidal zones, puddles on snow, subglacial ice, and glacial meltwater. All together, 32 *Penicillium* species were recorded. Most of these (78%) originated from different glacial ice samples. The following have been isolated in low to relatively large numbers: *P. bialowiezense*, *P. brasilianum*, *P. brevicompactum*, *P. chrysogenum*, *P. commune*, *P. corylophilum*, *P. decumbens*, *P. discolor*, *P. echinulatum*, *P. expansum*, *P. glabrum*, *P. gladioli*, *P. groenlandense*, *P. lanosum*, *P. nalgiovense*, *P. olsonii*, *P. palitans*, *P. persicinum*, *P. polonicum*, *P. resedanum*, the new species *P. svalbardense* (Sonjak et al., 2005, 2006, 2007a, 2007b), *P. solitum*, *P. roquefortii*, *P. thomii*, and *P. tulipae*. Greater numbers of isolates were seen for *P. bialowiezense*, *P. chrysogenum*, *P. commune*, *P. crustosum*, *P. expansum*, *P. nordicum*, *P. italicum*, *P. solitum*, and *P. westlingii*. In particular, *P. crustosum* dominated the subglacial ice (up to 2,500 CFU liter^{-1}). Of all of these species, only *P. brevicompactum*, *P. chrysogenum*, *P. crustosum*, *P. decumbens*, and *P. glabrum* have been isolated from the Arctic permafrost (Ozerskaya et al., 2009), and *P. chrysogenum*, *P. expansum*, and *P. lanosum* from the Arctic tundra soil (Frisvad et al., 2006; Kurek et al., 2007). Additional isolated penicillia from the Arctic permafrost and cryopegs within the permafrost sediments have included *P. aurantiogriseum*, *P. citrinum*, *P. granulatum*, *P. griseofulvum*, *P. melinii*, *P. miczynskii*, *P. minioluteum*, *P. puberulum*, *P. purpurogenum*, *P. restrictum*, *P. rugulosum*, *P. simplicissimum*, *P. variabile*, *P. verrucosum*, and *P. viridicatum* (Gilichinsky et al., 2005; Ozerskaya et al., 2008, 2009; Stakhov et al., 2008; Ivanushkina et al., 2005). Most of the mentioned *Penicillium* species that have been isolated from Arctic polar environments have also been reported from the Antarctic Peninsula and continental habitats, like soil, soil with high biotic influence, moss, lichens, algae, ice sheets, and ice cores. Additional species like *P. antarcticum*, *P. fellutanum*, *P. hirsutum*, *P. janthinellum*, *P. jensenii*, *P. ochrochloron*, *P. paxilli*, *P. spinulosum*, *P. waksmanii* (Abyzov, 1993; Azmi and Seppelt, 1998; McRae et al., 1999; Tosi et al., 2002; Loque et al., 2010), and others are listed in a database of nonlichenized fungi from Antarctica (http://www.antarctica.ac.uk/bas_research/data/access/fungi/Speciespublic2.html#Use). *P. adametzii*, *P. barcinonense*, *P. camemberti*, *P. canescens*, *P. charlesii*, *P. citreonigrum*, *P. citreoviride*, *P. clavigerum*, *P. crustaceum*, *P. dierckxii*, *P. diversum*, *P. donkii*, *P. duclauxii*, *P. frequentans*, *P. funiculosum*, *P. glandicola*, *P. hordei*, *P. lilacinum*, *P. marneffei*, *P. meleagrinum*, *P. montanense*, *P. notatum*, *P. purpurascens*, *P. roseopurpureum*, *P. soppii*, *P. steckii*, *P. strictum*, *P. viridicatum*, and *P. velutinum* have also been isolated from these habitats.

The extremely high diversity of the *Penicillium* species that have been isolated from polar environments has resulted in unknown species that remain to be described. Some of the recently described species include *P. antarcticum* from the Antarctic, which was later also found in other countries (McRae et al., 1999); *P. svalbardense* from Arctic glacial ice (Sonjak et al., 2007b); and *P. jamesonlandense* from Greenland soil (Frisvad et al., 2006).

In most cases, there are no obvious relationships among the *Penicillium* populations and their polar habitats, except that greater numbers and diversity have been isolated from anthropogenically influenced sites, and quite importantly, many of the isolated glacial species have been described among the very common food-borne penicillia (Sonjak et al., 2006; Frisvad, 2008). Only a few studies have been published that did not report on the isolation of penicillia. These organisms were not recovered from Arctic soils of the Arctic Franz Joseph Land (Bergero et al., 1999)

or from some Antarctic soil samples (Tosi et al., 2005). Additionally, their DNA was not among other fungal DNA that has been amplified from ancient permafrost samples (300,000 to 400,000 years old), suggesting that penicillia are modern-day contaminants (Lydolph et al., 2005). These studies further indicate that it is often difficult to separate the indigenous from the nonindigenous penicillia. However, extremely high numbers of certain *Penicillium* species have been isolated from Arctic glacial ice (Sonjak et al., 2006), with the isolation of new species (Sonjak et al., 2007b) and of genotypically and phenotypically different strains of certain species from a single glacier (Sonjak et al., 2005, 2006, 2007a). This thus indicates that glacial *Penicillium* species have become selectively enriched in these cold, oligotrophic environments with shifting osmotic pressures, and can as such be identified as indigenous.

THE GENUS *THELEBOLUS*

Species of the genus *Thelebolus* (*Thelebolaceae*, *Thelebolales*, *Leotiomycetidae*, *Leotiomycetes*, *Ascomycota*) are among the rare representatives of polar environments that have been assigned as psychrophilic (Wicklow and Malloch, 1971), as they develop their teleomorphic stage only at temperatures lower than 15°C. They have been retrieved mainly from Antarctic guano and the dung of animals, as well as from lakes, and occasionally they have been found also in temperate and tropical regions. The species *T. microsporus*, which occurs globally in boreal climate zones, has evolved into two endemic genotypes in the Antarctic. These have been described as the novel species *T. ellipsoideus* and *T. globosus* (de Hoog et al., 2005). The segregation of these new species has been explained by the loss of a bird vector and the subsequent life trapped under perennial ice, which resulted in their strongly reduced morphology and the degeneration of their asci. As they became unable to forcibly discharge spores, waterborne conidia were produced instead. These dramatic changes developed over a relatively short time frame, as indicated by multilocus sequence analyses (de Hoog et al., 2005).

CONCLUSIONS

Extremely cold polar environments are inhabited by extremophilic microorganisms that even today represent a largely unknown biodiversity. Classical culture-based methods, molecular techniques, and more recently also metagenomic analyses (Simon et al., 2009; Kennedy et al., 2010) have revealed the existence of new species, genera, and even phylogenetic lineages. Many of these microorganisms were previously considered mainly as entrapped living fossils, providing us with insight into the biological past. We now know that such microorganisms can live as metabolically active communities in sea ice, microbial mats on sea ice shelves, glacial inclusions, brine networks, permafrost, ephemeral cryoconite holes, and subglacial environments, at temperatures down to −20°C (Rivkina et al., 2000; Deming, 2002; Price and Sowers, 2004). These contribute to the local geobiochemistry, and when released they seed the immediate environment, whether it is air, soil, or water, with psychrotolerant propagules. The air and water currents can transport them globally to temperate zones, where they can find refuge either in refrigerated food or in other cold man-made environments.

The importance of psychrophilic and psychrotolerant microorganisms has increased due to the recent accelerated disappearance of the polar environments and due to the mounting evidence that the biosphere has experienced several extremely low-temperature periods, perhaps even during the earliest stages in the emergence and evolution of life. The earth has been covered in ice four times through the Snowball Earth periods. These episodes ranged from 3 million to 30 million years each (Bodiselitsch et al., 2005). The entire world ocean was frozen near the time of origin of the eukaryotic cell and the adaptive radiation of metazoa. The extensive polar microbial mat communities on the ice might provide insight into protected microhabitats that have allowed the survival, growth, and evolution of less tolerant biota, including multicellular eukaryotes, during periods of extensive glaciation. It has been hypothesized that the endemic descendants of primitive cold-adapted microorganisms can live today in rock, lakes, sediments, and subglacial ice. Since "microbial endemism" is currently a much debated issue in microbial ecology, the polar regions and the other extreme environments should be among the first places to examine the evolutionary processes that can give rise to microbial speciation (Vincent, 2000; Gostinčar et al., 2010).

The exciting field of polar microbiology is still today primarily dedicated to the study of prokaryotic microorganisms, due to the general belief that eukaryotes are not able to inhabit extremely stressed environments. This conviction was challenged when fungi were isolated from hypersaline waters (Gunde-Cimerman et al., 2000, 2005), from extremely acidic mine waters (Hölker et al., 2004), at 10 km below the surface of the oceans (Nagano et al., 2010), and from the surface of rocks in arid and cold climates (Sterflinger and Krumbein, 1997; Sterflinger et al., 1999; Gorbushina and Broughton, 2009). Although the fungi have reaffirmed themselves as one of the ecologically most successful eukaryotic lineages, the extent of fungal diversity in many polar habitats remains mostly unknown.

A limited number of species of fungi have a general xerophylic phenotype, which is determined primarily by the low water activity potential rather than by the type of solute dissolved. Since extracellular freezing and osmotic/xeric stress lead to cell dehydration, they can activate some of the common responses at the level of compatible solutes, ion transport across membranes, regulation of water efflux, composition and fluidity of cell membranes, nutrient uptake, electron transport, and environmental sensing. Xerotolerant and xerophilic fungi therefore have a physiological advantage in ice-bound polar environments. Studies of fungal polar biodiversity have indeed revealed the prevalence and diversity of pigmented, xerotolerant fungi, which represent a new world of eukaryotic extremophiles. These fungi have developed little-investigated adaptive strategies that are crucial for their successful survival in some of the harshest and most extreme environments on our planet.

REFERENCES

Abyzov, S. S. 1993. Microorganisms in the Antarctic ice, p. 265–295. In E. I. Friedmann (ed.), *Antarctic Microbiology*. Wiley-Liss, New York, NY.

Abyzov, S. S., R. B. Hoover, S. Imura, I. N. Mitskevich, T. Naganuma, M. N. Poglazova, and M. V. Ivanov. 2004. Use of different methods for discovery of ice-entrapped microorganisms in ancient layers of the Antarctic glacier. *Adv. Space Res.* **33**:1222–1230.

Azmi, O. R., and R. D. Seppelt. 1998. The broad-scale distribution of microfungi in the Windmill Islands region, continental Antarctica. *Polar Biol.* **19**:92–100.

Babjeva, I., and I. Reshetova. 1998. Yeast resources in natural habitats at polar circle latitude. *Food Technol. Biotechnol.* **36**:1–5.

Baublis, J. A., R. A. Wharton, and P. A. Volz. 1991. Diversity of micro-fungi in an Antarctic dry valley. *J. Basic Microbiol.* **31**:1–12.

Belzile, C., W. F. Vincent, J. A. Gibson, and P. V. Hove. 2001. Bio-optical characteristics of the snow, ice, and water column of a perennially ice-covered lake in the High Arctic. *Can. J. Fish. Aquat. Sci.* **58**:2405–2418.

Bergero, R., M. Girlanda, G. C. Varese, D. Intili, and A. M. Luppi. 1999. Psychrooligotrophic fungi from Arctic soils of Franz Joseph Land. *Polar Biol.* **21**:361–368.

Blehert, D. S., A. C. Hicks, M. Behr, C. U. Meteyer, B. M. Berlowski-Zier, E. L. Buckles, J. T. H. Coleman, S. R. Darling, A. Gargas, R. Niver, J. C. Okoniewski, R. J. Rudd, and W. B. Stone. 2009. Bat white-nose syndrome: an emerging fungal pathogen? *Science* **323**:227.

Bodiselitsch, B., C. Koeberl, S. Master, and W. U. Reimold. 2005. Estimating duration and intensity of Neoproterozoic snowball glaciations from Ir anomalies. *Science* **308**:239–242.

Branda, E., B. Turchetti, G. Diolaiuti, M. Pecci, C. Smiraglia, and P. Buzzini. 2010. Yeast and yeast-like diversity in the southernmost glacier of Europe (Calderone Glacier, Apennines, Italy). *FEMS Microbiol. Ecol.* **72**:354–369.

Brown, M. V., and J. P. Bowman. 2001. A molecular phylogenetic survey of sea-ice microbial communities (SIMCO). *FEMS Microbiol. Ecol.* **35**:267–275.

Burford, E. P., M. Fomina, and G. M. Gadd. 2003. Fungal involvement in bioweathering and biotransformation of rocks and minerals. *Mineral. Mag.* **67**:1127–1155.

Butinar, L., I. Spencer-Martins, and N. Gunde-Cimerman. 2007. Yeasts in high Arctic glaciers: the discovery of a new habitat for eukaryotic microorganisms. *Antonie van Leeuwenhoek* **91**:277–289.

Butinar, L., T. Strmole, and N. Gunde-Cimerman. 2011. Relative incidence of ascomycetous yeasts in Arctic coastal environments. *Microbiol Ecol.* **61**:832–843.

Cavicchioli, R., K. S. Siddiqui, D. Andrews, and K. R. Sowers. 2002. Low-temperature extremophiles and their applications. *Curr. Opin. Biotechnol.* **13**:253–261.

Chi, Z. M., F. Wang, Z. Chi, L. X. Yue, G. L. Liu, and T. Zhang. 2009. Bioproducts from *Aureobasidium pullulans*, a biotechnologically important yeast. *Appl. Microbiol. Biotechnol.* **82**:793–804.

Christner, B. C., E. Mosley-Thompson, L. G. Thompson, V. Zagorodnov, K. Sandman, and J. N. Reeve. 2000. Recovery and identification of viable bacteria immured in glacial ice. *Icarus* **144**:479–485.

de García, V., S. Brizzio, D. Libkind, P. Buzzini, and M. van Broock. 2007. Biodiversity of cold-adapted yeasts from glacial meltwater rivers in Patagonia, Argentina. *FEMS Microbiol. Ecol.* **59**:331–341.

de García, V., S. Brizzio, D. Libkind, C. A. Rosa, and M. van Broock. 2010a. *Wickerhamomyces patagonicus* sp. nov., an ascomycetous yeast species from Patagonia, Argentina. *Int. J. Syst. Evol. Microbiol.* **60**:1693–1696.

de García, V., S. Brizzio, G. Russo, C. A. Rosa, T. Boekhout, B. Theelen, D. Libkind, and M. van Broock. 2010b. *Cryptococcus spencermartinsiae* sp. nov., a basidiomycetous yeast isolated from glacial waters and apple fruits. *Int. J. Syst. Evol. Microbiol.* **60**:707–711.

de Hoog, G. S., E. Göttlich, G. Platas, O. Genilloud, G. Leotta, and J. van Brummelen. 2005. Evolution, taxonomy and ecology of the genus *Thelebolus* in Antarctica. *Stud. Mycol.* **51**:33–73.

de la Torre, J. R., B. M. Goebel, E. I. Friedmann, and N. R. Pace. 2003. Microbial diversity of cryptoendolithic communities from the McMurdo Dry Valleys, Antarctica. *Appl. Environ. Microbiol.* **69**:3858–3867.

D'Elia, T., R. Veerapaneni, V. Theraisnathan, and S. O. Rogers. 2009. Isolation of fungi from Lake Vostok accretion ice. *Mycologia* **101**:751–763.

de los Ríos, A., J. Wierzchos, L. G. Sancho, and C. Ascaso. 2003. Acid microenvironments in microbial biofilms of Antarctic endolithic microecosystems. *Environ. Microbiol.* **5**:231–237.

de los Ríos, A., L. G. Sancho, M. Grube, J. Wierzchos, and C. Ascaso. 2005. Endolithic growth of two lecidea lichens in granite from Continental Antarctica detected by molecular and microscopy techniques. *New Phytol.* **165:**181–189.

Deming, J. W. 2002. Psychrophiles and polar regions. *Curr. Opin. Microbiol.* **5:**301–309.

De Wit, R., P. Dyer, O. Genilloud, E. Goetlich, D. Hodgson, G. S. de Hoog, B. Jones, J. Laybourn-Parry, F. Marinelli, E. Stackebrandt, J. Swings, B. J. Tindall, W. Vyverman, and A. Wilmotte. 2003. Antarctic lakes—'hot spots' for microbial diversity and biotechnological screening, p. 228. *In* T. Avšič-Županc, A. van Belkum, C. Bruschi, I. Chet, J. Cole, D. Farr, W. Holzapfel, R. J. Koerner, A. Netrusov, J.-C. Piffaretti, E. Z. Ron, R. Rosselló-Mora, B. Schink, S. Spiro, B. J. Tindall, and H. G. Trüper (ed.), *1st FEMS Congress of European Microbiologists, Slovenia, Ljubljana, June 29–July 3*. Federation of European Microbiological Societies, Delft, The Netherlands.

Dmitriev, V. V., D. A. Gilichinsky, R. N. Faizutdinova, N. V. Ostroumova, W. I. A. Golubev, and V. I. Duda. 1997a. Yeasts in late Pleistocene–early Pleistocene Siberian permafrost. *Cryosphera Zemli* **1:**67–70. (In Russian.)

Dmitriev, V. V., D. A. Gilichinsky, R. N. Faizutdinova, I. N. Shershunov, W. I. A. Golubev, and V. I. Duda. 1997b. Occurrence of viable yeasts in 3-million-year-old permafrost in Siberia. *Mikrobiologiya* **66:**655–660. (In Russian.)

Domsch, K. H., W. Gams, and T. H. Anderson. 1980. *Compendium of Soil Fungi*. Academic Press, London, United Kingdom.

Dugan, F., K. Schubert, and U. Braun. 2004. Checklist of *Cladosporium* names. *Schlechtendalia* **11:**1–103.

Ekstrom, G., M. Nettles, and G. A. Abers. 2003. Glacial earthquakes. *Science* **302:**622–624.

Ellis-Evans, J. C. 1985. Fungi from maritime Antarctic freshwater environments. *Br. Antarct. Surv. Bull.* **68:**37–45.

Fahnestock, M. 2003. Geophysics: glacial flow goes seismic. *Science* **302:**578–579.

Faizutdinova, R. N., N. E. Suzina, V. I. Duda, L. E. Petrovskaya, and D. A. Gilichinsky. 2005. Yeasts isolated from ancient permafrost, p. 118–126. *In* J. D. Castello and S. O. Rogers (ed.), *Life in Ancient Ice*. Princeton University Press, Princeton, NJ.

Fell, J. W., G. Scorzetti, L. Connell, and S. Craig. 2006. Biodiversity of micro-eukaryotes in Antarctic Dry Valley soils with <5% soil moisture. *Soil Biol. Biochem.* **38:**3107–3119.

Foght, J., J. Aislabie, S. Turner, C. E. Brown, J. Ryburn, D. J. Saul, and W. Lawson. 2004. Culturable bacteria in subglacial sediments and ice

from two Southern Hemisphere glaciers. *Microb. Ecol.* **47:**329–340.

Freeman, K. R., A. P. Martin, D. Karki, R. C. Lynch, M. S. Mitter, A. F. Meyer, J. E. Longcore, D. R. Simmons, and S. K. Schmidt. 2009. Evidence that chytrids dominate fungal communities in high-elevation soils. *Proc. Natl. Acad. Sci. USA* **106:**18315–18320.

Friedmann, E. I. 1982. Endolithic microorganisms in the Antarctic cold desert. *Science* **215:**1045–1053.

Friedmann, E. I., and A. M. Koriem. 1989. Life on Mars: how it disappeared (if it was ever there). *Adv. Space Res.* **9:**167–172.

Frisvad, J. C. 2008. Fungi in cold ecosystems, p. 137–156. *In* R. Margesin, F. Schinner, J.-C. Marx, and C. Gerday (ed.), *Psychrophiles: from Biodiversity to Biotechnology*. Springer, Berlin, Germany.

Frisvad, J. C., T. O. Larsen, P. W. Dalsgaard, K. A. Seifert, G. Louis-Seize, E. K. Lyhne, B. B. Jarvis, J. C. Fettinger, and D. P. Overy. 2006. Four psychrotolerant species with high chemical diversity consistently producing cycloaspeptide A, *Penicillium jamesonlandense* sp nov., *Penicillium ribium* sp nov., *Penicillium soppii* and *Penicillium lanosum*. *Int. J. Syst. Evol. Microbiol.* **56:**1427–1437.

Gadd, G. M. 2007. Geomycology: biogeochemical transformations of rocks, minerals, metals and radionuclides by fungi, bioweathering and bioremediation. *Mycol. Res.* **111:**3–49.

Gaidos, E., B. Lanoil, T. Thorsteinsson, A. Graham, M. Skidmore, S. K. Han, T. Rust, and B. Popp. 2004. A viable microbial community in a subglacial volcanic crater lake, Iceland. *Astrobiology* **4:**327–344.

Gilichinsky, D., E. Rivkina, C. Bakermans, V. Shcherbakova, L. Petrovskaya, S. Ozerskaya, N. Ivanushkina, G. Kochkina, K. Laurinavichuis, S. Pecheritsina, R. Fattakhova, and J. M. Tiedje. 2005. Biodiversity of cryopegs in permafrost. *FEMS Microbiol. Ecol.* **53:**117–128.

Gilichinsky, D. A., G. S. Wilson, E. I. Friedmann, C. P. McKay, R. S. Sletten, E. M. Rivkina, T. A. Vishnivetskaya, L. G. Erokhina, N. E. Ivanushkina, G. A. Kochkina, V. A. Shcherbakova, V. S. Soina, E. V. Spirina, E. A. Vorobyova, D. G. Fyodorov-Davydov, B. Hallet, S. M. Ozerskaya, V. A. Sorokovikov, K. S. Laurinavichyus, A. V. Shatilovich, J. P. Chanton, V. E. Ostroumov, and J. M. Tiedje. 2007. Microbial populations in Antarctic permafrost: biodiversity, state, age, and implication for astrobiology. *Astrobiology* **7:**275–311.

Golubev, W. I. 1998. New species of basidiomycetous yeasts, *Rhodotorula creatinovora* and *R. yakutica*, isolated from permafrost soils of Eastern-Siberian Arctic. *Mykol. Phytopathol.* **32:**8–13. (In Russian.)

Golubic, S., E. I. Friedmann, and J. Schneider. 1981. The lithobiontic ecological niche, with special reference to microorganisms. *J. Sediment. Res.* **51:**475–478.

Gorbushina, A. A., and W. J. Broughton. 2009. Microbiology of the atmosphere-rock interface: how biological interactions and physical stresses modulate a sophisticated microbial ecosystem. *Annu. Rev. Microbiol.* **63:**431–450.

Gosink, J. J., R. L. Irgens, and J. T. Staley. 1993. Vertical distribution of bacteria in Arctic sea ice. *FEMS Microbiol. Lett.* **102:**85–90.

Gostinčar, C., M. Grube, S. de Hoog, P. Zalar, and N. Gunde-Cimerman. 2010. Extremotolerance in fungi: evolution on the edge. *FEMS Microbiol. Ecol.* **71:**2–11.

Grabińska-Łoniewska, A., T. Koniłłowicz-Kowalska, G. Wardzyńska, and K. Boryn. 2007. Occurrence of fungi in water distribution system. *Pol. J. Environ. Stud.* **16:**539–547.

Gunde-Cimerman, N., L. Butinar, S. Sonjak, M. Turk, V. Uršič, P. Zalar, and A. Plemenitaš. 2005. Halotolerant and halophilic fungi from coastal environments in the Arctics, p. 397–423. *In* N. Gunde-Cimerman, A. Orenand, and A. Plemenitaš (ed.), *Adaptation to Life at High Salt Concentrations in Archaea, Bacteria, and Eukarya.* Springer, Dordrecht, The Netherlands.

Gunde-Cimerman, N., S. Sonjak, P. Zalar, J. C. Frisvad, B. Diderichsen, and A. Plemenitaš. 2003. Extremophilic fungi in arctic ice: a relationship between adaptation to low temperature and water activity. *Phys. Chem. Earth B* **28:**1273–1278.

Gunde-Cimerman, N., P. Zalar, S. de Hoog, and A. Plemenitaš. 2000. Hypersaline waters in salterns—natural ecological niches for halophilic black yeasts. *FEMS Microbiol. Ecol.* **32:**235–240.

Hölker, U., J. Bend, R. Pracht, L. Tetsch, T. Müller, M. Höfer, and G. S. de Hoog. 2004. *Hortaea acidophila,* a new acid-tolerant black yeast from lignite. *Antonie van Leeuwenhoek* **86:**287–294.

Hughes, K. A., and B. Lawley. 2003. A novel Antarctic microbial endolithic community within gypsum crusts. *Environ. Microbiol.* **5:**555–565.

Ivanushkina, N. E., G. A. Kochkina, and S. M. Ozerskaya. 2005. Fungi in ancient permafrost sediments of the Arctic and Antarctic regions, p. 127–139. *In* J. D. Castello and S. O. Rogers (ed.), *Life in Ancient Ice.* Princeton University Press, Princeton, NJ.

Jones, G. E. B. 1976. *Recent Advances in Aquatic Mycology.* The Gresham Press, Old Woking, Surrey, United Kingdom.

Junge, K., F. Imhoff, T. Staley, and W. Deming. 2002. Phylogenetic diversity of numerically important Arctic sea-ice bacteria cultured at subzero temperature. *Microb. Ecol.* **43:**315–328.

Kennedy, J., B. Flemer, S. A. Jackson, D. P. H. Lejon, J. P. Morrissey, F. O'Gara, and A. D. W. Dobson. 2010. Marine metagenomics: new tools for the study and exploitation of marine microbial metabolism. *Mar. Drugs* **8:**608–628.

Kriss, A. E., I. N. Mitskevich, E. P. Rozanova, and L. K. Osnitskaia. 1976. Microbiological studies of the Wanda Lake (Antarctica). *Mikrobiologiya* **45:**1075–1081. (In Russian.)

Kurek, E., T. Korniłowicz-Kowalska, A. Słomka, and A. J. Melke. 2007. Characteristics of soil filamentous fungi communities isolated from various micro-relief forms in the high Arctic tundra (Bellsund region, Spitsbergen). *Pol. Polar Res.* **28:**57–73.

Libkind, D., S. Brizzio, A. Ruffini, M. Gadanho, M. van Broock, and J. P. Sampaio. 2003. Molecular characterization of carotenogenic yeasts from aquatic environments in Patagonia, Argentina. *Antonie van Leeuwenhoek* **84:**313–322.

Libkind, D., M. Gadanho, M. van Broock, and J. P. Sampaio. 2005. *Sporidiobolus longiusculus* sp. nov. and *Sporobolomyces patagonicus* sp. nov., novel yeasts of the Sporidiobolales isolated from aquatic environments in Patagonia, Argentina. *Int. J. Syst. Evol. Microbiol.* **55:**503–509.

Libkind, D., M. Moline, J. P. Sampaio, and M. van Broock. 2009. Yeasts from high-altitude lakes: influence of UV radiation. *FEMS Microbiol. Ecol.* **69:**353–362.

López-García, P., F. Rodríguez-Valera, C. Pedrós-Alió, and D. Moreira. 2001. Unexpected diversity of small eukaryotes in deep-sea Antarctic plankton. *Nature* **409:**603–607.

Loque, C. P., A. O. Medeiros, F. M. Pellizzari, E. C. Oliveira, C. A. Rosa, and L. H. Rosa. 2010. Fungal community associated with marine macroalgae from Antarctica. *Polar Biol.* **33:**641–648.

Ludley, K. E., and C. H. Robinson. 2008. "Decomposer" Basidiomycota in Arctic and Antarctic ecosystems. *Soil Biol. Biochem.* **40:**11–29.

Lydolph, M. C., J. Jacobsen, P. Arctander, M. T. Gilbert, D. A. Gilichinsky, A. J. Hansen, E. Willerslev, and L. Lange. 2005. Beringian paleoecology inferred from permafrost-preserved fungal DNA. *Appl. Environ. Microbiol.* **71:**1012–1017.

Ma, L. J., C. M. Catranis, W. T. Starmer, and S. O. Rogers. 1999. Revival and characterization of fungi from ancient polar ice. *Mycologist* **13:**70–73.

Ma, L. J., S. O. Rogers, C. M. Catranis, and W. T. Starmer. 2000. Detection and characterization of ancient fungi entrapped in glacial ice. *Mycologia* **92:**286–295.

Margesin, R. 2009. Effect of temperature on growth parameters of psychrophilic bacteria and yeasts. *Extremophiles* **13:**257–262.

Margesin, R., and J. W. Fell. 2008. *Mrakiella cryoconiti* gen. nov., sp. nov., a psychrophilic, anamorphic, basidiomycetous yeast from alpine and arctic habitats. *Int. J. Syst. Evol. Microbiol.* **58:**2977–2982.

Margesin, R., P. A. Fonteyne, F. Schinner, and J. P. Sampaio. 2007. *Rhodotorula psychrophila* sp. nov., *Rhodotorula psychrophenolica* sp. nov. and *Rhodotorula glacialis* sp. nov., novel psychrophilic basidiomycetous yeast species isolated from alpine environments. *Int. J. Syst. Evol. Microbiol.* **57:**2179–2184.

Margesin, R., G. Zacke, and F. Schinner. 2002. Characterization of heterotrophic microorganisms in alpine glacier cryoconite. *Arct. Antarct. Alp. Res.* **34:**88–93.

McLoughlin, N., M. D. Brasier, D. Wacey, O. R. Green, and R. S. Perry. 2007. On biogenicity criteria for endolithic microborings on early Earth and beyond. *Astrobiology* **7:**10–26.

McRae, C. F., A. D. Hocking, and R. D. Seppelt. 1999. *Penicillium* species from terrestrial habitats in the Windmill Islands, East Antarctica, including a new species, *Penicillium antarcticum. Polar Biol.* **21:**97–111.

Montes, M. J., C. Belloch, M. Galiana, M. D. Garcia, C. Andrés, S. Ferrer, J. M. Torres-Rodriguez, and J. Guinea. 1999. Polyphasic taxonomy of a novel yeast isolated from Antarctic environment; description of *Cryptococcus victoriae* sp. nov. *Syst. Appl. Microbiol.* **22:**97–105.

Morozova, D., D. Möhlmann, and D. Wagner. 2007. Survival of methanogenic archaea from Siberian permafrost under simulated Martian thermal conditions. *Orig. Life Evol. Biosph.* **37:**189–200.

Mueller, D. R., W. F. Vincent, W. H. Pollard, and C. H. Fritsen. 2001. Glacial cryoconite ecosystems: a bipolar comparison of algal communities and habitats. *Nova Hedwig. Beih.* **123:**173–197.

Nagano, Y., T. Nagahama, Y. Hatada, T. Nunoura, H. Takami, J. Miyazaki, K. Takai, and K. Horikoshi. 2010. Fungal diversity in deep-sea sediments—the presence of novel fungal groups. *Fungal Ecol.* **3:**316–325.

Nienow, J. A., and E. I. Friedmann. 1993. Terrestrial lithophytic (rock) communities, p. 343–412. *In* E. I. Friedmann (ed.), *Antarctic Microbiology.* Wiley-Liss, New York, NY.

Omelon, C. R., W. H. Pollard, and F. G. Ferris. 2006. Environmental controls on microbial colonization of high Arctic cryptoendolithic habitats. *Polar Biol.* **30:**19–29.

Omelon, C. R., W. H. Pollard, and F. G. Ferris. 2007. Inorganic species distribution and microbial diversity within high Arctic cryptoendolithic habitats. *Microb. Ecol.* **54:**740–752.

Onofri, S., D. Barreca, L. Selbmann, D. Isola, E. Rabbow, G. Horneck, J. P. P. de Vera, J. Hatton, and L. Zucconi. 2008. Resistance of Antarctic black fungi and cryptoendolithic communities to simulated space and Martian conditions. *Stud. Mycol.* **61:**99–109.

Onofri, S., M. Fenice, A. R. Cicalini, S. Tosi, A. Magrino, S. Pagano, L. Selbmann, L. Zucconi, H. S. Vishniac, R. Ocampo-Friedmann, and E. I. Friedmann. 2000. Ecology and biology of microfungi from Antarctic rocks and soils. *Ital. J. Zool.* **67:**163–167.

Onofri, S., S. Pagano, L. Zucconi, and S. Tosi. 1999. *Friedmanniomyces endolithicus* (Fungi, Hyphomycetes), anam.-gen. and sp. nov., from continental Antarctica. *Nova Hedwig. Beih.* **68:**175–181.

Onofri, S., L. Selbmann, L. Zucconi, and S. Pagano. 2004. Antarctic microfungi as models for exobiology. *Planet. Space Sci.* **52:**229–237.

Ozerskaya, S., G. Kochkina, N. Ivanushkina, and D. A. Gilichinsky. 2009. Fungi in permafrost, p. 85–95. *In* R. Margesin (ed.), *Permafrost Soils.* Springer, Berlin, Germany.

Ozerskaya, S. M., G. A. Kochkina, N. E. Ivanushkina, E. V. Knyazeva, and D. A. Gilichinskii. 2008. The structure of micromycete complexes in permafrost and cryopegs of the Arctic. *Microbiology* **77:**482–489.

Panikov, N. S. 2009. Microbial activity in frozen soils, p. 119–147. *In* R. Margesin (ed.), *Permafrost Soils.* Springer, Berlin, Germany.

Pathan, A., B. Bhadra, Z. Begum, and S. Shivaji. 2010. Diversity of yeasts from puddles in the vicinity of Midre Lovénbreen Glacier, Arctic and bioprospecting for enzymes and fatty acids. *Curr. Microbiol.* **60:**307–314.

Pennisi, E. 2003. Neither cold nor snow stops tundra fungi. *Science* **301:**1307.

Pitt, J. I., and A. D. Hocking. 1999. *Fungi and Food Spoilage.* Aspen Publishers, Gaithersburg, Maryland.

Pitt, J. I., R. A. Samson, and J. C. Frisvad. 2000. List of accepted species and their synonyms in the family Trichocomaceae, p. 9–47. *In* R. A. Samson and J. I. Pitt (ed.), *Integration of Modern Taxonomic Methods for* Penicillium *and* Aspergillus *Classification.* Harwood Academic Publishers, Amsterdam, The Netherlands.

Poglazova, M. N., I. N. Mitskevich, S. S. Abyzov, and M. V. Ivanov. 2001. Microbiological characterization of the accreted ice of subglacial Lake Vostok, Antarctica. *Microbiology* **70:**723–730.

Price, P. B. 2000. A habitat for psychrophiles in deep Antarctic ice. *Proc. Natl. Acad. Sci. USA* **97:**1247–1251.

Price, P. B., and T. Sowers. 2004. Temperature dependence of metabolic rates for microbial growth, maintenance, and survival. *Proc. Natl. Acad. Sci. USA* **101:**4631–4636.

Priscu, J. C., C. H. Fritsen, E. E. Adams, S. J. Giovannoni, H. W. Paerl, C. P. McKay,

P. T. Doran, D. A. Gordon, B. D. Lanoil, and J. L. Pinckney. 1998. Perennial Antarctic lake ice: an oasis for life in a polar desert. *Science* 280:2095–2098.

Reeve, J. N., B. C. Christner, B. H. Kvitko, E. Mosley-Thompson, and L. G. Thompson. 2002. Life in glacial ice, p. 27. *In* M. Rossi, S. Bartolucci, M. Ciaramellaand, and M. Moracci (ed.), *Extremophiles 2002, The Fourth International Congress on Extremophiles, Naples, Italy.* Institute of Protein Biochemistry and University of Naples "Federico II," Naples, Italy.

Rivkina, E. M., E. I. Friedmann, C. P. McKay, and D. A. Gilichinsky. 2000. Metabolic activity of permafrost bacteria below the freezing point. *Appl. Environ. Microbiol.* 66:3230–3233.

Rivkina, E., K. Laurinavichius, J. McGrath, J. Tiedje, V. Shcherbakova, and D. Gilichinsky. 2004. Microbial life in permafrost. *Adv. Space Res.* 33:1215–1221.

Robinson, C. H. 2001. Cold adaptation in Arctic and Antarctic fungi. *New Phytol.* 151:341–353.

Rohde, R. A., and P. B. Price. 2007. Diffusion-controlled metabolism for long-term survival of single isolated microorganisms trapped within ice crystals. *Proc. Natl. Acad. Sci. USA* 104:16592–16597.

Ruibal, C., C. Gueidan, L. Selbmann, A. A. Gorbushina, P. W. Crous, J. Z. Groenewald, L. Muggia, M. Grube, D. Isola, C. L. Schoch, J. T. Staley, F. Lutzoni, and G. S. de Hoog. 2009. Phylogeny of rock-inhabiting fungi related to *Dothideomycetes. Stud. Mycol.* 64:123–133.

Ruisi, S., D. Barreca, L. Selbmann, L. Zucconi, and S. Onofri. 2007. Fungi in Antarctica. *Rev. Environ. Sci. Biotechnol.* 6:127–141.

Russell, N. J., P. Harrisson, I. A. Johnston, R. Jaenicke, M. Zuber, F. Franks, and D. Wynn-Williams. 1990. Cold adaptation of microorganisms. *Philos. Trans. R. Soc. London B* 326:595–611.

Samson, R. A., E. S. Hoekstra, J. C. Frisvad, and O. Filtenborg. 2002. *Introduction to Food- and Airborne Fungi.* Centralbureau voor Schimmelcultures, Utrecht, The Netherlands.

Schadt, C. W., A. P. Martin, D. A. Lipson, and S. K. Schmidt. 2003. Seasonal dynamics of previously unknown fungal lineages in tundra soils. *Science* 301:1359–1361.

Schmidt, N., and M. Bölter. 2002. Fungal and bacterial biomass in tundra soils along an arctic transect from Taimyr Peninsula, central Siberia. *Polar Biol.* 25:871–877.

Schubert, K., J. Z. Groenewald, U. Braun, J. Dijksterhuis, M. Starink, C. F. Hill, P. Zalar, G. S. de Hoog, and P. W. Crous. 2007. Biodiversity in the *Cladosporium herbarum* complex (*Davidiellaceae, Capnodiales*), with standardisation of methods for *Cladosporium* taxonomy and diagnostics. *Stud. Mycol.* 58:105–156.

Scorzetti, G., I. Petrescu, D. Yarrow, and J. W. Fell. 2000. *Cryptococcus adeliensis* sp. nov., a xylanase producing basidiomycetous yeast from Antarctica. *Antonie van Leeuwenhoek* 77:153–157.

Selbmann, L., G. S. de Hoog, A. Mazzaglia, E. I. Friedmann, and S. Onofri. 2005. Fungi at the edge of life: cryptoendolithic black fungi from Antarctic desert. *Stud. Mycol.* 51:1–32.

Selbmann, L., G. S. de Hoog, L. Zucconi, D. Isola, S. Ruisi, A. H. van den Ende, C. Ruibal, F. De Leo, C. Urzi, and S. Onofri. 2008. Drought meets acid: three new genera in a dothidealean clade of extremotolerant fungi. *Stud. Mycol.* 61:1–20.

Siegert, M. J., J. C. Ellis-Evans, M. Tranter, C. Mayer, J.-R. Petit, A. Salamatin, and J. C. Priscu. 2001. Physical, chemical and biological processes in Lake Vostok and other Antarctic subglacial lakes. *Nature* 414:603–609.

Siegert, M. J., M. Tranter, J. C. Ellis-Evans, J. C. Priscu, and W. Berry Lyons. 2003. The hydrochemistry of Lake Vostok and the potential for life in Antarctic subglacial lakes. *Hydrol. Processes* 17:795–814.

Simon, C., A. Wiezer, A. W. Strittmatter, and R. Daniel. 2009. Phylogenetic diversity and metabolic potential revealed in a glacier ice metagenome. *Appl. Environ. Microbiol.* 75:7519–7526.

Skidmore, M. L., J. M. Foght, and M. J. Sharp. 2000. Microbial life beneath a high Arctic glacier. *Appl. Environ. Microbiol.* 66:3214–3220.

Sonjak, S., J. C. Frisvad, and N. Gunde-Cimerman. 2005. Comparison of secondary metabolite production by *Penicillium crustosum* strains, isolated from Arctic and other various ecological niches. *FEMS Microbiol. Ecol.* 53:51–60.

Sonjak, S., J. C. Frisvad, and N. Gunde-Cimerman. 2006. *Penicillium* mycobiota in Arctic subglacial ice. *Microb. Ecol.* 52:207–216.

Sonjak, S., J. C. Frisvad, and N. Gunde-Cimerman. 2007a. Genetic variation among *Penicillium crustosum* isolates from Arctic and other ecological niches. *Microb. Ecol.* 54:298–305.

Sonjak, S., V. Uršič, J. C. Frisvad, and N. Gunde-Cimerman. 2007b. *Penicillium svalbardense,* a new species from Arctic glacial ice. *Antonie van Leeuwenhoek* 92:43–51.

Stakhov, V., S. Gubin, S. Maksimovich, D. Rebrikov, A. Savilova, G. Kochkina, S. Ozerskaya, N. Ivanushkina, and E. Vorobyova. 2008. Microbial communities of ancient seeds derived from permanently frozen Pleistocene deposits. *Microbiology* 77:348–355.

Starmer, W., J. Fell, C. Catranis, V. Aberdeen, L. Ma, S. Zhou, and S. Rogers. 2005. Yeasts in the genus *Rhodotorula* recovered from the Greenland ice sheet, p. 181–195. *In* J. D. Castello and S. O. Rogers (ed.), *Life in Ancient Ice*. Princeton University Press, Princeton, NJ.

Sterflinger, K., G. S. de Hoog, and G. Haase. 1999. Phylogeny and ecology of meristematic ascomycetes. *Stud. Mycol.* **43:**5–22.

Sterflinger, K., and W. E. Krumbein. 1997. Dematiaceous fungi as a major agent for biopitting on Mediterranean marbles and limestones. *Geomicrobiol. J.* **14:**219–230.

Steven, B., R. Leveille, W. H. Pollard, and L. G. Whyte. 2006. Microbial ecology and biodiversity in permafrost. *Extremophiles* **10:**259–267.

Takano, Y., K. Kobayashi, K. Marumo, and Y. Ishikawa. 2004. Biochemical indicators and enzymatic activity below permafrost environment, p. 84. *In* F. Robb, M.W. Adams, K. Horikoshi, R.M. Kelly, J. Littlechild, K.E. Nelson, J. Reeve, R. Roberts, K.R. Sowers, and K. Stetter (ed.), *Extremophiles 2004, 5th International Conference on Extremophiles, September 19-23, Cambridge, Maryland.* American Society for Microbiology, Washington, DC.

Thomas, D. N., and G. S. Dieckmann. 2002. Antarctic sea ice—a habitat for extremophiles. *Science* **295:**641–644.

Thomas-Hall, S. R., B. Turchetti, P. Buzzini, E. Branda, T. Boekhout, B. Theelen, and K. Watson. 2010. Cold-adapted yeasts from Antarctica and the Italian Alps—description of three novel species: *Mrakia robertii* sp. nov., *Mrakia blollopis* sp. nov. and *Mrakiella niccombsii* sp. nov. *Extremophiles* **14:**47–59.

Thomas-Hall, S., and K. Watson. 2002. *Cryptococcus nyarrowii* sp. nov., a basidiomycetous yeast from Antarctica. *Int. J. Syst. Evol. Microbiol.* **52:**1033–1038.

Thomas-Hall, S., K. Watson, and G. Scorzetti. 2002. *Cryptococcus statzelliae* sp. nov. and three novel strains of *Cryptococcus victoriae*, yeasts isolated from Antarctic soils. *Int. J. Syst. Evol. Microbiol.* **52:**2303–2308.

Tosi, S., B. Casado, R. Gerdol, and G. Caretta. 2002. Fungi isolated from Antarctic mosses. *Polar Biol.* **25:**262–268.

Tosi, S., S. Onofri, M. Brusoni, L. Zucconi, and H. Vishniac. 2005. Response of Antarctic soil fungal assemblages to experimental warming and reduction of UV radiation. *Polar Biol.* **28:**470–482.

Turchetti, B., P. Buzzini, M. Goretti, E. Branda, G. Diolaiuti, C. D'Agata, C. Smiraglia, and A. Vaughan-Martini. 2008. Psychrophilic yeasts in glacial environments of Alpine glaciers. *FEMS Microbiol. Ecol.* **63:**73–83.

Vincent, W. F. 2000. Evolutionary origins of Antarctic microbiota: invasion, selection and endemism. *Antarct. Sci.* **12:**374–385.

Vincent, W. F., D. R. Mueller, and S. Bonilla. 2004. Ecosystems on ice: the microbial ecology of Markham Ice Shelf in the high Arctic. *Cryobiology* **48:**103–112.

Vishniac, H. S. 1993. The microbiology of Antarctic soils, p. 297–342. *In* E. I. Friedmann (ed.), *Antarctic Microbiology*. Wiley-Liss, New York, NY.

Vishniac, H. S. 2002. *Cryptococcus tephrensis*, sp. nov., and *Cryptococcus heimaeyensis*, sp. nov.; new anamorphic basidiomycetous yeast species from Iceland. *Can. J. Microbiol.* **48:**463–467.

Vishniac, H. S. 2006. Yeast biodiversity in the Antarctic, p. 419–440. *In* C. A. Rosaand and G. Péter (ed.), *Biodiversity and Ecophysiology of Yeasts*. Springer, Berlin, Germany.

Vishniac, H. S., and S. Onofri. 2003. *Cryptococcus antarcticus* var. *circumpolaris* var. nov., a basidiomycetous yeast from Antarctica. *Antonie van Leeuwenhoek* **83:**231–233.

Vorobyova, E., V. Soina, M. Gorlenko, N. Minkovskaya, N. Zalinova, A. Mamukelashvili, D. Gilichinsky, E. Rivkina, and T. Vishnivetskaya. 1997. The deep cold biosphere: facts and hypothesis. *FEMS Microbiol. Rev.* **20:**277–290.

Wagner, D. 2008. Microbial communities and processes in Arctic permafrost environments, p. 133–154. *In* P. Dionand and C. S. Nautiyal (ed.), *Microbiology of Extreme Soils*. Springer, Berlin, Germany.

Wagner, D., A. Lipski, A. Embacher, and A. Gattinger. 2005. Methane fluxes in permafrost habitats of the Lena Delta: effects of microbial community structure and organic matter quality. *Environ. Microbiol.* **7:**1582–1592.

Wallenstein, M. D., S. McMahon, and J. Schimel. 2007. Bacterial and fungal community structure in Arctic tundra tussock and shrub soils. *FEMS Microbiol. Ecol.* **59:**428–435.

Wicklow, D., and D. Malloch. 1971. Studies in the genus *Thelebolus*: temperature optima for growth and ascocarp development. *Mycologia* **63:**118–131.

Xiao, N., K. Suzuki, Y. Nishimiya, H. Kondo, A. Miura, S. Tsuda, and T. Hoshino. 2010. Comparison of functional properties of two fungal antifreeze proteins from *Antarctomyces psychrotrophicus* and *Typhula ishikariensis*. *FEBS J.* **277:**394–403.

Xin, M. X., and P. J. Zhou. 2007. *Mrakia psychrophila* sp. nov., a new species isolated from Antarctic soil. *J. Zhejiang Univ. Sci. B* **8:**260–265.

Zalar, P., G. S. de Hoog, H.-J. Schroers, P. W. Crous, J. Z. Groenewald, and N. Gunde-Cimerman. 2007. Phylogeny and ecology of the ubiquitous saprobe *Cladosporium sphaerospermum*, with descriptions of seven new species from hypersaline environments. *Stud. Mycol.* **58:**157–183.

Zalar, P., C. Gostinčar, G. S. de Hoog, V. Uršič, M. Sudhadham, and N. Gunde-Cimerman. 2008. Redefinition of *Aureobasidium pullulans* and its varieties. *Stud. Mycol.* **61:**21–38.

ADAPTATIONS AND PHYSIOLOGY OF COLD-ADAPTED MICROORGANISMS IN POLAR ENVIRONMENTS

GENERAL CHARACTERISTICS OF COLD-ADAPTED MICROORGANISMS

Shawn Doyle, Markus Dieser,
Erik Broemsen, and Brent Christner

5

INTRODUCTION

From a biological perspective, Earth is a cold planet. The majority of the ocean is deep water that never reaches temperatures above 5°C and a large portion of the biosphere is periodically or permanently frozen (i.e., the cryosphere). Specific examples of cold environments for life include the upper atmosphere (Amato et al., 2007), benthic marine waters and sediments (Li et al., 2008), permafrost (Liebner et al., 2009), polar deserts (Smith et al., 2006), snow (Carpenter et al., 2000), glaciers and subglacial lakes (Christner et al., 2000, 2006), and sea ice (Junge et al., 2011). Although interest in microbes inhabiting low-temperature environments has increased in recent years, significant gaps remain in our understanding of what makes certain microorganisms cold adapted. The thermal tolerance limits of cold-adapted species and their metabolic processes provide a general scheme for their classification. A psychrophile is generally classified as an organism that has a temperature for maximum growth rate (i.e., T_{opt}) of <25°C, meaning that its cellular processes function optimally at relatively low temperatures. Depending on its range of tolerance, a psychro-

phile can be subcategorized as being either a stenopsychrophile (i.e., growth is inhibited at temperatures >15°C) or a eurypsychrophile (capable of growth at low temperature but with a shorter generation time at mesophilic temperatures) (Feller and Gerday, 2003).

Organisms that live at low temperature are faced with a number of significant biochemical and physiological challenges, including reduced enzyme activity, protein denaturation and misfolding, and decreased membrane fluidity and transport efficiency (D'Amico et al., 2006). In environments that experience subzero temperatures, the cascade of events that occur during freezing concentrates the cells, together with salts, minerals, and gas impurities, into highly saline brines at particle and grain boundaries (Price, 2000; Mader et al., 2006). Hence, cellular survival during freezing requires mechanisms to offset the effect of physical and osmotic stress. Remarkably, recent data have indicated that some level of microbial metabolism remains functional under frozen conditions (Carpenter et al., 2000; Christner, 2002; Amato and Christner, 2009). Interest in the structural, biochemical, and physiological properties of psychrophilic microorganisms has motivated investigations to characterize the adaptations that maintain enzymatic reaction rates, macromolecular stability,

Shawn Doyle, Markus Dieser, Erik Broemsen, and Brent Christner, Department of Biological Sciences, Louisiana State University, 202 Life Sciences Building, Baton Rouge, LA 70803.

Polar Microbiology: Life in a Deep Freeze
Edited by Robert V. Miller and Lyle G. Whyte © 2012 ASM Press, Washington, DC

and homeostasis at cold temperature. This chapter provides an overview on the state of knowledge about adaptations that allow certain bacteria and archaea to persist in the coldest regions of the biosphere.

KINETIC AND BIOCHEMICAL CHALLENGES AT LOW TEMPERATURE

Reaction Kinetics

Chemical reaction rates are dependent on the kinetic energy of the substrate atoms and decrease exponentially with temperature. Thus, temperature places a significant thermodynamic constraint on enzyme reaction rates and limits the biochemical and physiological potential of a cellular system. The Arrhenius equation is a mathematical expression that can be used to approximate the relationship between temperature (T) and a given chemical reaction rate (k):

$$k = \kappa(k_B T/h)e^{-\Delta G*/RT}$$

where κ is the transmission coefficient, k_B is Boltzmann's constant, h is Planck's constant, $\Delta G*$ is the free energy of activation, and R is the universal gas constant. In equation form, the exponential effect that temperature has on the reaction rate can be easily appreciated. The observed ratio of growth or metabolic reaction rate for a 10°C change (i.e., the Q_{10} value) (Fig. 1) typically ranges between 2 and 3 (Takacs and Priscu, 1998; Lipson et al., 2002); however, Q_{10} values for bacterial growth and metabolism have been shown to increase at temperatures near 0°C (Sand-Jensen et al., 2007) (Fig. 1).

Catalytic Efficiency

The catalytic efficiency (k_{cat}/K_m) of an enzyme is defined by the ratio k_{cat} to K_m (the substrate concentration at which the reaction proceeds at half its maximum rate, or V_{max}). The k_{cat} describes the catalytic rate of an enzyme, and larger k_{cat} values indicate faster reaction catalysis.

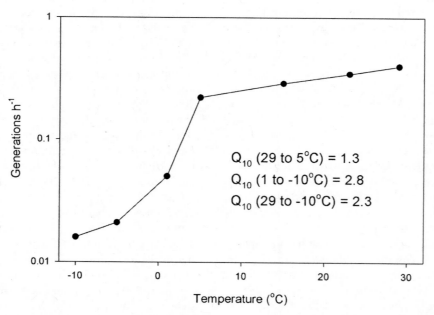

FIGURE 1 The effect of temperature on the growth rate of *Psychrobacter cryohalolentis* K5 in liquid culture. Q_{10} values for bacterial growth increase with decreasing temperature (Sand-Jensen et al., 2007). A Q_{10} of 2.3, similar to published values for complex communities (Takacs and Priscu, 1998), is obtained when using the high- and low-temperature data points. (Generation times are based on unpublished data [P. Amato] and the data for growth at −10°C are from Bakermans et al. [2003].)

TABLE 1 Examples of molecular adaptations in genes, proteins, and enzymes documented in psychrophilic prokaryotes

Organism	Cellular component	Adaptation	Reference(s)
Bacteria			
Antarctic bacterium DS2-3R	Citrate synthase	Predominantly negative surface potential; increased flexibility of the small domain relative to the large domain; reduced subunit interface interactions with no intersubunit ion-pair networks; polypeptide loops of increased length containing more charged and fewer proline residues; increase in solvent-exposed hydrophobic residues; increase in intramolecular ion pairs	Russell et al., 1998
Aquaspirillium arcticum	Malate dehydrogenase	Increased relative flexibility of active-site residues; favorable surface charge distribution for substrate and cofactor; reduced ion-pair interactions for the intersubunit	Kim et al., 1999
Colwellia maris	Isocitrate dehydrogenase	High catalytic activity at low temperatures; low thermostability	Watanabe et al., 2005
Colwellia psychrerythraea 34H	Aminopeptidase	Reduction in proline and residues involved in ion pairing; low hydrophobic residue content.	Huston et al., 2004
Desulfotalea psychrophila	Isocitrate dehydrogenase	Surface amino acid residues highly acidic; destabilizing charged residues prevalent; decreased number of amino acids which form ion pairs; smaller ionic networks	Fedøy et al., 2007
Pseudoalteromonas haloplanktis sp.	Genome	Upregulation of proteins involved in translation, protein folding, membrane integrity, and antioxidant activities	Piette et al., 2010
	α-Amylase	Reduction in the number of proline and arginine residues; increased number of nonpolar residues exposed to the solvent	Aghajari et al., 1998
	Xylanase	Fewer salt bridges; increased solvent exposure of nonpolar residues	Collins et al., 2005
	DNA ligase	High catalytic activity at low temperatures; low thermostability; decreased number of arginine and proline residues; fewer hydrogen bonds and salt bridges	Georlette et al., 2000
Psychrobacter cryohalolentis K5	Iron transport proteins	Upregulated expression at subzero temperatures	Bakermans et al., 2007
Psychrobacter immobilis sp.	β-Lactamase	Low T_{opt} and a low level of thermal stability.	Feller et al., 1995
	Lipase	Low proportion of arginine residues as compared to lysine; low content in proline residues; small hydrophobic core; small number of salt bridges and of aromatic-aromatic interactions	Arpigny et al., 1997
Shewanella livingstonensis Ac10	Proteome	Increased abundance of proteins involved in RNA and protein synthesis and folding, membrane transport, and motility	Kawamoto et al., 2007
Vibrio salmonicida	Uracil-DNA N-glycosylase	Decreased number of proline and arginine residues; lower Arg/(Arg + Lys) ratio; increased percentage of charged residues	Ræder et al., 2008

(*Continued*)

TABLE 1 Examples of molecular adaptations in genes, proteins, and enzymes documented in psychrophilic prokaryotes–*Continued*

Organism	Cellular component	Adaptation	Reference(s)
Bacteria			
Bacillus sp. strain TA39	Subtilisin	High catalytic activity at low temperatures; low thermostability; decreased number of residues that form salt bridges and aromatic interactions; increased interactions with the solvent due to large number of aspartic acid residues in the loops connecting secondary structures	Narinx et al., 1997
Pseudomonas sp. strain TACII18	Phosphoglycerate kinase	High catalytic activity at low temperatures; low thermostability	Mandelman et al., 2001
Vibrio sp. strain I5	3-Isopropylmalate dehydrogenase	Decreased number of residues that form salt bridges; reduction in aromatic-aromatic interactions; fewer proline residues and longer surface loops; substitutions of strictly conserved residues found in mesophilic and thermophilic counterparts	Wallon et al., 1997
Archaea			
Cenarchaeum symbiosum	DNA polymerase	Low thermostability	Schleper et al., 1997
Methanococcoides burtonii	Proteome	Higher number of the noncharged, polar amino acids (in particular glutamine and threonine); increase of hydrophobic residues in the solvent-accessible area; fewer charged residues	Saunders et al., 2003
		Differences in protein expression involved in metabolism, membrane transport, transcription/translation, and protein folding	Lim et al., 2000; Goodchild et al., 2005; Williams et al., 2010
	GTPase (EF-2)	Relatively high catalytic activity at low temperatures; reduced thermostability; low activation energy for GTP hydrolysis; decreased number of salt bridges; smaller number of residues comprising the hydrophobic core; lack of proline residues in loop structures; higher methionine content	Thomas and Cavicchioli, 1998, 2000, 2002; Siddiqui et al., 2002

The K_m is often used as an indicator of an enzyme's affinity for substrate (Koshland, 2002), with larger K_m values indicating a poorer substrate affinity. Thus, the ratio of an enzyme's catalytic rate to substrate affinity provides a quantitative description of how quickly an enzyme can generate product per unit concentration of substrate. With decreasing temperature, enzymes and substrates are less likely to collide and form the appropriate enzyme-substrate complex, which has the effect of reducing overall reaction rates. Under these circumstances, enzymes must be more catalytically efficient in order to generate enzymatic products at rates sufficient to maintain metabolic homeostasis. An enzyme can increase efficiency by increasing the substrate affinity (decreased K_m) and/or increasing the rate of catalysis (increased k_{cat}).

Molecular properties that increase enzyme catalytic efficiency have been most widely documented and studied in psychrophilic bacteria (Table 1). For instance, the α-amylase of *Pseudoalteromonas haloplanktis* has a higher k_{cat} relative to its mesophilic homologs (Feller et al., 1992; D'Amico et al., 2001) (Fig. 2). A

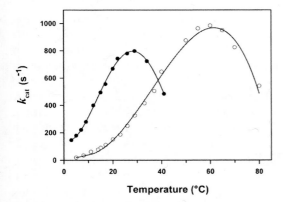

FIGURE 2 Thermodependence of enzyme activity. α-Amylase activity for *Pseudoalteromonas haloplanktis* (•) and *Bacillus amyloliquefaciens* (○), illustrating the greater k_{cat} of the psychrophilic organism's enzyme at low temperature. (Adapted from Feller et al. [1992].)

similar trend is found in the DNA ligase of *P. haloplanktis*, which has an ~10- and 2-fold higher k_{cat} and K_m, respectively, compared to *Escherichia coli* (Georlette et al., 2000). Whereas the α-amylase and DNA ligase from *P. haloplanktis* only show optimization of the k_{cat}, other enzymes appear to have alterations that affect the activity of both k_{cat} and K_m. For example, Hoyoux et al. (2001) and Bentahir et al. (2000) observed that k_{cat} values are higher and K_m values are lower in β-galactosidase and phosphoglycerate kinase from psychrophilic bacteria. While lower K_m values are also often associated with adaptations to lower substrate concentrations, they also decrease the k_{cat}, which is a tradeoff for a cold-adapted enzyme (Hochachka and Somero, 2002).

Similar biochemical adaptations have been discovered in psychrophilic archaea as well. The best-studied cold-adapted protein of *Methanococcoides burtonii* is the elongation factor 2 (EF-2) protein (Thomas and Cavicchioli, 1998, 2000, 2002; Thomas et al., 2001; Siddiqui et al., 2002). EF-2 is a GTPase involved in the translocation step of the ribosome during protein synthesis. Similar to characteristics observed in cold-adapted bacterial proteins (Table 1), the catalytic efficiency of the archaeal EF-2 protein is higher at low temperatures relative to a thermophilic homologue because of a re-

duced K_m (Thomas et al., 2001). Intriguingly, the GTP-binding affinity of the *M. burtonii* EF-2 at the temperature of its environmental source is approximately eightfold lower than values at warmer temperatures ($K_m < 10$ μM at T_{opt} of 23°C; $K_m = 81$ μM at 10°C) (Thomas et al., 2001). As such, the upregulation of EF-2 levels may be an important way in which *M. burtonii* compensates for reduced catalytic efficiency at low temperatures. Relative to different growth rates at 8 versus 23°C (0.013 versus 0.046 h^{-1}), EF-2 concentrations are about three times higher per cell when cultured at 8°C (Thomas et al., 2001). In addition to features described above, the *M. burtonii* EF-2 also possesses a relatively low activation energy for GTP hydrolysis and has poor thermostability (Thomas and Cavicchioli, 2000, 2002; Siddiqui et al., 2002).

The kinetics of reactions at low temperature are a challenge to physiological systems that must meet the energy demands of anabolism (energy-consuming reactions, biosynthesis) via catabolic processes (i.e., respiratory or fermentative generation of ATP). Strategies that compensate and improve the catalytic rate include increasing the enzyme and/or substrate concentration or expressing enzymes adapted for low temperatures (i.e., isozymes) that have increased catalytic efficiencies in the cold (k_{cat}/K_m; Table 1). Napolitano and Shain (2005) demonstrated that psychrophilic bacteria increased their total adenylate pool (ATP, ADP, and AMP) when cultured at the lowest range of their growth temperature (i.e., below the T_{opt}). Based on these results and the key role of ATP in numerous energy-requiring biochemical reactions, the authors deduced that elevation in the concentration of these key substrates may be a strategy to offset reduced enzymatic reaction rates, facilitating the catalysis of reactions involving adenylates at low temperature.

FLUIDITY AT COLD TEMPERATURES

Viscosity

The viscosity of a liquid is inversely related to temperature and the diffusion rate of compounds dissolved within the liquid. For example, viscosity increases by over twofold between 37 and 0°C (D'Amico et al., 2006) and a temperature drop from 30 to 10°C decreases the diffusion rate by approximately 50% (Lettinga et al., 2001). These temperature-dependant properties of water have the effect of reducing the rate at which microorganisms can catalyze biochemical reactions, import nutrients, and export waste products. Furthermore, the increase in water viscosity at low temperature imparts a physical constraint on mobility, and consequently, a microbe must exert more energy for motility in a cold, more viscous medium. Despite these challenges, motility has been observed in *Colwellia psychrerythraea* 34H at temperatures as low as −10°C in marine broth supplemented with up to 20% glycerol (Junge et al., 2003). Junge et al. (2003) also observed that association with surfaces was important for sustained subzero metabolism in brine inclusions and hypothesized that motility may be an important survival strategy in sea-ice brine channels. Specific microbiological adaptations that aid cells in combating increased viscosity at low temperature have not yet been identified and represent territory for future work.

Fluidity of Lipid Membranes

If a membrane transitions into a gel-crystalline state, physiological processes such as transport, energy generation, and cell division are not possible (Shivaji and Prakash, 2010). Low-temperature adjustment of the membrane fatty acid composition is a well-characterized adaptation that is used to increase membrane fluidity at cold temperatures (Los and Murata, 2004; Nichols et al., 2004). A common strategy for increasing membrane disorder is to introduce unsaturation into the fatty acid chains of membrane lipids (Russell, 2008). Genetic investigations of cyanobacteria (*Synechocystis* and *Synechococcus*), *E. coli*, and *Bacillus subtilis* have shown that the expression of genes encoding fatty acid desaturases is part of the well-characterized cold stress response (Sinensky, 1974; Cybulski et al., 2004; Los and Murata, 2004). Desaturases

commonly introduce double bonds into fatty acyl chains at the Δ9 position, although some bacteria also possess Δ5 and Δ10 desaturases (Russell, 2008). Membrane fluidity is also increased through incorporating branched-chain fatty acids into the lipid, shortening the fatty acid chain length, altering the size and charge of polar head groups, changing the *cis/trans* fatty acid ratio, and/or decreasing the sterol/phospholipid ratio (Shivaji and Prakash, 2010). Some of these mechanisms are conditional on the cell's physiological state; e.g., shortening of the fatty acid chain length is only possible if the cells are actively growing. The common outcome of all these molecular alterations is increased steric hindrance between adjacent membrane phospholipid molecules, which results in a membrane that has a lower gel-phase transition temperature.

Archaeal lipids are chemically unique from those found in the *Bacteria* and *Eukarya* and are composed of long-chain hydrocarbons (phytanyl or biphytanyl) bonded by an ether linkage to a glycerol backbone. Some hyperthermophilic archaea contain a membrane that is composed of a glycerol dibiphytanyl glycerol tetraether (i.e., a lipid monolayer) and contains branched hydrocarbon chains with cyclopentane rings, which densely packs and stabilizes the membrane at high temperatures (Gabriel and Chong, 2000). Pelagic crenarchaeota that inhabit marine waters near the freezing point of water contain membranes that are biochemically similar to their high-temperature relatives. The differences, however, are notable and include bicyclic biphytanyl chains that contain cyclohexane rings, which increase the fluidity of the membrane at low temperature (Sinninghe Damsté et al., 2002). Schouten et al. (2002) found a significant linear correlation between temperature and the number of cyclopentane rings present in membrane lipids from marine crenarchaeota, implying that the lipid composition of membranes in these archaea are adjusted based on the growth temperature.

The major phospholipids identified in *M. burtonii*, a model psychrophilic archaeon, are archaeol phosphatidylglycerol, archaeol phosphatidylinositol, hydroxyarchaeol phosphatidylglycerol, and hydroxyarchaeol phosphatidylinositol (Nichols et al., 2004). The ratio and composition of these phospholipids and their level of unsaturation are dependent on the growth temperature. For example, the proportion of unsaturated lipids in *M. burtonii* at 4°C is increased relative to 23°C (Nichols et al., 2004), a trend also found in bacteria and the cold-adapted archaeon *Halorubrum lacusprofundi* (Gibson et al., 2005). In *H. lacusprofundi*, the proportion of unsaturated diether lipids changes drastically with decreasing growth temperature and ranges from trace amounts at 25°C to ~65% of the total lipid fraction at 12°C (Gibson et al., 2005). The genome of *M. burtonii* contains open reading frames for a CDP-inositol transferase and CDP-glycerol transferase, as well as homologues of plant geranylgeranyl reductase; however, homologues of bacterial or eukaryotic desaturase genes have not been identified. Thus, it appears that the biosynthesis of unsaturated membrane lipids involves the direct incorporation of phospholipid intermediates into the glycerol backbone, rather than a bacterium-like desaturase mechanism, the latter of which can alter preexisting lipids to be cold tolerant (Nichols et al., 2004). The incomplete saturation of lipid precursors is also thought to be the biosynthetic pathway used to synthesize the unsaturated diether lipids of *H. lacusprofundi* (Gibson et al., 2005).

Role of Accessory Pigments

In addition to providing protection from UV radiation, carotenoid pigments also appear to have a role in membrane stabilization at cold temperatures (Chattopadhyay et al., 1997; Jagannadham et al., 2000; Fong et al., 2001). The production of these pigments increases in certain bacteria at low temperature (Chattopadhyay et al., 1997; Fong et al., 2001), and Jagannadham et al. (2000) suggested that polar carotenoid molecules may counterbalance the destabilizing effect between unsaturated and branched-chain fatty acids. Furthermore, differences in pigment polarity or configuration, or differences in the location or orientation of the pigments in the membrane, may affect the membrane properties (Britton, 1995). For example, the binding of carotenoid pigment bacterioruberin to membrane vesicles has been shown to increase the membrane rigidity (Strand et al., 1997). Recently, Dieser et al. (2010) demonstrated that carotenoid-possessing bacterial isolates from Antarctic environments were more resistant to episodic freezing compared to nonpigmented strains. This supports the concept that carotenoids contribute to membrane stability at low temperatures and suggests they may also have a role in freeze tolerance.

MACROMOLECULAR STABILITY AT LOW TEMPERATURE

Protein Stability

In general, the proteins of psychrophilic microorganisms must maintain flexibility to perform catalysis at low temperatures, whereas thermophilic proteins are rigid to protect them from thermal denaturation. Importantly, adaptations that enhance protein flexibility reduce the activation energy needed for the formation of the activated enzyme-substrate complex, resulting in enhanced catalytic activity at low temperature (Feller and Gerday, 2003). Distinct amino acid modifications are associated with increasing the conformational flexibility of a polypeptide (Thomas and Cavicchioli, 1998; Feller and Gerday, 2003; Georlette et al., 2004). General structural features that have been attributed to the higher flexibility of cold-active enzymes include (i) a reduced number of hydrogen bonds, (ii) fewer salt bridges, (iii) lower content of proline residues in loop regions, (iv) extended and highly charged surface loops, and (v) a low Arg/(Arg + Lys) content (Table 1). However, there are limits to the degree by which a protein can be modified to enhance flexibility (e.g., a very loosely packed hydrophobic core results in an unstable protein), and as such, conformations optimized to low temperature have inherent structural limitations. Some psychrophiles possess more than one genetic form of an enzyme (i.e., isozymes), at least one of which is expressed and has a higher catalytic efficiency at lower temperatures (e.g., Bakermans et al., 2007); however, genomic data suggest that this is not a widespread or universal strategy for cold adaptation in bacteria and archaea.

Trends in the amino acid composition and usage provide useful information on the mechanisms of protein thermal adaptation within psychrophilic microorganisms. The catalytic domains of psychrophilic enzymes contain fewer residues that limit polypeptide chain flexibility (e.g., proline) and favor residue clusters that increase local mobility (e.g., glycine), supporting the "flexibility concept" of thermal protein structure compensation (Georlette et al., 2004). Studies on the EF-2 protein of $M.$ $burtonii$ and its thermophilic archaeal relatives revealed that the protein lacks proline residues in loop structures, which is likely to affect the flexibility of the polypeptide backbone. The EF-2 protein also has a higher methionine composition compared with its thermophilic counterpart (41 versus 34 residues), implying that the absence of branched, charged, or dipolar R-groups associated with methionine increases protein fluidity; the presence of fewer salt bridges is also expected to enhance structural flexibility (Thomas and Cavicchioli, 1998). Thermolability of the EF-2 protein is consistent with the low free energy of activation (ΔG^*) for the unfolding of the protein and a decreased enthalpy (ΔH^*) (by 60%) for the transition state compared with the form in thermophilic archaea (Siddiqui et al., 2002). The reduced ΔH^* implies that fewer or weaker noncovalent bonds are broken to reach the transition state. Furthermore, protein unfolding can expose infolded hydrophobic residues and reduce the entropy (ΔS^*) of the transition state (Siddiqui et al., 2002). Assuming a more flexible structure of the catalytic region of the protein due to the decrease in the number of enthalpy-driven interactions, the protein is capable of occupying more conformational states in the ground state of the enzyme-substrate

complex, and subsequently, these states experience less disordering during activation (Siddiqui et al., 2002).

The DNA ligase of the psychrophile $P.$ $haloplanktis$ is similar in size and structure and is homologous to the DNA ligase possessed by $E.$ $coli$ and some thermophiles (Georlette et al., 2000). However, within the conserved regions found in all DNA ligases, proline was replaced by other amino acids in the $P.$ $haloplanktis$ homolog, and fewer amino acids participating in hydrogen bonding and salt bridge formation were found in the protein structure (Georlette et al., 2000). Structurally, this cold-adapted ligase had a significantly lower T_{opt} than the DNA ligases of $E.$ $coli$ and $Thermus$ $scotoductus$, but a shorter half-life at moderate temperatures; e.g., at 18 and 25°C, the $P.$ $haloplanktis$ ligase has a half-life of 24 and 11.5 min, respectively. This observation can be attributed to a paucity of macromolecular interactions, inferred from the primary structure of the enzyme, which results in a molecule with poor thermal stability. Structural studies have revealed that activity of the $P.$ $haloplanktis$ DNA ligase is lost before temperature-dependent protein unfolding, suggesting that the active site has even greater thermolability than the enzyme structure as a whole (Georlette et al., 2003). The thermolability of a cold-adapted DNA ligase has recently been exploited to develop a novel temperature-sensitive vaccine. Duplantis et al. (2010) genetically engineered a live vaccination for tularemia by substituting gene orthologs of DNA ligase from the pyschrophile $C.$ $psychrerythraea$ into the pathogenic mesophile $Francisella$ $tularensis$. Since the mutant strains of the pathogen were temperature sensitive, they were shown to be unable to proliferate at the core body temperature of mice. However, vaccination at a cooler region of the animal (e.g., the tail) induced protective immunity that protected the mice against the pathogenic effects observed in infections of the wild type. The use of essential genes from psychrophiles to tailor live vaccines has great potential in the biomedical sciences and could be equally valuable for any application that needs to control bacterial growth at moderate temperatures without using chemicals or antibiotics.

A comparison between the amino acid compositions in protein sequences of various psychrophilic and mesophilic bacteria shows fundamental differences in their proportions (Metpally and Reddy, 2009; also see Table 1). The amino acids serine, aspartic acid, threonine, and alanine are overrepresented in the coil regions of the secondary structure, whereas glutamic acid and leucine are underrepresented in the helical regions of proteins from psychrophiles. Furthermore, amino acids with aliphatic, basic, aromatic, and hydrophilic side chains have decreased usage in the proteins of psychrophilic bacteria (Metpally and Reddy, 2009). A significant portion of the *Psychrobacter arcticus* 273-4 proteome contains proteins with reduced usage of the acidic amino acids proline and arginine, which would result in polypeptides with increased flexibility at low temperatures (Ayala-del-Río et al., 2010). Similar molecular adaptations are found in archaea, and the number of expressed proteins containing noncharged, polar amino acids (in particular glutamine and threonine) is higher in psychrophilic archaea, whereas the content of hydrophobic amino acids (in particular leucine) is reduced (Saunders et al., 2003). In contrast, the proteins of hyperthermophiles are characterized by having a larger number of lysine, arginine, and glutamic acid residues. Structural models of proteins from cold-adapted archaea reveal a strong tendency in the solvent-accessible area for hydrophobic residues and fewer charged residues (Saunders et al., 2003). These characteristics most likely destabilize the surface of the proteins, and the higher number of glutamine and threonine residues may compensate for the general reduction in surface charged residues (Saunders et al., 2003).

The Cold Shock and Acclimation Response

Exposure to a temperature downshift induces a cold shock response in bacteria that involves the expression of cold shock proteins (Csps) and acclimation proteins (Phadtare and Inouye, 2008; reviewed in Chaikam and Karlson, 2010) and may also include the uptake or synthesis of compatible solutes (Ko et al., 1994), both of which significantly increase cold and freeze tolerance. Many of the proteins involved in the cold shock response are thought to function as molecular chaperones that destabilize RNA secondary structures and assist in the refolding of cold-denatured proteins (Phadtare and Inouye, 2008). In *E. coli*, 26 genes are upregulated subsequent to cold shock, with CspA being the major gene product, representing up to 13% of total cellular protein (Goldstein et al., 1990; Gualerzi et al., 2003).

Compared to their mesophilic counterparts, the synthesis of housekeeping proteins is not repressed upon cold shock in steno- and eurythermal psychrophilic bacteria (Hébraud and Potier, 1999). The synthesis of cold acclimation proteins (CAPs) coincides with continuous growth at low temperature in certain bacteria. Piette et al. (2010) demonstrated that the major CAP upregulated in *P. haloplanktis* TAC125 was a chaperone protein that was upregulated 37-fold at 4°C (compared to 18°C) and interacts with newly synthesized polypeptides on the ribosome. This molecular chaperone is involved in initial protein folding and may be the rate-limiting step for bacterial growth at low temperatures (Piette et al., 2010). While CAPs appear to be common in steno- and eurythermal psychrophiles, they are also expressed by some mesophilic bacterial species during growth at low temperatures (e.g., Panoff et al., 1997).

Csps are conspicuously absent from most archaeal genome sequences, and in particular from thermophilic and hyperthermophilic archaea (Lim et al., 2000; Saunders et al., 2003). It is important to note that not all CspA homologues have a role in cold acclimation and this family of molecular chaperones appears to have broader roles in RNA destabilization associated with physiological stress (Phadtare and Inouye, 2008). However, cold-induced Csps possess a unique and highly conserved cold shock domain (CSD) characteristic of nucleic acid–binding proteins (Graumann and Marahiel, 1998). Homologs of the Csps have been found in a small number of archaea, and strikingly, some of these archaeal Csps and their associated CSD folds have a high degree of sequence conservation and functional similarity to their bacterial counterparts (Giaquinto et al., 2007). Among other archaea (described in Giaquinto et al., 2007), Csp genes have been identified in the Antarctic psychrophilic archaea *Methanogenium frigidium* (Saunders et al., 2003) and *H. lacusprofundi* (Reid et al., 2006; Giaquinto et al., 2007). Unlike *M. frigidium*, *M. burtonii* (another cold-adapted archaeon isolated from Ace Lake, Antarctica) does not possess a Csp homologue (Saunders et al., 2003). However, two hypothetical proteins in the genome of *M. frigidium* are predicted to have a CSD fold (Saunders et al., 2003). The biological function of archaeal Csp and CSD homologs has not been thoroughly investigated, but they are hypothesized to be an important aspect of their capacity to function at low temperatures (Saunders et al., 2003; Reid et al., 2006; Giaquinto et al., 2007).

RNA Stability

The increased strength of hydrogen bonds between nucleic acid bases and enhanced formation of RNA secondary structures at low temperatures are examples of fundamental challenges that necessitate cold-adapted solutions for central cellular processes such as genome replication, transcription, and translation (Feller and Gerday, 2003). In the same sense that enzymes must reduce stabilizing interactions, small molecules such as tRNA must also be innately flexible to maintain their function at decreased kinetic energies. Dalluge et al. (1997) examined the nucleotide composition of tRNA from three psychrophilic bacterial species and found 40 to 70% more dihydrouridine than in tRNA pools characterized from in *E. coli*. Dihydrouri-

dine is a universal posttranscriptional tRNA modification; it is a nonplanar base that favors the C-2'-endo sugar conformation, and the number of modified nucleosides per tRNA has been shown to increase with temperature (Edmonds et al., 1991). The conformation of a tRNA with a dihydrouridine base is such that the accommodation required for molecular stacking involves rearrangements that propagate to the adjacent 5' residue, producing a structure that is more flexible by reducing the stabilizing effects of base stacking (Dalluge, 1997). In the hyperthermophilic archaeon *Stetteria hydrogenophila*, tRNA modification levels are as high as 13%, containing a structural diversity of 31 modified nucleosides (Noon et al., 2003). Conversely, the modification level in psychrophilic archaea is low and makes up less than 2% in *M. burtonii* (Noon et al., 2003). However, *M. burtonii* contains significant amounts of dihydrouridine (Noon et al., 2003). The incorporation of this modified tRNA nucleoside appears to enhance and maintain the molecular flexibility of RNA at low temperature (Dalluge et al., 1996). Unlike in *M. burtonii*, dihydrouridine in the tRNA of *S. hydrogenophila* (Noon et al., 2003) and other mesophilic or thermophilic archaea is low or below the level of detection (Edmonds et al., 1991).

Small RNA molecules can act as regulatory elements termed riboswitches (Vitreschak et al., 2004). Riboswitches bind with high affinity to metabolites and are often located in the 5' or 3' untranslated region (UTR) of mRNAs for operons encoding proteins in the metabolic pathways of the regulated metabolites (e.g., coenzyme B12) (Nahvi et al., 2002; Narberhaus et al., 2006). Most known riboswitches downregulate translation; however, there are also examples of riboswitches that operate by initiating translation (Rodionov et al., 2003). Some riboswitches are thought to act as cellular "thermometers" and may be used to up- and/or downregulate the translation of specific mRNAs in response to a change in temperature (Narberhaus et al., 2006). For example, the *E. coli cspA* mRNA

5' UTR forms a temperature-dependent secondary structure that exposes the Shine-Dalgarno sequence and AUG start codon at cold shock temperatures, whereas the region is conformationally hidden within a helical structure at warmer temperatures (Giuliodori et al., 2010).

WATER ACTIVITY AND FREEZING

During the phase transition to ice, dissolved and particulate impurities (including cells) are excluded from the ice matrix into interstitial aqueous channels at the ice-grain boundaries (Fig. 3). The chemical composition and percentage of the unfrozen water in the ice are a function of temperature (Fig. 4). For example, seawater freezes at −1.86°C and the ionic strength of the unfrozen water in the ice increases with decreasing temperature (Fig. 4B).

At −10°C, the ionic strength of the unfrozen water in saline ice (containing ~34 g of dissolved salts and ions per liter—mostly sodium, chloride, sulfate, magnesium, calcium, and potassium) is approximately fourfold higher (water activity [a_w] = 0.91) than the bulk salinity of seawater (a_w = 0.98); at −20°C, the salinity of the unfrozen water is approximately sevenfold higher (a_w = 0.82). Due to lower dissolved solute contents, the percentage of unfrozen water present at subzero temperatures in permafrost and glacial ice (0.45 and 0.19 to 0.0057 mole liter^{-1}, respectively; Fig. 4B) can be significantly lower than that of sea ice (Fig. 4A). Thus, since the a_w and chemistry of the unfrozen water are functions of temperature, one of the key differences between these very different icy substrates is the total amount of unfrozen water in the ice.

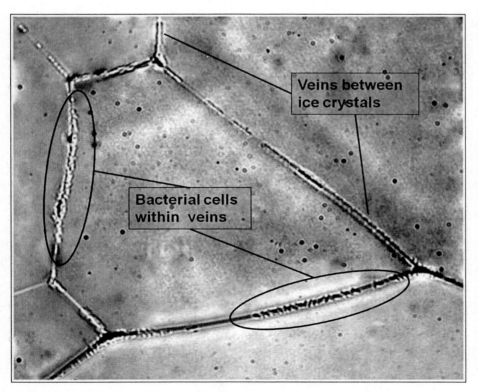

FIGURE 3 Bright field image of polycrystalline ice. The image shows narrow, intergranular veins and triple junctions formed between ice crystals. *Escherichia coli* cells that were frozen in LB broth at −20°C on a cryostage were excluded into the vein network during ice formation. The width of the image is 230 μm.

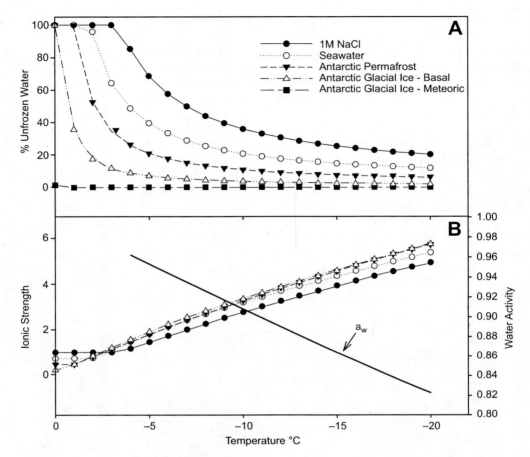

FIGURE 4 (A) Bulk percentage estimates of unfrozen water in various types of icy substrates. The inclusion of 1 M NaCl is for comparison. (B) Predicted ionic strength and a_w of the unfrozen water. (Water chemistry data: Seawater, Appelo and Postma [2005]; Antarctic Permafrost, McLeod et al. [2008]; Antarctic Glacial Ice, S. Montross [unpublished data]. Calculations were performed on dissolved major ions using FREZCHEM [v. 11.2; Mironenko et al., 1997].)

Production of Compatible Solutes

The formation of extracellular ice increases osmotic pressure and draws water from the cell. Biochemical reactions cannot occur without water, nor can catalytic (i.e., protein) or structural (e.g., lipid) macromolecules prevent their denaturation in its absence. In order to maintain a positive water balance at low a_w, microorganisms increase the cytoplasmic solute concentration by synthesizing or transporting an organic solute (i.e., a compatible solute) or inorganic ion. The genetic response to osmotic shock has been studied extensively (Poolman and Glaasker, 1998; Wood et al.,

2001), and when *E. coli* encounters high salinity, the cell balances the sudden increase in osmolarity by importing K^+; after compatible solutes have been imported or synthesized, the K^+ is actively transported out of the cell using a specific K^+ efflux system (Kempf and Bremer, 1998). In some cold-adapted bacteria, a decrease in temperature induces the uptake or synthesis of compatible solutes, which significantly increases their cold tolerance (Ko et al., 1994).

Extremely halophilic archaea use a "salt-in" strategy to create osmotic balance by maintaining K^+ (or Na^+ in some species) at multimolar intracellular concentrations (Empadinhas and da Costa, 2009). High salt concentrations destabilize proteins by disrupting ionic interactions and competing for water of hydration. To offset this effect, the cytoplasmic enzymes of halophilic archaea are composed of more acidic amino acids and coil structures, fewer hydrophobic amino acids and helix formations, and require high K^+ concentrations to function (Paul et al., 2008). In nonhalophiles, neutrally charged organic compounds such as proline, glycine betaine, and/or trehalose are synthesized or transported into the cell as a compatible solute. The highly water-soluble organic compounds that serve as compatible solutes in bacteria, plants, and animals can be zwitterionic, noncharged, or anionic solutes, and do not inhibit cellular functions at their physiological concentrations (Roberts, 2005). Compatible solutes have also been identified in psychrophilic, nonhalophilic archaea (Thomas et al., 2001; Costa et al., 2006). However, whereas typical compatible solutes have the effect of maintaining a positive water balance inside the cell, compatible solutes (i.e., potassium aspartate) reported for *M. burtonii* affect enzyme activity and stability. Potassium aspartate improves the affinity of EF-2 for GTP, and it has been speculated that this increase may offset the decreased substrate affinity associated with temperatures below optimum (23°C) (Thomas et al, 2001).

EPS Production

Exopolysaccharides (EPSs) are large sugar polymers made of residues that are synthesized by cells and secreted into the extracellular environment. EPS is thought to have a role in metal capture (i.e., Fe^{2+}, Zn^{2+}, Cu^{2+}, and Co^{2+}) and in retarding the diffusion of exoenzymes (Qin et al., 2007). However, it also appears to serve as a cryoprotectant (Krembs et al., 2002) by aiding microorganisms with water retention, the concentration of extracellular substrates, stabilizing extracellular enzymes, and/or freezing point depression (Mancuso Nichols et al., 2005). Junge et al. (2006) demonstrated significantly higher rates of protein synthesis in *C. psychrerythraea* 34H at temperatures between −1 and −20°C in the presence of EPS and suggested that it increased metabolic activity by trapping water molecules at the cell surface and/or stabilizing membrane-associated enzymes. EPS production is also a characteristic of cold-adapted archaea, and *M. burtonii* and *H. lacusprofundi* form multicellular aggregates at low temperatures that appear to be embedded in a network of EPS fibrils (Reid et al., 2006). Similarly, the euryarchaeon SM1 synthesizes a pili-like fiber during growth at ~10°C (Moissl et al., 2003). Although the full range of function for these extracellular, network-forming polymers has not been resolved, they may also mediate cell-cell and cell-surface interactions (Moissl et al., 2003) or facilitate the exchange of nutrients, membrane components, and/or genetic material (Reid et al., 2006).

Ice-Interacting Proteins That Influence Freezing and Ice Structure

Some microorganisms produce proteins at low temperature that affect ice nucleation, ice crystal structure, and/or the process of recrystalization. Proteins that bind to and inhibit the growth of ice crystals are referred to as ice-binding or antifreeze proteins (IBPs or AFPs). AFPs were first described in Antarctic notothenioid fish (DeVries and Wohlschlag, 1969), and more recently have been identified in four bacterial phyla (*Actinobacteria*, *Firmicutes*, *Bacteroidetes*, and *Proteobacteria*) (Walker et al., 2006; Raymond et al., 2007, 2008), snow molds (Hoshino et al., 2003), and sea ice diatoms (Janech et al., 2006). Three nonhomologous IBPs have been identified within the *Bacteria*, which correspond to genetic forms found with (i) *Pseudomonas putida* (Muryoi et al., 2004), (ii) *Marinomonas primoryensis* (Gilbert et al., 2005), and (iii) *Colwellia* sp. SLW05 (Raymond et al., 2007). The >1-MDa AFP of *M. primoryensis* is the only known true bacterial AFP and results in a noncolligative freeze point depression of >2°C (Garnham et al., 2008). Not all AFPs have an effect on freezing point depression (i.e., thermal hysteresis), and many only have an ice-interacting activity that inhibits ice crystal recrystallization. Such IBPs have been identified in bacteria isolated from winter soils (Walker et al., 2006), Antarctic sea ice (Raymond et al., 2007), glacier ice (Raymond et al., 2008), lakes (Gilbert et al., 2004), and cyanobacterial mats (Raymond and Fritsen, 2001). In addition, highly active intracellular AFPs have been reported in an Antarctic strain of *Flavobacterium xanthum* (Kawahara et al., 2007).

A variety of bacteria and fungi (as well as plant, lichen, and invertebrate species) produce proteinaceous compounds that serve as efficient ice nucleators. Hence, while AFPs function to depress the freezing point, ice-nucleating proteins (INPs) induce freezing at temperatures warmer than would occur spontaneously. The best-studied microorganisms that produce INPs are not particularly cold adapted and comprise certain culturable species of plant-associated bacteria (*Pseudomonas syringae*, *Pseudomonas viridiflava*, *Pseudomonas fluorescens*, *Pantoea agglomerans*, and *Xanthomonas campestris*), fungi (e.g., *Fusarium avenaceum*), and algae (*Chlorella minutissima*) (Lundheim, 2002). The ability to catalyze freezing at elevated temperatures has been hypothesized to benefit a plant-associated microorganism in several possible ways: by providing access to nutrients leaked from freeze-damaged plant tissue, aiding in infection, and enhancing survival by preventing damage resulting from low-temperature freezing (Hirano and Upper, 1995). However, the INPs possessed by some animal species have been shown to be important in freeze tolerance (Lundheim, 2002), a property that they may also confer to microbial species with the ice-nucleating phenotype.

SUBEUTECTIC METABOLISM: RESIDUAL REACTIONS OR SURVIVAL STRATEGY?

Challenges to Long-Term Survival
During dormancy, a microorganism accumulates genetic and cellular damage. If metabolically inert for extended time frames, a microbial population is at risk for extinction from the community. In icy environments, natural sources of ionizing radiation (i.e., from cosmic sources and the decay of radionuclides such as ^{40}K, ^{232}Th, and ^{238}U); L-amino acid racemization; and spontaneous reactions that oxidize, depurinate, and crosslink nucleic acids (Hansen et al., 2006) are all examples of mechanisms that place limits on long-term survival. Nevertheless, there are numerous reports documenting the persistence of viable microorganisms in ancient ice and permafrost hundreds of thousands to millions of years old (Abyzov et al., 1998; Brinton et al., 2002; Christner et al., 2003; Steven et al., 2006; Bidle et al., 2007; Johnson et al., 2007). Theoretical constraints on survival coupled with the observation of living microbes in these natural archives raise an intriguing question: are microorganisms in perennially frozen environments capable of metabolizing in situ?

It is well-known that planktonic microorganisms incorporated into sea ice during crystallization subsequently develop into spatially defined communities in and on the ice surface (reviewed by Junge et al., 2011). Microorganisms in the sea ice habitat experience very low ice temperatures during the winter, and some of the resident bacteria have the capacity to maintain their metabolic activity in brine inclusions to temperatures of −20°C (Junge et al., 2004). During the austral summer, the permanent ice covers of Antarctic Dry Valleys lakes develop pockets of liquid water that receive sufficient light and nutrients to support the growth of photosynthetically based microbial assemblages (Priscu et al., 1998). In contrast, it is less clear if active biological processes are widespread and important for survival in icy subsurface habitats such as glacial ice and permafrost. In essence, it is not known if the environments are simply natural archives, preserving cells to reanimate when liquid conditions are introduced, or specialized microbial habitats of the cryosphere. The crux of this debate is based on our lack of knowledge regarding the limits to metabolic activity under these physicochemical conditions.

Subzero Growth and Metabolism

Table 2 summarizes a variety of recent studies that report evidence for subzero metabolic activity (direct and indirect) in saline brines and otherwise frozen environments. Indirect evidence for subzero microbial activity is based on inferences on gas concentration and isotopic composition (i.e., CO_2, CH_4, N_2O) within glacial ice (Sowers, 2001; Campen et al., 2003) and gas emissions from permafrost soils (Panikov, 2009). In freshwater and saline ice, various studies have shown that cells are physically partitioned in the aqueous interstitial veins that exist at grain boundaries (Junge

TABLE 2 Reports documenting evidence for subzero metabolic activity

Source	T (°C)	Activity measurement	Technique[a]	Reference(s)
Bulk community measurements	−15	Respiration	[^{14}C]Glucose uptake	Gilichinsky et al., 2003
	−15	Respiration	[^{14}C]Glucose and [^{14}C]acetate mineralization	Steven et al., 2007
	−16.5	CH_4 production	[^{14}C]Bicarbonate and [^{14}C]acetate utilization	Rivkina et al., 2004
	−20	Lipid synthesis	[^{14}C]Acetate incorporation	Rivkina et al., 2000
	−39	Respiration	[^{14}C]Glucose uptake	Panikov et al., 2006
Tundra	−30	N mineralization and nitrification	Measurement of nitrate and ammonium	Schimel et al., 2004
Snow	−17	Macromolecular synthesis	[^{3}H]Thymidine and [^{3}H]leucine incorporation	Carpenter et al., 2000
Sea ice	−1	Protein synthesis	[^{3}H]Leucine incorporation	Grossmann, 1994
	−1.5	DNA synthesis	[^{3}H]Thymidine incorporation	Smith and Clement, 1990
	−20	Respiration	CTC reduction	Junge et al., 2004
Marine	−1.3	Protein synthesis	[^{14}C]Amino acid incorporation	Ritzrau, 1997

Laboratory studies of environmental isolates

Permafrost	−10	Reproduction	Plate counts, optical density	Bakermans et al., 2003
	−20	Metabolism	Resazurin reduction	Jakosky et al., 2003
	−35	Metabolism	$^{14}CO_2$ uptake	Panikov and Sizova, 2007
	−80	Metabolism	Adenylate pool	Amato and Christner, 2009
Sea ice	−10	Motility	Microscopy	Junge et al., 2003
	−12	Reproduction	Optical density	Breezee et al., 2004
	−196	Protein synthesis	[^{3}H]Leucine incorporation	Junge et al., 2006
Glacial ice	−15	Macromolecular synthesis	[^{3}H]Thymidine and [^{3}H]leucine incorporation	Christner, 2002; Amato et al., 2010
	−32	N_2O production	CF-IRMS	Miteva et al., 2007
Marine sediments	−1.7	Sulfate reduction	Reduction of [^{35}S] SO_4^{2-}	Knoblauch et al., 1999

Metabolism inferred from biogenic gases

Tundra	−5	N_2O production	Gas chromatography	Brooks et al., 1997
	−12	CO_2 production	Infrared gas analyzer	Mikan et al., 2002
Glacial ice	−10 to −40	Gas concentrations and isotopic composition	Gas chromatography and mass spectrometry	Sowers, 2001; Campen et al., 2003; Tung et al., 2006

[a]CF-IRMS, continuous flow isotope ratio monitoring mass spectrometry; CTC, 5-cyano-2,3-ditolyl tetrazolium chloride.

et al., 2004; Mader et al., 2006; Amato et al., 2009; see Fig. 3), supporting Price's (2000) hypothesis that ice-grain boundaries provide a microbial habitat. Physiological studies of permafrost bacteria and fungi indicate that cells are capable of remaining metabolically active to temperatures approaching −40°C (Panikov et al., 2006; Panikov and Sizova, 2007). Amato and Christner (2009) provided evidence for ATP generation via oxidation phosphorylation at temperatures as low as −80°C, and astonishingly, Junge et al. (2006) reported a measurable rate of leucine incorporation into protein at −196°C!

Price and Sowers (2004) compiled available subzero metabolic rates at temperatures down to −40°C, and based on thermodynamic constraints, found no evidence for a minimum temperature for metabolism. An Arrhenius plot of the data revealed that the subzero metabolic rates reported binned into three distinct categories that were identified as rates sufficient to support microbial growth, maintenance metabolism, and survival metabolism. Maintenance metabolism is the level of activity required for osmoregulation, maintaining the intracellular pH, the turnover of macromolecules, motility, etc., but is not sufficient for reproduction. Survival metabolism is a theoretical concept and is defined as the amount of energy expended to repair damage to DNA and offset amino acid racemization (Morita, 1988). The latter form of metabolism has never been directly measured and is inferred indirectly from geochemical measurements (Table 2).

Laboratory experiments demonstrate that bacteria and fungi can remain metabolically functional at very low subzero temperatures. This, coupled with geochemical evidence for in situ metabolism under frozen conditions, implies that permafrost and glacier ice may be active ecosystems. At warmer ice temperatures, the a_w is higher and there is a greater percentage of liquid water in the ice (Fig. 4), providing more favorable conditions for biological activity and an increased microenvironment for nutrient acquisition. Metabolism conducted by populations entrapped in ice or permafrost may be physiologically similar to the "starvation-survival" state described by Morita (1988). Although subnanometer-thin films of liquid water have been reported down to temperatures of −196°C (Pearson and Derbyshire, 1974), the increased ionic concentration of unfrozen water at decreasing temperatures in ice is likely a major constraint on biological activity, creating chemical conditions incompatible with a microorganism's physiology (Fig. 4). Hence, while the known limit for bacterial growth is −12°C (Breezee et al., 2004), the minimum temperature for maintaining metabolic reactions has yet to be defined. Deciphering the low temperature limits for life is of inherent scientific value and will also provide information that will aid in understanding the fundamental nature of microbial persistence and survival in the cryosphere. If such frozen matrices are habitats in a conventional sense, it would represent a significant extension in our view of the known boundaries for life.

CONCLUSION
Here we have discussed many of the most common and generally understood biochemical and physical adaptations that appear unique to the psychrophilic lifestyle. Psychrophilic microorganisms use a range of strategies to persist at low temperatures, including possessing catalytically efficient enzymes, synthesizing specialized lipids that increase membrane flexibility, and producing proteins that affect freezing and ice structure. However, our understanding of what precisely makes an organism psychrophilic is still in its infancy, and future investigations are needed to address basic science questions regarding cold adaptation. It is also worthy to note that the interesting properties of cold-adapted microorganisms have enormous biotechnological potential (Christner, 2010; Duplantis et al., 2010). The enzymes of psychrophiles are prime targets for industrial use because of their high efficiency and optimum catalysis at moderate to low temperatures. A better understanding of the properties that make a macromolecule

cold adapted might even be used as a rationale for designing molecules with specific activities at low temperature (e.g., the opposite of Dumon et al., 2008). We know much more about the adaptations that confer thermophily relative to psychrophily, which is ironic considering that most life in the biosphere exists at temperatures <5°C (i.e., deep cold ocean). Fortunately, efforts to discover novel psychrophiles and their adaptations have increased significantly over the last decade. Coupled with technological advances in high-throughput DNA sequencing and proteomics, we can expect that information on cold-adapted bacteria and archaea will increase in the future, as will our knowledge about nature's solutions to cold adaptation.

REFERENCES

Abyzov, S. S., I. N. Mitskevich, and M. N. Poglazova. 1998. Microflora of the deep glacier horizons of central Antarctica. *Microbiology* **67:**66–73.

Aghajari, N., G. Feller, C. Gerday, and R. Haser. 1998. Structures of the psychrophilic *Alteromonas haloplanctis* α-amylase give insights into cold adaptation at a molecular level. *Structure* **6:** 1503–1516.

Amato, P., and B. C. Christner. 2009. Energy metabolism response to low-temperature and frozen conditions in *Psychrobacter cryohalolentis*. *Appl. Environ. Microbiol.* **75:**711–718.

Amato, P., S. M. Doyle, J. R. Battista, and B. C. Christner. 2010. Implications of subzero metabolic activity on long-term microbial survival in terrestrial and extraterrestrial permafrost. *Astrobiology* **10:**789–798.

Amato, P., S. M. Doyle, and B. C. Christner. 2009. Macromolecular synthesis by yeasts under frozen conditions. *Environ. Microbiol.* **11:**589–596.

Amato, P., M. Parazols, M. Sancelme, P. Laj, G. Mailhot, and A. Delort. 2007. Microorganisms isolated from the water phase of tropospheric clouds at the Puy de Dôme: major groups and growth abilities at low temperatures. *FEMS Microbiol. Ecol.* **59:**242–254.

Appelo, C. A. J., and D. Postma. 2005. *Geochemistry, Groundwater and Pollution*, 2nd ed. A.A. Balkerma Publishers, Leiden, The Netherlands.

Arpigny, J. L., J. Lamotte, and C. Gerday. 1997. Molecular adaptation to cold of an Antarctic bacterial lipase. *J. Mol. Catal. B Enzym.* **3:**29–35.

Ayala-del-Río, H. L., P. S. Chain, J. J. Grzymski, M. A. Ponder, N. Ivanova, P. W. Bergholz, G. Di Bartolo, L. Hauser, M. Land, C. Bakermans, D. Rodrigues, J. Klappenbach, D. Zarka, F. Larimer, P. Richardson, A. Murray, M. Thomashow, and J. M. Tiedje. 2010. The genome sequence of *Psychrobacter arcticus* 273-4, a psychroactive Siberian permafrost bacterium, reveals mechanisms for adaptation to low-temperature growth. *Appl. Environ. Microbiol.* **76:**2304–2312.

Bakermans, C., S. L. Tollaksen, C. S. Giometti, C. Wilkerson, J. M. Tiedje, and M. F. Thomashow. 2007. Proteomic analysis of *Psychrobacter cryohalolentis* K5 during growth at subzero temperatures. *Extremophiles* **11:**343–354.

Bakermans, C., A. I. Tsapin, V. Souza-Egipsy, D. A. Gilichinsky, and K. H. Nealson. 2003. Reproduction and metabolism at −10°C of bacteria isolated from Siberian permafrost. *Environ. Microbiol.* **5:**321–326.

Bentahir, M., G. Feller, M. Aittaleb, J. Lamotte-Brasseur, T. Himri, J.-P. Chessa, and C. Gerday. 2000. Structural, kinetic, and calorimetric characterization of the cold-active phosphoglycerate kinase from the Antarctic *Pseudomonas* sp. TACII18. *J. Biol. Chem.* **275:**11147–11153.

Bidle, K. D., S. Lee, D. R. Marchant, and P. G. Falkowski. 2007. Fossil genes and microbes in the oldest ice on Earth. *Proc. Natl. Acad. Sci. USA* **104:**13455–13460.

Breezee, J., N. Cady, and J. T. Staley. 2004. Subfreezing growth of the sea ice bacterium "*Psychromonas ingrahamii*." *Microb. Ecol.* **47:**300–304.

Brinton, K. L. F., A. I. Tsapin, D. A. Gilichinsky, and G. D. McDonald. 2002. Aspartic acid racemization and age-depth relationships for organic carbon in Siberian permafrost. *Astrobiology* **2:**77–82.

Britton, G. 1995. Structure and properties of carotenoids in relation to function. *FASEB J.* **9:** 1551–1558.

Brooks, P. D., S. K. Schmidt, and M. W. Williams. 1997. Winter production of CO_2 and N_2O from alpine tundra: environmental controls and relationship to inter-system C and N fluxes. *Oecologia* **110:**403–413.

Campen, R. K., T. Sowers, and R. B. Alley. 2003. Evidence of microbial consortia metabolizing within a low-latitude mountain glacier. *Geology* **31:**231–234.

Carpenter, E. J., S. Lin, and D. G. Capone. 2000. Bacterial activity in South Pole snow. *Appl. Environ. Microbiol.* **66:**4514–4517.

Chaikam, V., and D. T. Karlson. 2010. Comparison of structure, function and regulation of plant cold shock domain proteins to bacterial and animal cold shock domain proteins. *BMB Rep.* **43:**1–8.

Chattopadhyay, M. K., M. V. Jagannadham, M. Vairamani, and S. Shivaji. 1997. Carotenoid pigments of an Antarctic psychrotrophic bacterium *Micrococcus roseus*: temperature dependent biosynthesis, structure, and interaction with synthetic membranes. *Biochem. Biophys. Res. Commun.* 239:85–90.

Christner, B. C. 2002. Incorporation of DNA and protein precursors into macromolecules by bacteria at −15°C. *Appl. Environ. Microbiol.* 68: 6435–6438.

Christner, B. C. 2010. Bioprospecting for microbial products that affect ice crystal formation and growth. *Appl. Microbiol. Biotechnol.* 85:481–489.

Christner, B. C., E. Mosley-Thompson, L. G. Thompson, and J. N. Reeve. 2001. Isolation of bacteria and 16S rDNAs from Lake Vostok accretion ice. *Environ. Microbiol.* 3:570–577.

Christner, B. C., E. Mosley-Thompson, L. G. Thompson, V. Zagorodnov, K. Sandman, and J. N. Reeve. 2000. Recovery and identification of viable bacteria immured in glacial ice. *Icarus* 144:479–485.

Christner, B. C., E. Mosley-Thompson, L. G. Thompson, and J. N. Reeve. 2003. Bacterial recovery from ancient ice. *Environ. Microbiol.* 5:433–436.

Christner, B. C., G. Royston-Bishop, C. F. Foreman, B. R. Arnold, M. Tranter, K. A. Welch, W. B. Lyons, A. I. Tsapin, M. Studinger, and J. C. Priscu. 2006. Limnological conditions in Subglacial Lake Vostok, Antarctica. *Limnol. Oceanogr.* 51:2485–2501.

Collins, T., C. Gerday, and G. Feller. 2004. Xylanases, xylanase families and extremophilic xylanases. *FEMS Microbiol. Rev.* 29:3–23.

Cybulski, L. E., G. Solar, P. O. Craig, M. Espinosa, and D. Mendoza. 2004. *Bacillus subtilis* DesR functions as a phosphorylation-activated switch to control membrane lipid fluidity. *J. Biol. Chem.* 279:39340–39347.

Dalluge, J. J., T. Hamamoto, K. Horikoshi, R. Y. Morita, K. O. Stetter, and J. A. McCloskey. 1997. Posttranscriptional modification of tRNA in psychrophilic bacteria. *J. Bacteriol.* 179:1918–1923.

Dalluge, J. J., T. Hashizume, A. E. Sopchik, J. A. McCloskey, and D. R. Davis. 1996. Conformational flexibility in RNA: the role of dihydrouridine. *Nucl. Acids Res.* 24:1073–1079.

D'Amico, S., T. Collins, J.-C. Marx, G. Feller, and C. Gerday. 2006. Psychrophilic microorganisms: challenges for life. *EMBO Rep.* 7:385–389.

D'Amico, S., C. Gerday, and G. Feller. 2001. Structural determinants of cold adaptation and stability in a large protein. *J. Biol. Chem.* 276: 25791–25796.

DeVries, A. L., and D. E. Wohlschlag. 1969. Freezing resistance in some Antarctic fishes. *Science* 163:1073–1075.

Dieser, M., M. Greenwood, and C. M. Foreman. 2010. Carotenoid pigmentation in Antarctic heterotrophic bacteria as a strategy to withstand environmental stresses. *Arct. Antarct. Alp. Res.* 42:396–405.

Dumon, C., A. Varvak, M. A. Wall, J. E. Flint, R. J. Lewis, J. H. Lakey, C. Morland, P. Luginbühl, S. Healey, T. Todaro, G. DeSantis, M. Sun, L. Parra-Gessert, X. Tan, D. P. Weiner, and H. J. Gilbert. 2008. Engineering hyperthermostability into a GH11 xylanase is mediated by subtle changes to protein structure. *J. Biol. Chem.* 283:22557–22564.

Duplantis, B., M. Osusky, C. L. Schmerk, D. R. Ross, C. M. Bosio, and F. E. Nano. 2010. Essential genes from Arctic bacteria used to construct stable, temperature-sensitive bacterial vaccines. *Proc. Natl. Acad. Sci. USA* 107:13456–13460.

Edmonds, C. G., P. F. Crain, R. Gupta, T. Hashizume, C. H. Hocart, J. A. Kowalak, S. C. Pomerantz, K. O. Stetter, and J. A. McCloskey. 1991. Posttranscriptional modification of tRNA in thermophilic archaea (archaebacteria). *J. Bacteriol.* 173:3138–3148.

Eisenthal, R., M. J. Danson, and D. W. Hough. 2007. Catalytic efficiency and k_{cat}/K_M: a useful comparator? *Trends Biotechnol.* 25:247–249.

Empadinhas, N., and M. S. da Costa. 2009. Diversity, distribution and biosynthesis of compatible solutes in prokaryotes. *Contrib. Sci.* 5:95–109.

Fedøy, A., N. Yang, A. Martinez, H. S. Leiros, and I. H. Steen. 2007. Structural and functional properties of isocitrate dehydrogenase from the psychrophilic bacterium *Desulfotalea psychrophila* reveal a cold-active enzyme with an unusual high thermal stability. *J. Mol. Biol.* 372:130–149.

Feller, G., and C. Gerday. 2003. Psychrophilic enzymes: hot topics in cold adaptation. *Nat. Rev. Microbiol.* 1:200–208.

Feller, G., T. Lonhienne, C. Deroanne, C. Libioulle, J. Van Beeumen, and C. Gerday. 1992. Purification, characterization, and nucleotide sequence of the thermolabile α-amylase from the Antarctic psychrotroph *Alteromonas haloplanctis* A23. *J. Biol. Chem.* 267:5217–5221.

Feller, G., P. Sonnet, and C. Gerday. 1995. The β-lactamase secreted by the Antarctic psychrophile *Psychrobacter immobilis* A8. *Appl. Environ. Microbiol.* 61:4474–4476.

Fong, N. J. C., M. L. Burgess, K. D. Barrow, and D. R. Glenn. 2001. Carotenoid accumulation in the psychrotrophic bacterium *Arthrobacter agilis* in response to thermal and salt stress. *Appl. Microbiol. Biotechnol.* 56:750–756.

Gabriel, J. L., and P. K. L. Chong. 2000. Molecular modeling of archaebacterial bipolar tetra-ether lipid membranes. *Chem. Phys. Lipids* 105:193–200.

Garnham, C. P., J. A. Gilbert, C. P. Hartman, R. L. Campbell, J. Laybourn-Parry, and P. L. Davies. 2008. A Ca^{2+} dependent bacterial antifreeze protein domain has a novel β-helical ice-binding fold. *Biochem. J.* 411:171–180.

Georlette, D., V. Blaise, T. Collins, S. D'Amico, E. Gratia, A. Hoyoux, J. C. Marx, G. Sonan, G. Feller, and C. Gerday. 2004. Some like it cold: biocatalysis at low temperatures. *FEMS Microbiol. Rev.* 28:25–42.

Georlette, D., B. Damien, V. Blaise, E. Depiereux, V. N. Uversky, C. Gerday, and G. Feller. 2003. Structural and functional adaptations to extreme temperatures in psychrophilic, mesophilic, and thermophilic DNA ligases. *J. Biol. Chem.* 278:37015–37023.

Georlette, D., Z. O. Jónsson, F. Van Petegem, J. Chessa, J. Van Beeumen, U. Hübscher, and C. Gerday. 2000. A DNA ligase from the psychrophile *Pseudoalteromonas haloplanktis* gives insights into the adaptation of proteins to low temperatures. *Eur. J. Biochem.* 267:3502–3512.

Giaquinto, L., P. M. G. Curmi, K. S. Siddiqui, A. Poljak, E. DeLong, S. DasSarma, and R. Cavicchioli. 2007. Structure and function of cold shock proteins in archaea. *J. Bacteriol.* 189:5738–5748.

Gibson, J. A. E., M. R. Miller, N. W. Davies, G. P. Neill, D. S. Nichols, and J. K. Volkman. 2005. Unsaturated diether lipids in the psychrotrophic archaeon *Halorubrum lacusprofundi*. *Syst. Appl. Microbiol.* 28:19–26.

Gilbert, J. A., P. L. Davies, and J. Laybourn-Parry. 2005. A hyperactive, Ca^{2+}-dependent antifreeze protein in an Antarctic bacterium. *FEMS Microbiol. Lett.* 245:67–72.

Gilbert, J. A., P. J. Hill, C. E. R. Dodd, and J. Laybourn-Parry. 2004. Demonstration of antifreeze protein activity in Antarctic lake bacteria. *Microbiology* 150:171–180.

Gilbert, M., T. Zuber, M. Bunce, R. Rønn, D. Gilichinsky, D. Froese, and E. Willerslev. 2007. Ancient bacteria show evidence of DNA repair. *Proc. Natl. Acad. Sci. USA* 36:14401–14405.

Gilichinsky, D., E. Rivkina, V. Shcherbakova, K. Laurinavichuis, and J. Tiedje. 2003. Supercooled water brines within permafrost—an unknown ecological niche for microorganisms: a model for astrobiology. *Astrobiology* 3:331–341.

Giuliodori, A. M., F. Di Pietro, S. Marzi, B. Masquida, R. Wagner, P. Romby, C. O. Gualerzi, and C. L. Pon. 2010. The *cspA* mRNA is a thermosensor that modulates translation of the cold-shock protein CspA. *Mol. Cell* 37:21–33.

Goldstein, J., N. S. Pollitt, and M. Inouye. 1990. Major cold shock protein of *Escherichia coli*. *Proc. Natl. Acad. Sci. USA* 87:283–287.

Goodchild, A., M. Raftery, N. F. W. Saunders, M. Guilhaus, and R. Cavicchioli. 2005. Cold adaptation of the Antarctic archaeon, *Methanococcoides burtonii* assessed by proteomics using ICAT. *J. Proteome Res.* 4:473–480.

Graumann, P. L., and M. A. Marahiel. 1998. A superfamily of proteins that contain the cold-shock domain. *Trends Biochem. Sci.* 23:286–290.

Grossmann, S. 1994. Bacterial activity in sea ice and open water of the Weddell Sea, Antarctica: a microautoradiographic study. *Microb. Ecol.* 28:1–18.

Gualerzi, C. O., A. M. Giuliodori, and C. L. Pon. 2003. Transcriptional and post-transcriptional control of cold-shock genes. *J. Mol. Biol.* 331:527–539.

Hansen, A. J., D. L. Mitchell, C. Wiuf, L. Paniker, T. B. Brand, J. Binladen, D. A. Gilichinsky, R. Rønn, and E. Willerslev. 2006. Crosslinks rather than strand breaks determine access to ancient DNA sequences from frozen sediments. *Genetics* 173:1175–1179.

Hébraud, M., and P. Potier. 1999. Cold shock response and low temperature adaptation in psychrotrophic bacteria. *J. Mol. Microbiol. Biotechnol.* 1:211–219.

Hirano, S. S., and C. D. Upper. 1995. Ecology of ice nucleation-active bacteria, p. 41–61. *In* R. E. Lee Jr., G. J. Warren, and L. V. Gusta (ed.), *Biological Ice Nucleation and Its Applications*. APS Press, St. Paul, MN.

Hochachka, P.W., and G. N. Somero. 2002. Temperature, p. 290–449. *In Biochemical Adaptation: Mechanism and Process in Physiological Evolution*. Oxford University Press, New York, NY.

Hoshino, T., M. Kiriaki, S. Ohgiya, M. Fujiwara, H. Kondo, Y. Nishimiya, I. Yumoto, and S. Tsuda. 2003. Antifreeze proteins from snow mold fungi. *C. J. Bot.* 81:1175–1181.

Hoyoux, A., I. Jennes, P. Dubois, S. Genicot, F. Dubail, J. M. François, E. Baise, G. Feller, and C. Gerday. 2001. Cold-Adapted β-galactosidase from the Antarctic psychrophile *Pseudoalteromonas haloplanktis*. *Appl. Environ. Microbiol.* 67:1529–1535.

Huston, A. L., B. Methe, and J. W. Deming. 2004. Purification, characterization, and sequencing of an extracellular cold-active aminopeptidase produced by marine psychrophile *Colwellia psychrerythraea* strain 34H. *Appl. Environ. Microbiol.* 70:3321–3328.

Jagannadham, M. V., M. K. Chattopadhyay, C. Subbalakshmi, M. Vairamani, K. Narayanan,

C. M. Rao, and S. Shivaji. 2000. Carotenoids of an Antarctic psychrotolerant bacterium, *Sphingobacterium antarcticus*, and a mesophilic bacterium, *Sphingobacterium multivorum*. *Arch. Microbiol.* 173:418–424.

Jakosky, B. M., K. H. Nealson, C. Bakermans, R. E. Ley, and M. T. Mellon. 2003. Subfreezing activity of microorganisms and the potential habitability of Mars' polar region. *Astrobiology* 3:343–350.

Janech, M. G., A. Krell, T. Mock, J. S. Kang, and J. A. Raymond. 2006. Ice-binding proteins from sea ice diatoms (Bacillariophyceae). *J. Phycol.* 42:410–416.

Johnson, S. S., M. B. Hebsgaard, T. R. Christensen, M. Mastepanov, R. Nielsen, K. Munch, T. Brand, M. T. P. Gilbert, M. T. Zuber, M. Bunce, R. Rønn, D. Gilichinsky, D. Froese, and E. Willerslev. 2007. Ancient bacteria show evidence of DNA repair. *Proc. Natl. Acad. Sci. USA* 104:14401–14405.

Junge, K., B. C. Christner, and J. T. Staley. 2011. Diversity of psychrophilic bacteria from sea ice—and glacial ice communities, p. 794–815. *In* K. Horikoshi, G. Antranikian, A. T. Bull, F. T. Robb, and K. O. Stetter (ed.), *Extremophiles Handbook*. Springer, Berlin, Germany.

Junge, K., H. Eicken, and J. W. Deming. 2003. Motility of *Colwellia psychrerythraea* strain 34H at subzero temperatures. *Appl. Environ. Microbiol.* 69:4282–4284.

Junge, K., H. Eicken, and J. W. Deming. 2004. Bacterial activity at −2 to −20°C in Arctic wintertime sea ice. *Appl. Environ. Microbiol.* 70:550–557.

Junge, K., H. Eicken, B. D. Swanson, and J. W. Deming. 2006. Bacterial incorporation of leucine into protein down to −20°C with evidence for potential activity in sub-eutectic saline ice formations. *Cryobiology* 52:417–429.

Kawahara, H., Y. Iwanaka, S. Higa, N. Muryoi, M. Sato, M. Honda, H. Omura, and H. Obata. 2007. A novel, intracellular antifreeze protein in an Antarctic bacterium, *Flavobacterium xanthum*. *Cryo Lett.* 28:39–49.

Kawamoto, J., T. Kurihara, M. Kitagawa, I. Kato, and N. Esaki. 2007. Proteomic studies of an Antarctic cold-adapted bacterium, *Shewanella livingstonensis* Ac10, for global identification of cold-inducible proteins. *Extremophiles* 11:819–826.

Kempf, B., and E. Bremer. 1998. Uptake and synthesis of compatible solutes as microbial stress responses to high-osmolarity environments. *Arch. Microbiol.* 170:319–330.

Kim, S. Y., K. Y. Hwang, S. H. Kim, H. C. Sung, Y. S. Han, and Y. Cho. 1999. Structural basis for cold adaptation. Sequence, biochemical properties, and crystal structure of malate dehydrogenase from a psychrophile *Aquaspirillium arcticum*. *J. Biol. Chem.* 274:11761–11767.

Knoblauch, C., B. B. Jørgensen, and J. Harder. 1999. Community size and metabolic rates of psychrophilic sulfate-reducing bacteria in Arctic marine sediments. *Appl. Environ. Microbiol.* 65:4230–4233.

Ko, R., L. T. Smith, and G. M. Smith. 1994. Glycine betaine confers enhanced osmotolerance and cryotolerance on *Listeria monocytogenes*. *J. Bacteriol.* 176:426–431.

Koshland, D. E., Jr. 2002. The application and usefulness of the ratio k_{cat}/K_M. *Bioorg. Chem.* 30:211–213.

Krembs, C., H. Eicken, K. Junge, and J. W. Deming. 2002. High concentrations of exopolymeric substances in Arctic winter sea ice: implications for the polar ocean carbon cycle and cryoprotection of diatoms. *Deep Sea Res. Part 1 Oceanogr. Res. Pap.* 49:2163–2181.

Lettinga, G., S. Rebac, and G. Zeeman. 2001. Challenge of psychrophilic anaerobic wastewater treatment. *Trends Biotechnol.* 19:363–370.

Li, Y., F. Li, X. Zhang, S. Qin, Z. Zeng., H. Dang., and Y. Qin. 2008. Vertical distribution of bacterial and archaeal communities along discrete layers of a deep-sea cold sediment sample at the East Pacific Rise (~13°N). *Extremophiles* 12:573–585.

Liebner, S., K. Rublack, T. Stuehrmann, and D. Wagner. 2009. Diversity of aerobic methanotrophic bacteria in a permafrost active layer soil of the Lena Delta, Siberia. *Microb. Ecol.* 57:25–35.

Lim, J., T. Thomas, and R. Cavicchioli. 2000. Low temperature regulated DEAD-box RNA helicase from the Antarctic archaeon, *Methanococcoides burtonii*. *J. Mol. Biol.* 297:553–567.

Lipson, D. A., C. W. Schadt, and S. K. Schmidt. 2002. Changes in microbial community structure and function in an alpine dry meadow following spring snow melt. *Microb. Ecol.* 43:307–314.

Los, D. A., and N. Murata. 2004. Membrane fluidity and its role in the perceptions of environmental signals. *Biochim. Biophys. Acta* 1666:142–147.

Lundheim, R. 2002. Physiological and ecological significance of biological ice nucleators. *Philos. Trans. R. Soc. Lond. B* 357:937–943.

Mader, H. M., M. E. Pettitt, J. L. Wadham, E. W. Wolff, and R. J. Parkes. 2006. Subsurface ice as a microbial habitat. *Geology* 34:169–172.

Mancuso Nichols, C. A., J. Guezenec, and J. P. Bowman. 2005. Bacterial exopolysaccharides from extreme marine environments with special consideration of the Southern Ocean, sea ice, and deep-sea hydrothermal vents: a review. *Mar. Biotechnol.* 7:253–271.

Mandelman, D., M. Bentahir, G. Feller, C. Gerday, and R. Haser. 2001. Crystallization and preliminary X-ray analysis of a bacterial psychrophilic enzyme, phosphoglycerate kinase. *Acta Crystallogr. D Biol. Crystallogr.* **57:**1666–1668.

McLeod, M., J. G. Bockheim, and M. R. Balks. 2008. Glacial geomorphology, soil development and permafrost features in central-upper Wright Valley, Antarctica. *Geoderma* **144:**93–103.

Metpally, R. P. R., and B. V. B. Reddy. 2009. Comparative proteome analysis of psychrophilic versus mesophilic bacterial species: insights into the molecular basis of cold adaptation of proteins. *BMC Genomics* **10:**11.

Mikan, C. J., J. P. Schimel, and A. P. Doyle. 2002. Temperature controls of microbial respiration in Arctic tundra soils above and below freezing. *Soil Biol. Biochem.* **34:**1785–1795.

Mironenko, M. V., S. A. Grant, G. M. Marion, and R. E. Farren. 1997. FREZCHEM2: a chemical thermodynamic model for electrolyte solutions at subzero temperatures. CRREL Spec. Rept. 97-5. USACRREL, Hanover, NH.

Miteva, V., T. Sowers, and J. Brenchley. 2007. Production of N_2O by ammonia oxidizing bacteria at subfreezing temperatures as a model for assessing the N_2O anomalies in the Vostok ice core. *Geomicrobiol. J.* **24:**451–459.

Moissl, C., C. Rudolph, R. Rachel, M. Koch, and R. Huber. 2003. In situ growth of the novel SM1 euryarchaeon from a string-of-pearls-like microbial community in its cold biotope, its physical separation and insights into its structure and physiology. *Arch. Microbiol.* **180:**211–217.

Morita, R. Y. 1988. Bioavailability of energy and its relationship to growth and starvation survival in nature. *Can. J. Microbiol.* **34:**436–441.

Muryoi, N., M. Sato, S. Kaneko, H. Kawahara, H. Obata, M. W. Yaish, M. Griffith, and B. R. Glick. 2004. Cloning and expression of *afpA*, a gene encoding an antifreeze protein from the Arctic plant growth-promoting rhizobacterium *Pseudomonas putida* GR12–2. *J. Bacteriol.* **186:**5661–5671.

Nahvi, A., N. Sudarsan, M. S. Ebert, X. Zou, K. L. Brown, and R. R. Breaker. 2002. Genetic control by a metabolite binding mRNA. *Chem. Biol.* **9:**1043–1049.

Napolitano, M. J., and D. H. Shain. 2005. Distinctions in adenylate metabolism among organisms inhabiting temperature extremes. *Extremophiles* **9:**93–98.

Narberhaus, F., T. Waldminghaus, and S. Chowdhury. 2006. RNA thermometers. *FEMS Microbiol. Rev.* **30:**3–16.

Narinx, E., E. Baise, and C. Gerday. 1997. Subtilisin from psychrophilic Antarctic bacteria: characterization and site-directed mutagenesis of residues

possibly involved in the adaptation to cold. *Protein Eng.* **10:**1271–1279.

Nichols, D. S., M. R. Miller, N. W. Davies, A. Goodchild, M. Raftery, and R. Cavicchioli. 2004. Cold adaptation in the Antarctic archaeon *Methanococcoides burtonii* involves membrane lipid unsaturation. *J. Bacteriol.* **186:**8508–8515.

Noon, K. R., R. Guymon, P. F. Crain, J. A. McCloskey, M. Thomm, J. Lim, and R. Cavicchioli. 2003. Influence of temperature on tRNA modification in archaea: *Methanococcoides burtonii* (optimum growth temperature [T_{opt}], 23°C) and *Stetteria hydrogenophila* (T_{opt}, 95°C). *J. Bacteriol.* **185:**5483–5490.

Panikov, N. S. 2009. Microbial activity in frozen soils. p. 119–147. *In* R. Margesin (ed.), *Permafrost Soils*. Springer, Berlin, Germany.

Panikov, N. S., P. W. Flanagan, W. C. Oechel, M. A. Mastepanov, and T. R. Christensen. 2006. Microbial activity in soils frozen to below −39°C. *Soil Biol. Biochem.* **38:**785–794.

Panikov, N. S., and M. V. Sizova. 2007. Growth kinetics of microorganisms isolated from Alaskan soil and permafrost in solid media frozen down to 35°C. *FEMS Microbiol. Ecol.* **59:**500–512.

Panoff, J. M., D. Corroler, B. Thammavongs, and P. Boutibonnes. 1997. Differentiation between cold shock proteins and cold acclimation proteins in a mesophilic gram-positive bacterium, *Enterococcus faecalis* JH2-2. *J. Bacteriol.* **179:**4451–4454.

Paul, S., S. K. Bag, S. Das, E. T. Harvill, and C. Dutta. 2008. Molecular signature of hypersaline adaptation: insights from genome and proteome composition of halophilic prokaryotes. *Genome Biol.* **9:**R70.

Pearson, R. T., and W. Derbyshire. 1974. NMR studies of water adsorbed on a number of silica surfaces. *J. Colloid Interface Sci.* **46:**232–248.

Phadtare, S., and M. Inouye. 2008. Cold-shock proteins, p. 191–209. *In* R. Margesin, F. Schinner, J.-C. Marx, and C. Gerday (ed.), *Psychrophiles: from Biodiversity to Biotechnology*. Springer, Berlin, Germany.

Piette, F., S. D'Amico, C. Struvay, G. Mazzucchelli, J. Renaut, M. L. Tutino, A. Danchin, P. Leprince, and G. Feller. 2010. Proteomics of life at low temperatures: trigger factor is the primary chaperone in the Antarctic bacterium *Pseudoalteromonas haloplanktis* TAC125. *Mol. Microbiol.* **76:**120–132.

Poolman, B., and E. Glaasker. 1998. Regulation of compatible solute accumulation in bacteria. *Mol. Microbiol.* **29:**397–407.

Price, P. B. 2000. A habitat for psychrophiles in deep Antarctic ice. *Proc. Natl. Acad. Sci. USA* **97:**1247–1251.

Price, P. B., and T. Sowers. 2004. Temperature dependence of metabolic rates for microbial growth, maintenance, and survival. *Proc. Natl. Acad. Sci. USA* **101**:4631–4636.

Priscu, J. C., C. H. Fritsen, E. E. Adams, S. J. Giovannoni, H. W. Paerl, C. P. McKay, P. T. Doran, D. A. Gordon, B. D. Lanoil, and J. L. Pinckney. 1998. Perennial Antarctic lake ice: an oasis for life in a polar desert. *Science* **280**:2095–2098.

Qin, G., L. Zhu, X. Chen, P. G. Wang, and Y. Zhang. 2007. Structural characterization and ecological roles of a novel exopolysaccharide from the deep-sea psychrotolerant bacterium *Pseudoalteromonas* sp. SM9913. *Microbiology* **153**:1566–1572.

Ræder, I. L. U., I. Leiros, N. P. Willassen, A. O. Smalas, and E. Moe. 2008. Uracil-DNA N-glycosylase (UNG) from the marine pyschrophilic bacterium *Vibrio salmonicida* shows cold adapted features: a comparative analysis to *Vibrio cholerae* uracil-DNA N-glycosylase. *Enzyme Microb. Technol.* **42**:594–600.

Raymond, J. A., B. C. Christner, and S. C. Schuster. 2008. A bacterial ice-binding protein from the Vostok ice core. *Extremophiles* **12**:713–717.

Raymond, J. A., and C. H. Fritsen. 2001. Semi-purification and ice recrystallization inhibition activity of ice-active substances associated with Antarctic photosynthetic organisms. *Cryobiology* **43**:63–70.

Raymond, J. A., C. Fritsen, and K. Shen. 2007. An ice-binding protein from an Antarctic sea ice bacterium. *FEMS Microbiol. Ecol.* **61**:214–221.

Reid, I. N., W. B. Sparks, S. Lubow, M. McGrath, M. Livio, J. Valenti, K. R. Sowers, H. D. Shukla, S. MacAuley, T. Miller, R. Suvanasuthi, R. Belas, A. Colman, F. T. Robb, P. DasSarma, J. A. Müller, J. A. Coker, R. Cavicchioli, F. Chen, and S. DasSarma. 2006. Terrestrial models for extraterrestrial life: methanogens and halophiles at Martian temperatures. *Int. J. Astrobiol.* **5**:89–97.

Ritzrau, W. 1997. Pelagic microbial activity in the Northeast Water Polynya, summer 1992. *Polar Biol.* **17**:259–267.

Rivkina, E. M., E. I. Friedmann, C. P. McKay, and D. A. Gilichinsky. 2000. Metabolic activity of permafrost bacteria below the freezing point. *Appl. Environ. Microbiol.* **66**:3230–3233.

Rivkina, E., K. Laurinavichius, J. McGrath, J. Tiedje, V. Shcherbakova, and D. Gilichinsky. 2004. Microbial life in permafrost. *Adv. Space Res.* **33**:1215–1221.

Roberts, M. F. 2005. Organic compatible solutes of halotolerant and halophilic microorganisms. *Saline Systems* **1**:5.

Rodionov, D. A., A. G. Vitreschak, A. A. Mironov, and M. S. Gelfand. 2003. Regulation of lysine biosynthesis and transport genes in bacteria: yet another RNA riboswitch? *Nucl. Acids Res.* **31**:3–16.

Russell, N. J. 2008. Membrane components and cold sensing, p. 177–190. *In* R. Margesin, F. Schinner, J.-C. Marx, and C. Gerday (ed.), *Psychrophiles: from Biodiversity to Biotechnology*. Springer, Berlin, Germany.

Russell, R. J. M., U. Gerike, M. J. Danson, D. W. Hough, and G. L. Taylor. 1998. Structural adaptations of the cold-active citrate synthase from an Antarctic bacterium. *Structure* **6**:351–361.

Sand-Jensen, K., N. L. Pedersen, and M. Søndergaard. 2007. Bacterial metabolism in small temperate streams under contemporary and future climates. *Freshw. Biol.* **52**:2340–2353.

Saunders, N. F. W., T. Thomas, P. M. G. Curmi, J. S. Mattick, E. Kuczek, R. Slade, J. Davis, P. D. Franzmann, D. Boone, K. Rusterholtz, R. Feldman, C. Gates, S. Bench, K. Sowers, K. Kadner, A. Aerts, P. Dehal, C. Detter, T. Glavina, S. Lucas, P. Richardson, F. Larimer, L. Hauser, M. Land, and R. Cavicchioli. 2003. Mechanisms of thermal adaptation revealed from the genomes of the Antarctic archaea *Methanogenium frigidum* and *Methanococcoides burtonii*. *Genome Res.* **13**:1580–1588.

Schimel, J. P., C. Bilbrough, and J. M. Welker. 2004. Increased snow depth affects microbial activity and nitrogen mineralization in two Arctic tundra communities. *Soil Biol. Biochem.* **36**:217–227.

Schleper, C., R. V. Swanson, E. J. Mathur, and E. F. DeLong. 1997. Characterization of a DNA polymerase from the uncultivated psychrophilic archaeon *Cenarchaeum symbiosum*. *J. Bacteriol.* **179**:7803–7811.

Schouten, S., E. C. Hopmans, E. Schefuss, and J. S. Sinninghe Damsté. 2002. Distributional variations in marine crenarchaeotal membrane lipids: a new tool for reconstructing ancient sea water temperatures? *Earth Planet. Sci. Lett.* **204**:265–274.

Shivaji, S., and J. S. S. Prakash. 2010. How do bacteria sense and respond to low temperature? *Arch. Microbiol.* **192**:85–95.

Siddiqui, K. S., R. Cavicchioli, and T. Thomas. 2002. Thermodynamic activation properties of elongation factor 2 (EF-2) proteins from psychrotolerant and thermophilic Archaea. *Extremophiles* **6**:143–150.

Sinensky, M. 1974. Homeoviscous adaptation—a homeostatic process that regulates the viscosity of membrane lipids in *Escherichia coli*. *Proc. Natl. Acad. Sci. USA* **71**:522–525.

Sinninghe Damsté, J. S., S. Schouten, E. C. Hopmans, A. C. T. van Duin, and J. A. J. Geenevasen. 2002. Crenarchaeol: the characteristic core glycerol dibiphytanyl glycerol tetraether membrane lipid of cosmopolitan pelagic crenarchaeota. *J. Lipid Res.* **43:**1641–1651.

Smith, J. J., L. A. Tow, W. Stafford, C. Cary, and D. Cowan. 2006. Bacterial diversity in three different Antarctic cold desert mineral soils. *Microb. Ecol.* **51:**413–421.

Smith, R. E. H., and P. Clement. 1990. Heterotrophic activity and bacterial productivity in assemblages of microbes from sea ice in the high Arctic. *Polar Biol.* **10:**351–357.

Sowers, T. 2001. The N_2O record spanning the penultimate deglaciation from the Vostok ice core. *J. Geophys. Res.* **106:**903–931.

Steven, B., R. Leveille, W. H. Pollard, and L. G. Whyte. 2006. Microbial ecology and biodiversity in permafrost. *Extremophiles* **10:**259–267.

Steven, B., W. H. Pollard, C. W. Greer, and L. G. Whyte. 2007. Microbial diversity and activity through a permafrost/ground ice core profile from the Canadian high Arctic. *Environ. Microbiol.* **10:**3388–3403.

Strand, A., S. Shivaji, and S. Liaaen-Jensen. 1997. Bacterial carotenoids 55. C_{50}-carotenoids 25. Revised structures of carotenoids associated with membranes in psychrotrophic *Micrococcus roseus. Biochem. Syst. Ecol.* **25:**547–552.

Takacs, C. T., and J. C. Priscu. 1998. Bacterioplankton dynamics in the McMurdo Dry Valley lakes: production and biomass loss over four seasons. *Microb. Ecol.* **36:**239–250.

Thomas, T., and R. Cavicchioli. 1998. Archaeal cold-adapted proteins: structural and evolutionary analysis of the elongation factor 2 proteins from psychrophilic, mesophilic and thermophilic methanogens. *FEBS Lett.* **439:**281–286.

Thomas, T., and R. Cavicchioli. 2000. Effect of temperature on stability and activity of elongation factor 2 proteins from Antarctic and thermophilic methanogens. *J. Bacteriol.* **182:**1328–1332.

Thomas, T., and R. Cavicchioli. 2002. Cold adaptation of archaeal elongation factor 2 (EF-2) proteins. *Curr. Prot. Pept. Sci.* **3:**223–230.

Thomas, T., N. Kumar, and R. Cavicchioli. 2001. Effects of ribosomes and intracellular solutes on activities and stabilities of elongation factor 2 proteins from psychrotolerant and thermophilic methanogens. *J. Bacteriol.* **183:**1974–1982.

Tung, H. C., P. B. Price, N. E. Bramall, and G. Vrdoljak. 2006. Microorganisms metabolizing on clay grains in 3-km-deep Greenland basal ice. *Astrobiology* **6:**69–86.

Vitreschak, A. G., D. A. Rodionov, A. A. Mironov, and M. S. Gelfand. 2004. Riboswitches: the oldest mechanism for the regulation of gene expression? *Trends Genet.* **20:**44–50.

Walker, V. K., G. R. Palmer, and G. Voordouw. 2006. Freeze-thaw tolerance and clues to winter survival of a soil community. *Appl. Environ. Microbiol.* **72:**1784–1792.

Wallon, G., S. T. Lovett, C. Magyar, A. Svingor, A. Szilagyi, P. Zàvodszky, D. Ringe, and G. A. Petsko. 1997. Sequence and homology model of 3-isopropylmalate dehydrogenase from the psychrotrophic bacterium *Vibrio* sp. I5 suggest reasons for thermal instability. *Protein Eng.* **10:**665–672.

Watanabe, S., Y. Yasutake, I. Tanaka, and Y. Takada. 2005. Elucidation of stability determinants of cold-adapted monomeric isocitrate dehydrogenase from a psychrophilic bacterium, *Colwellia maris*, by construction of chimeric enzymes. *Microbiology* **151:**1083–1094.

Williams, T. J., D. W. Burg, M. J. Raftery, A. Poljak, M. Guilhaus, O. Pilak, and R. Cavicchioli. 2010. Global proteomic analysis of the insoluble, soluble, and supernatant fractions of the psychrophilic archaeon *Methanococcoides burtonii.* Part I: The effect of growth temperature. *J. Proteome Res.* **9:**640–652.

Wood, J. M., E. Bremer, L. N. Csonka, R. Kraemer, B. Poolman, T. van der Heide, and L. T. Smith. 2001. Osmosensing and osmoregulatory compatible solute accumulation by bacteria. *Comp. Biochem. Physiol. A Mol. Integr. Physiol.* **130:**437–460.

Yamashita, Y., N. Nakamura, K. Omiya, J. Nishikawa, H. Kawahara, and H. Obata. 2002. Identification of an antifreeze lipoprotein from *Moraxella* sp. of Antarctic origin. *Biosci. Biotechnol. Biochem.* **66:**239–247.

GENOMIC AND EXPRESSION ANALYSES OF COLD-ADAPTED MICROORGANISMS

Corien Bakermans, Peter W. Bergholz,
Debora F. Rodrigues, Tatiana A. Vishnivetskaya,
Héctor L. Ayala-del-Río, and James M. Tiedje

6

INTRODUCTION

Cold adaptation of microorganisms is the result of intrinsic genome-wide changes that facilitate growth at temperatures of 5°C or below (Feller and Gerday, 2003; Cavicchioli and Siddiqui, 2004; Rodrigues and Tiedje, 2008). These adaptations can be detected via genomic analyses and may only be apparent at the genome level (rather than at the level of individual genes) due to differences between organisms resulting from genetic drift or the specific environment. For example, biases in amino acid abundance of the genomes of hyperthermophiles have been reported and reflect adaptations to living at high temperatures (Singer and Hickey, 2003). In addition, examination of gene content has been used to better understand the metabolic capabilities of microorganisms with the smallest genomes like *Mycoplasma genitalium*, *Chlamydia*, and the insect symbionts, e.g., *Carsonella ruddii* (Fraser et al., 1995; Read et al., 2000; McCutcheon and Moran, 2007).

Genomic analysis does not have to stop at the sequence level—complex metabolic changes in the whole organism can be elucidated using postgenomic technologies. For example, microarrays can be used to study the transcriptome (all the genes expressed during specific culture conditions) and have been used extensively to examine cold shock (Phadtare and Inouye, 2004; Gao et al., 2006; Beckering et al., 2002). However, there are very few studies of the transcriptome or proteome (all the proteins expressed during specific conditions) during reproductive growth at cold temperatures, especially at temperatures below 0°C, which dominate in polar environments (Budde et al., 2006). Examination of the transcriptome and proteome enables the investigation of the underlying gene and protein expression, respectively, that results in cold adaptation and ultimately permits the successful colonization of cold environments by cold-adapted microorganisms.

Corien Bakermans, Division of Mathematics and Natural Sciences, Altoona College, Pennsylvania State University, 3000 Ivyside Park, Altoona, PA 16601. *Peter W. Bergholz,* Department of Food Science, Cornell University, Stocking Hall, Room 412, Ithaca, NY 14853. *Debora F. Rodrigues,* Department of Civil and Environmental Engineering, University of Houston, N107 Engineering Building 1, Houston, TX 77204-4003. *Tatiana A. Vishnivetskaya,* Center for Environmental Biotechnology, The University of Tennessee, 676 Dabney-Buehler Hall, 1416 Circle Drive, Knoxville, TN 37996-1605, and Biosciences Division, Oak Ridge National Laboratory, Oak Ridge, TN 37831-6038. *Héctor L. Ayala-del-Río,* Department of Biology, University of Puerto Rico at Humacao, Humacao, PR 00791. *James M. Tiedje,* Center for Microbial Ecology, 540 Plant and Soil Sciences Building, Michigan State University, East Lansing, MI 48824-1325.

Polar Microbiology: Life in a Deep Freeze
Edited by Robert V. Miller and Lyle G. Whyte © 2012 ASM Press, Washington, DC

Comparative studies of cold-adapted organisms are beginning to reveal which adaptations are common to all psychrophiles and which are specific to the particular environment each psychrophile inhabits or the particular family of organisms it represents. To date, 47 cold-adapted microorganisms have been completely sequenced (Casanueva et al., 2010), accounting for a mere 3% of all prokaryotic genomes sequenced (~1,500). Approximately half (25) of these cold-adapted organisms have been isolated from polar regions (Table 1), while the majority (40) have been isolated from cold marine environments, which are distinctly different from cold terrestrial environments like permafrost. Marine environments tend to have high thermal stability as well as stable solute concentrations, while terrestrial environments do not. Consequently, genomic analysis of microorganisms isolated from marine environments likely reflects adaptations to stable low temperatures and homogenous environmental chemistry, while genomic analysis of microorganisms isolated from terrestrial environments may reveal unique mechanisms of cold adaptation related to heterogenous environmental chemistry due to extreme temperature fluctuations and its effect on the extent and composition of thin films of liquid water within a highly constrained physical environment (Bock and Eicken, 2005; Rivkina et al., 2000).

For many years, it has been known that cold-adapted organisms can be classified into two overlapping groups based on their cardinal growth temperatures: psychrophiles, which have a maximum growth temperature (T_{max}) of <20°C; and psychrotrophs (or psychrotolerants), which have a T_{max} of >20°C (Morita, 1975). More recently, the terms "stenopsychrophile" and "eurypsychrophile" have been favored (Feller and Gerday, 2003; Cavicchioli, 2006; Bakermans and Nealson, 2004) for cold-adapted microorganisms with a narrow or a wide tolerance to low temperatures, respectively. Notably, stenopsychrophiles (also known as true psychrophiles) are isolated most frequently from thermally stable cold marine

environments, while eurypsychrophiles (also known as psychrotrophs) dominate in cold terrestrial ecosystems (Helmke and Weyland, 2004). Historically, the term "psychrotolerant" or "psychrotroph" was applied to many eurypsychrophiles, but these terms are inaccurate when applied to polar microbes, because these organisms are often isolated only from habitats with temperatures of <4°C, indicating that they are not only viable in cold habitats, but better able to compete against other species in the cold (Feller and Gerday, 2003). The genomic basis of differences in cold adaptation between stenopsychrophiles and eurypsychrophiles has yet to be thoroughly investigated. Here, we use the term "psychrophile" to describe both groups of cold-adapted organisms.

Genomics can be used to investigate cold adaptation at the level of whole genes by examining gene content, gene expression, protein expression, and other unique features, while at the molecular level, genomic analyses may identify trends in amino acid composition, codon usage, and nucleotide content that result from cold adaptation. In this chapter we discuss (i) use of ecological information to discern cold-adapted microorganisms, (ii) unique gene- and protein-expression adaptations for coping with cold environment stresses, (iii) sequence adaptations that facilitate protein function at low temperature, and (iv) a case study comparing cold-adapted and warm-adapted species of the genus *Exiguobacterium*.

ECOLOGICAL EVIDENCE OF BACTERIAL ADAPTATION TO COLD

Earth's cold environments have temperatures that are close to or below the freezing point of water and constitute some 75% of the biosphere (Table 2). Permanently cold environments include a variety of polar and alpine habitats, the deep ocean, and permafrost, while seasonally and artificially cold environments include snow, lake ice, surface soils frozen in winter, and refrigerated appliances and products. These environments are inhabited by a variety of cold-adapted (psychrophilic) organisms such as bacteria, archaea, yeasts,

TABLE 1 Genome sequences of cold-adapted microbes from polar regions

Microorganism	Habitat	Growth temp range (°C)	Reference(s)
Terrestrial			
Exiguobacterium sibiricum 255-15	Siberian permafrost, 2 million–3 million years old	−5 to 39	Rodrigues et al., 2008
Psychrobacter arcticus 273-4	Siberian permafrost, 20,000–40,000 years old	−10 to 28	Ayala-del-Río et al., 2010
Psychrobacter cryohalolentis K5	Cryopeg in 40,000-year-old Siberian permafrost	−10 to 30	NP[a]
"*Planococcus halocryophilus*" Or1	Permafrost, Axel Heiberg Island, Nunavut, Canada		L. G. Whyte, personal communication[b]
Acidobacteria species Terriglobus saanensis SP1PR4 "*Granulicella mallensis*" MP5ACTX8 "*Granulicella tundricola*" MP5ACTX9	Alpine soils, Kilpisjärvi region in northern Finland		M. M. Häggblom, personal communication[b]
Polar lakes			
Halorubrum lacusprofundi ATCC 49239	Deep Lake, Antarctica	0 to 42	NP
Marinobacter sp. ELB17	East Lobe Lake Bonney, Antarctica	(T_{opt}^c = 12)	NP
Methanococcoides burtonii DSM 6242	Ace Lake, Antarctica	<4 to 29	Allen et al., 2009; Saunders et al., 2003
Methanogenium frigidum Ace-2	Ace Lake, Antarctica	0 to 18	Saunders et al., 2003
Polar sea ice			
Psychromonas ingrahamii 37	Sea ice off Point Barrow, Alaska	−12 to 10	Riley et al., 2008
Shewanella frigidimarina NCIMB 400	Sea ice, Antarctica	<0 to 28	NP
Polar seas			
Agreia sp. PHSC20c1	Coastal Antarctica, western Antarctic Peninsula		NP
Octadecabacter antarcticus 307	McMurdo Sound, Antarctica	<0 to <19	NP
Octadecabacter antarcticus 238	350 km offshore off Deadhorse, Alaska	<0 to <19	NP
Polaribacter filamentus 215	Surface seawater in Arctic Ocean	<4 to 19	NP
Polaribacter irgensii 23-P	Penola Strait, western Antarctic Peninsula	−1.5 to 10	NP
Pseudoalteromonas haloplanktis TAC125	Antarctic coastal seawater	<0 to 20	Medigue et al., 2005
Psychroflexus torquis ATCC 700755	Prydz Bay, Antarctica	<0 to 19	NP
Sphingopyxis alaskensis RB2256	Deep fjord of Gulf of Alaska	4 to 44	Allen et al., 2009
Polar seas, animal associated			
Aliivibrio salmonicida LFI1238	Fish pathogen of Atlantic cod from Hammerfest, Norway	1 to 22	Hjerde et al., 2008
Pseudoalteromonas tunicate D2	From adult tunicate *Ciona intestinalis*, 10-m depth, Gullmarsjorden, Sweden	4 to <37 (T_{opt} = 28)	NP

(Continued)

TABLE 1 (*Continued*)

Microorganism	Habitat	Growth temp range (°C)	Reference(s)
Polar marine sediments			
Colwellia psychrerythraea 34H	Arctic marine sediments	−1 to 10	Methe et al., 2005
Desulfotalea psychrophila LSv54	Arctic sediments off the coast of Svalbard	−1.8 to 19 (T_{opt} = 10)	Rabus et al., 2004

[a]NP, not published.
[b]These sequences have been completed but not yet released.
[c]T_{opt}, optimal temperature for maximum growth rate.

algae, insects, marine and terrestrial invertebrates, fish, and plants, but are dominated by microorganisms in terms of species diversity and biomass (Feller and Gerday, 2003). Psychrophiles are particularly abundant in the cold deep oceans that cover ~70% of the surface of the planet.

Cold temperature plays a critical role in the selection and survival of the microorganisms inhabiting these cold environments via its direct impact on organisms as well as through its influence on other environmental parameters such as reduction of water availability, elevated salt concentrations, inaccessibility of organic and inorganic nutrients, and long-term influence of gamma radiation from ^{40}K in soil minerals. Despite the numerous challenges of cold environments, some cold-adapted microbes grow in the 0 to 10°C range, some grow at temperatures down to −10°C, and some are able

TABLE 2 Cold environments on Earth

Environment	Definition and dominant features	Area occupied (km^2)
Permafrost and frozen ground	Perennially or seasonally frozen ground High organic content (Arctic), congelation ice, $T < 0$°C	54 million
Cryopegs	Lenses of "overcooled brine" found within permafrost Moderate to high salinity, $T < 0$°C	Negligible
Snow	Precipitation within the atmosphere, crystalline water ice Meteoric ice, high porosity, seasonal, $T < 0$°C	47 million
Lake and river ice	Ice that forms in response to seasonal cooling Low to moderate salinity, congelation ice, seasonal, $T < 0$°C	Negligible, highly variable
Glaciers and ice sheets	Ice masses that rest on solid land Meteoric ice, low nutrient, $T < 0$°C	>50,000
Arctic Ocean	Water between North America and Eurasia at the North Pole Moderate salinity (~3.5%), stable T (−2 to 4°C), largely covered by sea ice	14.0 million
Antarctic Ocean	Water surrounding the Antarctic continent Moderate salinity (~3.5%), stable T (−2 to 4°C)	20.3 million
Antarctic sea ice	Formed from seawater that freezes in the polar oceans Moderate to high salinity, congelation ice, seasonal, $T < 0$°C	3 million–4 million (summer minimum) 17 million–20 million (winter maximum)
Deep ocean	Lowest layer of the ocean at depths below 1,500–1,800 m High pressure (up to 110 MPa), saline, stable T (1 to 5°C)	(90% of the volume of the oceans)

to metabolize in snow and ice at temperatures down to −20°C (Junge et al., 2004; Carpenter et al., 2000). Some models predict the ability of microorganisms to metabolize at −40°C and survive at −45°C (Cavicchioli, 2006; Price and Sowers, 2004).

Characteristics of Cold Environments and Their Implications for Microbial Ecology

Important environmental factors affected by temperature that require specific gene- and protein-based adaptations by psychrophiles are the properties of water, nutrient availability, pH, and salinity.

The most relevant changes in the properties of water at lower temperatures (Table 3) are reduced dissociation constant (K_w), reduced diffusion, and increased viscosity (Angell, 1982). The most dramatic effect, at least from a biochemistry viewpoint, is the degree of ionization of water (measured as K_w), which decreases with decreasing temperature (Hepler and Woolley, 1973). Given that H^+ and OH^- are involved in most biochemical reactions, the decrease in ionization will affect enzyme equilibrium and kinetics at low temperature (Franks et al., 1990). Reduced diffusion rates of water (and chemicals dissolved in water) are a direct result of the increase in viscosity of the water at cold temperatures.

Reduced diffusion rates at low temperature cause a decrease in the availability of substrates necessary for growth (i.e., organic carbon, phosphates, sulfates, and nitrogenous compounds). Therefore, microbial growth and survival in cold environments depend upon obtaining and sequestering these sparse resources more successfully than competing species. The efficiency with which nutrients can be obtained through active transport will depend on the affinity of uptake mechanisms for their substrates (Nedwell, 1999). The ability of an organism to sequester substrate (the specific affinity, a_A) can be described by a Monod-type saturation curve relating maximum specific growth rate (μ_{max}) to the concentration of the rate-limiting substrate (Gottschal, 1985) as follows: $a_A = \mu_{max}/K_s$, where K_s is the substrate affinity constant (Nedwell and Rutter, 1994). The general consensus among studies is that when environmental temperatures decrease below the optimum for growth there is concomitant decrease in a_A (Ponder et al., 2005; Herbert and Bell, 1977; Nedwell and Rutter, 1994), which applies to the uptake of both organic and inorganic substrates. Hence, the combination of reduced diffusion of and affinity for substrates presents a formidable challenge to microorganisms in cold environments, which are likely starved for nutrients (Nedwell, 1999).

Temperature affects the balance of carbonate and bicarbonate ions in water or soil solution, and these differences can result in variable pH of the environment (Thomas, 1996). The pH per se also influences availability of nutrients,

TABLE 3 Physical properties of liquid water at −25 and +25°C[a]

Water properties	Result at temp of:	
	−25°C	+25°C
Density (g cm^{-3})	0.987	0.996
Heat capacity, C_p (J mol^{-1} K^{-1})	80	75
Isothermal compressibility (10^6 MPa^{-1})	720	440
Hypersonic sound velocity (m s^{-1})	1,220	1,480
Dielectric permittivity	102	79
Self-diffusion coefficient (10^5 cm^2 s^{-1})	0.321	2.23
Viscosity (mPa s)	6.5	0.89
pK_w	17.3	14.0

[a]Adapted from Angell (1982) with permission from Springer Science + Business Media B.V.

carbon, essential trace elements, and enzymatic reactions, and therefore has a great impact on bacterial growth. Hence, even minor changes in the pH resulting from temperature change can alter the selective advantage of different microbial community members. Indeed, a recent examination of microbial biogeography in Arctic soils identified pH as the primary factor affecting phylogenetic composition of the communities (Chu et al., 2010).

At temperatures below 0°C, the salinity of liquid water increases significantly. As water freezes, salt is excluded because salt has a different crystalline structure: salt forms cubic crystals, whereas ice is hexagonal. The resulting brine is denser and has a depressed freezing point. It is common to find pockets or channels of brine forming within ice or frozen soil, particularly when initial solute concentrations are high, as in seawater (Golden, 2001; Collins et al., 2008). The increase in salinity affects microorganisms through osmotic stress, which leads to the movement of water out of cells through osmosis. Osmotic stress also inhibits transport of substrates and cofactors into the cell, which can lead to growth inhibition (Walter et al., 1987). Because of the effects of low temperatures on salinity, microorganisms from cold environments frequently exhibit adaptations similar to those of halotolerant and halophilic microorganisms (Leblanc et al., 2003; Ponder et al., 2005).

Besides cold temperatures, psychrophilic microorganisms must cope with the above unique environmental stresses generated by low temperatures. These environmental factors would be expected to play an important role in the selection and success of psychrophiles, due to the need to combine unique physiological adaptations for survival in cold environments.

Ecological Adaptation in *Exiguobacterium* spp. and *Psychrobacter* spp.

Key evidence for successful adaptation to cold comes from ecological information, namely that cold-adapted organisms would be expected to preferentially occupy, or be more

numerous in, cold environments than similar warm environments. The genera *Exiguobacterium* and *Psychrobacter* represent gram-positive and gram-negative bacteria, respectively, that the authors have studied as examples of organisms that show ecological evidence of adaptation to cold (Rodrigues and Tiedje, 2007; Rodrigues et al., 2009). These microorganisms have repeatedly been isolated from geological strata within Siberian permafrost that has been frozen for 20,000 to 3 million years, as well as from Antarctica, sea ice, and the Himalayan Mountains (Vishnivetskaya et al., 2005; Shravage et al., 2007; Shivaji et al., 2005; Chaturvedi and Shivaji, 2006; Bowman et al., 1997; Rodrigues et al., 2006; Bakermans et al., 2006). Furthermore, isolates of both genera grow at subzero temperatures, a direct indicator of cold adaptation. Strains of these two genera were among the first psychrophile genomes sequenced and are used in this chapter, along with other examples, to illustrate various aspects of cold adaptation.

Both *Exiguobacterium* spp. and *Psychrobacter* spp. are more abundant in Antarctica and Siberian permafrost than in temperate or tropical sites, as demonstrated by quantitative real-time PCR (Rodrigues et al., 2009). In warmer environments, *Psychrobacter* spp. populations were patchy, and when present they were in very low density (Rodrigues and Tiedje, 2007; Rodrigues et al., 2009). The distribution of *Exiguobacterium* spp. was also patchy in warmer habitats, though when present their abundance was similar to that found in colder habitats. The abundance and distribution of these microorganisms were correlated with specific environmental factors including salinity, pH, and K^+ and can be explained by their physiology. *Psychrobacter* spp. and *Exiguobacterium* spp. isolated from cold environments can tolerate increased osmolarity and show associated changes in membrane composition, cell morphology, and size (Ponder et al., 2005; Rodrigues et al., 2006; Shivaji et al., 2005; Crapart et al., 2007; Romanenko et al., 2004). Additionally, the Siberian permafrost isolates *Psychrobacter arcticus* sp. 273-4 and *Exiguobacterium sibiricum* sp. 255-15 both maintain membrane potential, electron transport capabilities, and protein synthesis capabilities when incubated in high-osmolarity media (Ponder et al., 2008). These data suggest that these genera are adapted to increased salinity to successfully survive in the permafrost, and could also explain their occurrence in saltier environments that may not be cold. The preferred environmental pH for these genera seems to be between 6 and 8. The maintenance of pH homeostasis is very important for cell survival, and potassium is the principal cation used in bacteria to maintain pH homeostasis (White, 2000). Elevated K levels would be expected in soil solution (brine) of younger permafrost and seasonally frozen soils due to its concentration in the brine by freezing and may explain the correlation with K as well as with salinity. These ecological and physiological results clearly show that psychrophiles have developed specific physiological responses to environmental factors generated by cold temperatures.

A latitudinal gradient in diversity at the species level is also evident for both genera. Clone libraries of *Psychrobacter* spp. and *Exiguobacterium* spp. 16S rRNA genes in samples from Antarctica (polar), Michigan (temperate), and Puerto Rico (tropical) showed a higher diversity of operational taxonomic units (OTUs) in Antarctic samples than in samples from Michigan and Puerto Rico (Fig. 1, top). Additionally, OTUs frequently found in Antarctica were less frequently found in environments from more equatorial latitudes (i.e., Michigan and Puerto Rico) (Fig. 1, bottom). The higher abundance and diversity of these genera at the poles than in the other regions suggest that *Psychrobacter* and *Exiguobacterium* have cold, polar environments as their preferred habitats, and hence are cold adapted by ecological criteria.

GENE EXPRESSION RESPONSES TO THE COLD

The multifaceted stress on microbial cells at low temperatures has led cold-adapted microorganisms to develop unique mechanisms to detect and respond using cellular subsystems.

Cells have some mechanisms to detect low temperature directly, such as RNA "thermometers," which permit the cell to modify translation or RNA half-life when low temperature induces secondary structures in RNA (Eriksson et al., 2002). However, gene expression responses are more often targeted to the myriad impacts of low temperature, including loss of membrane fluidity, water loss through osmosis, starvation for macronutrients due to thermal limits on nutrient incorporation rates, decreased enzymatic reaction rates, and denaturation of enzymes.

Fundamentals of Gene Expression Responses to Cold

To date, five prominent eurypsychrophiles have been subjected to functional genomics experimentation at low temperature, the aquatic methanogen *Methanococcoides burtonii*, the permafrost γ-proteobacteria *Psychrobacter arcticus* 273-4 and *Psychrobacter cryohalolentis* K-5, the permafrost firmicute *E. sibiricum* 255-15, and the marine α-proteobacterium *Sphingopyxis alaskensis*. Findings from studies with these organisms will be presented in this section, with reference to other psychrophilic and mesophilic microbes where appropriate.

In order to acclimate for low-temperature growth, there are three principal gene expression responses available to eurypsychrophiles (Georlette et al., 2004; Bergholz et al., 2009; Bakermans et al., 2007). A eurypsychrophile may acquire mutations for local disorder in its protein structures that permit low-temperature function of an enzyme while also maintaining function at warm temperatures (see "Protein Adaptations to Cold," p. 138) (Feller and Gerday, 2003; Feller, 2007; Georlette et al., 2004; Ayala-del-Río et al., 2010). Because these genes are optimized for function over a wide thermal range, their expression should be dictated by the metabolic demands of the cell. Second, a eurypsychrophile may compensate for decreased reaction rates in the cold by increasing the number of active sites catalyzing a reaction through expression. This response seems intuitive but may incur great

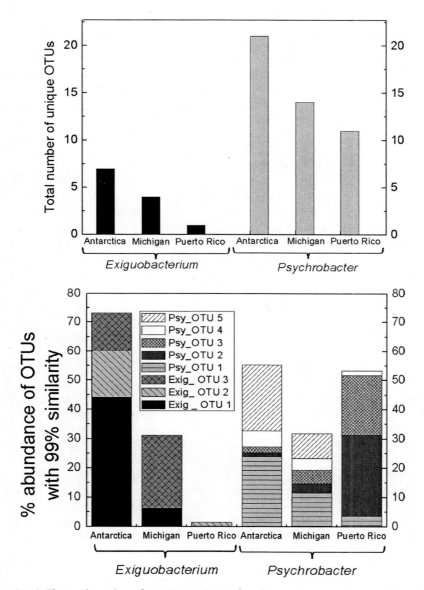

FIGURE 1 (Top) The total number of 16S rRNA OTUs found in each indicated region for each genus. The results show that both *Exiguobacterium* and *Psychrobacter* have higher diversity (higher number of OTUs) in cold environments than in warm. (Bottom) Graph representing the percentages of *Psychrobacter* and *Exiguobacterium* OTUs in each region that share 99% similarity. Similar OTUs found more frequently in cold habitats are not abundant in warmer habitats and vice versa.

cost, because protein synthesis is the largest energy expense for cells. Third, the cell may differentially express isozymes (i.e., enzymes with a similar function, but adapted to dif-

ferent thermal ranges), relying on a cold-adapted isozyme to catalyze reactions at low temperature and a warm-adapted isozyme at warmer temperatures (Iost and Dreyfus, 2006;

Maki et al., 2006). Moreover, isozymes may be specialized for differing substrate pools; for example, cold-adapted DEAD-box helicases (e.g., CsdA in *Escherichia coli*) interact with many RNA-processing enzymes, leading to activity on the bulk RNA pool, but the warm-temperature-adapted DEAD-box helicase RhlB is specialized for interaction with the RNA degradosome while DbpA is specific for ribosome biogenesis (Iost and Dreyfus, 2006). The existence of temperature-specialized isozymes in microbes has been documented in a few cases (Ishii et al., 1987; Bergholz et al., 2009); isozymes are logical strategies for proteins with shared evolutionary history but adapted for function in different environments.

Acclimation for Life in Cold Habitats

TRANSLATION AND CHAPERONE PROTEINS

The expression of both RNA and protein chaperones is a conserved feature of growth at suboptimal temperatures for both eurypsychrophiles and mesophiles, because low temperature changes the strength of intramolecular interactions (Russell, 2000), especially secondary structure formation. Cold shock RNA chaperone proteins prevent secondary structure formation in RNA transcripts and are commonly elevated during initial exposure to and during growth at low temperature in all model organisms, except *P. arcticus* and *Bacillus subtilis*, which produce high levels of *cspA* mRNA at all temperatures (Budde et al., 2006; Bergholz et al., 2009; Giuliodori et al., 2004). DEAD-box RNA helicases support the unwinding of RNA secondary structures. DEAD-box helicase isozymes have been observed in *P. arcticus* 273-4, *E. sibiricum* 255-15, *E. coli*, and *B. subtilis* over changing growth temperature (Fig. 2A) (Rodrigues et al., 2008; Beckering et al., 2002; Bergholz et al., 2009; Iost and Dreyfus, 2006). RNA degradation appears to play an important and highly conserved role during growth in the cold; the RNA degradosome may act to recycle ribo-

nucleotides to new RNA molecules (Ting et al., 2010; Bergholz et al., 2009).

Data on protein chaperones are less consistent across organisms. Protein chaperones are commonly elevated in mesophiles and psychrotolerant pathogens, and a strain of *E. coli* expressing a psychrophilic Cpn60/10 chaperone was reported to grow at 4°C (Phadtare and Inouye, 2004; Budde et al., 2006). *S. alaskensis*, isolated from 10°C waters, has undergone chaperone gene duplication and elevates the expression of protein chaperones at low temperature, while *M. burtonii* and *P. arcticus* only elevate the expression of Cpn60/10 and the *groEL/ES* transcripts at high growth rates or higher-than-optimal growth temperatures (Ting et al., 2010; Bergholz et al., 2009; Campanaro et al., 2010). The current data suggest that the rescue of denatured proteins is essential to expand the growth temperature range of mesophiles, but psychrophiles tend to increase expression of the ClpB protein–disaggregating chaperone and the oxidative stress–associated Hsp33 (Bergholz et al., 2009; Campanaro et al., 2010). Peptide degradation functions are also commonly elevated at low temperature in psychrophiles, allowing for the recycling of amino acids from putative aggregated, denatured peptides into new proteins (Ting et al., 2010; Campanaro et al., 2010; Bergholz et al., 2009). Peptidyl-prolyl *cis/trans* isomerases are responsible for folding nascent polypeptides and are also downregulated in some psychrophiles, in contrast with results from mesophiles (Phadtare and Inouye, 2004; Budde et al., 2006; Bergholz et al., 2009; Goodchild et al., 2004; Thomashow, 1999; Campanaro et al., 2010). Psychrophilic microbes generally upregulate ribosomal proteins at higher growth rates, although interesting exceptions have been noted, especially in *M. burtonii*, where ribosomal proteins involved in channeling mRNA and nascent peptides through the ribosome were upregulated at low temperature (Campanaro et al., 2010), and the ribosomes of *Listeria monocytogenes* have been shown to become thermolabile upon cold acclimation (Bayles et al., 2000).

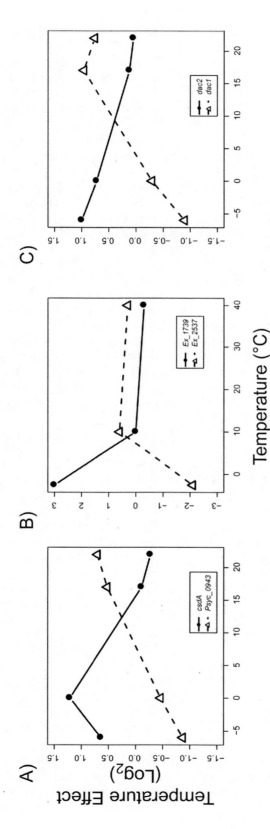

FIGURE 2 Expression of isozymes in *Psychrobacter arcticus* and *Exiguobacterium sibiricum* at different temperatures. (A) DEAD-box helicase isozyme expression in *P. arcticus*; (B) α-amylase isozyme expression in *E. sibiricum*; (C) D-alanyl-D-alanine carboxypeptidase isozyme expression in *P. arcticus*.

CARBON AND ENERGY METABOLISM

Although the general response is decreased expression of carbon-metabolizing enzymes at lower temperatures, specific enzymes are often reported as compensated at low temperature through either increased expression or isozyme exchange (Ting et al., 2010; Rodrigues et al., 2008; Bergholz et al., 2009). For example, *E. sibiricum* expressed two different α-amylase isozymes at −2.5 and 39°C (Fig. 2B). Similarly, *P. arcticus* increased expression of one acetyl-coenzyme A synthesis pathway while downregulating a second when cultivated in a simple medium at low temperature (Bergholz et al., 2009). Compensation can also occur by using a different growth substrate at low temperatures and has been documented in both *E. sibiricum* and *P. arcticus* when grown in complex media (Rodrigues et al., 2008; Ponder et al., 2005).

Energy metabolism data have presented conflicting physiological inferences of the cell at low temperature (Rodrigues et al., 2008; Goodchild et al., 2005; Bakermans et al., 2007; Bergholz et al., 2009; Campanaro et al., 2010). Transcriptome analysis indicates that psychrophiles are responding to slow growth at cold temperatures by generally decreasing gene expression for central and energy metabolism. In contrast, some proteome analyses detected increased levels of ATP-generating enzymes and tricarboxylic acid cycle enzymes. It is possible that low temperature increases the half-life of carbon- and energy-metabolizing enzymes while slow growth rates decrease the instantaneous ATP requirement of the cell, leading to downregulation of transcription. It should be noted that there is increased energy required per unit biomass synthesized during growth at low temperature (Bakermans and Nealson, 2004).

AMINO ACID BIOSYNTHESIS

Every functional genomics experiment performed at suboptimal growth temperatures has indicated increases in certain amino acid biosynthesis enzymes, and these expression changes are often targeted toward those amino acids that are energetically costly to synthesize (Akashi and Gojobori, 2002). *E. sibiricum* upregulated expression of glutamate, histidine, arginine, lysine, and serine biosynthetic enzymes, but *P. arcticus* increased expression of methionine, tryptophan, and proline while it decreased expression of genes for lysine and arginine synthesis and glycine cleavage (Rodrigues et al., 2008; Bergholz et al., 2009). At low temperature, *S. alaskensis* increased protein expression for histidine, tryptophan, and proline, but decreased expression for arginine, aspartate, branched-chain amino acids, and serine (Ting et al., 2010). Mesophilic organisms and other psychrophiles display similar levels of species-to-species variation in amino acid biosynthesis at low temperature. Therefore, it appears that low temperature influences amino acid metabolism, but each eurypsychrophile has unique amino acid biosynthesis responses. The composition of culture medium might impact these responses as well.

COMPATIBLE SOLUTES

Compatible solutes are produced by bacteria to maintain the turgor pressure necessary for growth while preventing toxic concentrations of inorganic solutes in the cytoplasm (Poolman and Glaasker, 1998). Compatible solutes act globally and nonspecifically, allowing microorganisms to occupy broader thermal niches and to react quickly to changes in temperature (Chin et al., 2010). Glycine betaine and trehalose are compatible solutes whose synthesis and transport are known to be induced in a variety of bacteria during growth at suboptimal temperature, including *L. monocytogenes* and *E. coli* (Phadtare and Inouye, 2004; Angelidis et al., 2002). Genes for the conversion of choline to betaine, the transport of betaine, and the symport of proline/Na$^+$ are commonly found in the genomes of psychrophiles (Methe et al., 2005; Bowman, 2008). For example, *Psychromonas ingrahamii* contains the genes for the synthesis of glycine betaine from choline and for the uptake of both choline and glycine betaine (Riley et al.,

2008). Polar microbes have also demonstrated expression of compatible solute biosynthesis during growth at low temperature. *M. burtonii* upregulates glycine betaine transporter genes during growth at 4°C (Campanaro et al., 2010). *E. sibiricum* upregulated transporters for glycine betaine, carnitine, and choline during growth at −2.5°C (Rodrigues et al., 2008). *P. arcticus* was observed to downregulate glycine betaine and carnitine transport and synthesis genes at low temperature, but the conditions of the experiment included 7.5% salt (wt/vol) at the optimal growth temperature, suggesting that increased compatible solute synthesis and uptake at higher temperatures was to support increased growth rate in the presence of salt (Bergholz et al., 2009).

MEMBRANE FLUIDITY

While partitioning the cell from its environment, the phospholipid bilayer also acts as a solvent system for respiratory chain enzymes and transport proteins. The fluidity of membrane lipids decreases at cold temperature and may strongly influence the overall metabolic rate of the cell in spite of genetic adaptations in membrane proteins for low-temperature activity. The loss of membrane fluidity can be sensed by membrane-bound histidine kinases such as the DesK protein in *B. subtilis* (Weber and Marahiel, 2003), and changes made to ensure homeoviscous adaptation and continued function at low temperature. Fatty acid biosynthesis is often modified at low temperatures to produce shorter lipid chain length and branched-chain lipids, while desaturase enzymes are produced to increase the proportion of unsaturated fatty acids in polar eurypsychrophiles at low temperature (Rodrigues et al., 2008; Bergholz et al., 2009; Ponder et al., 2005). Interestingly, two gram-negative psychrophiles, *P. arcticus* and *S. alaskensis*, increased expression of phosphatidylserine decarboxylase, which produces phosphatidylethanolamine lipids. While the potential of membrane lipid head groups in modifying membrane fluidity has been discussed, the role of this polar head group in low-temperature growth is unclear (Chintalapati et al., 2004).

THE CELL WALL AT LOW TEMPERATURE

In bacteria, the peptidoglycan cell wall is an elastic material that expands by stretching according to turgor pressure applied by the expanding cytoplasmic membrane. The resistance of peptidoglycan to stretching is partially determined by the density of peptide cross-linkages: less cross-linkage yields greater deformability (Yao et al., 1999). Two functional genomics studies have detected cell wall biosynthesis changes in eurypsychrophiles growing in the cold (Rodrigues et al., 2008; Bergholz et al., 2009). Two D-alanyl-D-alanine carboxypeptidases (Dac) are present in the *P. arcticus* genome and responsible for preventing the formation of peptide cross-linkages in peptidoglycan (Scheffers and Pinho, 2005). Deletion of the cold-upregulated locus, *dac2*, resulted in a lower growth rate at $T < 4°C$ compared to the wild type, while deletion of the warm-temperature-upregulated locus, *dac1*, lowered the growth rate at 17°C. The expression of autolysins was also upregulated during growth of *P. arcticus* at suboptimal temperature. Together, these results suggest that *P. arcticus* decreases the resistance of the cell wall to stretching during growth at low temperature in minimal cultivation medium with 7.5% salt (Bergholz et al., 2009). In contrast, *E. sibiricum* was cultivated in complex medium and upregulated the expression of cell wall biosynthesis transcripts at low temperature. The expression response in *E. sibiricum* was coupled with a marked difficulty in lysing cells after cultivation at low temperature (Rodrigues et al., 2008). It remains unclear whether the composition of the cultivation media or some basic difference in the cell wall composition of these two organisms (i.e., gram-negative versus gram-positive) is responsible for the contrasting cell wall responses.

TRANSPORTERS

Low temperature decreases diffusion rates in solutions, but freezing concentrates solutes into more limited volume while also limiting access to solute pools with physical barriers. Transport protein systems adapted for increased substrate affinity can help the cell cope with limited nutrient availability, and cells can increase expression of transporter genes to increase substrate uptake. During growth at subzero temperatures, *P. arcticus* increased expression of sulfate uptake transporters while the expression of ammonium and phosphate transporters was unchanged (Bergholz et al., 2009). *M. burtonii* substrate-binding enzymes for glycine betaine and molybdate ABC transporters were upregulated at 4°C, but the substrate-binding enzymes for the phosphate ABC transporter were downregulated at 4°C (Campanaro et al., 2010). In *E. sibiricum*, all the genes involved in the glucose phosphotransferase transport system were upregulated at low temperatures, while glucose permeases were upregulated at higher temperatures. These results suggested that *E. sibiricum* requires active transport of nutrients at lower temperature to increase substrate uptake (Rodrigues et al., 2008). These data suggest specific mechanisms that cold-adapted microorganisms may use to overcome nutrient limitation due to decreased diffusion rates and substrate affinity at low temperatures and corroborate previous whole-cell studies (Nedwell and Rutter, 1994; Ponder et al., 2005; Herbert and Bell, 1977).

GENOME TOPOLOGY

Under cold conditions, DNA becomes more negatively supercoiled (Golovlev, 2003; Rodrigues et al., 2008) due to the increased strength of hydrogen bonding interactions (Mizushima et al., 1997) and must be stabilized in a more functional conformation. For example, nucleoid-associated proteins such as gyrase A, integration host factor, and histonelike nucleoid structuring protein are suggested to be necessary for the relaxation of DNA during cold stress (Scherer and Neuhaus, 2002; Gualerzi et al., 2003; Weber and Marahiel, 2003). However, the potential roles of repeated sequences in DNA topology at low temperature have received very limited treatment. The *P. arcticus* genome contains numerous transposase genes and a

highly abundant interspersed repeat sequence that accounts for 1.92% of the total genome sequence (Ayala-del-Río et al., 2010). This repeat sequence is similar to *nemis* elements in *Neisseria meningitidis*, which bind DNA-binding proteins in vitro and are cleaved by RNase III when they appear in transcripts (Buisine et al., 2002; De Gregorio et al., 2003). The potential role of such interspersed repeated sequences in regulating DNA topology at low temperature remains unknown. Alternately, repeated sequences and transposases might play a role in recombination and contribute to genomic evolution. Interestingly, inactivation of transposes in the psychrophile *Photobacterium profundum* SS9 leads to cold sensitivity (Lauro et al., 2008). While few data are available on genome evolution in psychrophiles via transposases, numerous transposases are present in the genomes of *P. arcticus* and *E. sibiricum*, and they are overrepresented in the *M. burtonii* genome and metagenomic data from the cold deep ocean (Allen et al., 2009; DeLong et al., 2006; Vishnivetskaya and Kathariou, 2005).

PROTEIN ADAPTATIONS TO COLD

The Low-Temperature Challenge

Low temperatures inhibit the rates of biochemical reactions catalyzed by enzymes. During enzyme catalysis, conformational motions of the enzyme facilitate bringing together chemically reactive species and cofactors to form new bonds (Fields, 2001). As temperatures decrease there is insufficient kinetic energy in the reactants and in the surrounding medium to satisfy the thermal energy requirements of the enzyme, and catalysis inevitably slows down.

Kinetic modeling of temperature effects in enzymatic reaction rates, using Arrhenius and other equations, predicts that for every 10°C drop in growth temperature a two- to fourfold decrease in enzymatic rates will occur (Georlette et al., 2004). For mesophilic enzymes with an optimal temperature of 37°C, a drop to 0°C results in a 20- to 250-fold reduction in activity (Feller and Gerday, 2003). Although a decrease in enzymatic activity will slow down many cellular processes, microbial growth has been documented at −10°C in species of the genus *Psychrobacter* (Bakermans et al., 2006; Ponder et al., 2005) and at −12°C in *P. ingrahamii* (Breezee et al., 2004). The cold-adapted α-amylase from *Alteromonas haloplanktis* A23 is up to 10 times more active than mesophilic homologs at moderate temperatures (20 to 30°C), demonstrating that psychrophiles possess enzymes that are very efficient at low temperatures (Feller et al., 1992).

The Stability-Activity Relationship

Although there are many factors that contribute to cold growth (described above), detailed structural analyses at the enzyme level have assisted the examination of the efficient catalysis of cold-adapted enzymes. From a structural point of view, the prevailing hypothesis is that psychrophilic enzymes have low levels of conformational stability to compensate for the reduction in thermal energy (Casanueva et al., 2010; Feller and Gerday, 2003; Siddiqui and Cavicchioli, 2006; Feller, 2010). At the active site of enzymes there is an inverse relationship between stability and enzymatic activity (Shoichet et al., 1995). As enzymes increase their stability by amino acid selection and orientation, the active site becomes more rigid, limiting interactions with the substrate. To be active, enzymes cannot be the rigid structures depicted in figures, but rather must be a group of closely related conformations in constant motion. This means that many cold-active enzymes increase the molecular flexibility of the protein structure to facilitate the interaction between the substrate and the active site and thus reduce the activation energy of the reaction.

Structural Features of Cold-Adapted Enzymes

Considerable efforts have been made to identify which amino acid residues are responsible for low-temperature activity of proteins and their location in protein structure. Studies on the low stability and thermolability of psychro-

philic enzymes reveal that all structural factors that increase protein stability are weakened in psychrophilic proteins (Siddiqui and Cavicchioli, 2006; Smalås et al., 2000). In particular, intramolecular interactions that occur between residues of the protein seem to be reduced in cold-active enzymes, providing the needed flexibility for catalysis at low temperatures. Some of the structural adaptations that increase protein flexibility are summarized below. For detailed reviews of all known structural adaptations present in psychrophilic proteins, see Siddiqui and Cavicchioli (2006) and Smalås et al. (2000).

HYDROPHOBIC INTERACTIONS

Hydrophobic interactions are important noncovalent interactions between nonpolar amino acids that stabilize the three-dimensional structure of proteins (Nelson and Cox, 2004). To reduce the entropy of the system, i.e., increase the stability of the protein, nonpolar regions of proteins group together to minimize their contact with the surrounding water. This interaction will favor nonpolar amino acids in the protein core while minimizing nonpolar residues in solvent-exposed regions, i.e., the surface. Cold-active proteins increase their flexibility by reducing the hydrophobicity of the core while making the surface regions more hydrophobic (Feller et al., 1994; Aghajari et al., 1998). The genomes of archaeal psychrophiles *Methanogenium frigidum* and *M. burtonii* show a strong tendency toward more hydrophobic residues in solvent-accessible areas of the proteins (Saunders et al., 2003). In the *P. arcticus* 273-4 genome there is a general preference for hydrophobic amino acids, suggesting that hydrophobicity is a key component of cold-active enzymes at the genome level.

ELECTROSTATIC INTERACTIONS

Electrostatic interactions, intramolecular noncovalent interactions between charged groups, are extremely important for maintaining secondary and tertiary structure of proteins (Whitford, 2005). Polar and charged amino acids are the major contributors to electrostatic interactions, resulting in H-bonding, salt bridges, and aromatic interactions.

Arginine. Arginine, a polar, positively charged amino acid, likely interacts with the side chains of negatively charged amino acids such as aspartate and glutamate to increase protein stability via formation of hydrogen bonds and salt bridges (Siddiqui and Cavic-chioli, 2006; Adekoya et al., 2006). Cold-adapted enzymes tend to have less arginine in order to reduce structural stability. In the *P. arcticus* 273-4 genome there was a reduction in arginine content and increased lysine within genes in the replication, recombination, and repair categories (Ayala-del-Río et al., 2010). When comparing the amino acid substitutions between six psychrophilic genomes and six mesophilic genomes, psychrophilic proteins contained fewer basic residues (argine, lysine, and histidine) (Metpally and Reddy, 2009). A reduction in arginine was also observed in the genome sequences of the marine psychrophiles *Shewanella halifaxensis* and *Shewanella sediminis* (Zhao et al., 2010) and in genome fragments of six Antarctic marine bacteria (Grzymski et al., 2006), suggesting it is an important substitution for cold adaptation.

Acidic Residues. Acidic amino acids contribute to protein structure in different ways. In helix structures acidic amino acids, among other charged residues, form ion pairing and hydrogen bonds, thus playing an important role in protein stabilization (Gianese et al., 2002; Xiao and Honig, 1999; Vogt et al., 1997). A reduction in acidic residues could increase protein flexibility, an advantage in psychrophilic enzymes. In *P. arcticus* 273-4 approximately 400 genes had fewer acidic amino acids compared to mesophilic homologs (Ayala-del-Río et al., 2010). This adaptation was observed in gene categories essential for cell growth, e.g., replication, transcription, translation, envelope biogenesis, and lipid and amino acid biosynthesis. In the genomes of *M. frigidum* and *M. burtonii* there was a reduction in the contribution of charged amino acids to both accessible and inaccessible areas of the protein (Saunders

et al., 2003). The removal of acidic amino acids has also been observed in genome fragments from Antarctic marine bacteria (Grzymski et al., 2006) and in the systematic analysis of 21 psychrophilic proteins (Gianese et al., 2002). However, psychrophilic proteins may also have increased negative charge on their surface (Feller and Gerday, 2003; Siddiqui and Cavicchioli, 2006; Smalås et al., 2000). Negative charges on protein surfaces improve solvent interaction, i.e., hydration of polar residues, and thereby reduce compactness of the protein (Makhatadze and Privalov, 1994; Davail et al., 1994). In addition, formation of clusters of negatively charged amino acids gives rise to repulsive interactions that desta-bilize protein structure (Siddiqui and Cavic-chioli, 2006; Saunders et al., 2003). Although trends of increased acidic residues in psychro-philic proteins have not been observed at the whole-genome level, it is a relevant mecha-nism detected in detailed structural analyses of psychrophilic enzymes that should be consid-ered when studying cold adaptation (de Backer et al., 2002; Violot et al., 2005).

Structural Elements

α-HELICES AND β-SHEETS

α-Helices and β-sheets are the main com-ponents of secondary structure that define protein folding and thus affect function. The stability of both β-sheets and α-helices is sensitive to changes in amino acid sequences since it depends on hydrogen bonds between adjacent strands and on internal hydrogen bonds and side chain-side chain interactions, respectively (Siddiqui and Cavicchioli, 2006). Turns (loops) connect the elements of sec-ondary structure and are much more flexible than α-helices and β-sheets. They can repre-sent up to 30% of the protein and could rep-resent a large fraction of the solvent-exposed area (Whitford, 2005). Since the structures of both α-helices and β-sheets are sensitive to changes in amino acid sequence, only minor changes are allowed to occur in those regions. Turns tend to be more flexible, and because they are in solvent-exposed areas they represent points of major interest when evaluating protein stability.

PROLINE AND GLYCINE
Proline and glycine are both nonpolar aliphatic amino acids with unique effects on protein conformation. Proline is the only amino acid possessing a side chain that forms a covalent bond with the backbone nitrogen atom, a cyclic pyrrolidine ring (Whitford, 2005). The presence of a cyclic ring restricts the rotation to the N-Cα bond, reducing the conformational flexibility of the protein in that location. As expected, increased abundance of proline residues has been related to increased protein stability (Fields, 2001; Reiersen and Rees, 2001). A reduction in proline residues was detected in the genome sequences of *P. arcticus* 273-4 (Ayala-del-Río et al., 2010), *S. halifaxensis*, and *S. sediminis* (Zhao et al., 2010); in genome fragments of six Antarctic marine bacteria (Grzymski et al., 2006); and in a multigenome comparison study (Metpally and Reddy, 2009), suggesting that fewer prolines is a very important adaptation to cold in psychrophiles.

Glycine is a small amino acid with only two hydrogen atoms attached to the Cα and no side group, which confers a high degree of conformational flexibility. Although a glycine preference seems an obvious adaptation in psychrophilic proteins, no trend in glycine was identified until recently (Metpally and Reddy, 2009). In psychrophiles, clustering of glycine residues or their location close to functional regions of the protein (e.g., active site) increases conformational flexibility without changes in global amino acid content (Feller, 2010).

DISORDERED REGIONS
Disordered regions of proteins do not form a stable three-dimensional structure (Dunker, 2007) and hence are often not evident in high-resolution structural determinations (Bloomer et al., 1978). Many disordered regions are located at the N or C terminus, forming random coils. Although there are many possible functions for the disordered regions (Dunker

et al., 2002), functional contribution to increased protein flexibility has been recently documented in psychrophilic proteins (Violot et al., 2005). DD-Peptidase and a DEAD-box RNA helicase of *P. arcticus* show evidence of disordered regions that were beneficial at low temperatures (Fig. 3) (Bergholz et al., 2009). For both genes, there were two open reading frames (ORFs) annotated in the genome (i.e., isozymes), but one copy was expressed at optimal temperature while the other was expressed at low temperature. In both cases, analysis using DisEMBL 1.5 predicted two- to fivefold more disordered amino acid residues in the low-temperature-expressed locus (Linding et al., 2003). An exchange in expression between α-amylase isozymes from -2.5 to 39°C was recently documented in *E. sibiricum* 255-15 (Rodrigues et al., 2008). Although disordered regions were not analyzed, the cold-expressed α-amylase was larger in size and had some of the characteristics of cold-adapted proteins described above. Similar results were observed in the genome fragments of six Antarctic marine bacteria (Grzymski et al., 2006), suggesting that disordered regions could be an understudied but important factor in the flexibility of psychrophilic proteins.

COMPARISON OF COLD- AND WARM-ADAPTED *EXIGUOBACTERIUM* STRAINS
Bacteria of the genus *Exiguobacterium* are low-GC, gram-positive facultative anaerobes that cluster closely to *Bacillales* family XII, *Incertae sedis*, within the *Firmicutes*. Members of the genus *Exiguobacterium* have been isolated from remarkably diverse habitats including frozen permafrost sediments and glacial ice as well as thermal springs. Interestingly, analysis of the 16S rRNA gene sequences from 41 *Exiguobacterium* isolates revealed the presence of two distinct clusters: isolates from cold and temperate environments in one cluster and isolates from thermal and marine environments in the second cluster (Vishnivetskaya et al., 2009). This clustering suggests adaptation and speciation events that may be related to environmental temperatures.

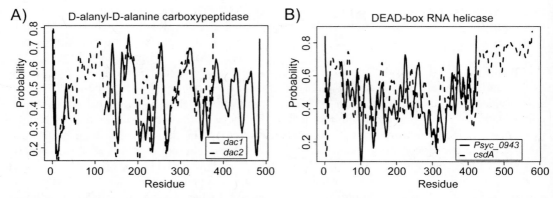

FIGURE 3 Disordered region predictions from the primary sequence of isozyme loci. Dashed lines are putative low-temperature-adapted loci, and solid lines are predicted from genes upregulated at the optimal temperature for maximum growth rate (see Fig. 2). The vertical axis is the probability that an amino acid residue is in a coil, as predicted by DisEMBL 1.5. (A) Aligned sequences of D-alanyl-D-alanine carboxypeptidase isozymes in *P. arcticus*. (B) Aligned sequences of DEAD-box RNA helicase isozymes in *P. arcticus*.

Phylogeny Reflects Adaptations to Environmental Conditions

To further explore what aspect of the environment most affected evolution and speciation of *Exiguobacterium* species, additional 16S rRNA gene sequences were obtained from the NCBI database and analyzed. Of the 660 entries for *Exiguobacterium* (including 201 uncultured clones), 373 sequences that contained 400 bp from the beginning of the 16S rRNA gene (position 100 to 500 of *E. coli* 16S rRNA gene) were analyzed using Uni-Frac (Lozupone et al., 2006; Lozupone and Knight, 2005). The sequences represented *Exiguobacterium* isolates or environmental clones derived from 19 different habitats including permafrost and hot springs. Principal component analysis separated samples into two groups (A and B) along axis P1, explaining 65.6% of the variation (Fig. 4). Group A combined sequences derived from permafrost; various soils and sediments; industrial processes; vegetation; biomes of shrimp, larvae, and oysters; and clinical samples. Group B included sequences from aquatic environments including hot springs and marine habitats, glacier ice, atmospheric air, and polluted sites. The second axis variable, P2, explained 11.2% of variations observed within groups;

for example, sequences from Antarctica were located closely to permafrost *Exiguobacterium*, whereas cave sediments were the most distant from permafrost. In contrast, analysis of 422 sequences from the end of 16S rRNA gene (position 1000 to 1300 of *E. coli* 16S rRNA) placed all *Exiguobacterium* entries into the same cluster independent of environmental source. These data suggest that environmental temperature is not the main factor structuring *Exiguobacterium* diversity, but that the main factors are physical (water or soil) and chemical (degree of pollution) states of the environments (Fig. 4). Not surprisingly, the growth temperature ranges of *Exiguobacterium* isolates are related to environmental temperature, with permafrost strains growing at −3 to 40°C and isolates from thermal springs from 15 to 55°C (Vishnivetskaya et al., 2005). The members of group A and B could be further distinguished by the chromosomal location of rRNA operons and hybridization patterns of stress response genes (Vishnivetskaya et al., 2009).

Genomic Comparison of Two Strains

The genomes of two *Exiguobacterium* strains, one cold adapted and the other warm adapted, were sequenced by the Joint Genome Institute

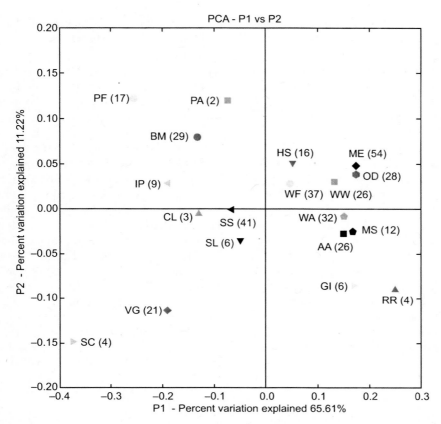

FIGURE 4 Results of principal component analysis (PCA) obtained with weighted and normalized UniFrac using the 16S rRNA sequences of *Exiguobacterium* isolates and clones from different environments. PF, permafrost; PA, Antarctica; SS, soil or sediments; SL, sludge; SC, sediments from cave; IP, industrial processes; VG, vegetation; BM, biome of shrimp/larvae/oyster; CL, clinical samples; HS, hot springs; WF, freshwater; WA, water; WW, wastewater; ME, marine environments; MS, marine sediments; OD, oil or other polluted sites; AA, atmospheric air; RR, rhizosphere; GI, glacier ice. The number of sequences used for analyses is given in brackets.

and are available through the Integrated Microbial Genomes website (http://img.jgi.doe.gov). *E. sibiricum* 255-15 was isolated from a two-million- to three-million-year-old permafrost sample and *Exiguobacterium* sp. strain AT1b from a Yellowstone hot spring (Vishnivetskaya et al., 2000). Their very different growth temperature ranges and optima are shown in Fig. 6, which make them an attractive pair to compare at the genome level for insight into temperature adaptation. The average nucleotide identity between the whole-genome sequences of the two species was 69.2%—as determined by using the in silico DNA-DNA

hybridization method of the JSpecies software (http://www.imedea.uib-csic.es/jspecies/about.html)—and is significantly lower than the recommended cutoff point of 94% for species delineation (Konstantinidis and Tiedje, 2005; Goris et al., 2007; Richter and Rosselló-Móra, 2009), and near the genus boundary of 65 to 73% (Konstantinidis and Tiedje, 2007). The BLAST options (-x, 150; -q, -1; -F, F; -e, 1e-15; and -a, 2) and average nucleotide identity calculation settings (identity, ≥30%; alignment, ≥70%; and length, 1,020 nucleotides) were used. The tetra-nucleotide distribution regression coefficient was 0.86, further demonstrating that the DNA from these two strains varies greatly and that they represent distinct species.

Features of both genomes are described in Table 4. The genome of *E. sibiricum* 255-15 contains one circular chromosome that encodes 3,054 putative proteins and two small plasmids that encode 8 hypothetical genes; while *Exiguobacterium* sp. strain AT1b contains a single replication unit, which contains 3,043 putative proteins. In both strains, genes were evenly distributed between the forward (~49.5%) and reverse (~50.5%) strands. The GC content was constant across both genomes, although they also contained AT-rich fragments that encode uncharacterized proteins. The AT-rich fragments in *E. sibiricum* 255-15 include a restriction endonuclease, an ATP-dependent protease, a glycosyltransferase, and a CDP-glycerophosphotransferase; while the *Exiguobacterium* sp. strain AT1b fragment contains a sulfurtransferase and a sulfur modification protein.

Exiguobacterium species typically have nine rRNA operons, as shown by Southern hybridization of EcoRI-digested genomic DNA (Vishnivetskaya et al., 2009). Four *Exiguobacterium* strains had fewer (seven or eight) hybridization bands evident, which might be attributed to genes present on multiple fragments of a similar size or multiple genes present on the same fragment. Within the genome of *E. sibiricum* 255-15 the rRNA operons were located primarily on the positive (eight copies) strand rather than the negative (one copy) strand, while in *Exiguobacterium* sp. strain AT1b rRNA operons were more evenly distributed between the positive (four copies) and negative (five copies) strands (Fig. 5). The average GC content of the rRNA operons (54.95% for strain 255-15 and 56.12% for strain AT1b) was higher than the average GC content of the genome.

While distinct in many ways, the overall content and organization of the two *Exiguobacterium* genomes share a high degree of

TABLE 4 Genome features of *Exiguobacterium* strains

Feature	*Exiguobacterium sibericum* 255-15	*Exiguobacterium* sp. strain AT1b
Size (bp)	3,034,136	2,999,895
Genes, total no.	3,151	3,138
G+C content (%)	47.70	48.46
Protein-coding genes (%)	3,054 (96.92)	3,043 (96.97)
Average gene size (bp)	886	891
Coding bases (%)	87.8	89.6
Protein coding with function prediction (%)	2,177 (69.09)	2,157 (68.74)
Protein coding without function prediction (%)	877 (27.83)	886 (28.24)
Transposases	27	3
Pseudogenes (%)	39 (1.24)	23 (0.73)
rRNA operons	9	9
tRNAs	69	68
Other rRNAs (misc-RNA)	1	0
Extrachromosomal elements	2 (4,885 bp; 1,765 bp)	0
Cold shock genes	6	2
Phage protein genes	1	10
Nucleases	34	35

similarity. A total of 99.2 and 97.7% of the forward and reverse strands, respectively, of the *Exiguobacterium* sp. strain AT1b genome align with the *E. sibiricum* 255-15 genome (Fig. 5). The number of blocks of identical sequences (minimum 5 bp; maximum 2,757 bp) on the forward strand is 1,970, with a total sequence length of 164,300 bp that comprises 5.4% of the *E. sibiricum* 255-15 and 5.5% of the *Exiguobacterium* sp. strain AT1b genome length. On the reverse strand, there are 1,447 blocks of identical sequences comprising 119,425 bp (3.9% of the genome lengths). A TaxPlot comparison based on the protein sequences encoded by the *Exiguobacterium* genomes showed that 2,487 (81.7%) proteins of *Exiguobacterium* sp. strain AT1b are similar to proteins of *E. sibiricum* 255-15.

Transposases usually catalyze site-specific DNA rearrangements and horizontal gene transfer in bacteria, and may play a role in evolutionary and adaptation processes. Complete insertion sequence (IS) elements are present in the genome of *E. sibiricum* 255-15, suggesting that the IS elements could be active (Vishnivetskaya and Kathariou, 2005). Many transposases (27) were predicted in the genome of *E. sibiricum* 255-15, while fewer (3) were identified in *Exiguobacterium* sp. strain AT1b. Five distinct families of transposases were identified in *E. sibiricum* 255-15 and included IS*605 orfB* (13 elements), IS*3* (5 elements), IS*4* (5 elements), IS*200* (3 elements), and IS*891*/IS*1136*/IS*1341* (1 element). Family IS*605 orfB* transposases were present in both genomes. Transposases homologous to those identified in the genome of *E. sibiricum* 255-15 were found in 24 *Exiguobacterium* strains by Southern hybridization, with the putative transposase IS*605 orfB* detected in all strains (Vishnivetskaya et al., 2009).

Exiguobacterium strains displayed temperature-dependent nuclease activity, with *E. sibiricum* 255-15 showing higher DNase activity at 4°C, whereas DNase production by *Exiguobacterium* sp. strain AT1b was 55.2 to 67.7% higher at 37°C than at lower temperatures. Genome sequences showed the presence of restriction endonucleases, ribonucleases, and exonucleases in both strains (Table 1). Increased nuclease activity and hence degradation of nucleic acids at low temperature supports the idea that cells need to recycle nucleotides to new molecules of nucleic acids to continue growth at low temperatures.

Exiguobacterium species are facultative anaerobes expected to grow both aerobically and anaerobically (Rodrigues et al., 2006). The genome analyses revealed the presence of nitrate reductase in the genome of the thermophile *Exiguobacterium* sp. strain AT1b but not in the genome of the eurypsychrophile *E. sibiricum* 255-15. The presence of nitrate reductase in the thermophilic bacterium suggests that it is able to produce energy by reduction of nitrate (NO_3^-) to nitrite (NO_2^-) by anaerobic respiration. The genome of oxidase-positive *E. sibiricum* 255-15 contained the complete tricarboxylic acid cycle and a branched aerobic respiratory chain, which consists of monomeric NADH-quinone oxidoreductase, menaquinol-cytochrome *c* reductase, and three terminal oxidases (Rodrigues et al., 2006, 2008). *E. sibiricum* 255-15 can therefore utilize oxygen for energy production within an electron transfer chain. *Exiguobacterium* sp. strain AT1b contained an NADH-quinone oxidoreductase and cytochrome oxidase genes as well, but a menaquinol-cytochrome *c* reductase was not identified. All tested *Exiguobacterium* strains isolated from permafrost, soils, and vegetation were nitrate reductase negative, whereas species from hot spring, marine, atmospheric, and wastewater habitats were positive. Furthermore, nitrate reductase-negative strains were oxidase positive, while nitrate reductase-positive strains, with the exception of *Exiguobacterium mexicanum*, were oxidase negative (Vishnivetskaya et al., 2009). These data support the idea that the clustering of the genus into two groups has been accompanied by specific physiological attributes related to adaptation to specific environments.

An important question is whether the psychrotrophic lifestyle of certain *Exiguobacterium*

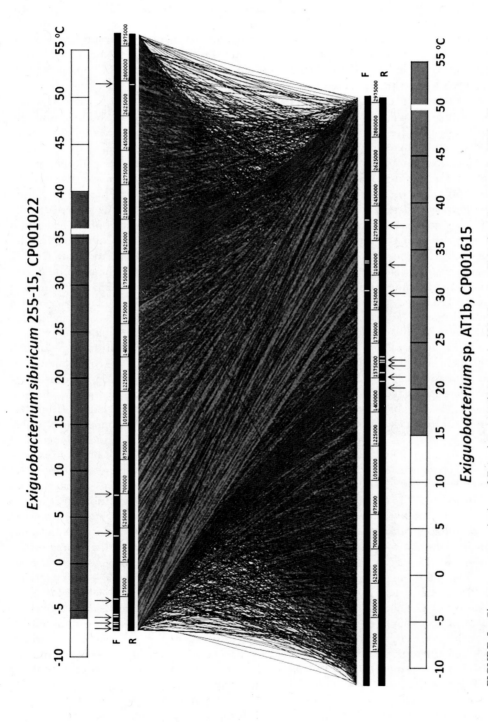

FIGURE 5 Chromosome organization of *Exiguobacterium sibiricum* strain 255-15 versus *Exiguobacterium* sp. strain AT1b. Chromosomes were compared using Artemis Comparison Tool (www.sanger.ac.uk/resources/software/act/). Black bars symbolize chromosomes. Gray lines connect homologous regions present in the same orientation, while black lines connect regions of inverted orientation. Localization of the rRNA operons is indicated by white bars (also highlighted by arrows). Temperature bars are shown at the top and bottom, indicating the growth range and optima of the two compared strains.

spp. may be conferred by a unique set of genes that are absent in thermophilic members of the genus. For example, upregulation of DNA topoisomerases in *E. sibiricum* 255-15 was observed at growth temperature extremes: gyrase B was upregulated at −2.5 and 39°C but gyrase A only at −2.5°C (Rodrigues et al., 2008). Homologous gyrase B, but not gyrase A, genes were found in the genomes of *Exiguobacterium* isolates from thermophilic, marine, and polluted sites by Southern hybridization. The gyrase B from *Exiguobacterium* sp. strain AT1b has 75% identity with the gyrase B from *E. sibiricum* 255-15, while gyrase A has only 65.8% identity (below the detection limit of Southern hybridization). Further physiological studies of thermophilic *Exiguobacterium* are needed to verify what changes in the genome content and organization may be a result of adaptations to environmental temperature conditions.

SUMMARY AND FUTURE DIRECTIONS

Microorganisms are the oldest and most widespread form of life on Earth. Their success in colonizing harsh environments has been due to their capacity to adapt. Microbes in cold habitats face conflicting signals from the environment. While the nutrient concentrations in ice-bound brines and cultivation media may be large enough to support exponential growth, low temperature limits the rate at which an organism may exploit nutrients and changes in water activity limit the ability of cells to expand and divide. The integration of these conflicting signals appears to result in gene expression to increase transport of specific nutrients and biosynthesis of specific building blocks, combined with responses to break down nonfunctional proteins and RNAs into their subunits. Meanwhile, the bulk of energy and carbon metabolism enzymes are unaffected by temperature, and gene expression is decreased in these pathways to slow activity to levels dictated by physical limitations on other metabolic pathways in the cell (Fig. 6).

The physical limitations imposed by low temperature on biomolecules can only be overcome by increasing the disorder of the cellular system. This principle has been demonstrated through the use of chaotropic solutes to increase growth rates at low temperature (Chin et al., 2010). Adaptations at the protein level have allowed psychrophilic microorganisms to grow at temperatures below zero, where the kinetic energy is not sufficient for mesophiles. Although proteins adapt to the cold via different mechanisms, the goal is the same: increase protein flexibility to increase disorder and decrease the activation energy requirements. It is the combination of all the mechanisms mentioned above distributed, many times unevenly, across the proteome that confers upon psychrophiles the ability to grow at low temperatures. These genetic adaptations permit psychrophiles to increase the disorder of many cellular systems, including changes in DEAD-box helicases and RNA chaperones, changes in ribosomal protein composition, and increases in protein-disaggregating chaperones, fatty acid desaturases, and cell wall-degrading enzymes. Expression of these genes during growth at low temperature permits the maintenance of molecular motion in spite of decreased heat in the cell. The role of disordered regions in proteins and temperature-specialized isozymes in cold growth represents an understudied but important factor in cold adaptation. Further studies are necessary to determine the frequency of disordered regions and isozymes between psychrophilic and mesophilic genomes.

Functional genomics experiments present a systemwide accounting of transcript and protein abundance in the cell, but represent physiological changes very coarsely and often leave a large number of new hypotheses to be tested. Future experiments to examine physiological questions resulting from transcriptome and proteome data on psychrophiles could include (i) a comparison of the instantaneous and growth-normalized energy requirements of psychrophiles; (ii) the effect of growth temperature on cellular polymers,

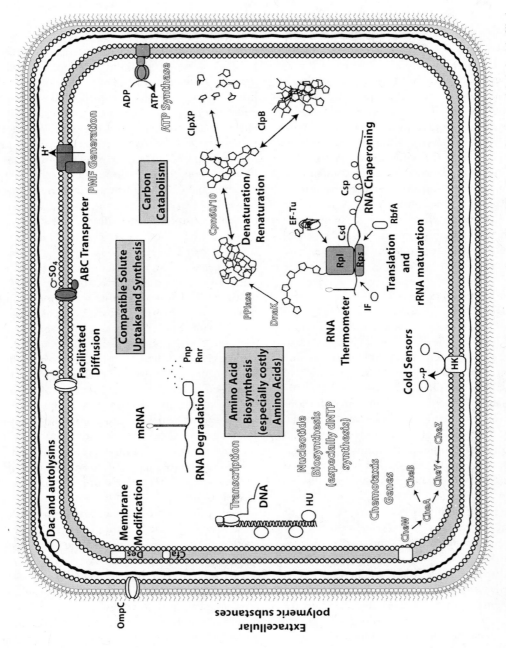

FIGURE 6 Gene expression during cold-acclimated growth in a gram-negative psychrophile. Depicted genes are labeled with filled text when transcription is upregulated and outlined text when downregulated. Proteins and pathways with conflicting results across species or across proteome and transcriptome datasets are shaded in gray.

including exopolysaccharides and peptidogly-can; (iii) metabolomic analysis of eurypsychro-philes at different growth temperatures, with particular attention to amino acid and compat-ible solute pools; (iv) the impact of cold tem-peratures on the proteome, with a focus on amino acid biosynthesis; and (v) an assessment of the relative impacts of posttranscriptional and posttranslational regulation on enzyme ac-tivity at cold temperature.

Experiments that examine the longitudinal trends, not only in gene expression but also in nutrient flux in frozen environments, are merited. The temporal and spatial dynamics of nutrient availability and dispersal in ice is unknown and is likely to have a major impact on metabolism. Enzyme-labeled fluorescent substrates may be ideal for determining if cells obtain nutrients in bursts or at a more con-stant rate over time, as the fluorescent marker precipitates from water when cleaved from its carrier molecule. Similar techniques could be used to examine the spatial relationship be-tween extracellular enzyme activity in ice and microbial cells. Other measures of incorpora-tion such as temporal integration and spatial analysis of radionucleotide dispersal and incor-poration processes after freezing could provide a much needed "microbial perspective" on life in the cold.

Many indicators of cold adaptation have been identified from high-resolution analy-sis of single proteins. Although genome se-quences have expanded our view on the generality of the previously identified adap-tations, more work is needed to analyze all genomes employing similar techniques, given their varying sensitivity in identifying trends. For example, more sensitive techniques were required to identify genome-wide evidence of cold adaptation in proteins of *P. arcticus* 273-4. These "psychrophilic" signatures may have been less evident in *P. arcticus* 273-4 be-cause it is a eurypsychrophile that evolved many adaptations primarily while living in the "active layer," an environment with a wide temperature range and frequent temperature swings across the freezing point. Thus, other trends seen in *P. arcticus* 273-4 may also be hallmarks of a eurypsychrophile lifestyle and may include the use of isozymes, efficient use of resources, and efficient growth at low tem-peratures. Alternatively, these trends may be unique to *Psychrobacter* or terrestrial psychro-philes. The idea that terrestrial psychrophiles may have unique adaptations is supported by analysis of *Exiguobacterium* species, which suggests that a major factor structuring diver-sity may be the physical nature of the habi-tat (terrestrial versus aquatic; see "Phylogeny Reflects Adaptations to Environmental Con-ditions," p. 142).

Unfortunately, limited data hamper our ability to make further conclusions. More genome sequences from psychrophiles, espe-cially psychrophiles from terrestrial habitats, are needed for comparative studies—particu-larly given the current emphasis being placed on understanding permafrost in the face of global climate change. The availability of ge-nome sequences of closely related microor-ganisms with different temperature optima, temperature ranges, and habitats will facili-tate a more comprehensive characterization of features that facilitate growth at low tem-perature. A comprehensive characterization of these features in a variety of psychrophiles is crucial to discerning the impact of envi-ronmental factors, including temperature, on gene content, gene sequences, and genome structure.

ACKNOWLEDGMENTS
The authors' research on permafrost microorganisms was supported as part of NASA's Astrobiology Insti-tute and the National Science Foundation's Life in Extreme Environments (LExEn) program. The U.S. Department of Energy's funding of the Joint Genome Institute for sequencing five strains of *Psychrobacter* and *Exiguobacterium*, as well as several other cold-adapted microorganisms mentioned herein, was critical and is gratefully acknowledged.

REFERENCES
Adekoya, O. A., R. Helland, N. P. Willassen, and I. Sylte. 2006. Comparative sequence and structure analysis reveal features of cold adaptation

of an enzyme in the thermolysin family. *Proteins* **62**:435–449.

Aghajari, N., G. Feller, C. Gerday, and R. Haser. 1998. Structures of the psychrophilic *Alteromonas haloplanctis* α-amylase give insights into cold adaptation at a molecular level. *Structure* **6**:1503–1516.

Akashi, H., and T. Gojobori. 2002. Metabolic efficiency and amino acid composition in the proteomes of *Escherichia coli* and *Bacillus subtilis*. *Proc. Natl. Acad. Sci. USA* **99**:3695–3700.

Allen, M. A., F. M. Lauro, T. J. Williams, D. Burg, K. S. Siddiqui, D. De Francisci, K. W. Y. Chong, O. Pilak, H. H. Chew, M. Z. De Maere, L. Ting, M. Katrib, C. Ng, K. R. Sowers, M. Y. Galperin, I. J. Anderson, N. Ivanova, E. Dalin, M. Martinez, A. Lapidus, L. Hauser, M. Land, T. Thomas, and R. Cavicchioli. 2009. The genome sequence of the psychrophilic archaeon, *Methanococcoides burtonii*: the role of genome evolution in cold adaptation. *ISME J.* **3**:1012–1035.

Angelidis, A. S., L. T. Smith, L. M. Hoffman, and G. M. Smith. 2002. Identification of OpuC as a chill-activated and osmotically activated carnitine transporter in *Listeria monocytogenes*. *Appl. Environ. Microbiol.* **68**:2644–2650.

Angell, C. A. 1982. Supercooled water, p. 1–82. *In* F. Franks (ed.), *Water: a Comprehensive Treatise*. Plenum Press, New York, NY.

Ayala-del-Río, H. L., P. S. Chain, J. J. Grzymski, M. A. Ponder, N. Ivanova, P. W. Bergholz, G. Di Bartolo, L. Hauser, M. Land, C. Bakermans, D. Rodrigues, J. Klappenbach, D. Zarka, F. Larimer, P. Richardson, A. Murray, M. Thomashow, and J. M. Tiedje. 2010. The genome sequence of *Psychrobacter arcticus* 273-4, a psychroactive Siberian permafrost bacterium, reveals mechanisms for adaptation to low-temperature growth. *Appl. Environ. Microbiol.* **76**:2304–2312.

Bakermans, C., H. L. Ayala-del-Río, M. A. Ponder, T. Vishnivetskaya, D. Gilichinsky, M. F. Thomashow, and J. M. Tiedje. 2006. *Psychrobacter cryohalolentis* sp. nov. and *Psychrobacter arcticus* sp. nov., isolated from Siberian permafrost. *Int. J. Syst. Evol. Microbiol.* **56**:1285–1291.

Bakermans, C., and K. H. Nealson. 2004. Relationship of critical temperature to macromolecular synthesis and growth yield in *Psychrobacter cryopegella*. *J. Bacteriol.* **186**:2340–2345.

Bakermans, C., S. L. Tollaksen, C. S. Giometti, C. Wilkerson, J. M. Tiedje, and M. F. Thomashow. 2007. Proteomic analysis of *Psychrobacter cryohalolentis* K5 during growth at subzero temperatures. *Extremophiles* **11**:343–354.

Bayles, D. O., M. H. Tunick, T. A. Foglia, and A. J. Miller. 2000. Cold shock and its

effect on ribosomes and thermal tolerance in *Listeria monocytogenes*. *Appl. Environ. Microbiol.* **66**:4351–4355.

Beckering, C. L., L. Steil, M. H. Weber, U. Volker, and M. A. Marahiel. 2002. Genomewide transcriptional analysis of the cold shock response in *Bacillus subtilis*. *J. Bacteriol.* **184**:6395–6402.

Bergholz, P. W., C. Bakermans, and J. M. Tiedje. 2009. *Psychrobacter arcticus* 273-4 uses resource efficiency and molecular motion adaptations for subzero temperature growth. *J. Bacteriol.* **191**:2340–2352.

Bloomer, A. C., J. N. Champness, G. Bricogne, R. Staden, and A. Klug. 1978. Protein disk of tobacco mosaic virus at 2.8Å resolution showing the interactions within and between subunits. *Nature* **276**:362–368.

Bock, C., and H. Eicken. 2005. A magnetic resonance study of temperature-dependent microstructural evolution and self-diffusion of water in Arctic first-year sea ice. *Ann. Glaciol.* **40**:179–184.

Bowman, J. P. 2008. Genomic analysis of psychrophilic prokaryotes, p. 265–284. *In* R. Margesin, F. Schinner, J.-C. Marx, and C. Gerday (ed.), *Psychrophiles: from Biodiversity to Biotechnology*. Springer, Berlin, Germany.

Bowman, J. P., D. S. Nichols, and T. A. McMeekin. 1997. *Psychrobacter glacincola* sp. nov., a halotolerant, psychrophilic bacterium isolated from Antarctic sea ice. *Syst. Appl. Microbiol.* **20**:209–215.

Breezee, J., N. Cady, and J. T. Staley. 2004. Subfreezing growth of the sea ice bacterium "*Psychromonas ingrahamii*." *Microb. Ecol.* **47**:300–304.

Budde, I., L. Steil, C. Scharf, U. Völker, and E. Bremer. 2006. Adaptation of *Bacillus subtilis* to growth at low temperature: a combined transcriptomic and proteomic appraisal. *Microbiology* **152**:831–853.

Buisine, N., C. M. Tang, and R. Chalmers. 2002. Transposon-like Correia elements: structure, distribution and genetic exchange between pathogenic *Neisseria* sp. *FEBS Lett.* **522**:52–58.

Campanaro, S., T. J. Williams, D. W. Burg, D. De Francisci, L. Treu, F. M. Lauro, and R. Cavicchioli. 2011. Temperature-dependent global gene expression in the Antarctic archaeon *Methanococcoides burtonii*. *Environ Microbiol.* **13**:2018–2038.

Carpenter, E. J., S. Lin, and D. G. Capone. 2000. Bacterial activity in South Pole snow. *Appl. Environ. Microbiol.* **66**:4514–4517.

Casanueva, A., M. Tuffin, C. Cary, and D. A. Cowan. 2010. Molecular adaptations to psychrophily: the impact of "omic" technologies. *Trends Microbiol.* **18**:374–381.

Cavicchioli, R., and K. S. Siddiqui. 2004. Cold adapted enzymes, p. 615–638. *In* A. Pandey,

C. Webb, C. R. Soccol, and C. Larroche (ed), *Enzyme Technology*. Asiatech Publishers Inc., New Delhi, India.

Chaturvedi, P., and S. Shivaji. 2006. *Exiguobacterium indicum* sp. nov., a psychrophilic bacterium from the Hamta glacier of the Himalayan mountain ranges of India. *Int. J. Syst. Evol. Microbiol.* 56:2765–2770.

Chin, J. P., J. Megaw, C. L. Magill, K. Nowotarski, J. P. Williams, P. Bhaganna, M. Linton, M. F. Patterson, G. J. Underwood, A. Y. Mswaka, and J. E. Hallsworth. 2010. Solutes determine the temperature windows for microbial survival and growth. *Proc. Natl. Acad. Sci. USA* 107:7835–7840.

Chintalapati, S., M. D. Kiran, and S. Shivaji. 2004. Role of membrane lipid fatty acids in cold adaptation. *Cell. Mol. Biol. (Noisy-le-grand)* 50:631–642.

Chu, H., N. Fierer, C. L. Lauber, J. G. Caporaso, R. Knight, and P. Grogan. 2010. Soil bacterial diversity in the Arctic is not fundamentally different from that found in other biomes. *Environ. Microbiol.* 12:2998–3006.

Collins, R. E., S. D. Carpenter, and J. W. Deming. 2008. Spatial heterogeneity and temporal dynamics of particles, bacteria, and pEPS in Arctic winter sea ice. *J. Mar. Syst.* 74:902–917.

Crapart, S., M. Fardeau, J. Cayol, P. Thomas, C. Sery, B. Ollivier, and Y. Cambet-Blanc. 2007. *Exiguobacterium profundum* sp. nov., a moderately thermophilic, lactic acid-producing bacterium isolated from a deep-sea hydrothermal vent. *Int. J. Syst. Evol. Microbiol.* 57:287–292.

Davail, S., G. Feller, E. Narinx, and C. Gerday. 1994. Cold adaptation of proteins. Purification, characterization, and sequence of the heat-labile subtilisin from the Antarctic psychrophile *Bacillus* TA41. *J. Biol. Chem.* 269:17448–17453.

de Backer, M., S. McSweeney, H. B. Rasmussen, B. W. Riise, P. Lindley, and E. Hough. 2002. The 1.9 Å crystal structure of heat-labile shrimp alkaline phosphatase. *J. Mol. Biol.* 318:1265–1274.

De Gregorio, E., C. Abrescia, M. S. Carlomagno, and P. P. Di Nocera. 2003. Ribonuclease III-mediated processing of specific *Neisseria meningitidis* mRNAs. *Biochem. J.* 374:799–805.

DeLong, E. F., C. M. Preston, T. Mincer, V. Rich, S. J. Hallam, N. U. Frigaard, A. Martinez, M. B. Sullivan, R. Edwards, B. R. Brito, S. W. Chisholm, and D. M. Karl. 2006. Community genomics among stratified microbial assemblages in the ocean's interior. *Science* 311:496–503.

Dunker, A. K. 2007. *Disordered Proteins*. John Wiley & Sons Ltd., London, United Kingdom.

Dunker, A. K., C. J. Brown, J. D. Lawson, L. M. Iakoucheva, and Z. Obradovic. 2002. Intrinsic disorder and protein function. *Biochemistry* 41:6573–6582.

Eriksson, S., R. Hurme, and M. Rhen. 2002. Low-temperature sensors in bacteria. *Philos. Trans. R. Soc. Lond. B Biol. Sci.* 357:887–893.

Feller, G. 2007. Life at low temperatures: is disorder the driving force? *Extremophiles* 11:211–216.

Feller, G. 2010. Protein stability and enzyme activity at extreme biological temperatures. *J. Phys. Condens. Matter* 22:323101.

Feller, G., and C. Gerday. 2003. Psychrophilic enzymes: hot topics in cold adaptation. *Nat. Rev. Microbiol.* 1:200–208.

Feller, G., T. Lonhienne, C. Deroanne, C. Libioulle, J. Van Beeumen, and C. Gerday. 1992. Purification, characterization, and nucleotide sequence of the thermolabile α-amylase from the Antarctic psychrotroph *Alteromonas haloplanctis* A23. *J. Biol. Chem.* 267:5217–5221.

Feller, G., F. Payan, F. Theys, M. Qian, R. Haser, and C. Gerday. 1994. Stability and structural analysis of α-amylase from the antarctic psychrophile *Alteromonas haloplanctis* A23. *Eur. J. Biochem.* 222:441–447.

Fields, P. A. 2001. Review: Protein function at thermal extremes: balancing stability and flexibility. *Comp. Biochem. Physiol. A Mol. Integr. Physiol.* 129:417–431.

Franks, F., S. F. Mathias, and R. H. Hatley. 1990. Water, temperature and life. *Philos. Trans. R. Soc. Lond. B Biol. Sci.* 326:517–531; discussion 531–533.

Fraser, C. M., J. D. Gocayne, O. White, M. D. Adams, R. A. Clayton, R. D. Fleischmann, C. J. Bult, A. R. Kerlavage, G. Sutton, J. M. Kelley, J. L. Fritchman, J. F. Weidman, K. V. Small, M. Sandusky, J. Fuhrmann, D. Nguyen, T. R. Utterback, D. M. Saudek, C. A. Phillips, J. M. Merrick, J. F. Tomb, B. A. Dougherty, K. F. Bott, P. C. Hu, T. S. Lucier, S. N. Peterson, H. O. Smith, C. A. Hutchison, and J. C. Venter. 1995. The minimal gene complement of *Mycoplasma genitalium*. *Science* 270:397–403.

Gao, H. C., Z. M. K. Yang, L. Y. Wu, D. K. Thompson, and J. Z. Zhou. 2006. Global transcriptome analysis of the cold shock response of *Shewanella oneidensis* MR-1 and mutational analysis of its classical cold shock proteins. *J. Bacteriol.* 188:4560–4569.

Georlette, D., V. Blaise, T. Collins, S. D'Amico, E. Gratia, A. Hoyoux, J. C. Marx, G. Sonan, G. Feller, and C. Gerday. 2004. Some like it cold: biocatalysis at low temperatures. *FEMS Microbiol. Rev.* 28:25–42.

Gianese, G., F. Bossa, and S. Pascarella. 2002. Comparative structural analysis of psychrophilic and meso- and thermophilic enzymes. *Proteins* **47**:236–249.

Giuliodori, A. M., A. Brandi, C. O. Gualerzi, and C. L. Pon. 2004. Preferential translation of cold-shock mRNAs during cold adaptation. *RNA* **10**:265–276.

Golden, K. M. 2001. Brine percolation and the transport properties of sea ice. *Ann. Glaciol.* **33**:28–36.

Golovlev, E. L. 2003. Bacterial cold shock response at the level of DNA transcription, translation and chromosome dynamics. *Mikrobiologiia* **72**:5–13. (In Russian.)

Goodchild, A., M. Raftery, N. F. Saunders, M. Guilhaus, and R. Cavicchioli. 2004. Biology of the cold adapted archaeon, *Methanococcoides burtonii* determined by proteomics using liquid chromatography-tandem mass spectrometry. *J. Proteome Res.* **3**:1164–1176.

Goodchild, A., M. Raftery, N. F. Saunders, M. Guilhaus, and R. Cavicchioli. 2005. Cold adaptation of the Antarctic archaeon, *Methanococcoides burtonii* assessed by proteomics using ICAT. *J. Proteome Res.* **4**:473–480.

Goris, J., K. T. Konstantinidis, J. A. Klappenbach, T. Coenye, P. Vandamme, and J. M. Tiedje. 2007. DNA-DNA hybridization values and their relationship to whole-genome sequence similarities. *Int. J. Syst. Evol. Microbiol.* **57**:81–91.

Gottschal, J. C. 1985. Some reflections on microbial competitiveness among heterotrophic bacteria. *Antonie van Leeuwenhoek* **51**:473–494.

Grzymski, J. J., B. J. Carter, E. F. DeLong, R. A. Feldman, A. Ghadiri, and A. E. Murray. 2006. Comparative genomics of DNA fragments from six Antarctic marine planktonic bacteria. *Appl. Environ. Microbiol.* **72**:1532–1541.

Gualerzi, C. O., A. M. Giuliodori, and C. L. Pon. 2003. Transcriptional and post-transcriptional control of cold-shock genes. *J. Mol. Biol.* **331**:527–539.

Helmke, E., and H. Weyland. 2004. Psychrophilic versus psychrotolerant bacteria—occurrence and significance in polar and temperate marine habitats. *Cell. Mol. Biol. (Noisy-le-grand)* **50**:553–561.

Hepler, L. G., and E. M. Woolley. 1973. Hydration effects and acid-base equilibria, p. 145–172. *In* F. Franks (ed.), *Water: a Comprehensive Treatise.* Plenum Press, New York, NY.

Herbert, R. A., and C. R. Bell. 1977. Growth characteristics of an obligately psychrophilic *Vibrio* sp. *Arch. Microbiol.* **113**:215–220.

Hjerde, E., M. S. Lorentzen, M. T. Holden, K. Seeger, S. Paulsen, N. Bason, C. Churcher, D. Harris, H. Norbertczak, M. A. Quail, S. Sanders, S. Thurston, J. Parkhill, N. P. Wil-lassen, and N. R. Thomson. 2008. The genome sequence of the fish pathogen *Aliivibrio salmonicida* strain LFI1238 shows extensive evidence of gene decay. *BMC Genomics* **9**:616.

Iost, I., and M. Dreyfus. 2006. DEAD-box RNA helicases in *Escherichia coli*. *Nucleic Acids Res.* **34**:4189–4197.

Ishii, A., T. Ochiai, S. Imagawa, N. Fukunaga, S. Sasaki, O. Minowa, Y. Mizuno, and H. Shiokawa. 1987. Isozymes of isocitrate dehydrogenase from an obligately psychrophilic bacterium, *Vibrio* sp. strain ABE-1: purification, and modulation of activities by growth-conditions. *J. Biochem.* **102**:1489–1498.

Junge, K., H. Eicken, and J. W. Deming. 2004. Bacterial activity at −2 to −20°C in Arctic wintertime sea ice. *Appl. Environ. Microbiol.* **70**:550–557.

Konstantinidis, K. T., and J. M. Tiedje. 2005. Genomic insights that advance the species definition for prokaryotes. *Proc. Natl. Acad. Sci. USA* **102**:2567–2572.

Konstantinidis, K. T., and J. M. Tiedje. 2007. Prokaryotic taxonomy and phylogeny in the genomic era: advancements and challenges ahead. *Curr. Opin. Microbiol.* **10**:504–509.

Lauro, F. M., K. Tran, A. Vezzi, N. Vitulo, G. Valle, and D. H. Bartlett. 2008. Large-scale transposon mutagenesis of *Photobacterium profundum* SS9 reveals new genetic loci important for growth at low temperature and high pressure. *J. Bacteriol.* **190**:1699–1709.

Leblanc, L., C. Leboeuf, F. Leroi, A. Hartke, and Y. Auffray. 2003. Comparison between NaCl tolerance response and acclimation to cold temperature in *Shewanella putrefaciens*. *Curr. Microbiol.* **46**:157–162.

Linding, R., L. J. Jensen, F. Diella, P. Bork, T. J. Gibson, and R. B. Russell. 2003. Protein disorder prediction: implications for structural proteomics. *Structure* **11**:1453–1459.

Lozupone, C., M. Hamady, and R. Knight. 2006. UniFrac—an online tool for comparing microbial community diversity in a phylogenetic context. *BMC Bioinformatics* **7**:371.

Lozupone, C., and R. Knight. 2005. UniFrac: a new phylogenetic method for comparing microbial communities. *Appl. Environ. Microbiol.* **71**:8228–8235.

Makhatadze, G. I., and P. L. Privalov. 1994. Hydration effects in protein unfolding. *Biophys. Chem.* **51**:291–309.

Maki, S., M. Yoneta, and Y. Takada. 2006. Two isocitrate dehydrogenases from a psychrophilic bacterium, *Colwellia psychrerythraea*. *Extremophiles* **10**:237–249.

McCutcheon, J. P., and N. A. Moran. 2007. Parallel genomic evolution and metabolic interdependence in an ancient symbiosis. *Proc. Natl. Acad. Sci USA* **104:**19392–19397.

Medigue, C., E. Krin, G. Pascal, V. Barbe, A. Bernsel, P. N. Bertin, F. Cheung, S. Cruveiller, S. D'Amico, A. Duilio, G. Fang, G. Feller, C. Ho, S. Mangenot, G. Marino, J. Nilsson, E. Parrilli, E. P. C. Rocha, Z. Rouy, A. Sekowska, M. L. Tutino, D. Vallenet, G. von Heijne, and A. Danchin. 2005. Coping with cold: the genome of the versatile marine Antarctica bacterium *Pseudoalteromonas haloplanktis* TAC125. *Genome Res.* **15:**1325–1335.

Methe, B. A., K. E. Nelson, J. W. Deming, B. Momen, E. Melamud, X. Zhang, J. Moult, R. Madupu, W. C. Nelson, R. J. Dodson, L. M. Brinkac, S. C. Daugherty, A. S. Durkin, R. T. DeBoy, J. F. Kolonay, S. A. Sullivan, L. Zhou, T. M. Davidsen, M. Wu, A. L. Huston, M. Lewis, B. Weaver, J. F. Weidman, H. Khouri, T. R. Utterback, T. V. Feldblyum, and C. M. Fraser. 2005. The psychrophilic lifestyle as revealed by the genome sequence of *Colwellia psychrerythraea* 34H through genomic and proteomic analyses. *Proc. Natl. Acad. Sci. USA* **102:**10913–10918.

Metpally, R. P., and B. V. Reddy. 2009. Comparative proteome analysis of psychrophilic versus mesophilic bacterial species: insights into the molecular basis of cold adaptation of proteins. *BMC Genomics* **10:**11.

Mizushima, T., K. Kataoka, Y. Ogata, R. Inoue, and K. Sekimizu. 1997. Increase in negative supercoiling of plasmid DNA in *Escherichia coli* exposed to cold shock. *Mol. Microbiol.* **23:**381–386.

Morita, R. Y. 1975. Psychrophilic bacteria. *Bacteriol. Rev.* **39:**144–167.

Nedwell, D. B. 1999. Effect of low temperature on microbial growth: lowered affinity for substrates limits growth at low temperature. *FEMS Microbiol. Ecol.* **30:**101–111.

Nedwell, D. B., and M. Rutter. 1994. Influence of temperature on growth rate and competition between two psychrotolerant Antarctic bacteria: low temperature diminishes affinity for substrate uptake. *Appl. Environ. Microbiol.* **60:**1984–1992.

Nelson, D. L., and M. M. Cox. 2004. *Lehninger Principles of Biochemistry*, 4th ed. W. H. Freeman & Co., New York, NY.

Phadtare, S., and M. Inouye. 2004. Genome-wide transcriptional analysis of the cold shock response in wild-type and cold-sensitive, quadruple-*csp*-deletion strains of *Escherichia coli*. *J. Bacteriol.* **186:**7007–7014.

Ponder, M. A., S. J. Gilmour, P. W. Bergholz, C. A. Mindock, R. Hollingsworth, M. F. Thomashow, and J. M. Tiedje. 2005. Characterization of potential stress responses in ancient

Siberian permafrost psychroactive bacteria. *FEMS Microbiol. Ecol.* **53:**103–115.

Ponder, M., M. Thomashow, and J. Tiedje. 2008. Metabolic activity of Siberian permafrost isolates, *Psychrobacter arcticus* and *Exiguobacterium sibiricum*, at low water activities. *Extremophiles* **12:**481–490.

Poolman, B., and E. Glaasker. 1998. Regulation of compatible solute accumulation in bacteria. *Mol. Microbiol.* **29:**397–407.

Price, P. B., and T. Sowers. 2004. Temperature dependence of metabolic rates for microbial growth, maintenance, and survival. *Proc. Natl. Acad. Sci. USA* **101:**4631–4636.

Rabus, R., A. Ruepp, T. Frickey, T. Rattei, B. Fartmann, M. Stark, M. Bauer, A. Zibat, T. Lombardot, I. Becker, J. Amann, K. Gellner, H. Teeling, W. D. Leuschner, F. O. Glöckner, A. N. Lupas, R. Amann, and H. P. Klenk. 2004. The genome of *Desulfotalea psychrophila*, a sulfate-reducing bacterium from permanently cold Arctic sediments. *Environ. Microbiol.* **6:**887–902.

Read, T. D., R. C. Brunham, C. Shen, S. R. Gill, J. F. Heidelberg, O. White, E. K. Hickey, J. Peterson, T. Utterback, K. Berry, S. Bass, K. Linher, J. Weidman, H. Khouri, B. Craven, C. Bowman, R. Dodson, M. Gwinn, W. Nelson, R. DeBoy, J. Kolonay, G. McClarty, S. L. Salzberg, J. Eisen, and C. M. Fraser. 2000. Genome sequences of *Chlamydia trachomatis* MoPn and *Chlamydia pneumoniae* AR39. *Nucleic Acids Res.* **28:**1397–1406.

Reiersen, H., and A. R. Rees. 2001. The hunchback and its neighbours: proline as an environmental modulator. *Trends Biochem. Sci.* **26:**679–684.

Richter, M., and R. Rosselló-Móra. 2009. Shifting the genomic gold standard for the prokaryotic species definition. *Proc. Natl. Acad. Sci. USA* **160:**19126–19131.

Riley, M., J. T. Staley, A. Danchin, T. Z. Wang, T. S. Brettin, L. J. Hauser, M. L. Land, and L. S. Thompson. 2008. Genomics of an extreme psychrophile, *Psychromonas ingrahamii*. *BMC Genomics* **9:**210.

Rivkina, E. M., E. I. Friedmann, C. P. McKay, and D. A. Gilichinsky. 2000. Metabolic activity of permafrost bacteria below the freezing point. *Appl. Environ. Microbiol.* **66:**3230–3233.

Rodrigues, D. F., E. da C. Jesus, H. L. Ayala-Del-Río, V. H. Pellizari, D. Gilichinsky, L. Sepulveda-Torres, and J. M. Tiedje, (2009) Biogeography of two cold-adapted genera: *Psychrobacter* and *Exiguobacterium*. *ISME J.* **3:**658–665.

Rodrigues, D. F., J. Goris, T. Vishnivetskaya, D. Gilichinsky, M. F. Thomashow, and J. M. Tiedje. 2006. Characterization of *Exiguobacterium* isolates from the Siberian permafrost. Description of *Exiguobacterium sibiricum* sp. nov. *Extremophiles* **10**:285–294.

Rodrigues, D. F., N. Ivanova, Z. He, M. Huebner, J. Zhou, and J. M. Tiedje. 2008. Architecture of thermal adaptation in an *Exiguobacterium sibiricum* strain isolated from 3 million year old permafrost: a genome and transcriptome approach. *BMC Genomics* **9**:547.

Rodrigues, D. F., and J. M. Tiedje. 2007. Multilocus real-time PCR for quantitation of bacteria in the environment reveals *Exiguobacterium* to be prevalent in permafrost. *FEMS Microbiol. Ecol.* **59**:489–499.

Romanenko, L. A., A. M. Lysenko, M. Rohde, V. V. Mikhailov, and E. Stackebrandt. 2004. *Psychrobacter maritimus* sp. nov. and *Psychrobacter arenosus* sp. nov., isolated from coastal sea ice and sediments of the Sea of Japan. *Int. J. Syst. Evol. Microbiol.* **54**:1741–1745.

Russell, N. J. 2000. Toward a molecular understanding of cold activity of enzymes from psychrophiles. *Extremophiles* **4**:83–90.

Saunders, N. F., T. Thomas, P. M. Curmi, J. S. Mattick, E. Kuczek, R. Slade, J. Davis, P. D. Franzmann, D. Boone, K. Rusterholtz, R. Feldman, C. Gates, S. Bench, K. Sowers, K. Kadner, A. Aerts, P. Dehal, C. Detter, T. Glavina, S. Lucas, P. Richardson, F. Larimer, L. Hauser, M. Land, and R. Cavicchioli. 2003. Mechanisms of thermal adaptation revealed from the genomes of the Antarctic *Archaea Methanogenium frigidum* and *Methanococcoides burtonii*. *Genome Res.* **13**:1580–1588.

Scheffers, D. J., and M. G. Pinho. 2005. Bacterial cell wall synthesis: new insights from localization studies. *Microbiol. Mol. Biol. Rev.* **69**:585–607.

Scherer, S., and K. Neuhaus. 2006. Life at low temperatures, p. 210–262. *In* M. Dworkin (ed.), *The Prokaryotes*. Springer-Verlag, New York, NY.

Shivaji, S., G. S. Reddy, K. Suresh, P. Gupta, S. Chintalapati, P. Schumann, E. Stackebrandt, and G. I. Matsumoto. 2005. *Psychrobacter vallis* sp. nov. and *Psychrobacter aquaticus* sp. nov., from Antarctica. *Int. J. Syst. Evol. Microbiol.* **55**:757–762.

Shoichet, B. K., W. A. Baase, R. Kuroki, and B. W. Matthews. 1995. A relationship between protein stability and protein function. *Proc. Natl. Acad. Sci. USA* **92**:452–456.

Shravage, B. V., K. M. Dayananda, M. S. Patole, and Y. S. Shouche. 2007. Molecular microbial diversity of a soil sample and detection of ammonia oxidizers from Cape Evans, McMurdo Dry Valley, Antarctica. *Microbiol. Res.* **162**:15–25.

Siddiqui, K. S., and R. Cavicchioli. 2006. Cold-adapted enzymes. *Annu. Rev. Biochem.* **75**:403–433.

Singer, G. A. C., and D. A. Hickey. 2003. Thermophilic prokaryotes have characteristic patterns of codon usage, amino acid composition and nucleotide content. *Gene* **317**:39–47.

Smalås, A. O., H. K. Leiros, V. Os, and N. P. Willassen. 2000. Cold adapted enzymes. *Biotechnol. Annu. Rev.* **6**:1–57.

Thomas, G. W. 1996. Soil pH and soil acidity, p. 475–490. *In* J. M. Bigham (ed.), *Methods of Soil Analysis*. Soil Science Society of America and American Society of Agronomy, Madison, WI.

Thomashow, M. F. 1999. Plant cold acclimation: freezing tolerance genes and regulatory mechanisms. *Annu. Rev. Plant Physiol. Plant Mol. Biol.* **50**:571–599.

Ting, L., T. J. Williams, M. J. Cowley, F. M. Lauro, M. Guilhaus, M. J. Raftery, and R. Cavicchioli. 2010. Cold adaptation in the marine bacterium, *Sphingopyxis alaskensis*, assessed using quantitative proteomics. *Environ. Microbiol.* **12**:2658–2676.

Violot, S., N. Aghajari, M. Czjzek, G. Feller, G. K. Sonan, P. Gouet, C. Gerday, R. Haser, and V. Receveur-Bréchot. 2005. Structure of a full length psychrophilic cellulase from *Pseudoalteromonas haloplanktis* revealed by X-ray diffraction and small angle X-ray scattering. *J. Mol. Biol.* **348**:1211–1224.

Vishnivetskaya, T. A., and S. Kathariou. 2005. Putative transposases conserved in *Exiguobacterium* isolates from ancient Siberian permafrost and from contemporary surface habitats. *Appl. Environ. Microbiol.* **71**:6954–6962.

Vishnivetskaya, T., S. Kathariou, J. McGrath, D. Gilichinsky, and J. M. Tiedje. 2000. Low-temperature recovery strategies for the isolation of bacteria from ancient permafrost sediments. *Extremophiles* **4**:165–173.

Vishnivetskaya, T. A., S. Kathariou, and J. M. Tiedje. 2009. The *Exiguobacterium* genus: biodiversity and biogeography. *Extremophiles* **13**:541–555.

Vishnivetskaya, T., R. Ramley, D. F. Rodrigues, J. M. Tiedje, and S. Kathariou. 2005. *Exiguobacterium* from frozen subsurface sediments (Siberian permafrost) and from other sources have growth temperature ranges reflective of the environmental thermocline of their origin, p. 139. *In Joint International Symposia for Subsurface Microbiology (ISSM 2005) and Environmental Biogeochemistry (ISEB XVII), Jackson Hole, Wyoming*. American Society for Microbiology, Washington, DC.

Vogt, G., S. Woell, and P. Argos. 1997. Protein thermal stability, hydrogen bonds, and ion pairs. *J. Mol. Biol.* **269**:631–643.

Walter, R. P., J. G. Morris, and D. B. Kell. 1987. The roles of osmotic stress and water activity in the inhibition of the growth, glycolysis and glucose phosphotransferase system of *Clostridium pasteurianum. J. Gen. Microbiol.* **133:**259–266.

Weber, M. H., and M. A. Marahiel. 2003. Bacterial cold shock responses. *Sci. Prog.* **86:**9–75.

White, D. 2000. *The Physiology and Biochemistry of Prokaryotes,* p. 565. Oxford University Press, New York, NY.

Whitford, D. 2005. *Proteins: Structure and Function.* John Wiley & Sons Ltd., Chichester, United Kingdom.

Xiao, L., and B. Honig. 1999. Electrostatic contributions to the stability of hyperthermophilic proteins. *J. Mol. Biol.* **289:**1435–1444.

Yao, X., M. Jericho, D. Pink, and T. Beveridge. 1999. Thickness and elasticity of gram-negative murein sacculi measured by atomic force microscopy. *J. Bacteriol.* **181:**6865–6875.

Zhao, J.-S., Y. Deng, D. Manno, and J. Hawari. 2010. *Shewanella* spp. genomic evolution for a cold marine lifestyle and *in-situ* explosive biodegradation. *PLoS One* **5:**e9109.

METAGENOMIC ANALYSIS OF POLAR ECOSYSTEMS

Etienne Yergeau and Charles W. Greer

7

INTRODUCTION

Metagenomics

The word "metagenomics" was first coined by Jo Handelsman (Handelsman et al., 1998) and can be described as the application of modern genomics techniques to the study of microbial communities directly in their natural environments, bypassing the need for isolation and lab cultivation of individual species. In the present chapter we use this definition as broadly as possible, including studies based on large-scale 16S rRNA gene libraries, environmental microarrays, large-insert clone libraries, and shotgun genome sequencing (Color Plate 6). Small-scale 16S rRNA gene libraries are covered in Chapters 1 and 2 and will not be discussed here in the context of metagenomics.

NEXT-GENERATION SEQUENCING

The world of sequencing was revolutionized with the technological advances that were brought by next-generation technology, the first of which was released by the 454 Life Science Corporation (Margulies et al., 2005). Two major innovations allowed this giant leap in sequencing capacity: (i) the parallel-

Etienne Yergeau and Charles W. Greer, National Research Council of Canada, Biotechnology Research Institute, Montreal, QC H4P 2R2, Canada.

ization of the reactions in plates containing hundreds of thousands of individual wells of picoliter capacity; and (ii) the clonal amplification of DNA molecules via emulsion PCR, eliminating the need for tedious clone libraries (Rothberg and Leamon, 2008).

Shortly after, other next-generation sequencing technologies were released, based on different sequencing chemistry and library preparation, but all with the same basic concept of massive parallelization and cloneless library preparation (Shendure and Ji, 2008). Because of their longer read length, most metagenomic studies to date have used 454 sequencing technology. However, the improvement in read length of other available technologies (e.g., Illumina, Solexa, SOLiD, Helicos) and in data analysis tools that can correctly assign shorter reads (Dalevi et al., 2008; Krause et al., 2008; Liu et al., 2007, 2008) now make it now feasible to use higher-throughput technology that generates shorter reads for metagenomic studies (e.g., Illumina) (Lazarevic et al., 2009; Qin et al., 2010).

SHOTGUN SEQUENCING APPROACHES

Shotgun sequencing involves the fragmentation of the total environmental genomic material and subsequent separation and sequencing. Early shotgun sequencing work used cloning

Polar Microbiology: Life in a Deep Freeze
Edited by Robert V. Miller and Lyle G. Whyte © 2012 ASM Press, Washington, DC

approaches (Tyson et al., 2004; Venter et al., 2004). These remarkable sequencing efforts led the way for other early shotgun metagenomic studies. The advent of novel cloneless sequencing technologies has democratized this approach and made it possible for individual researchers to perform large-scale metagenomic sequencing projects that were previously only accessible to large sequencing centers. Since then numerous shotgun sequencing projects have been carried out on a large range of contrasting environments, including polar environments.

AMPLICON SEQUENCING

The principal gene amplified and sequenced in the context of metagenomic studies is the 16S rRNA gene. These 16S rRNA gene studies were pioneered by Norman Pace and coworkers (Pace et al., 1985; Schmidt et al., 1991), and more than 20 years later this gene is still the gold standard for studying bacterial diversity in environmental samples. Until recently, 16S rRNA gene studies involved the amplification and cloning of 16S rRNA genes directly from environmental samples. The increase in sequencing capacity and the elimination of cloning steps in next-generation sequencing increased the depth of studies from a few hundred sequences to tens of thousands of sequences per sample, providing an unprecedented view of microbial diversity. In fact, 16S rRNA gene libraries that were partly abandoned for other higher-throughput techniques (e.g., denaturing gradient gel electrophoresis, microarrays) are now making a comeback with next-generation sequencing (Tringe and Hugenholtz, 2008). In addition to the 16S rRNA gene, other genes could be subjected to amplicon sequencing, like functional genes of interest in a particular environment or other housekeeping marker genes, like *cpn60* (Schellenberg et al., 2009). However, gaps in databases make most other genes hard to relate to a known organism.

LARGE-INSERT LIBRARIES

Large-insert libraries (fosmid, cosmid, and bacterial artificial chromosome) contain large fragments (in the kbp range) of environmental genomic DNA and are generally screened for characteristics of interest. One way is to screen the library for a gene of interest using PCR and to sequence positive clones. Another way is to functionally screen the library, using specific growth media or growth-related assays. The main advantage of large-insert libraries is to reduce the sequencing effort needed when the only interest resides in particular enzymes or organisms. Because these libraries contain complete genes in their genetic context, they are also advantageous because they can provide insights and a functional annotation that shotgun libraries can rarely offer. On the downside, some genes might be overlooked because they cannot be expressed in the host cell. Large-insert libraries are normally Sanger sequenced, since it offers a better scalability than next-generation sequencing.

ENVIRONMENTAL MICROARRAYS

The rapid increase in genetic databases has facilitated the development of comprehensive platforms encompassing the known range of bacterial and archaeal diversity based upon 16S rRNA gene sequences (e.g., the PhyloChip) (Brodie et al., 2006; DeSantis et al., 2007) or the known range of functional genes involved in a variety of biogeochemical cycles (e.g., Geo-Chip 3.0) (He et al., 2010). However, it is still commonly accepted that public databases contain only a small part of the diversity present in environmental samples. Environment-specific microarrays targeting genes and microorganisms that are known to be important in an environment could be a way to circumvent this important limitation (Yergeau et al., 2009a). Anonymous microarray approaches, based on unsequenced genomic fragments isolated from the environment, have also been shown to be useful in metatranscriptomic studies (Yergeau et al., 2010b).

Polar Environments

The description of polar environments is outside the scope of this chapter, and therefore only the most relevant characteristics will be mentioned here. Although they are sometimes treated together as "polar environments," Arctic and Antarctic environments are very different in certain aspects. For instance, Antarctic environments are subjected to far more severe climatic conditions at similar latitudes (Convey, 2001). Furthermore, the Antarctic continent has been isolated from other landmasses for over 25 million years by the Southern Ocean and the South Polar Air Vortex, whereas the Arctic is well connected with neighboring regions. However, both Arctic and Antarctic environments are subjected to many of the same constraints. One of the most obvious constraints for microbial growth is the low average temperature. This affects microorganisms directly and indirectly through decreases in primary production. Less obvious are the large diel variations in temperature and frequent freeze-thaw cycles in surface environments, which are thought to be more important than stable low temperatures (Yergeau and Kowalchuk, 2008). Furthermore, some environments are nutrient limited and terrestrial environments can be water limited, especially during winter, when all the available water is frozen.

Challenges of Metagenomic Studies in Polar Environments

Polar environments are particularly challenging to study using metagenomics. One of the main limitations is the relatively low microbial biomass present in most polar environments, resulting in low recoveries of microbial DNA, which can limit the use of some technologies. Improvements in protocols are helping to reduce the amount of DNA required. For instance, the amount of DNA required for shotgun library preparation prior to 454 sequencing recently dropped from microgram quantities to around 500 ng. Whole-genome amplification, like multiple displacement amplification, is another way to solve this issue, but recent evidence shows that this technique can create large representational biases in the amplified DNA (Bodelier et al., 2009; Yergeau et al., 2010a), which is clearly undesirable for most metagenomic studies. Another challenge is that

polar environments, because of their remoteness, are less well studied than other environments, which makes it more difficult to obtain significant matches for sequences in databases and makes microarrays more difficult to apply.

Application of Metagenomics to Polar Environments

In our opinion, two major ecological themes could greatly benefit from the application of metagenomic methods. The first one is the adaptation and response of microorganisms to ongoing climatic changes. Indeed, the polar regions are subjected to the fastest warming rates on Earth (IPCC, 2007). Of particular interest are changes in microbial community composition, in diversity and functional redundancy. Specific genes or pathways that might be important to monitor in view of their importance in greenhouse gas emission include carbon degradation genes and methane production and oxidation genes. Laboratory and field experiments together with sampling of natural habitats along latitudinal/altitudinal gradients could be interesting ways to gather information about microbial responses to climate change.

The second research theme that could be enhanced by the application of metagenomics is bioremediation. Partly because of global warming, human activity is steadily increasing in polar regions, increasing the risks of accidental contamination. Metagenomics of contaminated polar environments can provide a basic understanding of the ecology of pollutant-degrading microorganisms and help devise bioremediation treatments. Metagenomics could also be used to link organisms/genes to the efficiency of different bioremediation treatments. Furthermore, studies of contaminated environments could lead to the discovery of novel cold-adapted bioremediation enzymes.

Other than ecological research, metagenomics in polar environments can be used to further mine for cold-adapted enzymes useful for different biotechnological applications (Ferrer et al., 2007). This type of study is normally better served by a focused sequencing effort on large-insert clones that were prescreened for a particular activity. This provides the researcher with a complete gene in its genomic context, something that would be hardly feasible using shotgun metagenomic sequencing even in the simplest of environments.

Although the field of metagenomics is currently growing at an explosive rate, there are very few studies focused on polar environments (Tables 1 and 2). As of April 2010, a search using (arctic* AND metagenom*) in Web of Science only yielded eight results, of which only three were true metagenomic works in Arctic environments. A similar search using (antarctic* AND metagenom*) resulted in 16 records, of which 10 were true metagenomic works in the Antarctic. Only one and two microarray studies were retrieved for Arctic and Antarctic environments, respectively. This

is quite surprising since polar environments are ideal testing grounds for new techniques, because of their limited complexity. Furthermore, polar environments are of considerable interest because of their relevance to the environmental problems mentioned above. The following sections describe studies that are applying metagenomic techniques to characterize and understand polar ecosystems.

MARINE ECOSYSTEMS

Seawater and Sediments

Bacteria and archaea communities of five regions of the Arctic Ocean were recently studied using 454 GS 20 sequencing (Galand et al., 2009a). The main focus of the study was to observe rare microbes in these environments,

TABLE 1 Summary of Arctic metagenomic studies

Environment	Method	No. of targets	Notes	Reference
Alaskan soil	Large-insert clone libraries	714,000 clones screened, 14 clones sequenced	Screened for β-lactamase activity	Allen et al., 2009
68 sites in four oceanic regions (Arctic: 16 sites)	Shotgun 454 GS FLX sequencing	181 Mbp (Arctic: 69 Mbp)	Viruses	Angly et al., 2006
Five geographical regions of the Arctic Ocean	16S rRNA amplicon GS 20 sequencing	545,246 bacterial and 195,107 archaeal 16S rRNA gene sequences (~70 bp)		Galand et al., 2009a
Eight samples from the western Arctic Ocean	16S rRNA amplicon GS 20 sequencing	195,107 archaeal 16S rRNA gene sequences (~70 bp)		Galand et al., 2009b
13 samples from three basins of the Arctic Ocean	16S rRNA amplicon GS 20 sequencing	313,827 bacterial 16S rRNA gene sequences (~70 bp)		Galand et al., 2010
Seashore sediments of the Kongsfjorden, Spitsbergen, Norway	Large-insert clone libraries	60,132 clones screened, 2 clones sequenced	Screened for esterase/ lipase activity	Jeon et al., 2009
11 samples from the western Arctic Ocean	16S rRNA amplicon GS 20 sequencing	250,555 bacterial 16S rRNA gene sequences (~70 bp)		Kirchman et al., 2010
Soil bioremediation experiments at two sites in the Canadian High Arctic	Functional and 16S rRNA gene microarrays	159 bacterial and archaeal taxa; 127 functional genes		Yergeau et al., 2009a
Canadian High Arctic, 2-m deep permafrost and active-layer soil	Shotgun 454 GS FLX Titanium sequencing	853 Mbp		Yergeau et al., 2010a

TABLE 2 Summary of Antarctic metagenomic studies

Environment	Method	No. of targets	Notes	Reference
Unvegetated soil on Pointe Geologie archipelago	Large-insert clone libraries	10,000 clones screened, 1 clone sequenced	Screened for glycosyl hydrolase activity	Berlemont et al., 2009
Debris-covered alpine glacier ice from the Dry Valleys	Large-insert and 16S rRNA gene clone libraries	258 16S rRNA and 559 bacterial artificial chromosome clones sequenced		Bidle et al., 2007
Penguin rookery soil, King George Island	Large-insert clone libraries	85,000 clones screened, 1 clone sequenced	Screened for MTA phosphorylase activity	Cieslinski et al., 2009b
Soil in the Miers Valley	Large-insert clone libraries	10,000 clones screened, 1 clone sequenced	Screened for alkaliphilic esterase activity	Heath et al., 2009
Ultra-oligotrophic lake, Livingston Island	Shotgun 454 GS FLX Titanium sequencing	20.6 Mbp		López-Bueno et al., 2009
Four sampling sites in different oceanic regions	Large-insert clone libraries	74,181 clones screened (Antarctic: 9,307 clones), 22 clones sequenced (Antarctic:1 clone)	Screened for archaeal 16S rRNA genes by PCR	Martin-Cuadrado et al., 2008
Antarctic water, 500 m deep	Large-insert clone libraries	6,107 clones screened, 1 clone sequenced	Screened for archaeal 16S rRNA genes by PCR	Moreira et al., 2004
Antarctic water, 500 m deep	Large-insert clone libraries	3,200 clones screened, 2 clones sequenced	Screened for bacterial 16S rRNA genes by PCR	Moreira et al., 2006
Rhizosphere soil, Admiralty Bay	16S rRNA amplicon GS FLX sequencing	~30,000 16S rRNA gene sequences		Teixeira et al., 2010
Vegetated and bare soils at five different sites	Functional gene microarray "Geochip"	>10,000 genes in >150 functional groups		Yergeau et al., 2007a
Vegetated and bare soils at five different sites	16S rRNA gene microarray "Phylochip"	8,741 bacterial and archaeal taxa		Yergeau et al., 2009b
Control and warmed soils at three different sites along the Peninsula	16S rRNA amplicon GS FLX Titanium sequencing	~150,000 16S rRNA gene sequences (~400 bp)		E. Yergeau et al., 2011

which showed the same distribution patterns as the more abundant community members.

The archaeal part of this dataset was reused in another publication, to more precisely describe the archaea of the Arctic Ocean (Galand et al., 2009b). Generally, the libraries were dominated by the Marine Group I *Crenarchaeota*, but depth had a strong influence on community composition.

Some of the bacterial libraries were also reused in a subsequent publication, where it was shown that bacteria in the Arctic Ocean have a biogeography that could be explained by hydrography (Galand et al., 2010). The bacterial libraries were also reanalyzed in another publication, which indicated that the bacterial communities were stable over time (winter versus summer sampling) and among oceanic regions (three different regions) (Kirchman et al., 2010). Interestingly, the little variation that was observed in bacterial communities was due to abundant community members rather than the rare ones.

A recent survey of the viral metagenome of four different oceanic regions showed that the Arctic Ocean contained a particular viral assemblage not observed in the other regions. The Arctic was much richer in prophage-like sequences (Angly et al., 2006).

Three studies have reported results from a large-insert library that was used to access the metagenomic information of Antarctic seawater (Martin-Cuadrado et al., 2008; Moreira et al., 2004, 2006). PCR targeting the 16S rRNA genes of archaea or bacteria was used to screen the libraries, and the sequence of each positive clone was determined. This analysis revealed novel genes and the metabolic potential of different organisms. Jeon et al. (2009) screened a large-insert library constructed from Arctic seashore sediments for esterase activity. They then characterized two esterase genes and their associated enzymes, which were similar to temperate enzymes but with lower activation energies, indicating a certain cold adaptation.

TERRESTRIAL ECOSYSTEMS

Soil

Several metagenomic studies have been carried out at sites that range from the Falkland Islands all the way to the base of the Antarctic Peninsula. Using a 16S rRNA gene microarray, the PhyloChip (DeSantis et al., 2007), Yergeau et al. (2009b) showed that bacterial diversity decreased with increasing latitude but only in plots devoid of vegetation cover. These data confirmed previous results from 16S rRNA gene clone libraries (Yergeau et al., 2007b). Another microarray study, using the GeoChip

2.0 (He et al., 2007), at the same sites revealed a wide metabolic potential that varied with the plant cover and the geographical location (Yergeau et al., 2007a).

A recent study at the same sites used 454 GS FLX Titanium sequencing of the 16S rRNA gene to compare the bacterial community structure and diversity in warmed versus control plots (Yergeau et al., 2011). This study revealed that soil warming induced significant shifts in the major soil bacterial groups like *Acidobacteria* and α-*Proteobacteria*, which in turn led to changes in soil respiration. These shifts could lead to rapid changes in biogeochemical cycles in Antarctic soils following global warming, which would influence ecosystem processes.

A recent study used 454 GS FLX sequencing targeting 16S rRNA gene amplicons to observe the bacterial diversity of the rhizosphere of two native Antarctic vascular plants at different sites around Admiralty Bay, Antarctica (Teixeira et al., 2010). Both plants had similar bacterial communities, with no clear plant rhizosphere effect. However, bacterial diversity was quite different from that found in temperate soils and other polar soils. *Bifidobacterium* (*Actinobacteria*), *Arcobacter* (*Proteobacteria*), and *Faecalibacterium* (*Firmicutes*) were the dominant genera.

A recent shotgun metagenomic study in the Canadian High Arctic revealed the metabolic potential of the polar soils of this region. This study sequenced a permafrost sample and its overlying active-layer soil and focused particularly on genes that might be important for greenhouse gas emissions following permafrost thaw (Yergeau et al., 2010a). Metagenomic sequencing of permafrost soils revealed that they contain a wide range of decomposition and methane-related genes that could lead to rapid emissions of greenhouse gases following permafrost thaw.

Several studies have used large-insert clone libraries to screen Arctic or Antarctic soils for particular functions, with the goal of potentially using the recovered genes to produce enzymes useful in biotechnological applications. The targeted enzymes included esterase (Heath et al., 2009), glycosyl hydrolase (Berlemont et al., 2009), lipase (Cieslinski et al., 2009a), and methylthioadenosine phosphorylase (Cieslinski et al., 2009b). Another study screened large-insert libraries constructed from DNA extracted from pristine Alaskan soil for β-lactamase activity. This study showed that Arctic soils are a reservoir of β-lactamase genes, which could have repercussions on bacterial resistance to antibiotics (Allen et al., 2009).

Freshwater and Ice

The only metagenomic study of polar freshwater ecosystems published to date looked at the viral communities of an Antarctic lake (López-Bueno et al., 2009). This study revealed that in contrast to other environments, a large proportion of the viral sequences retrieved from the lake were from eukaryotic viruses and not from bacteriophages. The authors also observed seasonal changes in the viral community, which they attributed to shifts in host organisms. Partial sequencing of hundreds of large-insert clones constructed from DNA extracted from ancient ice in the Dry Valleys of Antarctica revealed a wide range of functional genes (Bidle et al., 2007). The authors hypothesized that the genetic material contained in such ancient ice could serve as a source of genes for microbes upon ice thawing.

CONCLUSIONS

Polar environments can be among the simplest environments on Earth, making them excellent models to understand ecosystem processes and the responses of microorganisms to environmental perturbations. This vulnerability and the rate of the ongoing change make polar environments a priority in climate change and bioremediation studies. However, as this chapter highlights, few studies have yet used modern metagenomic techniques to investigate polar environments. The next logical move would be to apply existing metagenomic tools to a wider range of polar environments to further the understanding of their basic microbiology.

The next step for the analysis of polar environments would be to move toward metatranscriptomic studies. This type of analysis poses particular challenges, especially for polar environments. They are often characterized by low biomass. In combination with the cold temperatures, this leads to very low activity rates. Under these circumstances, it is very difficult to retrieve enough material to conduct meaningful metatranscriptomic studies. The remoteness of polar regions also makes it difficult to acquire and properly preserve samples in a state where the transcriptomic information will be protected. Another challenge that is not specific to polar environments, but common to all metatranscriptomic studies, is related to the overwhelming amount of rRNA as compared to mRNA in RNA extracts. Some commercial kits are available to enrich RNA extracts in mRNA, but generally they only partially remove the rRNA in environmental samples. Even with these difficulties, recent work in this field is showing some promise (Stewart et al., 2010).

Taking it one step further would be to utilize new techniques or to combine existing methods in novel approaches. Other next-generation sequencing techniques (e.g., Helicos, Illumina, SOLiD) could be used to increase the amount of sequencing data retrieved from these environments. However, the shorter read length makes these techniques better suited for 16S rRNA amplicon sequencing or for shotgun sequencing of already well-characterized environments. Emerging technologies could also be interesting to apply to polar environments. For instance, bioremediation studies could benefit from metabolomics (Turnbaugh and Gordon, 2008), proteomics (Verberkmoes et al., 2009), and newly developed reactome arrays (Beloqui et al., 2009). Other strategies might include combining metagenomics with complementary techniques like microfluidics and cell sorting (Warnecke and Hugenholtz, 2007). Third-generation sequencing techniques that promise sequence reads in the kbp range (Branton et al., 2008; Eid et al., 2009) will lead to another revolution in the field of metagenomics and will probably yield even greater insight into polar environments.

Nevertheless, even with the best technology and the most relevant samples, the reach of the conclusions from small unreplicated datasets will remain anecdotal. Well-planned, large-scale projects with a solidly replicated sampling design are needed to gain in-depth knowledge of the environmental genomic and transcriptomic information on crucially important polar environments. The multiplexing possibility and the depth of coverage of new metagenomics approaches are now making this a realistic option. As polar microbiologists are slowly grasping the vast possibilities brought about by the recent sequencing shakeup, it is now time to move from mastering techniques to utilizing them in large-scale experiments, to finally shedding a major beam of light into the proverbial microbial black box.

REFERENCES

Allen, H. K., L. A. Moe, J. Rodbumrer, A. Gaarder, and J. Handelsman. 2009. Functional metagenomics reveals diverse β-lactamases in a remote Alaskan soil. *ISME J.* 3:243–251.

Angly, F. E., B. Felts, M. Breitbart, P. Salamon, R. A. Edwards, C. Carlson, A. M. Chan, M. Haynes, S. Kelley, H. Liu, J. M. Mahaffy, J. E. Mueller, J. Nulton, R. Olson, R. Parsons, S. Rayhawk, C. A. Suttle, and F. Rohwer. 2006. The marine viromes of four oceanic regions. *PLoS Biol.* 4:e368.

Beloqui, A., M. E. Guazzaroni, F. Pazos, J. M. Vieites, M. Godoy, O. V. Golyshina, T. N. Chernikova, A. Waliczek, R. Silva-Rocha, Y. Al-Ramahi, V. La Cono, C. Mendez, J. A. Salas, R. Solano, M. M. Yakimov, K. N. Timmis, P. N. Golyshin, and M. Ferrer. 2009. Reactome array: forging a link between metabolome and genome. *Science* 326:252–257.

Berlemont, R., M. Delsaute, D. Pipers, S. D'Amico, G. Feller, M. Galleni, and P. Power. 2009. Insights into bacterial cellulose biosynthesis by functional metagenomics on Antarctic soil samples. *ISME J.* 3:1070–1081.

Bidle, K. D., S. Lee, D. R. Marchant, and P. G. Falkowski. 2007. Fossil genes and microbes in the oldest ice on Earth. *Proc. Natl. Acad. Sci. USA* 104:13455–13460.

Bodelier, P. L. E., M. Kamst, M. Meima-Franke, N. Stralis-Pavese, and L. Bodrossy. 2009. Whole-community genome amplification (WCGA) leads to compositional bias in methane-oxidizing communities as assessed by *pmoA*-based microarray analyses and QPCR. *Environ. Microbiol. Rep.* 1:434–441.

Branton, D., D. W. Deamer, A. Marziali, H. Bayley, S. A. Benner, T. Butler, M. Di Ventra, S. Garaj, A. Hibbs, X. Huang, S. B. Jovanovich, P. S. Krstic, S. Lindsay, X. S. Ling, C. H. Mastrangelo, A. Meller, J. S. Oliver, Y. V. Pershin, J. M. Ramsey, R. Riehn, G. V. Soni, V. Tabard-Cossa, M. Wanunu, M. Wiggin, and J. A. Schloss. 2008. The potential and challenges of nanopore sequencing. *Nat. Biotechnol.* 26:1146–1153.

Brodie, E. L., T. Z. DeSantis, D. C. Joyner, S. M. Baek, J. T. Larsen, G. L. Andersen, T. C. Hazen, P. M. Richardson, D. J. Herman, T. K. Tokunaga, J. M. M. Wan, and M. K. Firestone. 2006. Application of a high-density oligonucleotide microarray approach to study bacterial population dynamics during uranium reduction and reoxidation. *Appl. Environ. Microbiol.* 72:6288–6298.

Cieslinski, H., A. Bialkowska, K. Tkaczuk, A. Dlugolecka, J. Kur, and M. Turkiewicz. 2009a. Identification and molecular modeling of a novel lipase from an Antarctic soil metagenomic library. *Pol. J. Microbiol.* 58:199–204.

Cieslinski, H., A. Dlugolecka, J. Kur, and M. Turkiewicz. 2009b. An MTA phosphorylase gene discovered in the metagenomic library derived from Antarctic top soil during screening for lipolytic active clones confers strong pink fluorescence in the presence of rhodamine B. *FEMS Microbiol. Lett.* 299:232–240.

Convey, P. 2001. Antarctic ecosystems, p. 171–184. *In* S. A. Levin (ed.), *Encyclopedia of Biodiversity.* Academic Press, San Diego, CA.

Dalevi, D., N. N. Ivanova, K. Mavromatis, S. D. Hooper, E. Szeto, P. Hugenholtz, N. C. Kyrpides, and V. M. Markowitz. 2008. Annotation of metagenome short reads using proxygenes. *Bioinformatics* 24:I7–I13.

DeSantis, T. Z., E. L. Brodie, J. P. Moberg, I. X. Zubieta, Y. M. Piceno, and G. L. Andersen. 2007. High-density universal 16S rRNA microarray analysis reveals broader diversity than typical clone library when sampling the environment. *Microb. Ecol.* 53:371–383.

Eid, J., A. Fehr, J. Gray, K. Luong, J. Lyle, G. Otto, et al. 2009. Real-time DNA sequencing from single polymerase molecules. *Science* 323:133–138.

Ferrer, M., O. Golyshina, A. Beloqui, and P. N. Golyshin. 2007. Mining enzymes from extreme environments. *Curr. Opin. Microbiol.* 10:207–214.

Galand, P. E., E. O. Casamayor, D. L. Kirchman, and C. Lovejoy. 2009a. Ecology of the rare microbial biosphere of the Arctic Ocean. *Proc. Natl. Acad. Sci. USA* 106:22427–22432.

Galand, P. E., E. O. Casamayor, D. L. Kirchman, M. Potvin, and C. Lovejoy. 2009b. Unique archaeal assemblages in the Arctic Ocean unveiled by massively parallel tag sequencing. *ISME J.* **3:**860–869.

Galand, P. E., M. Potvin, E. O. Casamayor, and C. Lovejoy. 2010. Hydrography shapes bacterial biogeography of the deep Arctic Ocean. *ISME J.* **4:**564–576.

Handelsman, J., M. R. Rondon, S. F. Brady, J. Clardy, and R. M. Goodman. 1998. Molecular biological access to the chemistry of unknown soil microbes: a new frontier for natural products. *Chem. Biol.* **5:**R245–R249.

He, Z., Y. Deng, J. D. Van Nostrand, Q. Tu, M. Xu, C. L. Hemme, X. Li, L. Wu, T. J. Gentry, Y. Yin, J. Liebich, T. C. Hazen, and J. Zhou. 2010. GeoChip 3.0 as a high-throughput tool for analyzing microbial community composition, structure and functional activity. *ISME J.* **4:**1167–1179.

He, Z. L., T. J. Gentry, C. W. Schadt, L. Wu, J. Liebich, S. C. Chong, W. Wu, P. Jardine, C. Criddle, and J. Z. Zhou. 2007. GeoChip: a comprehensive microarray for investigating biogeochemical, ecological and environmental processes. *ISME J.* **1:**67–77.

Heath, C., X. P. Hu, S. C. Cary, and D. Cowan. 2009. Identification of a novel alkaliphilic esterase active at low temperatures by screening a metagenomic library from Antarctic desert soil. *Appl. Environ. Microbiol.* **75:**4657–4659.

IPCC. 2007. *Climate Change 2007: the Physical Science Basis. Contribution of Working Group I to the Fourth Assessment Report of the Intergovernmental Panel on Climate Change.* Cambridge University Press, Cambridge, United Kingdom and New York, NY.

Jeon, J. H., J. T. Kim, S. G. Kang, J. H. Lee, and S. J. Kim. 2009. Characterization and its potential application of two esterases derived from the Arctic sediment metagenome. *Mar. Biotechnol.* **11:**307–316.

Kirchman, D. L., M. T. Cottrell, and C. Lovejoy. 2010. The structure of bacterial communities in the western Arctic Ocean as revealed by pyrosequencing of 16S rRNA genes. *Environ. Microbiol.* **12:**1132–1143.

Krause, L., N. N. Diaz, A. Goesmann, S. Kelley, T. W. Nattkemper, F. Rohwer, R. A. Edwards, and J. Stoye. 2008. Phylogenetic classification of short environmental DNA fragments. *Nucleic Acids Res.* **36:**2230–2239.

Lazarevic, V., K. Whiteson, S. Huse, D. Hernandez, L. Farinelli, M. Osteras, J. Schrenzel, and P. Francois. 2009. Metagenomic study of the oral microbiota by Illumina high-throughput sequencing. *J. Microbiol. Methods* **79:**266–271.

Liu, Z., T. Z. DeSantis, G. L. Andersen, and R. Knight. 2008. Accurate taxonomy assignments from 16S rRNA sequences produced by highly parallel pyrosequencers. *Nucleic Acids Res.* **36:**e120.

Liu, Z., C. Lozupone, M. Hamady, F. D. Bushman, and R. Knight. 2007. Short pyrosequencing reads suffice for accurate microbial community analysis. *Nucleic Acids Res.* **35:**e120.

López-Bueno, A., J. Tamames, D. Velázquez, A. Moya, A. Quesada, and A. Alcamí. 2009. High diversity of the viral community from an Antarctic lake. *Science* **326:**858–861.

Margulies, M., M. Egholm, W. E. Altman, S. Attiya, J. S. Bader, L. A. Bemben, et al. 2005. Genome sequencing in microfabricated high-density picolitre reactors. *Nature* **437:**376–380.

Martin-Cuadrado, A. B., F. Rodríguez-Valera, D. Moreira, J. C. Alba, E. Ivars-Martínez, M. R. Henn, E. Talla, and P. López-García. 2008. Hindsight in the relative abundance, metabolic potential and genome dynamics of uncultivated marine archaea from comparative metagenomic analyses of bathypelagic plankton of different oceanic regions. *ISME J.* **2:**865–886.

Moreira, D., F. Rodríguez-Valera, and P. López-García. 2004. Analysis of a genome fragment of a deep-sea uncultivated Group II euryarchaeote containing 16S rDNA, a spectinomycin-like operon and several energy metabolism genes. *Environ. Microbiol.* **6:**959–969.

Moreira, D., F. Rodríguez-Valera, and P. López-García. 2006. Metagenomic analysis of mesopelagic Antarctic plankton reveals a novel deltaproteobacterial group. *Microbiology* **152:**505–517.

Pace, N. R., D. A. Stahl, D. J. Lane, and G. J. Olsen. 1985. The analysis of natural microbial populations by ribosomal RNA sequences. *ASM News* **51:**4–12.

Qin, J., R. Li, J. Raes, M. Arumugam, K. S. Burgdorf, C. Manichanh, T. Nielsen, N. Pons, F. Levenez, T. Yamada, D. R. Mende, J. Li, J. Xu, S. Li, D. Li, J. Cao, B. Wang, H. Liang, H. Zheng, Y. Xie, J. Tap, P. Lepage, M. Bertalan, J. M. Batto, T. Hansen, D. Le Paslier, A. Linneberg, H. B. Nielsen, E. Pelletier, P. Renault, T. Sicheritz-Ponten, K. Turner, H. Zhu, C. Yu, S. Li, M. Jian, Y. Zhou, Y. Li, X. Zhang, S. Li, N. Qin, H. Yang, J. Wang, S. Brunak, J. Doré, F. Guarner, K. Kristiansen, O. Pedersen, J. Parkhill, J. Weissenbach, MetaHIT Consortium, P. Bork, S. D. Ehrlich, and J. Wang. 2010. A human gut microbial gene catalogue established by metagenomic sequencing. *Nature* **464:**59–65.

Rothberg, J. M., and J. H. Leamon. 2008. The development and impact of 454 sequencing. *Nat. Biotechnol.* **26:**1117–1124.

Schellenberg, J., M. G. Links, J. E. Hill, T. J. Dumonceaux, G. A. Peters, S. Tyler, T. B. Ball, A. Severini, and F. A. Plummer. 2009. Pyrosequencing of the chaperonin-60 universal target as a tool for determining microbial community composition. *Appl. Environ. Microbiol.* **75:**2889–2898.

Schmidt, T. M., E. F. DeLong, and N. R. Pace. 1991. Analysis of a marine picoplankton community by 16S rRNA gene cloning and sequencing. *J. Bacteriol.* **173:**4371–4378.

Shendure, J., and H. Ji. 2008. Next-generation DNA sequencing. *Nat. Biotechnol.* **26:**1135–1145.

Stewart, F. J., E. A. Ottesen, and E. F. DeLong. 2010. Development and quantitative analyses of a universal rRNA-subtraction protocol for microbial metatranscriptomics. *ISME J.* **4:**896–907.

Teixeira, L. C., R. S. Peixoto, J. C. Cury, W. J. Sul, V. H. Pellizari, J. M. Tiedje, and A. S. Rosado. 2010. Bacterial diversity in rhizosphere soil from Antarctic vascular plants of Admiralty Bay, maritime Antarctica. *ISME J.* **4:**989–1001.

Tringe, S. G., and P. Hugenholtz. 2008. A renaissance for the pioneering 16S rRNA gene. *Curr. Opin. Microbiol.* **11:**442–446.

Turnbaugh, P. J., and J. I. Gordon. 2008. An invitation to the marriage of metagenomics and metabolomics. *Cell* **134:**708–713.

Tyson, G. W., J. Chapman, P. Hugenholtz, E. E. Allen, R. J. Ram, P. M. Richardson, V. V. Solovyev, E. M. Rubin, D. S. Rokhsar, and J. F. Banfield. 2004. Community structure and metabolism through reconstruction of microbial genomes from the environment. *Nature* **428:**37–43.

Venter, J. C., K. Remington, J. F. Heidelberg, A. L. Halpern, D. Rusch, J. A. Eisen, D. Y. Wu, I. Paulsen, K. E. Nelson, W. Nelson, D. E. Fouts, S. Levy, A. H. Knap, M. W. Lomas, K. Nealson, O. White, J. Peterson, J. Hoffman, R. Parsons, H. Baden-Tillson, C. Pfannkoch, Y. H. Rogers, and H. O. Smith. 2004. Environmental genome shotgun sequencing of the Sargasso Sea. *Science* **304:**66–74.

Verberkmoes, N. C., A. L. Russell, M. Shah, A. Godzik, M. Rosenquist, J. Halfvarson, M. G. Lefsrud, J. Apajalahti, C. Tysk, R. L. Hettich, and J. K. Jansson. 2009. Shotgun metapro-

teomics of the human distal gut microbiota. *ISME J.* **3:**179–189.

Warnecke, F., and P. Hugenholtz. 2007. Building on basic metagenomics with complementary technologies. *Genome Biol.* **8:**231.

Yergeau, E., M. Arbour, R. Brousseau, D. Juck, J. R. Lawrence, L. Masson, L. G. Whyte, and C. W. Greer. 2009a. Microarray and real-time PCR analyses of the responses of high Arctic soil bacteria to hydrocarbon pollution and bioremediation treatments. *Appl. Environ. Microbiol.* **75:**6258–6267.

Yergeau, E., S. Bokhorst, S. Kang, J. Zhou, C. W. Greer, R. Aerts, and G. A. Kowalchuk. 2011. Shifts in soil microorganisms in response to warming are consistent across a range of Antarctic environments. *ISME J.* doi:10.1038/ismej.2011.124.

Yergeau, E., H. Hogues, L. G. Whyte, and C. W. Greer. 2010a. The functional potential of high Arctic permafrost revealed by metagenomic sequencing, qPCR, and microarray analyses. *ISME J.* **4:**1206–1214.

Yergeau, E., S. Kang, Z. He, J. Zhou, and G. A. Kowalchuk. 2007a. Functional microarray analysis of nitrogen and carbon cycling genes across an Antarctic latitudinal transect. *ISME J.* **1:**163–179.

Yergeau, E., and G. A. Kowalchuk. 2008. Responses of Antarctic soil microbial communities and associated functions to temperature and freeze-thaw cycle frequency. *Environ. Microbiol.* **10:**2223–2235.

Yergeau, E., J. R. Lawrence, D. R. Korber, M. J. Waiser, and C. W. Greer. 2010b. Metatranscriptomic analysis of the response of river biofilms to pharmaceutical products using anonymous DNA microarrays. *Appl. Environ. Microbiol.* **76:**5432–5439.

Yergeau, E., K. K. Newsham, D. A. Pearce, and G. A. Kowalchuk. 2007b. Patterns of bacterial diversity across a range of Antarctic terrestrial habitats. *Environ. Microbiol.* **9:**2670–2682.

Yergeau, E., S. A. Schoondermark-Stolk, E. L. Brodie, S. Déjean, T. Z. DeSantis, O. Gonçalves, Y. M. Piceno, G. L. Andersen, and G. A. Kowalchuk. 2009b. Environmental microarray analyses of Antarctic soil microbial communities. *ISME J.* **3:**340–351.

POLAR MICROORGANISMS
AND BIOTECHNOLOGY

Georges Feller and Rosa Margesin

8

INTRODUCTION

Among the various organisms thriving in extreme environments on Earth, psychrophiles (cold-loving organisms) are the most abundant in terms of biomass, diversity, and distribution. If a psychrophile is defined as an organism living permanently at temperatures close to the freezing point of water in thermal equilibrium with the medium (without entering the debate on classification), this definition includes de facto a large range of representatives from all three domains: *Bacteria*, *Archaea*, and *Eukarya* (including yeasts, algae, marine invertebrates, and polar fish). Such biodiversity partly correlates with a less detrimental effect of low temperatures on cellular structures as compared with high temperatures or extreme pH, for instance, but it also reflects the fact that most of Earth's biotopes are cold and have been successfully colonized by diverse organisms (Margesin and Schinner, 1999b; Deming, 2002; Margesin et al., 2002, 2008; D'Amico et al., 2006; Gerday and Glansdorff, 2007). Extreme psychrophiles have been traditionally sampled from Antarctic and Arctic polar regions, as-

suming that low temperatures persisting over a geological time scale have promoted deep and efficient adaptations to freezing conditions. More recently, Arctic permafrost, representing more than 20% of terrestrial soils, has revealed an unsuspected biodiversity in cryopegs, i.e., salty water pockets that have remained liquid for about 100,000 years at −10°C (Gilichinsky et al., 2005). High-altitude mountains, glaciers, and natural caves are additional sources of cold-adapted organisms. However, the largest psychrophilic reservoir is provided by oceans, covering 70% of our planet, which have a constant temperature of 4°C below a depth of 1,000 m, irrespective of latitude. Furthermore, deep-sea sediments, previously considered as abiotic, apparently host a considerable microbial biomass that remains almost uncharacterized as a result of technical difficulties in sampling and culturing (Leigh Mascarelli, 2009).

Such high abundance of psychrophiles evidently offers a huge potential for biomining using culture-based techniques, recombinant protein expression, and metagenomic approaches (Leary, 2008; Lohan and Johnston, 2005). Incidentally, the first cold-adapted enzymes from Antarctic bacteria that have been cloned, sequenced, and expressed in a recombinant form were lipases, subtilisins, and

Georges Feller, Laboratory of Biochemistry, Centre for Protein Engineering, University of Liège, B-4000 Liège, Belgium. *Rosa Margesin*, Institute of Microbiology, University of Innsbruck, Technikerstrasse 25, A-6020 Innsbruck, Austria.

Polar Microbiology: Life in a Deep Freeze
Edited by Robert V. Miller and Lyle G. Whyte © 2012 ASM Press, Washington, DC

α-amylase, i.e., well-known representatives of industrial enzymes. This illustrates that besides the fundamental research on biocatalysis in the cold, the biotechnological potential of psychrophilic enzymes was already put into perspective in the early 1990s. Since then, numerous possible applications based on psychrophiles have been described and patenting in this field is growing (Hoag, 2008). By contrast, the number of known or proven current applications remains modest. It should be stressed that confidentiality accompanying commercial products frequently obscures the possible psychrophilic origin of compounds and, accordingly, the number of current applications is certainly much higher than those summarized below. We present here an overview of the biotechnological uses of psychrophiles and of their biomolecules using selected examples. Previous reviews should be consulted for a complete coverage of this topic (Russell, 1998; Margesin and Schinner, 1999a; Gerday et al., 2000; Allen et al., 2002; Cavicchioli et al., 2002; Marx et al., 2007; Margesin et al., 2008; Margesin and Feller, 2010).

ADVANTAGES OF POLAR MICROORGANISMS IN BIOTECHNOLOGY

While the growth of most microorganisms is stopped or at least severely inhibited in a refrigerator, psychrophiles actively divide at these temperatures. As shown in Fig. 1, some wild-type psychrophilic bacteria display doubling times at 4°C comparable to those of fast-growing *Escherichia coli* laboratory strains grown at 37°C. The latter fail to grow exponentially below 8°C (Strocchi et al., 2006), whereas psychrophilic bacteria maintain doubling times as low as 2 to 3 h at 4°C. This is primarily achieved by a weak effect of low temperature on growth rates of psychrophiles as compared with mesophiles. This efficiency is mainly determined by cold adaptation of the enzymatic machinery (see below). Although contamination of cold-room facilities by psychrophilic or psychrotolerant species is a concern for food spoilage, their growth properties can be advantageously exploited in human-driven operations. Psychrophilic strains, as cell factories, can be grown at tap-water temperatures, avoiding heating of fermentation units,

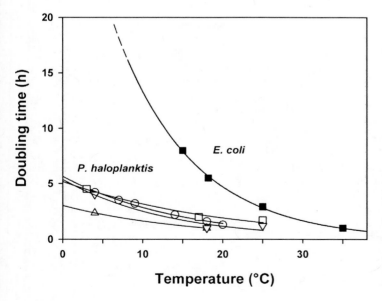

FIGURE 1 Temperature dependence of growth (expressed by the generation time) of *Escherichia coli* and of representative psychrophilic gram-negative bacteria (data from various Antarctic isolates of *Pseudoalteromonas haloplanktis*), displaying high growth rates at low temperatures for the latter.

or at lower temperatures to produce heat-labile compounds and aggregation-prone proteins (Table 1). A promising application also lies in their use in wastewater or soil bioremediation during winter in temperate regions or as bio-additives to relieve pollution (e.g., oil spills) in cold ecosystems that have an inherently lower propensity to recover.

Microbial adaptation to low temperatures of course requires a vast array of metabolic and structural adjustments at nearly all organization levels of the cell, which have begun

TABLE 1 Selected applications of cold-active biomolecules in biotechnology[a]

Application field	Advantage(s)	Involved biomolecule(s)
Gene expression	Recombinant protein expression at low temperature	Chaperonins
Detergents	Washing at low temperature (energy saving, and applicable to synthetic fibers), contact lens cleaning	Protease, lipase, amylase, cellulase, oxygenase
Food industry	Reduced incubation time for lactose hydrolysis in milk and dairy products	β-Galactosidase
	Improved bread quality	Xylanase
	Improved juice clarification, increased juice yield	Pectinase, cellulase
	Efficient and gentle removal of fish skin, meat tenderization	Protease, carbohydrase
	Cold pasteurization, food preservation	Catalase, lysozyme, glucose oxidase, antifreeze proteins
	Improved taste and texture of ice cream	Antifreeze proteins
	Improved taste and aroma of fermentation products (e.g., cheese, dry sausages, alcoholic beverages)	Enzymes involved in fermentation and ripening
Organic synthesis	Synthesis of volatile and heat-sensitive compounds (e.g., flavors and fragrances) Synthesis of acrylamide	Lipase, esterase, protease, etc. Nitrile hydratase
	Organic-phase biocatalysis (increased solvent choice, product yield, and biocatalysis stability)	Enzymes operating under low-water conditions
Molecular biology	Mild heat inactivation of enzymes without interference with subsequent reactions	Various enzymes
	Efficient low-temperature ligation	DNA ligase
	Prevention of carryover contamination in PCR	Uracil-DNA glycosylase
	Rapid 5′ end labeling of nucleic acids	Alkaline phosphatase
	Efficient protoplast formation	Cellulase, xylanase, etc.
Pharmaceuticals, cosmetics	Debridement of necrotic tissue, digestion promotion, chemonucleolytic agents Scar treatment, cosmetic creams	Multienzyme systems, proteases Glycoproteins (antifreeze)
Textiles	Improved quality after desizing, biopolishing, and stone-washing of fabrics	Amylase, laccase, cellulase
Biosensors	Selective, sensitive, and rapid online monitoring of low-temperature processes; quality control	Various enzymes
Environment	In situ/on-site bioremediation of organic contaminants	Enzymes involved in biodegradation
	Low-energy wastewater treatment	Enzymes involved in mineralization, nitrification, denitrification, etc.
	Low-energy anaerobic wastewater treatment	Enzymes involved in anaerobic biodegradation
	Low-temperature biogas (methane) production	Enzymes involved in anaerobic biodegradation
	Low-temperature composting	Enzymes involved in litter degradation

[a]Adapted from Margesin et al. (2007) with permission of the publisher.

to be understood thanks to the availability of genome sequences and of proteomic approaches (Gerday and Glansdorff, 2007; Margesin et al., 2008). A survey of these data shows that the main upregulated functions for growth at low temperatures are protein synthesis (transcription, translation), RNA and protein folding, membrane homeostasis, antioxidant activities, and regulation of specific metabolic pathways. However, the few common features shared by all these psychrophilic genomes and proteomes have suggested that cold adaptation superimposes on preexisting cellular organization, and accordingly, the adaptive strategies may differ between various microorganisms.

BIOPROSPECTING THE POLAR GENETIC RESOURCES

Two reports of the United Nations University Institute of Advanced Studies (UNU-IAS) have described the bioprospecting activities in the Antarctic (Lohan and Johnston, 2005) and in the Arctic (Leary, 2008). The former report stressed that the absence of clear rules governing the use and ownership of genetic resources from Antarctica, resulting from the peculiar international status of the continent, obviously inhibits commercially oriented research and information exchanges. Such concerns have been frequently debated (Williams, 2004). On the other hand, the latter report devoted to the Arctic provides an extensive survey of companies active in this field and of the patents and products derived from Arctic organisms. This survey clearly demonstrates the potential of psychrophiles in an unsuspected wide range of applications and the intense commercial activity in the field. More recently, UNU-IAS has launched Bioprospector, an online database (http://www.bioprospector.org/bioprospector/) surveying patents, commercial products, and companies involved in applied research using genetic resources from both the Antarctic and the Arctic. This excellent initiative, accompanied by relevant publications, is currently the most updated survey of biotechnological applications based on psychrophiles.

The main areas of interest in terms of investigation, patenting, and commercial products are ranked below (Lohan and Johnston, 2005; Leary, 2008), and some relevant examples are given in the next sections (see also Table 1).

- Enzymes: their use in a wide range of industrial processes including food technology, as well as laboratory reagents in molecular biology or medical research.
- Biomolecules: generally as food additives such as dietary supplements for use in aquaculture, livestock, and human diets, with special focus on polyunsaturated fatty acids and on antifreeze proteins.
- Pharmaceutical and medical uses: mainly focused on screening for new antibiotics or anticancer drugs (Biondi et al., 2008), but also on cosmetics and nutraceuticals.
- Bioremediation: applying biostimulation or bioaugmentation to degrade pollutants with cold-adapted microorganisms following accidental spills or to address past waste disposal practices.

POLAR BACTERIA AS CELL FACTORIES

To facilitate biotechnological applications of psychrophiles and of their products, recombinant protein secretion systems efficiently working at low temperature are indispensable. The production level of cold-active (heat-labile) proteins by wild-type strains is usually too low for production on an industrial scale. The first recombinant production of a cold-active enzyme (α-amylase from Antarctic *Pseudoalteromonas haloplanktis*) in an Antarctic host bacterium of the same species was described in 2001 (Tutino et al., 2001). The cold gene-expression system was further developed and optimized for the recombinant extracellular secretion of heterologous proteins in *P. haloplanktis*, with enzymes originating from various Antarctic *P. haloplanktis* strains and a mesophilic yeast (Cusano et al., 2006; Papa et al., 2007). The simultaneous secretion of proteolytic enzymes that degraded the recombinant products could be considerably reduced by inactivating the secretion system with the use of a gene insertion strategy; the mutant strain still secreted the cold-active enzyme (α-amylase) as efficiently as the wild type and in a stable form (Parrilli et al., 2008).

Another recombinant protein expression system working at low temperature was developed by using an Antarctic *Shewanella* sp. strain (Miyake et al., 2007) and was based on the selection of a suitable promoter and a broad-host-range plasmid. High yields of β-lactamase were produced in the *Shewanella* sp. strain at 4°C; the enzyme yield produced at 4°C was 64% of that obtained at 18°C. The efficiency of the system was demonstrated by the production of foreign proteins (putative peptidases and a glucosidase) from the psychrophile *Desulfotalea psychrophila*.

Site-specific mutants of psychrophiles are a useful tool to study cold adaptation and expression of cold-active enzymes at low temperature (Bakermans et al., 2009). For example, the role of a substrate-binding subunit of a specific transporter in a Siberian psychrophile *Psychrobacter arcticus* strain in the transport of several substrates at low temperatures could be elucidated.

Contrary to the above-mentioned works, which focused on the construction of cold expres-

sion systems for *Proteobacteria*, an expression system was developed for high-G+C gram-positive bacteria that are known to occur frequently in cold environments (Miteva et al., 2008). Seven psychrophilic isolates from Greenland ice cores belonging to various genera of the class *Actinobacteria* (*Arthrobacter*, *Microbacterium*, *Curtobacterium*, and *Rhodoglobus*) were transformed with a shuttle vector that was constructed by using a small cryptic plasmid from a psychrophile *Arthrobacter agilis* strain and conferred antibiotic resistance. In some isolates, plasmid stability was higher at 5 than at 25°C, which points to the efficiency of the expression system within a restricted low-temperature range.

COLD-ACTIVE ENZYMES IN BIOTECHNOLOGY

In contrast to these variable cellular adjustments, most enzymes from psychrophiles are cold active, and this peculiarity provides the basis for the main physiological adaptation to low temperatures. Indeed, cold-active enzymes allow the persistence of metabolic fluxes compatible with sustained growth at freezing temperatures. As shown in Fig. 2, psychrophilic

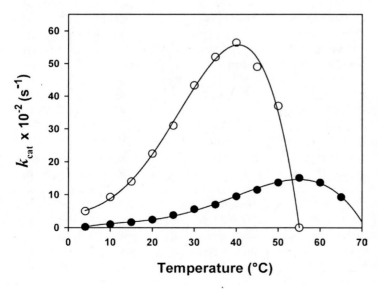

FIGURE 2 Temperature dependence of the activity of psychrophilic (O) and mesophilic (●) enzymes recorded at various temperatures illustrates the main properties of cold-adapted enzymes: cold activity and heat lability (data for cellulases from the Antarctic bacterium *Pseudoalteromonas haloplanktis* and from the mesophile *Erwinia chrysanthemi*). (Adapted from D'Amico et al. [2006] with permission of the publisher.)

enzymes can be up to 10 times more active at low and moderate temperatures as compared with their mesophilic homologues. Furthermore, psychrophilic enzymes are heat labile and are frequently inactivated at temperatures that are not detrimental for their mesophilic counterparts. These specific traits are responsible for the three main advantages of cold-active enzymes in biotechnology: (i) as a result of their high activity, a lower concentration of the enzyme catalyst is required to reach a given activity, therefore reducing the amount of costly enzyme preparation in a process; (ii) as a result of their cold activity, they remain efficient at tap-water or ambient temperature, therefore avoiding heating during a process, either at domestic (e.g., washing machine) or industrial levels; and (iii) as a result of heat lability, they can be efficiently and sometimes selectively inactivated after a process by moderate heat input. Besides these traits specifically linked to temperature adaptation, an additional important aspect has to be mentioned: enzymes from organisms endemic to cold environments can be a valuable source of new catalysts possessing useful enzymological characteristics such as novel substrate specificities or product properties, as exemplified by lipases from the yeast *Candida antarctica* or by the xylanase from the bacterium *P. haloplanktis* (see below).

The conformations and three-dimensional structures of psychrophilic proteins are not markedly different from their mesophilic homologues, and furthermore all amino acid side chains that are essential for the catalytic mechanism are strictly identical. It was found, however, that cold-active enzymes maintain the appropriate flexibility and dynamics of the active site at temperatures at which their mesophilic and thermophilic counterparts have severely restricted molecular motions (D'Amico et al., 2003b; Feller and Gerday, 2003; Tehei et al., 2004). This is achieved by the disappearance of discrete stabilizing interactions either in the whole molecule or at least in structures adjacent to the active site. Among these destabilizing factors, the most relevant are a reduced number of proline residues and of electrostatic

interactions (ion pairs, H-bonds, aromatic interactions), a weakening of the hydrophobic effect, the strategic location of glycine residues, an improved interaction of surface side chains with the solvent, and an improved charge-induced interaction with substrates and cofactors (Smalås et al., 2000; Siddiqui and Cavicchioli, 2006). This adaptive destabilization of psychrophilic enzymes has been demonstrated to be responsible for both cold activity and low thermal stability (D'Amico et al., 2001, 2003a).

INDUSTRIAL ENZYMES FROM POLAR MICROORGANISMS

At the industrial level, the best-known representative of polar microorganisms is certainly the yeast *Candida antarctica*, as its species name unambiguously refers to the sampling origin. This yeast produces two lipases, A and B, the latter being sold, for instance, as Novozym 435 by Novozymes (Bagsvaerd, Denmark). Although the moderate heat stability of this lipase in aqueous solutions can be of concern, this enzyme is stabilized in its immobilized form. As a result of its substrate and stereospecificity, lipase B is involved in a very large number of organosynthesis applications related to food/feed processing, pharmaceuticals, and cosmetics (Babu et al., 2008). In a survey of patents related to Antarctica (Lohan and Johnston, 2005), it was shown that lipases from *C. antarctica* by far dominate the number of process- or product-based patents. This is a significant example of the potential for novel catalysts from genetic resources in cold environments.

The market for enzymes used in detergents represents 30 to 40% of all enzymes produced worldwide. Among these enzymatic cleaning agents, subtilisin (an alkaline serine protease predominantly produced by *Bacillus* species) dominates this market. At the domestic level, however, the current trend is to use detergents at lower washing temperatures because of the associated reductions in energy consumption and costs, as well as to protect the texture and colors of the fabrics. Accordingly, cold-active subtilisins are required for optimal washing results at tap-water temperatures, and the current advertisements for cold-active detergents indicate that this goal has been reached. The first psychrophilic subtilisins isolated from Antarctic *Bacillus* species have been extensively characterized to comply with this requirement (Davail et al., 1994; Narinx et al., 1997). However, they suffered from a low heat stability that can compromise their storage, and also from a low chemical stability in the detergent components. Therefore, subtilisins currently incorporated in cold-active detergents are engineered enzymes that combine storage stability, alkaline stability, and activity and cold activity. Although psychrophilic subtilisins are not components per se of cold-active detergents, they have largely contributed to the advancement of this economically attractive concept.

The xylanase from the Antarctic bacterium *P. haloplanktis* is a good example of the successful biotechnological transfer from academic research to industry. Xylanases are glycoside hydrolases that degrade the polysaccharide β-1,4-xylan, thus breaking down hemicellulose, one of the major components of plant cell walls. Xylanases are also a key ingredient of industrial dough conditioners used to improve bread quality. It was found that the Antarctic enzyme belonged to a new class of xylanases as both its amino acid sequence and fold were distinct from previously characterized xylanases. The psychrophilic enzyme was therefore subjected to intensive investigations aimed at elucidating the structural origins of its high cold activity and weak stability as well as understanding its enzymological mode of action (Collins et al., 2002, 2003, 2005; Van Petegem et al., 2003; De Vos et al., 2006). Furthermore, baking trials have revealed that the psychrophilic xylanase is very effective in improving dough properties and final bread quality, with, for instance, a positive effect on loaf volume (Collins et al., 2006). This efficiency appears to be related to the high activity of the psychrophilic xylanase at cool temperatures required for dough resting and to its specific mode of xylan hydrolysis. Fol-

lowing careful production optimization of this peculiar xylanase, the product is now sold by Puratos (Groot-Bijgaarden, Belgium). This is apparently the psychrophilic enzyme produced at the highest weight amounts to date.

β-Galactosidase, or lactase, is also a glycoside hydrolase that specifically hydrolyzes the milk sugar lactose into galactose and glucose. It should be stressed that 75% of the world population suffers from lactose intolerance arising from deficient synthesis of intestinal lactase in adults and resulting in digestive disorders due to fermentation of lactose by enteric bacteria. In this context, a cold-active lactase from an Antarctic bacterium has been patented for its capacity to hydrolyze lactose during milk storage at low temperatures (Hoyoux et al., 2001). It is worth mentioning that commercially available lactases require milk heating to become active. This heating step has, however, detrimental effects on milk quality as it alters the aspect, taste, and texture (by Maillard reactions, activation of proteases, coagulation, etc.). Although the psychrophilic lactase is apparently not used for this specific application, it is expected that it will be produced soon in large quantities by Nutrilab (Bekkevoort, Belgium) to hydrolyze lactose (a by-product of the dairy industry) in the processing of the high-value sweetener D-tagatose, a natural monosaccharide with low caloric value and glycemic index.

POLAR PROTEINS IN MOLECULAR BIOLOGY AND COSMETICS

In pioneering work, Kobori et al. (1984) purified and characterized a heat-labile alkaline phosphatase from an Antarctic bacterium isolated in seawater samples from McMurdo Sound. Alkaline phosphatases are mainly used in molecular biology for the dephosphorylation of DNA vectors prior to cloning to prevent recircularization, for the dephosphorylation of 5′-nucleic acid termini before 5′ end labeling by polynucleotide kinase, or for removal of dNTPs and pyrophosphate from PCR reactions. However, the phosphatase has to be carefully removed after dephosphorylation to avoid interference with the subsequent steps. Furthermore, *E. coli* and calf intestinal alkaline phosphatase (which was the preferred enzyme for these applications) are heat stable and require detergent addition for inactivation. It follows that heat-labile alkaline phosphatases are excellent alternatives as they are inactivated by moderate heat treatment, allowing performance of the subsequent steps in the same test tube and minimizing nucleic acid losses. While the scientific report of Kobori et al. (1984) specifically stressed the usefulness of their heat-labile alkaline phosphatase as a new tool in molecular biology, this interesting finding was apparently not turned into a marketed product, possibly because gene cloning and heterologous expression were not well established at that time. Fifteen years later, the group of Bouriotis isolated an alkaline phosphatase from another Antarctic bacterium and cloned its gene in *E. coli* (Rina et al., 2000), solved its crystal structure (Wang et al., 2007), and also showed that its properties can be further improved by directed evolution in terms of high activity and heat lability (Koutsioulis et al., 2008). This heat-labile alkaline phosphatase, sold as Antarctic phosphatase, is now marketed by New England Biolabs (Ipswich, MA).

Cold-active chaperones have also found very useful application in the production of recombinant proteins. High-level expression of heterologous proteins in *E. coli* can result in the production of large amounts of incorrectly folded proteins, generating aggregates of inactive protein generally in the form of inclusion bodies. To circumvent this insolubility problem, low-temperature cultivation of *E. coli* represents a classical strategy and coexpression of chaperones also frequently improves the recovery of soluble proteins. Chaperones are a ubiquitous class of proteins that assist the folding of nascent polypeptides, preventing or even repairing misfolding. In this context, the chaperonins Cpn10 and Cpn60 (homologous to GroES and GroEL in *E. coli*) from the Antarctic bacterium *Oleispira antarctica* were shown to improve the growth of *E. coli* at low temperatures and to remain optimally active as folding catalysts at these low temperatures (Ferrer et al., 2003). Taking advantage of these properties, the ArcticExpress *E. coli* cells from Agilent Technologies (Santa Clara, CA) have been engineered to coexpress the cold-active chaperonins with the recombinant protein of interest, therefore improving protein processing at low temperatures and increasing the yield of active, soluble recombinant protein.

To conclude this section with beauty, a cosmetic additive is worth mentioning. Antarcticine-NF3 is a glycoprotein with antifreeze properties produced by the bacterium *Pseudoalteromonas antarctica*, which has been patented by Spanish researchers (Parenta Duena et al., 2006). It was found that Antarcticine is effective for scar treatment and re-epithelialization of wounds. This glycoprotein is now included in some cosmetic regeneration creams (sometimes under the name Antarctilyne). It is also proposed for use in association with edelweiss extract: this is of course reminiscent of the peculiar resistance to harsh conditions of both the Antarctic bacterium and the Alpine flower.

HYDROCARBON BIOREMEDIATION IN POLAR ENVIRONMENTS

The capacity of a broad spectrum of microorganisms to utilize hydrocarbons as the sole source of carbon and energy (biodegradation) was recognized by Zobell in 1946 (Zobell, 1946) and was the basis for the development of biological remediation methods. Bioremediation attempts to accelerate natural biodegradation rates through the optimization of limiting environmental conditions and is an ecologically and economically effective method; nonetheless, it has its limitations. A number of studies have shown the feasibility of bioremediation in polar regions; however, more field studies are needed to confirm that the desired cleanup levels can be reached (Atlas, 2010).

Low-temperature biodegradation of organic contaminants in cold ecosystems is a result of the degradation capacity of the indigenous microbial populations. They transform or mineralize organic pollutants into less harmful, nonhazardous substances, which are then integrated into natural biogeochemical cycles. Most studies on hydrocarbon bioremediation in polar regions have focused on the treatment of petroleum hydrocarbons, since increased petroleum exploration increases the risk of accidental oil release. The bioremediation potential of microorganisms (Bej et al., 2010) and bioremediation strategies to treat contaminants in polar environments (Filler et al., 2008) have recently been described in detail. Evidence for the biodegradation activity of indigenous microorganisms (Margesin, 2007; Brakstad, 2008; Filler et al., 2008; Bej et al., 2010) in contaminated cold environments is provided by high numbers and activities of hydrocarbon degraders, the prevalence of genotypes with catabolic pathways for the degradation of a wide range of hydrocarbons (Yergeau et al., 2009; Panicker et al., 2010), and high mineralization potentials. The majority of these studies are based on polar bacteria; the use of polar fungi for bioremediation (mycoremediation) requires further research (Bej et al., 2010).

Hydrocarbon degraders in polar regions are confronted with special challenges, which—besides reduced enzymatic reaction rates—include increased viscosity of liquid hydrocarbons, reduced volatility of toxic compounds, low levels of nutrients, limited bioavailability of contaminants, and sometimes extremes of pH and salinity (Margesin, 2004; Aislabie et al., 2006; Filler et al., 2008; Bej et al., 2010). Until recently, frozen soils have been considered to be a practically impermeable barrier to pollutants. Meanwhile, studies have confirmed that hydrocarbons can penetrate even into ice-saturated soils (Barnes and Chuvilin, 2009). Microbial activities have been measured at temperatures close to the freezing point of water and in marine ice at temperatures lower than −10°C, indicating that slow hydrocarbon biodegradation occurs in oil-contaminated ice (Brakstad, 2008).

Several remediation schemes have been implemented successfully at petroleum-contaminated sites in the Arctic during the past decade (Mar-gesin, 2004; Aislabie and Foght, 2008; Filler et al., 2009; Bej et al., 2010). Successful on-site treatments include biopiles (Thomassin-Lacroix et al., 2002) and landfarming, which is now well developed for cold regions and offers low-cost treatment of petroleum-contaminated soils (Paudyn et al., 2008; Walworth and Ferguson, 2008). Engineered bioremediation implies the use of mechanized systems (e.g., forced aeration, heating and insulation systems) and allows remediation of large volumes of petroleum-contaminated soils to cleanup standards within two to three treatment seasons in Alaska (Filler et al., 2006, 2009) and lengthening of the usual short Arctic bioremediation season (June to September) by 3 months (to May to November).

The most widely used bioremediation procedure in polar soils is biostimulation of the indigenous microorganisms by supplementation of appropriate nutrients (and optimization of other limiting factors, such as oxygen content, pH, and temperature); however, care has to be taken to avoid inhibition of biodegradation due to overfertilization (Walworth et al., 2007). Bioaugmentation by inoculating allochthonous hydrocarbon degraders has been used as a bioremediation option to treat petroleum-contaminated sites in Alaska, Canada, Greenland, and Norway. This strategy generally underperformed or gave no better results than fertilization (Margesin, 2004; Filler et al., 2009; Bej et al., 2010). In addition, such inocula are more expensive than commercially available fertilizers. Bioaugmentation with nonindigenous or genetically modified/engineered microorganisms is banned in Antarctica, Norway, Iceland, and Sweden (Filler et al., 2009). The construction of psychrophiles with specific degradative capabilities was reported 20 years ago (Kolenc et al., 1988) and was based on the transfer of the TOL plasmid from the mesophile Pseudomonas putida by conjugation to a psychrophile of the same species; the transconjugant degraded toluene at temperatures as low as 0°C. Recently, the gene coding for a monooxygenase involved in the degradation of aromatic hydrocarbons from the mesophile *Pseudomonas stutzeri* was recombinantly expressed in the Antarctic *P. haloplanktis* (Siani et al., 2006). However, the performance of such strains has still to be proven. Preconditions of a successful application of bioaugmentation are the expression of the biodegrading activities in the polluted environment and the survival of the inoculated strains at least for the time necessary for bioremediation.

WASTEWATER TREATMENT IN COLD ENVIRONMENTS

Bacteria and fungi able to degrade high amounts of organic compounds within a short time at low temperatures represent a promising source as inocula for accelerated wastewater treatment. For example, an Antarctic *Arthrobacter psychrolactophilus* strain displayed all the features necessary for its use as microbial starter, from the viewpoint of both biosafety and production. At 10°C, the strain induced a complete clarification of a synthetic wastewater turbid medium; it hydrolyzed proteins, starch, and lipids and improved the biodegradability of organic compounds in the wastewater (Gratia et al., 2009). Arctic and Antarctic cyanobacteria have been also positively evaluated for wastewater treatment (Chevalier et al., 2000). Another example is low-temperature degradation of phenol, which is the most common representative of aromatic toxic pollutants in a wide variety of wastewaters. Psychrophilic bacteria (*Rhodococcus* spp.) and yeasts (e.g., *Rhodotorula psychrophenolica*) fully degraded up to 12.5 to 15 mM phenol at 10°C under fed-batch cultivation; with some strains phenol degradation occurred even at temperatures as low as 1°C (Margesin et al., 2005). Immobilization may improve phenol degradation by psychrophilic yeast strains (Krallish et al., 2006). Fixed-biofilm reactors inoculated with bacterial consortia were used for the degradation of chlorophenols, and high removal efficiencies were obtained at temperatures down to 4°C (Zilouei et al., 2006). Interestingly, a temperature decrease resulted in a relative increase of the γ-*Proteobacteria* within the consortia. The

biodegradation of polychlorinated biphenyls in polar environments has been reported; however, further studies are needed to successfully implement the treatment of polychlorinated biphenyl-polluted areas (Lo Giudice et al., 2010). A further compound currently of interest is methyl *tert*-butyl ether (MTBE); its extensive use as both an octane enhancer and as an oxygenate in unleaded gasoline over the past decades has led to widespread pollution of surface water and groundwater. An aerobic mixed bacterial culture capable of utilizing MTBE and growing at from 3 to 30°C could be suitable for the cleanup of MTBE-contaminated aquifers. When inoculated into groundwater samples, it degraded MTBE simultaneously with other volatile organic compounds (Zaitsev et al., 2007).

POLAR PLANTS AND ANIMALS IN BIOTECHNOLOGY

Polar plants and animals have also found diverse applications and are worth citing in the context of the present survey. A special mention should be made of the very large number of products derived from the Arctic shrimp *Pandalus borealis* and the Antarctic krill *Euphausia superba*, mainly by Scandinavian and Japanese companies, in research, medical, and pharmaceutical applications (alkaline phosphatase, hydrolase for prevention of immune rejection reactions, hypotensor peptide, anti-inflammatory agent); in food, feed, and beverage processes; and in cosmetics and nutraceuticals (krill oils rich in omega-3 fatty acids). For instance, two shrimp enzymes are marketed for molecular biology applications that take advantage of their heat-labile properties. The heat-labile alkaline phosphatase from the Arctic shrimp *P. borealis* is available from Biotec Pharmacon (Tromsø, Norway) and GE Healthcare Life Sciences (Little Chalfont, United Kingdom) for applications similar to that of the above-mentioned Antarctic phosphatase. Shrimp nuclease selectively degrades double-stranded DNA; it is used for the removal of carryover contaminants in PCR mixtures, and is heat inactivated prior to addition of the template.

This enzyme is produced in recombinant form in *Pichia pastoris* and is available from Biotec Pharmacon, USB Corporation (Santa Clara, CA), and Thermo Scientific (Waltham, MA).

Heat-labile uracil-DNA *N*-glycosylase from Atlantic cod (*Gadus morhua*), which presents typical cold-adaptation features (Leiros et al., 2003), is also used to remove DNA contaminants in sequential PCR reactions. When PCR is performed with dUTP instead of dTTP, PCR products become distinguishable from target DNA, and can be selectively degraded by uracil-DNA *N*-glycosylase. Following degradation of contaminants, the enzyme is completely and irreversibly inactivated after heat treatment. Heat-labile uracil-DNA *N*-glycosylase, produced in recombinant form in *E. coli*, is available from Biotec Pharmacon.

Among polar plants, extracts of the Antarctic alga *Durvillaea antarctica* are included in cosmetics to improve skin vitality, such as in the Extra Firming Day Cream, a top seller of Clarins (Neuilly-sur-Seine, France).

A last example will give a sweet taste for psychrophilic proteins. Most antifreeze proteins are small glycoproteins, first discovered in polar fish (Fletcher et al., 2001), that allow the animals to thrive at subzero temperatures without freezing. As a result of their peculiar surface properties, antifreeze proteins bind to ice crystal seeds and inhibit growth of ice in body fluids that would otherwise be fatal. The precise mechanism of ice adsorption inhibition by antifreeze proteins remains poorly understood because of the complex water–ice interface and the structural diversity of these proteins. Besides this interesting finding, the gene of an antifreeze protein from the ocean pout, an eel-like fish found in northern and Arctic oceans, has been cloned and expressed in the baker's yeast *Saccharomyces cerevisiae*. This recombinant protein is now included in several edible ice cream brands from Unilever (Rotterdam, The Netherlands and London, United Kingdom) under the name of ice-structuring protein. It controls ice recrystallization following thawing-freezing cycles that otherwise drastically reduces taste and texture quality.

Furthermore, the antifreeze protein enables the production of healthier ice creams that are lower in fat and added sugar and with fewer additives. Approval for the use of this technology has been granted by regulatory administrations in many countries.

CONCLUSIONS

Among the extremophiles thriving at extreme biological temperatures, thermophiles have been generally considered as the most promising source of biotechnological innovations. However, recent developments based on cold-adapted organisms and their biomolecules, such as those mentioned here, have clearly demonstrated the huge potential of psychrophiles. This potential appears to be even larger than for thermophiles, considering both the broader psychrophilic biodiversity that encompasses microorganisms, plants, and animals (Margesin et al., 2007) and the broader fields of application. Last but not least, most biotechnological applications of psychrophiles are environmentally friendly and contribute to energy savings, both aspects being of increasing significance.

ACKNOWLEDGMENTS

Research at the authors' laboratories was supported by the European Union, the Région wallonne (Belgium), the Fonds National de la Recherche Scientifique (Belgium), the Autonome Provinz Bozen/Südtirol, and the Universities of Liège and Innsbruck. The facilities offered by the Institut Polaire Français are also acknowledged.

REFERENCES

Aislabie, J., and J. M. Foght. 2008. Hydrocarbon-degrading bacteria in contaminated soils, p. 69–83. *In* D. M. Filler, I. Snape, and D. L. Barnes (ed.), *Bioremediation of Petroleum Hydrocarbons in Cold Regions.* Cambridge University Press, Cambridge, Cambridge, United Kingdom.

Aislabie, J., D. J. Saul, and J. M. Foght. 2006. Bioremediation of hydrocarbon-contaminated polar soils. *Extremophiles* **10:**171–179.

Allen, D., A. L. Huston, L. E. Weels, and J. W. Deming. 2002. Biotechnological use of psychrophiles, p. 1–17. *In* G. Bitton (ed.), *Encyclopedia of Environmental Microbiology.* John Wiley and Sons, New York, NY.

Atlas, R. M. 2010. Microbial bioremediation in polar environments: current status and future directions, p. 255–275. *In* A. K. Bej, J. Aislabie, and R. M. Atlas (ed.), *Polar Microbiology: the Ecology, Biodiversity and Bioremediation Potential of Microorganisms in Extremely Cold Environments.* CRC Press, Boca Raton, FL.

Babu, J., P. W. Ramteke, and G. Thomas. 2008. Cold active microbial lipases: some hot issues and recent developments. *Biotechnol. Adv.* **26:**457–470.

Bakermans, C., R. E. Sloup, D. G. Zarka, J. M. Tiedje, and M. F. Thomashow. 2009. Development and use of genetic system to identify genes required for efficient low-temperature growth of *Psychrobacter arcticus* 273-4. *Extremophiles* **13:**21–30.

Barnes, D. L., and E. Chuvilin. 2009. Migration of petroleum in permafrost-affected regions, p. 263–278. *In* R. Margesin (ed.), *Permafrost Soils.* Springer, Berlin, Germany.

Bej, A. K., J. Aislabie, and R. M. Atlas (ed.). 2010. *Polar Microbiology: the Ecology, Biodiversity and Bioremediation Potential of Microorganisms in Extremely Cold Environments.* CRC Press, Boca Raton, FL.

Biondi, N., M. R. Tredici, A. Taton, A. Wilmotte, D. A. Hodgson, D. Losi, and F. Marinelli. 2008. Cyanobacteria from benthic mats of Antarctic lakes as a source of new bioactivities. *J. Appl. Microbiol.* **105:**105–115.

Brakstad, O. G. 2008. Natural and stimulated biodegradation of petroleum in permafrost-affected regions, p. 389–407. *In* R. Margesin, F. Schinner, J.-C. Marx, and C. Gerday (ed.), *Psychrophiles: from Biodiversity to Biotechnology.* Springer, Berlin, Germany.

Cavicchioli, R., K. S. Siddiqui, D. Andrews, and K. R. Sowers. 2002. Low-temperature extremophiles and their applications. *Curr. Opin. Biotechnol.* **13:**253–261.

Chevalier, P., D. Proulx, P. Lessard, W. F. Vincent, and J. de la Noue. 2000. Nitrogen and phosphorus removal by high latitude mat-forming cyanobacteria for potential use in tertiary wastewater treatment. *J. Appl. Phycol.* **12:**105–112.

Collins, T., D. De Vos, A. Hoyoux, S. N. Savvides, C. Gerday, J. Van Beeumen, and G. Feller. 2005. Study of the active site residues of a glycoside hydrolase family 8 xylanase. *J. Mol. Biol.* **354:**425–435.

Collins, T., A. Hoyoux, A. Dutron, J. Georis, B. Genot, T. Dauvrin, F. Arnaut, C. Gerday, and G. Feller. 2006. Use of glycoside hydrolase family 8 xylanases in baking. *J. Cereal Sci.* **43:**79–84.

Collins, T., M. A. Meuwis, C. Gerday, and G. Feller. 2003. Activity, stability and flexibility in glycosidases adapted to extreme thermal environments. *J. Mol. Biol.* **328:**419–428.

Collins, T., M. A. Meuwis, I. Stals, M. Claeyssens, G. Feller, and C. Gerday. 2002. A novel family 8 xylanase, functional and physicochemical characterization. *J. Biol. Chem.* **277:**35133–35139.

Cusano, A. M., E. Parrilli, A. Duilio, G. Sannia, G. Marino, and M. L. Tutino. 2006. Secretion of psychrophilic α-amylase deletion mutants in *Pseudoalteromonas haloplanktis* TAC125. *FEMS Microbiol. Lett.* **258:**67–71.

D'Amico, S., T. Collins, J. C. Marx, G. Feller, and C. Gerday. 2006. Psychrophilic microorganisms: challenges for life. *EMBO Rep.* **7:**385–389.

D'Amico, S., C. Gerday, and G. Feller. 2001. Structural determinants of cold adaptation and stability in a large protein. *J. Biol. Chem.* **276:**25791–25796.

D'Amico, S., C. Gerday, and G. Feller. 2003a. Temperature adaptation of proteins: engineering mesophilic-like activity and stability in a cold-adapted α-amylase. *J. Mol. Biol.* **332:**981–988.

D'Amico, S., J.-C. Marx, C. Gerday, and G. Feller. 2003b. Activity-stability relationships in extremophilic enzymes. *J. Biol. Chem.* **278:**7891–7896.

Davail, S., G. Feller, E. Narinx, and C. Gerday. 1994. Cold adaptation of proteins. Purification, characterization, and sequence of the heat-labile subtilisin from the Antarctic psychrophile *Bacillus* TA41. *J. Biol. Chem.* **269:**17448–17453.

De Vos, D., T. Collins, W. Nerinckx, S. N. Savvides, M. Claeyssens, C. Gerday, G. Feller, and J. Van Beeumen. 2006. Oligosaccharide binding in family 8 glycosidases: crystal structures of active-site mutants of the β-1,4-xylanase pXyl from *Pseudoalteromonas haloplanktis* TAH3a in complex with substrate and product. *Biochemistry* **45:**4797–4807.

Deming, J. W. 2002. Psychrophiles and polar regions. *Curr. Opin. Microbiol.* **5:**301–309.

Feller, G., and C. Gerday. 2003. Psychrophilic enzymes: hot topics in cold adaptation. *Nat. Rev. Microbiol.* **1:**200–208.

Ferrer, M., T. N. Chernikova, M. M. Yakimov, P. N. Golyshin, and K. N. Timmis. 2003. Chaperonins govern growth of *Escherichia coli* at low temperatures. *Nat. Biotechnol.* **21:**1266–1267.

Filler, D. M., C. M. Reynolds, I. Snape, A. J. Daugulis, D. L. Barnes, and P. J. Williams. 2006. Advances in engineered remediation methods for use in the Arctic and Antarctica. *Polar Rec.* **42:**111–120.

Filler, D. M., I. Snape, and D. L. Barnes (ed.). 2008. *Bioremediation of Petroleum Hydrocarbons in Cold Regions.* Cambridge University Press, Cambridge, United Kingdom.

Filler, D. M., D. R. van Stempvoort, and M. B. Leigh. 2009. Remediation of frozen ground contaminants with petroleum hydrocarbons: feasibility and limits, p. 279–301. In R. Margesin (ed.), Permafrost Soils. Springer, Berlin, Germany.

Fletcher, G. L., C. L. Hew, and P. L. Davies. 2001. Antifreeze proteins of teleost fishes. Annu. Rev. Physiol. 63:359–390.

Gerday, C., M. Aittaleb, M. Bentahier, J. P. Chessa, P. Claverie, T. Collins, S. D'Amico, J. Dumont, G. Garsoux, D. Georlette, A. Hoyoux, T. Lonhienne, M.-A. Meuwis, and G. Feller. 2000. Cold-adapted enzymes: from fundamentals to biotechnology. Trends Biotechnol. 18:103–107.

Gerday, C., and N. Glansdorff. 2007. Physiology and Biochemistry of Extremophiles. ASM Press, Washington, DC.

Gilichinsky, D., E. Rivkina, C. Bakermans, V. Shcherbakova, L. Petrovskaya, S. Ozerskaya, N. Ivanushkina, G. Kochkina, K. Laurinavichuis, S. Pecheritsina, R. Fattakhova, and J. M. Tiedje. 2005. Biodiversity of cryopegs in permafrost. FEMS Microbiol. Ecol. 53:117–128.

Gratia, E., F. Weekers, R. Margesin, S. D'Amico, P. Thonart, and G. Feller. 2009. Selection of a cold-adapted bacterium for bioremediation of wastewater at low temperatures. Extremophiles 13:763–768.

Hoag, H. 2008. Polar biotech. Nat. Biotechnol. 26:1204.

Hoyoux, A., I. Jennes, P. Dubois, S. Genicot, F. Dubail, J. M. Francois, E. Baise, G. Feller, and C. Gerday. 2001. Cold-adapted β-galactosidase from the Antarctic psychrophile Pseudoalteromonas haloplanktis. Appl. Environ. Microbiol. 67:1529–1535.

Kobori, H., C. W. Sullivan, and H. Shizuya. 1984. Heat-labile alkaline phosphatase from Antarctic bacteria: rapid 5' end labelling of nucleic acids. Proc. Natl. Acad. Sci. USA 81:6691–6695.

Kolenc, R. J., W. E. Inniss, B. R. Glick, C. W. Robinson, and C. I. Mayfield. 1988. Transfer and expression of mesophilic plasmid-mediated degradative capacity in a psychrotrophic bacterium. Appl. Environ. Microbiol. 54:638–641.

Koutsioulis, D., E. Wang, M. Tzanodaskalaki, D. Nikiforaki, A. Deli, G. Feller, P. Heikinheimo, and V. Bouriotis. 2008. Directed evolution on the cold adapted properties of TAB5 alkaline phosphatase. Protein Eng. Des. Sel. 21:319–327.

Krallish, I., S. Gonta, L. Savenkova, P. Bergauer, and R. Margesin. 2006. Phenol degradation by immobilized cold-adapted yeast strains of Cryptococcus terreus and Rhodotorula creatinivora. Extremophiles 10:441–449.

Leary, D. 2008. Bioprospecting in the Arctic. UNU-IAS report. United Nations University Institute of Advanced Studies, Yokohama, Japan.

Leigh Mascarelli, A. 2009. Geomicrobiology: low life. Nature 459:770–773.

Leiros, I., E. Moe, O. Lanes, A. O. Smalås, and N. P. Willassen. 2003. The structure of uracil-DNA glycosylase from Atlantic cod (Gadus morhua) reveals cold-adaptation features. Acta Crystallogr. D Biol. Crystallogr. 59:1357–1365.

Lo Giudice, A., V. Bruni, and L. Michaud. 2010. Potential for microbial biodegradation of polychlorinated biphenyls in polar environments, p. 373–391. In A. K. Bej, J. Aislabie, and R. M. Atlas (ed.), Polar Microbiology: the Ecology, Biodiversity and Bioremediation Potential of Microorganisms in Extremely Cold Environments. CRC Press, Boca Raton, FL.

Lohan, D., and S. Johnston. 2005. Bioprospecting in Antarctica. UNU-IAS report. United Nations University of Advanced Studies, Yokohama, Japan.

Margesin, R. 2004. Bioremediation of petroleum hydrocarbon-polluted soils in extreme temperature environments, p. 215–234. In A. Singh and O. P. Ward (ed.), Applied Bioremediation and Phytoremediation. Springer, Berlin, Germany.

Margesin, R. 2007. Alpine microorganisms: useful tools for low-temperature bioremediation. J. Microbiol. 45:281–285.

Margesin, R., and G. Feller. 2010. Biotechnological applications of psychrophiles. Environ. Technol. 31:835–844.

Margesin, R., G. Feller, C. Gerday, and N. J. Russell. 2002. Cold-adapted microorganisms: adaptation strategies and biotechnological potential, p. 871–885. In G. Bitton (ed.), Encyclopedia of Environmental Microbiology. John Wiley and Sons, New York, NY.

Margesin, R., P. A. Fonteyne, and B. Redl. 2005. Low-temperature biodegradation of high amounts of phenol by Rhodococcus spp. and basidiomycetous yeasts. Res. Microbiol. 156:68–75.

Margesin, R., G. Neuner, and K. B. Storey. 2007. Cold-loving microbes, plants, and animals—fundamental and applied aspects. Naturwissenschaften 94:77–99.

Margesin, R., and F. Schinner (ed.). 1999a. Biotechnological Applications of Cold-Adapted Organisms. Springer, Berlin, Germany.

Margesin, R., and F. Schinner (ed.). 1999b. Cold-Adapted Organisms: Ecology, Physiology, Enzymology and Molecular Biology. Springer, Berlin, Germany.

Margesin, R., F. Schinner, J.-C. Marx, and C. Gerday (ed.). 2008. Psychrophiles: from Biodiversity to Biotechnology. Springer, Berlin, Germany.

Marx, J.-C., T. Collins, S. D'Amico, G. Feller, and C. Gerday. 2007. Cold-adapted enzymes from marine Antarctic microorganisms. *Mar. Biotechnol. (NY)* **9**:293–304.

Miteva, V., S. Lantz, and J. Brenchley. 2008. Characterization of a cryptic plasmid from a Greenland ice core *Arthrobacter* isolate and construction of a shuttle vector that replicates in psychrophilic high G+C Gram-positive recipients. *Extremophiles* **12**:441–449.

Miyake, R., J. Kawamoto, Y. L. Wei, M. Kitagawa, I. Kato, T. Kurihara, and N. Esaki. 2007. Construction of a low-temperature protein expression system using a cold-adapted bacterium, *Shewanella* sp. strain Ac10, as the host. *Appl. Environ. Microbiol.* **73**:4849–4856.

Narinx, E., E. Baise, and C. Gerday. 1997. Subtilisin from psychrophilic Antarctic bacteria: characterization and site-directed mutagenesis of residues possibly involved in the adaptation to cold. *Protein Eng.* **10**:1271–1279.

Panicker, G., N. Mojib, J. Aislabie, and A. K. Bej. 2010. Detection, expression and quantitation of the biodegradative genes in Antarctic microorganisms using PCR. *Antonie van Leeuwenhoek* **97**:275–287.

Papa, R., V. Rippa, G. Sannia, G. Marino, and A. Duilio. 2007. An effective cold inducible expression system developed in *Pseudoalteromonas haloplanktis* TAC125. *J. Biotechnol.* **127**:199–210.

Parente Duena, A., J. Garces Garces, J. Guinea Sanchez, J. M. Garcia Anton, R. Casaroli Marano, M. Reina Del Pozo, and S. Vilaro Coma. April 2006. Use of a glycoprotein for the treatment and re-epithelialisation of wounds. U.S. patent 7, 022,668.

Parrilli, E., D. De Vizio, C. Cirulli, and M. L. Tutino. 2008. Development of an improved *Pseudoalteromonas haloplanktis* TAC125 strain for recombinant protein secretion at low temperature. *Microb. Cell Fact.* **7**:2.

Paudyn, K., A. Rutter, R. K. Rowe, and J. S. Poland. 2008. Remediation of hydrocarbon contaminated soils in the Canadian Arctic by landfarming. *Cold Reg. Sci. Technol.* **53**:102–114.

Rina, M., C. Pozidis, K. Mavromatis, M. Tzanodaskalaki, M. Kokkinidis, and V. Bouriotis. 2000. Alkaline phosphatase from the Antarctic strain TAB5. Properties and psychrophilic adaptations. *Eur. J. Biochem.* **267**:1230–1238.

Russell, N. J. 1998. Molecular adaptations in psychrophilic bacteria: potential for biotechnological applications. *Adv. Biochem. Eng. Biotechnol.* **61**:1–21.

Siani, L., R. Papa, A. Di Donato, and G. Sannia. 2006. Recombinant expression of toluene

o-xylene monooxygenase (ToMO) from *Pseudomonas stutzeri* OX1 in the marine Antarctic bacterium *Pseudoalteromonas haloplanktis* TAC125. *J. Biotechnol.* **126**:334–341.

Siddiqui, K. S., and R. Cavicchioli. 2006. Cold-adapted enzymes. *Annu. Rev. Biochem.* **75**:403–433.

Smalås, A. O., H. K. Leiros, V. Os, and N. P. Willassen. 2000. Cold adapted enzymes. *Biotechnol. Annu. Rev.* **6**:1–57.

Strocchi, M., M. Ferrer, K. N. Timmis, and P. N. Golyshin. 2006. Low temperature-induced systems failure in *Escherichia coli*: insights from rescue by cold-adapted chaperones. *Proteomics* **6**:193–206.

Tehei, M., B. Franzetti, D. Madern, M. Ginzburg, B. Z. Ginzburg, M. T. Giudici-Orticoni, M. Bruschi, and G. Zaccai. 2004. Adaptation to extreme environments: macromolecular dynamics in bacteria compared in vivo by neutron scattering. *EMBO Rep.* **5**:66–70.

Thomassin-Lacroix, E. J., M. Eriksson, K. J. Reimer, and W. W. Mohn. 2002. Biostimulation and bioaugmentation for on-site treatment of weathered diesel fuel in Arctic soil. *Appl. Microbiol. Biotechnol.* **59**:551–556.

Tutino, M. L., A. Duilio, R. Parrilli, E. Remaut, G. Sannia, and G. Marino. 2001. A novel replication element from an Antarctic plasmid as a tool for the expression of proteins at low temperature. *Extremophiles* **5**:257–264.

Van Petegem, F., T. Collins, M. A. Meuwis, C. Gerday, G. Feller, and J. Van Beeumen. 2003. The structure of a cold-adapted family 8 xylanase at 1.3 Å resolution. Structural adaptations to cold and investigation of the active site. *J. Biol. Chem.* **278**:7531–7539.

Walworth, J. L., and S. Ferguson. 2008. Landfarming, p. 170–189. *In* D. M. Filler, I. Snape, and D. L. Barnes (ed.), *Bioremediation of Petroleum Hydrocarbons in Cold Regions*. Cambridge University Press, Cambridge, United Kingdom.

Walworth, J. L., A. Pond, I. Snape, J. Rayner, S. Ferguson, and P. Harvey. 2007. Nitrogen requirements for maximizing petroleum bioremediation in a sub-Antarctic soil. *Cold Reg. Sci. Technol.* **48**:84–91.

Wang, E., D. Koutsioulis, H. K. Leiros, O. A. Andersen, V. Bouriotis, E. Hough, and P. Heikinheimo. 2007. Crystal structure of alkaline phosphatase from the Antarctic bacterium TAB5. *J. Mol. Biol.* **366**:1318–1331.

Williams, N. 2004. Chill wind over Antarctic biodiversity. *Curr. Biol.* **14**:R169–R170.

Yergeau, E., M. Arbour, R. Brousseau, D. Juck, J. R. Lawrence, L. Masson, L. G. Whyte, and C. W. Greer. 2009. Microarray and real-time PCR analyses of the responses of high Arctic soil bacteria to hydrocarbon pollution and bioremediation treatments. *Appl. Environ. Microbiol.* **75:** 6258–6267.

Zaitsev, G. M., J. S. Uotila, and M. M. Häggblom. 2007. Biodegradation of methyl *tert*-butyl ether by cold-adapted mixed and pure bacterial cultures. *Appl. Microbiol. Biotechnol.* **74:** 1092–1102.

Zilouei, H., A. Soares, M. Murto, B. Guieysse, and B. Mattiasson. 2006. Influence of temperature on process efficiency and microbial community response during the biological removal of chlorophenols in a packed-bed bioreactor. *Appl. Microbiol. Biotechnol.* **72:**591–599.

Zobell, C. E. 1946. Action of microorganisms on hydrocarbons. *Bacteriol. Rev.* **10:**1–49.

ECOLOGY AND BIOGEOCHEMICAL CYCLING OF POLAR MICROBIOLOGY COMMUNITIES

MICROBIAL CARBON CYCLING IN PERMAFROST

Tatiana A. Vishnivetskaya,
Susanne Liebner, Roland Wilhelm,
and Dirk Wagner

9

INTRODUCTION

With the greenhouse gases methane and carbon dioxide contributing to rising global temperatures, a comprehensive understanding of the global carbon cycle has become a worldwide policy imperative. Terrestrial and submarine permafrost is identified as one of the most vulnerable carbon pools on Earth (Osterkamp, 2001; Zimov et al., 2006). About one-third of the global soil carbon is preserved in northern latitudes in huge layers of frozen ground, which underlay around 24% of the exposed land area of the Northern Hemisphere (Gorham, 1991; Zhang et al., 1999). This carbon reservoir is of global climatic importance due to the disproportionately higher rate of warming in the circumpolar north predicted by climate change research (IPCC, 2007).

The term "permafrost" refers to a soil, sediment, or rock that has been frozen for a period

greater than two consecutive years and is used to describe both mineral and organic soils. Permafrost may extend deep beneath Earth's surface, in some places over a kilometer (Schurr et al., 2008), and is governed by the balance of surface temperature and geothermal gradients. Soil overlying permafrost undergoes seasonal freeze-thaw and is known as the active layer. Collectively, the two soil/sediment layers are referred to as cryosol (Canada, Europe), Gelisol (United States), or Cryozem (Russia), depending on the soil classification system (Jones et al., 2009). Permafrost may also exist at the surface, such as in protrusions through the sides of embankments or cliff faces, and exists in more southern latitudes in sporadic, isolated areas (also termed "discontinuous" permafrost).

The permafrost table, separating permafrost-affected soil from permafrost, is characterized by an impermeable ice cover that impedes the diffusion of gases and movement of liquids (Gilichinsky, 2002). As a result, permafrost is isolated from allochthonous inputs and is primarily a reducing, anaerobic environment, though microaerophilic zones are presumed to exist based on the diversity of cultured aerobes (Steven et al., 2006). An abundance of cells are observed in permafrost-affected soil at the permafrost table, resulting from the buildup of translocated cells unable to pass through the ice barrier, and possibly due to the

Tatiana A. Vishnivetskaya, Center for Environmental Biotechnology, The University of Tennessee, 676 Dabney-Buehler Hall, 1416 Circle Drive, Knoxville, TN 37996-1605, and Biosciences Division, Oak Ridge National Laboratory, Oak Ridge, TN 37831-6038. Susanne Liebner, Department of Arctic and Marine Biology, University of Tromsø, 9037 Tromsø, Norway. Roland Wilhelm, Department of Natural Resource Sciences, McGill University, Ste.-Anne-de-Bellevue, QC H9X 3V9, Canada. Dirk Wagner, Alfred Wegener Institute for Polar and Marine Research, Telegrafenberg A43, 14473 Potsdam, Germany.

Polar Microbiology: Life in a Deep Freeze
Edited by Robert V. Miller and Lyle G. Whyte © 2012 ASM Press, Washington, DC

infiltration of nutrients to a more stable, albeit colder, environment (Gilichinsky et al., 2008).

Temperatures in permafrost have been measured as low as −18°C in the Arctic and −27°C in the Antarctic (Vorobyova et al., 1997), with a typical temperature profile of permafrost from the High Arctic as follows: 5 m, −18°C with 3 to 4°C annual variation; 12 m, −17°C with 1.5°C annual variation; and 15 m, −17.5°C with less than 1°C annual variation (Steven et al., 2008). The extremity of cold affects the abundance of liquid water and therefore conditions suitable for life in permafrost. Films of liquid occur in close proximity to soil particles at subzero temperatures, where ~15 nm of liquid was observed at −1.5°C, and ~5 nm at −10°C (Gilichinsky et al., 1993). The size of a film is a function of the particle size and chemical composition. For instance, in sandy soils the amount of liquid water is negligible, while in a loam soil at the same temperature up to 10% of water remains unfrozen, adhered to soil particles. These thin layers of water, as well as pockets of brine and ice veins, are microhabitats where cells can be protected from damage-causing water crystallization and may attain the best possible osmotic balance.

Within permafrost there exists a variety of common features that provide habitats for microbial colonization (Color Plate 7). Taliks are discrete areas within permafrost, or adjacent to permafrost, that retain sufficient heat to remain perennially unfrozen. They are produced when warm surface waters infiltrate to deeper soils, creating a reservoir of thermal energy; they also occur in the active layer. In some areas, permafrost comprises upwards of 80% ice in the form of large features, such as massive ice sheets many kilometers in length; or on smaller scales, such as ice wedges and ice lenses, and as ice that fills soil pore space. Arctic polygon surface features lead to the formation of subsurface ice wedges, which are meter-wide, V-shaped ice features, capable of being several meters deep. They result from water accumulating in cracks produced by the freeze-thaw of surface soils. Lastly, residual pockets of seawater, from the subsidence of the polar ocean, exist as saturated, salt-rich permafrost environments known as salt lenses or cryopegs. All of these permafrost features sustain microbial communities that contribute to carbon cycling in polar regions. Whether their subzero contributions to the global carbon cycle are as significant as, for instance, the overall thawing of permafrost is unlikely. However, their significance as reservoirs of biodiversity is only beginning to be known.

Thawing of permafrost could release large quantities of previously frozen organic matter. Permafrost degradation through environmental change, such as erosion and soil subsidence, is considered to have a stronger impact on organic carbon decomposition rates than the direct effect of temperature rise alone (Eugster et al., 2000). The degradation of permafrost is associated with the release of climate-relevant trace gases from intensified microbial carbon turnover that may further increase global warming and transform the Arctic tundra ecosystems from a carbon sink to a carbon source (Oechel et al., 1993). Trace gas fluxes from permafrost ecosystems are influenced by a number of biotic and abiotic parameters. The decomposition of soil organic matter and the generation of greenhouse gases result from microbial activity that is affected by habitat characteristics (soil parameters) and by climate-related properties (e.g., soil temperature and moisture). The way in which gas is released from permafrost, i.e., the rate and pathway, determines the ratio of methane and carbon dioxide emitted to the atmosphere. However, the exact processes affecting carbon release, and the spatial distribution of involved microorganisms as well as their climate dependency, are not yet adequately quantified and understood.

The worldwide wetland area has been estimated at approximately 5.5×10^6 km^2. About half of it is located in high latitudes of the Northern Hemisphere (above 50°N) (Aselmann and Crutzen, 1989). The atmospheric input of methane from tundra soils of northern regions was estimated to vary between 7.2% (Denman et al., 2007) and 8.1% (Wuebbles and Hayhoe, 2002) of the total global CH$_4$ emissions.

In anaerobic wetland environments, the mineralization of organic matter is achieved stepwise by specialized microorganisms of the so-called anaerobic food chain (Schink and Stams, 2006). Important intermediates of organic matter decomposition under anaerobic conditions are polysaccharides, low-molecular-weight organic acids, phenolic compounds, and sugar monomers (Guggenberger et al., 1994; Kaiser et al., 2001). The fermentation of carbon by microorganisms runs much slower than the oxidative respiration; thus, an increase in water-saturated soils, as predicted for Arctic environments, resulting from thermokarst erosion (Grosse et al., 2006), might be accompanied by qualitative and quantitative alterations within the turnover of organic matter (Color Plate 8).

Although microorganisms are the drivers of carbon mineralization, the structure of the microbial community and its influence on carbon dynamics and ecosystem stability in Arctic permafrost-affected soils remain poorly understood (Wagner, 2008). Although several studies have investigated the diversity of methanogens (Høj et al., 2005, 2006; Ganzert et al., 2007), methanotrophs (Kaluzhnaya et al., 2002; Wartiainen et al., 2003; Liebner et al., 2008), and the entire microbial community (Zhou et al., 1997; Kobabe et al., 2004; Neufeld and Mohn, 2005; Hansen et al., 2007; Steven et al., 2007), very little is known about how microbial communities in Arctic regions will respond to ecosystem-specific changes induced by global warming.

This review first describes the different carbon pools, carbon fluxes, and freeze-thaw stresses related to microbial activities. It then examines methane-cycling communities in Arctic active-layer and permafrost environments.

CARBON TURNOVER IN ARCTIC TERRESTRIAL ECOSYSTEMS

Different Carbon Pools Including Cellular Carbon

Tundra soils store approximately 15% (191.8 × 10^{15} g, or 191.8 Pg) of the global terrestrial carbon (Post et al., 1982) in the form of poorly decomposed plant detritus and cellular biomass of soil organisms. Taking into account the carbon pools in permafrost deposits deeper than

3 m, the total organic C in northern regions was estimated to be approximately 1,672 Pg (Tarnocai et al., 2009). Carbon in the form of soil organic matter, humus, or peat accumulates in tundra soils because cold temperatures and water saturation slow decomposition relative to the rate of input of organic material from surface vegetation. Furthermore, organic matter of differing molecular weight and composition experiences varying rates of decomposition and turnover time. Labile soil carbon pools constituted of simple carbohydrates, amino acids, and other simple organic compounds are known to be easily depleted also at low temperatures, while little is known about the contribution of bacteria and fungi to decomposition of more stable, recalcitrant carbon pools. The turnover time for labile soil carbon, easily accessible by soil microbes, is often less than a few years, compared to recalcitrant carbon, with a turnover time of up to several thousand years (Knorr et al., 2005).

On Earth, the total number of prokaryotes is equal to 4×10^{30} to 6×10^{30} cells and the mass of their cellular C is 350 to 550 Pg (Whitman et al., 1998), with approximately 25% of this accumulation occurring in cold regions. The total bacterial cell counts in the tundra soils are comparable to temperate soils and reach as high as 10^8 to 10^9 cells g^{-1} in active-layer soils (Kobabe et al., 2004; Hansen et al., 2007; Wagner et al., 2009); in deeper permafrost, cell counts are lower (10^6 to 10^8 cells g^{-1}) (Vishnivetskaya et al., 2000; Steven et al., 2007). Strong positive correlation was detected throughout the soil profile between abundance of bacteria and organic C content ($R^2 = 0.951$, $P < 0.0001$, $N = 9$) (Kobabe et al., 2004). The composition of the microbial community was shown to be strongly linked to the quality of soil organic matter (Camill, 1999; Eskelinen et al., 2009).

Carbon Fluxes and Their Connection to Microbial Activity

Several studies have observed increased C emissions from climate-induced changes in Arctic ecosystems (Oechel and Vourlitis, 1994;

Walker et al., 2006a). The cumulative CO_2 production during a growing season, as estimated in central Alaska, ranged from 177 to 270 g CO_2-C m^{-2}, with the lowest value occurring in moist acidic tundra and the highest in sites where thawing of permafrost and thermokarst was most pronounced (Lee et al., 2010). Greenhouse gas fluxes measured over the season in a sub-Arctic wetland near Abisko in northern Sweden ranged as follows: 0.2 and 36.1 mg CH_4 m^{-2} h^{-1} net ecosystem exchange; −1,000 and 1,250 mg CO_2 m^{-2} h^{-1} (negative values meaning a sink of atmospheric CO_2); and dark respiration 110 and 1,700 mg CO_2 m^{-2} h^{-1} net ecosystem exchange (Ström and Christensen, 2007). The contribution of heterotrophic microbial respiration to the total CO_2 efflux from tundra soil in Scandinavia was roughly estimated as one-half (Sjögersten and Wookey, 2009). Photosynthetic rate and CO_2 and CH_4 emissions were affected by the plant species and rhizosphere composition, depth of water table, substrate availability, activity of methane-producing bacteria, soil pH, soil temperature, etc. These key determinants also define tundra microbial community composition, and variations in these parameters will result in microbial community changes.

Temperature was defined as a major driver of in situ decomposition, and positive regressions between soil temperature and soil respiration were found in tundra (Sjögersten and Wookey, 2009) and other soils (Trumbore et al., 1996; Knorr et al., 2005). Investigations have been made into how a changing climate would alter carbon turnover and microbial communities in high northern latitude soils. At Scandinavia tundra sites the warming increased the CO_2 efflux from the system by 33 to 50% compared to control (no warming) sites (Sjögersten and Wookey, 2009). Jonasson et al. (2004) found higher respiration rates, but lower microbial biomass carbon content resulting from temperature increase and litter addition. Microbial biomass N was found to decrease as temperature increased, resulting in a high mobilization of inorganic N. A long-term study in a sub-Arctic heath

ecosystem revealed that soil microbial community composition was slightly altered by warming (10- to 13-year study period), while the activity of cultivable bacterial and fungal communities was significantly increased upon warming (Rinnan et al., 2007). After 15 years of climate change simulations, soil microbial communities were significantly altered (hypothetically due to changes in plant community), while less than 10 years of simulated climate change (temperature, nutrient, and light manipulations) left the microbial biomass largely unaffected (Rinnan et al., 2007, 2009). Oechel et al. (2000) observed that continued warming and drying of two Arctic ecosystems between 1960 and 1998 has, after an initial increase, resulted in diminished CO_2 efflux and in some cases the presence of a summer CO_2 sink. This indicates the capacity for ecosystems to metabolically adjust to long-term (decadal or longer) climate changes, likely as a result of changes in nutrient cycling, physiological acclimation, and community reorganization. Nevertheless, the Arctic ecosystems are annual net sources of CO_2 to the atmosphere of at least 40 g C m^{-2} per year, due to winter release of CO_2, implying that further climate change may still exacerbate CO_2 emissions from Arctic ecosystems (Oechel et al., 2000). Finally, Oelbermann et al. (2008), using soils from Canada's Northwest Territories in laboratory incubations, monitored respiration rates in three phases after increasing the temperature from 14 to 21°C and observed a rapid increase in respiration, followed by a peak and finally a new equilibrium lower than in phase two. Here the temperature increases also led to higher metabolic diversity expressed as microbial richness through the number of oxidized C substrates in the Biolog Ecoplate.

Different mineralization rates of recalcitrant compounds have been attributed to differences in temperature, where more recalcitrant compounds are preferentially respired by Arctic microbes at higher temperatures. The main source of CO_2 in tundra originated from microbial oxidation of soil organic matter. The isotopic signatures of respired CO_2 and phospholipid fatty acid profiles showed significant differences in microbial community structure at 2, 12, and 24°C, indicating that recalcitrant compounds are preferentially respired by Arctic microbes at higher temperatures (Biasi et al., 2005). Therefore, large portions of tundra soil organic matter are potentially mineralizable.

This is a cause for concern, since the main source of CO_2 from tundra is hypothesized to originate from microbial oxidation of soil organic matter (Zak and Kling, 2006). Differences in energy partitioning and carbon balance were examined between acidic and nonacidic tundra, characterized by structural differences in plant growth forms, moss cover, soil pH, summer thaw, and CO_2 and CH_4 fluxes. Moist nonacidic tundra has greater heat flux, deeper summer thaw (active layer), is less of a carbon sink, and is a smaller source of methane than moist acidic tundra (Walker et al., 2006b). In moist acidic tundra of the Brooks Range, Alaska, net CO_2 uptake was 26% more (with 34% greater ecosystem respiration and 30% greater gross primary production) than the values obtained from moist nonacidic tundra (Eugster et al., 2005). Decomposition levels were two times faster in older, acidic tundra than in nonacidic tundra, which was attributable to a greater abundance of soil fungi and higher soil N availability in the acidic tundra (Gough et al., 2000). Fungal biomass in tundra soils is similar to that in temperate forest soils and correlates strongly with soil organic matter and soil N availability (Hobbie and Gough, 2004; Hobbie et al., 2009). A high proportion of fungi were found in alpine tundra sites with dwarf, shrub-rich vegetation characterized by high C-to-N and low soluble N-to-phenolics ratios in soil organic matter, while sites rich in herbaceous flowering plants (low C-to-N and high soluble N-to-phenolics ratios) were found to contain high proportions of bacteria (Eskelinen et al., 2009). Different groups of the microbial community are responsible for substrate utilization; for example, in a tundra heath in Abisko, northern Sweden, glycine and especially starch were mainly used up by bacteria and not fungi

(Rinnan et al., 2009). Psychrotolerant bacteria isolated from tundra soils showed cellulase, amylase, protease, and lipase activities, and they possessed numerous metabolic pathways and were able to utilize various carbon compounds including cellulose, xylan, glycogen, sugars, amino acids, organic acids, polyalcohols, and certain aromatic compounds as primary carbon and energy sources (Berestovskaia et al., 2006; Männistö and Häggblom, 2006; Vasilyeva et al., 2006; Nelson et al., 2009).

In tundra soils, nitrogen availability is the main constraint on primary production, so soil microbes and symbiotic fungi play a major role in N turnover and retention. A number of bacteria capable of nitrogen fixation were isolated from tundra soils and sphagnum bogs (Dedysh et al., 2004; Belova et al., 2006). Addition of labeled glycine-N, NH_4^+-N, and NO_3^--N to active-layer tundra soil showed that up to 14% of nitrogen was recovered in the plant-soil system, whereas the major remaining part of the ^{15}N was immobilized by microbes (Clemmensen et al., 2008).

Freeze–Thaw Stresses and Their Influence on Microbial Activity

The top 1 m of surface soil in polar regions is characterized by extensive periods of frozen inactivity, annual freeze-thaw cycling, and waterlogging during periods of thaw. These layers can be either mineral soils or humus rich and often show disrupted horizons, cracks, or patterned surface features such as frost mounds, caused by the physical actions of ice formation and thaw. The process of freezing, thawing, and refreezing of wet soil is responsible for many of the characteristics of the Arctic tundra landscape. Various global climate models predict the strongest future warming in the Northern Hemisphere high latitudes by the end of this century (Anisimov and Reneva, 2006; Tarnocai, 2006), and as a result the depth of seasonal thawing may increase on average by 15 to 25% or more (Anisimov and Reneva, 2006).

Ongoing global warming is predicted to cause an increase in the recurrence of soil freeze-thaw cycles in polar and high-latitude regions (Kreyling et al., 2010). Rapid freeze-thaw cycles often occur during the early spring period directly after snowmelt when soil temperatures are close to 0°C, and this period is characterized by strong stability in microbial and soil C and N pool sizes (Buckeridge et al., 2010). In field experiments, more frequent freeze-thaw cycles changed nitrate availability, which positively correlated with increased plant biomass production in grassland and decreased biomass production in heath sites (Kreyling et al., 2010). The activity of denitrifiers in the early spring period may be stimulated by N addition, causing increases in N_2O fluxes, suggesting that the soils are nitrogen limited and that in situ denitrifying activity is low in early spring (Buckeridge et al., 2010).

Naturally occurring processes such as deep snow cover during winter or increased soil frost (temperature at a depth of 15 cm below the surface decreased by 5°C in comparison to a control site) also affect microbial communities. Deep snow increased microbial N release and its availability to plants during spring, while soil pools of dissolved organic N and C and bacterial counts decreased (Buckeridge et al., 2010). Fungal mass and hyphal lengths were not affected by snow cover. However, fungi from under deepened snow did respond to substrate availability (Buckeridge et al., 2010). Soil frost triggers a change in the composition of the microbial community, leading to a considerable reduction of heterotrophic respiration, a decrease of 14% during the soil frost period itself and 63% during the following summer, which was only partly compensated for by a slight increase of rhizosphere respiration (Muhr et al., 2009). It is interesting to note that the differences between frost-induced and control plots occurred in organic and not in mineral horizons. Repetitive freeze-thawing has been seen to cause only minor changes in the bacterial community (Männistö and Häggblom, 2006).

Tundra soils are dominated by diverse microorganisms including bacteria, archaea, yeasts, algae, and mycelial fungi (D'Amico et al., 2006). Typical members of the active-layer tundra microbial communities are gram-negative α-, β-, and γ-Proteobacteria; the Cytophaga-Flavobacterium-Bacteroides group; Cyanobacteria; and gram-positive Actinobacteria and Firmicutes (Parinkina, 1989; Dobrovolskaya et al., 1996; Belova et al., 2006; Männistö and Häggblom, 2006; Liebner et al., 2008). The bacterial community of underlying permafrost sediments from different locations is characterized by varying abundances of Actinobacteria, Firmicutes, α- and γ-Proteobacteria, and Cytophaga-Flavobacterium-Bacteroides group members (Steven et al., 2007, 2008; Vorobyova et al., 1997). These microorganisms possess a wide range of adaptive strategies that allow them to effectively metabolize at low temperatures as well as tolerate and rapidly adapt to quick changes in environmental conditions. Repeated freeze-thawing studied in Arctic sediments from Svalbard showed that microbial activity in the frozen state decreased to 0.25% of initial levels at 4°C, but activity resumed without delay following thaw, reaching >60% of initial activity (Sawicka et al., 2010). Exposure of the sediments to successive large temperature changes (−20 versus 10°C) decreased [^{35}S]sulfate reduction rates by 80% of the initial activity, demonstrating that a fraction of the bacterial community rapidly recovers from extreme temperature fluctuations. 16S rRNA gene-based denaturing gradient gel electrophoresis profiles revealed persistence of dominant microbial taxa following repeated freeze-thaw cycles (Sawicka et al., 2010). In samples from northern Norway, microbial heterotrophic activity started directly after the spring thaw, increasing cumulative soil respiration (Bölter et al., 2005). Under laboratory conditions, freeze-thaw cycling showed different patterns for active versus nonactive bacteria, suggesting that freeze-thaw may disproportionately damage a subpopulation of active-layer communities (Bölter et al., 2005). Terminal restriction fragment length polymorphism (T-RFLP) analysis of the bacterial communities after three to five freeze-thaw cycles (ranging from 5 to −2, −5, and −10°C) showed that the dominant T-RFLP peaks did not change, though slight variation in their amplitude was detected (Männistö et al., 2009).

Freezing and thawing of soils affect the turnover of soil organic matter and thus the losses of C and N from soils. Repeated freezing and thawing influences the microbial processes in sediments and soils and facilitates sediment carbon cycling at high latitudes affected by global warming. The fast recovery of the microbial activity during spring suggests that carbon mineralization in thawing Arctic sediment may rapidly respond to warming, resulting in substantial changes in carbon cycling and growth of microbial populations.

METHANE-CYCLING MICROBIAL COMMUNITIES

The biological formation and consumptive reactions involving methane are carried out by specialized microorganisms. The production of methane, or methanogenesis, is possible from a single clade from the archaeal phylum *Euryarchaeota* (Table 1), also referred to as methanogens. Another more taxonomically diverse functional group of organisms are capable of consuming methane. These are known as methanotrophs, and they comprise obligate aerobic members of the phyla *Proteobacteria* (Bowman, 1999) and *Verrucomicrobia* (Dunfield et al., 2007; Pol et al., 2007), as well as anaerobic methane-oxidizing *Archaea* in marine habitats (Boetius et al., 2000). Further, there are *Bacteria* of a yet unknown taxon carrying out methane oxidation in the presence of very high nitrate and methane concentration in freshwater habitats (Raghoebarsing et al., 2006). The dominant methane-consuming microorganisms in permafrost-affected soils are those of the *Proteobacteria* phylum (Table 2). Because of the pronounced distribution of methanogenic *Archaea* and methanotrophic *Proteobacteria* in Arctic permafrost environments (reviewed by Wagner, 2008) and their significance for the global methane budget, these two groups are of particular concern for carbon cycling under anaerobic conditions.

Methane Production in Permafrost-Affected Tundra

Methanogenic *Archaea* represent a small group of strictly anaerobic microorganisms (Hedderich and Whitman, 2006). They can be found in temperate habitats like paddy fields (Grosskopf et al., 1998), lakes (Jurgens et al., 2000; Keough et al., 2003), and freshwater sediments (Chan et al., 2005); in the gastrointestinal tract of animals (Lin et al., 1997); or in extreme habitats such as hydrothermal vents (Jeanthon et al., 1999), hypersaline habitats (Mathrani and Boone, 1995), active-layer and permafrost-affected soils, and sediments (Kobabe et al., 2004; Rivkina et al., 1998). Methane represents the driving intermediate in permafrost carbon cycling under anaerobic conditions. It is produced via two main pathways: (i) the reduction of CO_2 to CH_4 using H_2 as a reductant, and (ii) the fermentation of acetate to CH_4 and CO_2 (Conrad, 2005). However, only a few species of methanogenic *Archaea* from low-temperature terrestrial environments have been characterized so far (Franzmann et al., 1992, 1997; Kotelnikova et al., 1998; Lomans et al., 1999; Shlimon et al., 2004; Krivushin et al., 2010). At present, 34 genera are described within the group of methanogenic *Archaea*, with an increasing number of cultivated methanogenic species (Table 1).

Although permafrost environments are characterized by extreme climate conditions, it was recently shown that the abundance and composition of methanogenic populations in permafrost-affected soils are similar to that of communities of comparable temperate ecosystems (Wagner et al., 2005). The highest cell counts of methanogenic *Archaea* have been detected in the active layer of permafrost, with numbers of up to 3×10^8 cells g^{-1} soil by fluorescence in situ hybridization (Kobabe et al., 2004), and in late Pleistocene permafrost deposits, with up to 9×10^8 cells g^{-1} sediment by quantitative PCR (our unpublished data). Methanogenic *Archaea* represented between 0.5 and 22.4% of the total cell counts. Phylogenetic analyses revealed a high diversity of

TABLE 1 Taxonomy of all previously described methanogenic *Archaea*[a]

Phylum	Class	Order	Family	Genus
Euryarchaeota	Methanobacteria	Methanobacteriales	Methanobacteriaceae	Methanobacterium
				Methanobrevibacter
				Methanosphaera
				Methanothermobacter
			Methanothermaceae	Methanothermus
	Methanococci	Methanococcales	Methanocaldococcaceae	Methanocaldococcus
				Methanotorris
			Methanococcaceae	Methanococcus
				Methanothermococcus
	Methanomicrobia	Methanocellales	Methanocellaceae	Methanocella
		Methanomicrobiales	Methanocorpusculaceae	Methanocorpusculum
			Methanomicrobiaceae	Methanoculleus
				Methanofollis
				Methanogenium
				Methanolacinia
				Methanomicrobium
				Methanoplanus
			Methanospirillaceae	Methanospirillum
			Incertae sedis	Methanocalculus
				Methanolinea
				Methanoregula
				Methanosphaerula
		Methanosarcinales	Methanosaetaceae	Methanosaeta
				Methanothrix
			Methanosarcinaceae	Methanomicrococcus
				Methanococcoides
				Methanohalobium
				Methanohalophilus
				Methanolobus
				Methanomethylovorans
				Methanosalsum
				Methanosarcina
			Methermicrococcaceae	Methermicrococcus
	Methanopyri	Methanopyrales	Methanopyraceae	Methanopyrus

Reference	Kluyver and van Niel 1936, Zentralbl. Bakteriol. Parasitenkd. Infektionskr. Hyg.; Balch et al. 1979, Microbiol. Rev.; Miller and Wolin 1985, Arch. Microbiol.; Wasserfallen et al. 2000, IJSEM	Stetter 1981, Zentralbl. Bakteriol. Parasitenkd. Infektionskr. Hyg.	Whitman 2002, Bergey's Manual of Systematic Bacteriology, 2nd ed., vol. 1; Whitman 2001, Bergey's Manual of Systematic Bacteriology, 2nd ed., vol. 1	Kluyver and van Niel 1936, Zentralbl. Bakteriol. Parasitenkd. Infektionskr. Hyg.; Whitman 2001, Bergey's Manual of Systematic Bacteriology, 2nd ed., vol. 1	Sakai et al. 2008, IJSEM	Zeller et al. 1987, Arch. Microbiol.	Mestrojuán et al. 1990, IJSEM; Zellner et al. 1999, IJSEM; Romesser et al. 1979, Arch. Mikrobiol.; Zellner et al. 1989, J. Gen. Appl. Microbiol.; Balch and Wolfe 1979, Microbiol. Rev.; Wildengruber 1982, Arch. Microbiol.	Ferry et al. 1974, IJSEM	Ollivier et al. 1998, IJSEM; Imachi et al. 2008, IJSEM; Cadillo-Quiroz et al. 2009, IJSEM	Patel and Sprott 1990, IJSEM; Huser et al. 1982, Arch. Microbiol.	Sprenger at al. 2000, IJSEM; Sowers and Ferry 1983, Appl. Environ. Microbiol.; Zhilina and Zavarzin 1987, Dokl. Akad. Nauk.; Paterek and Smith 1988, IJSEM; Koenig and Stetter 1982, Zentralbl. Bakteriol. Parasitenkd. Infektionskr. Hyg.; Lomans et al. 1999, Appl. Environ. Microbiol.; Boone and Baker 2001, Bergey's Manual of Systematic Bacteriology, 2nd ed., vol. 1; Kluyver and van Niel 1936, Zentralbl. Bakteriol. Parasitenkd. Infektionskr. Hyg.	Kurr et al. 1991, Arch. Microbiol.

Cheng et al. 2007, IJSEM

aThe references cited in the table may be found in the *List of Prokaryotic Names with Standing in Nomenclature* (http://www.bacterio.cict.fr/index.html).

TABLE 2 Taxonomy of methane-oxidizing *Proteobacteria*[a]

Phylum	*Proteobacteria*			
Subphylum	γ-*Proteobacteria* (Type I)		α-*Proteobacteria* (Type II)	
Order	*Methylococcales*		*Rhizobiales*	
Family	*Methylococcaceae*	*Crenotrichaceae*	*Methylocystaceae*	*Beijerinckiaceae*
Genus	*Methylobacter* *Methylomonas* *Methylomicrobium* *Methylococcus* *Methylosphaera* *Methylosarcina* *Methylocaldum* *Methylohalobius* *Methylothermus* *Methylosoma* *Methylovolum*	*Crenothrix*	*Methylosinus* *Methylocystis*	*Methylocella* *Methylocapsa* *Methyloferula*
Reference	Bowman et al. 1993, *IJSB*; Bowman et al. 1995, *IJSB*; Bowman et al. 1997, *Microbiology*; Wise et al. 2001, *IJSEM*; Heyer et al. 2005, *IJSEM*; Tsubota et al. 2005, *IJSEM*; Rahalker et al. 2007, *IJSEM*; Iguchi et al. 2010, *IJSEM*	Stoecker et al. 2006, *PNAS*	Browman et al. 1993, *IJSB*	Dedysh et al. 2000, *IJSEM*; 2002, *IJSEM*; Vorobev et al. 2010, *IJSEM*

[a]The references cited in the table may be found in the *List of Prokaryotic Names with Standing in Nomenclature* (http://www.bacterio.cict.fr/index.html).

methanogens in the active layer, with species belonging to the families *Methanobacteriaceae, Methanomicrobiaceae, Methanosarcinaceae,* and *Methanosaetaceae* (Høj et al., 2005; Metje and Frenzel, 2007; Ganzert et al., 2007; Koch et al., 2009). Other sequences detected were affiliated with the euryarchaeotal Rice Clusters II and V (Hales et al., 1996; Grosskopf et al., 1998; Ramakrishnan et al., 2001), as well as with Marine Group I 1.3b of the uncultured *Crenarchaeota* (nonmethanogenic *Archaea*) (Ochsenreiter et al., 2003). Environmental sequences from the Laptev Sea coast consist of four specific permafrost clusters (Ganzert et al., 2007). It was hypothesized that these clusters comprise methanogenic *Archaea* with a specific physiological potential to survive under harsh environmental conditions. The phylogenetic

affiliation of the sequences recovered by Ganzert et al. (2007) indicated that both hydrogenotrophic and acetoclastic methanogenesis exist in permafrost-affected soils. Recent studies on permafrost deposits from the Lena Delta (Siberia) revealed significant amounts of methane, which could be attributed to in situ activity of methanogenic *Archaea* (Wagner et al., 2007). A first study on submarine permafrost of the Laptev Sea shelf demonstrated that intact DNA was extractable from late Pleistocene permafrost deposits with an age of up to 111,000 years. At the depth where methane concentrations were at maximum, the lowest isotopic ratios for methane were measured, indicating active methanogenesis in the frozen sediments (Koch et al., 2009). Samples with high methane concentrations were dominated by sequences affiliated with the methylotrophic genera *Methanosarcina* and *Methanococcoides* as well as with uncultured *Archaea*. Another study on permafrost on Ellesmere Island reported an archaeal community composed of 61% *Euryarchaeota* (including methane-producing *Archaea*) and 39% *Crenarchaeota*, suggesting the presence of a diverse archaeal population also in perennially frozen sediments (Steven et al., 2007).

"*Candidatus* Methanosarcina gelisolum" strain SMA-21, which is closely related to *Methanosarcina mazei*, was recently isolated from a Siberian active-layer soil in the Lena Delta (our unpublished results). This strain grows well at 28°C and slowly at low temperatures (4 and 10°C) with H_2/CO_2 (80:20, vol/vol; pressurized at 150 kPa) as substrate. The cells grow as cocci, with a diameter of 1 to 2 μm, and cell aggregates were regularly observed. Previous strains of "*Ca.* Methanosarcina gelisolum" have demonstrated extreme tolerance to very low temperatures (−78.5°C), high salinity (up to 6 M), starvation, desiccation, and oxygen exposure (Morozova and Wagner, 2007). In a separate study, "*Ca.* Methanosarcina gelisolum" survived for 3 weeks under simulated thermophysical Martian conditions (Morozova et al., 2007). During the simulation experiment the diurnal temperature profile varied between −75 and 20°C and the water activity—dependent on the daily pressure variations around the mean value of 6 mbar—ranged between 0 and 1.

Methanogenesis was observed at low in situ temperatures, with rates of up to 39 nmol CH_4 h^{-1} g^{-1} soil from active-layer soils (Wagner et al., 2003; Høj et al., 2005; Metje and Frenzel, 2007). Interestingly, the highest activities were measured in the coldest zones of the profiles. Furthermore, it could be shown that methane production is limited more by the quality of soil organic carbon than by the in situ temperature (Wagner et al., 2005; Ganzert et al., 2007). Another important factor affecting methanogenic communities in permafrost-affected soils is the water regime. Along a natural soil moisture gradient, changes in archaeal community composition were observed, im-

plicating the differences in these communities in the large-scale variations in methane emissions related to changes in soil hydrology (Høj et al., 2006).

Methane Consumption in Active-Layer Soils

The biological consumption of methane is carried out by *Bacteria* termed methanotrophs. In active-layer soils, methanotrophy occurs at the oxic/anoxic soil interface, in the rhizosphere of aerenchymatous vascular plants such as *Carex* spp., and in association with submerged mosses (Color Plate 9).

The more the active-layer community of methane consumers is studied, the more complex this niche appears to be. Currently, this community is known to comprise obligate aerobic members of the phyla *Proteobacteria* (Bowman, 1999) and *Verrucomicrobia* (Dunfield et al., 2007; Pol et al., 2007), as well as anaerobic methane-oxidizing *Archaea* and *Bacteria* in marine (Boetius et al., 2000) and freshwater (Raghoebarsing et al., 2006) habitats. However, in permafrost-affected soils only *Bacteria* of the *Proteobacteria* phylum have been identified to date.

Methane-oxidizing *Proteobacteria* (MOP) utilize the enzyme methane monooxygenase, which enables them to grow on methane as single carbon and energy sources. MOP are affiliated with the γ- and α-*Proteobacteria*, referred to as type I and type II MOP, respectively. Type I and type II MOP can not only be distinguished based on their phylogeny but also with regard to their carbon assimilation pathway, the structure of their intracytoplasmic membranes, their resting stages, their GC content, the constitution of their methane monooxygenase, and their major phospholipid fatty acids. At the moment, 17 genera are described within the group of MOP (Table 2), and there is an increasing number of cultivated methanotrophic species.

MOP in polar Arctic active-layer soils were detected in Siberia (Vorkuta, Chukotka, Lena River Delta), on Svalbard, and in the Canadian High Arctic (Eureka). Viable methane oxidizers were also detected in deep Siberian permafrost sediments with ages of 1,000 to 100,000 years (Khmelenina et al., 2002). Cold-adapted methanotrophs have previously been cultivated from active-layer soils. *Methylobacter tundripaludum* (Wartiainen et al., 2006a) and *Methylocystis rosea* (Wartiainen et al., 2006b) were obtained from Arctic wetland soils on Svalbard, Norway. Both species can grow between 5 and 30°C, with optimum growth temperatures at 23 and 27°C, respectively. *Methylobacter psychrophilus*, a close relative of *M. tundripaludum*, was isolated from Siberian tundra and is the only true psychrophilic methanotroph reported (Omelchenko et al., 1996). Psychrophiles are organisms having optimum growth temperatures of <15°C and maximum growth temperatures of <20°C (Russell, 2000). Unfortunately, the culture of *M. psychrophilus* is no longer viable. The type II methanotroph *Methylocella tundrae* was obtained from an acidic *Sphagnum* tundra peatland in northern Russia (Dedysh et al., 2004). *M. tundrae* can grow at pH values between 4.2 and 7.5 and temperatures between 5 and 30°C, and has a temperature optimum of 15°C.

There are increasing data indicating that type I methanotrophs of the genera *Methylobacter* and *Methylosarcina* dominate Arctic permafrost-affected soils that have moderate pH values (Berestovskaya et al., 2002; Wartiainen et al., 2003; Wagner et al., 2005; Liebner and Wagner, 2007; Martineau et al., 2010). Type I, in contrast to type II methanotrophs, were also detected in the first metagenomic study on active-layer and permafrost samples (Yergeau et al., 2010). In particular, *M. tundripaludum* is believed to be a key player within the active methanotrophic community in pH-neutral Arctic wetland soils (Graef et al., 2011), which made the genome of *M. tundripaludum* a candidate for sequencing. Specific clusters of MOP closely related to *M. psychrophilus* and *M. tundripaludum* were also detected in active-layer samples from the Siberian Lena Delta (Liebner et al., 2008).

The diversity and composition of MOP in active-layer soils has only begun to be characterized. It remains unknown whether psychrophilic or psychrotrophic methanotrophs are responsible for methane oxidation at low and subzero temperatures in active-layer or permafrost sediments (Trotsenko and Khmelenina, 2005). However, according to the temperature response of methanotrophic activity and the organisms' lipid composition, it is currently hypothesized that constantly very low temperatures favor a more psychrophilic and/or psychrotrophic methanotrophic community, while pronounced temperature fluctuations, as seen occurring in active-layer soils, seem to promote mesophilic methanotrophs (Liebner and Wagner, 2007; Mangelsdorf et al., 2009).

REFERENCES

Anisimov, O., and S. Reneva. 2006. Permafrost and changing climate: the Russian perspective. *Ambio* **35:**169–175.

Aselmann, I., and J. Crutzen. 1989. Global distribution of natural freshwater wetlands and rice paddies, their net primary productivity, seasonality and possible methane emissions. *J. Atmos. Chem.* **8:**307–358.

Belova, S. E., T. A. Pankratov, and S. N. Dedysh. 2006. Bacteria of the genus *Burkholderia* as a typical component of the microbial community of sphagnum peat bogs. *Microbiology* **75:**90–96.

Berestovskaia, I., A. M. Lysenko, T. P. Turova, and L. V. Vasil'eva. 2006. A psychrotolerant *Caulobacter* sp. from Russian polar tundra soil. *Mikrobiologiia* **75:**377–382. (In Russian.)

Berestovskaya, Y. Y., L. V. Vasil'eva, O. V. Chestnykh, and G. A. Zavarzin. 2002. Methanotrophs of the psychrophilic microbial community of the Russian Arctic tundra. *Microbiology* **71:**460–466.

Biasi, C., O. Rusalimova, H. Meyer, C. Kaiser, W. Wanek, P. Barsukov, H. Junger, and A. Richter. 2005. Temperature-dependent shift from labile to recalcitrant carbon sources of arctic heterotrophs. *Rapid Commun. Mass Spectrom.* **19:**1401–1408.

Boetius, A., K. Ravenschlag, C. J. Schubert, D. Rickert, F. Widdel, A. Gieseke, R. Amann, B. B. Jørgensen, U. Witte, and O. Pfannkuche. 2000. A marine microbial consortium apparently mediating anaerobic oxidation of methane. *Nature* **407:**623–626.

Bölter, M., N. Soethe, R. Horn, and C. Uhlig. 2005. Seasonal development of microbial activity in soils of northern Norway. *Pedosphere* **15:**716–727.

Bowman, J. P. 1999. The methanotrophs—the families *Methylococcaceae* and *Methylocystaceae*. In M. Dworkin (ed.), *The Prokaryotes*. Springer-Verlag, New York, NY.

Buckeridge, K. M., Y. P. Cen, D. B. Layzell, and P. Grogan. 2010. Soil biogeochemistry during the early spring in low arctic mesic tundra and the impacts of deepened snow and enhanced nitrogen availability. *Biogeochemistry* 99:127–141.

Camill, P. 1999. Patterns of boreal permafrost peatland vegetation across environmental gradients sensitive to climate warming. *Can. J. Bot.* 77:721–733.

Chan, O. C., P. Claus, P. Casper, A. Ulrich, T. Lueders, and R. Conrad. 2005. Vertical distribution of the methanogenic archaeal community in Lake Dagow sediment. *Environ. Microbiol.* 7:1139–1149.

Clemmensen, K. E., P. L. Sorensen, A. Michelsen, S. Jonasson, and L. Ström. 2008. Site-dependent N uptake from N-form mixtures by arctic plants, soil microbes and ectomycorrhizal fungi. *Oecologia* 155:771–783.

Conrad, R. 2005. Quantification of methanogenic pathways using stable carbonisotopic signatures: a review and a proposal. *Organ. Geochem.* 36:739–752.

D'Amico, S., T. Collins, J.-C. Marx, G. Feller, and C. Gerday. 2006. Psychrophilic microorganisms: challenges for life. *EMBO Rep.* 7:385–389.

Dedysh, S. N., Y. Y. Berestovskaya, L. V. Vasylieva, S. E. Belova, V. N. Khmelenina, N. E. Suzina, Y. A. Trotsenko, W. Liesack, and G. A. Zavarzin. 2004. *Methylocella tundrae* sp nov., a novel methanotrophic bacterium from acidic tundra peatlands. *Int. J. Syst. Evol. Microbiol.* 54:151–156.

Denman, K. L., G. Brasseur, A. Chidthaisong, P. Ciais, P. M. Cox, R. E. Dickinson, D. Hauglustaine, C. Heinze, E. Holland, D. Jacob, U. Lohmann, S. Ramachandran, P. L. Da Silva Dias, S. C. Wofsy, and X. Zhang. 2007. Couplings between changes in the climate system and biogeochemistry. *In* IPCC, *Climate Change 2007: the Physical Science Basis. Contribution of Working Group I to the Fourth Assessment Report of the Intergovernmental Panel on Climate Change*. Cambridge University Press, Cambridge, United Kingdom, and New York, NY.

Dobrovolskaya, T. G., L. V. Lysak, and D. G. Zvyagintsev. 1996. Soils and microbial diversity. *Euras. Soil Sci.* 29:630–634.

Dunfield, P. F., A. Yuryev, P. Senin, A. V. Smirnova, M. S. Stott, S. Hou, B. Ly, J. H. Saw, Z. Zhou, Y. Ren, J. Wang, B. W. Mountain, M. A. Crowe, T. M. Weatherby, P. L. E. Bodelier, W. Liesack, L. Feng, L.

Wang, and M. Alam. 2007. Methane oxidation by an extremely acidophilic bacterium of the phylum Verrucomicrobia. *Nature* 450:879–882.

Eskelinen, A., S. Stark, and M. Männistö. 2009. Links between plant community composition, soil organic matter quality and microbial communities in contrasting tundra habitats. *Oecologia* 161:113–123.

Eugster, W., W. R. Rouse, R. A. Pielke, J. P. McFadden, D. D. Baldocchi, T. G. F. Kittel, F. S. Chapin III, G. Liston, P. L. Vidale, E. Vaganov, and S. Chambers. 2000. Land-atmosphere energy exchange in Arctic tundra and boreal forest: available data and feedbacks to climate. *Global Change Biol.* 6:84–115.

Eugster, W., J. P. McFadden, and F. S. Chapin. 2005. Differences in surface roughness, energy, and CO_2 fluxes in two moist tundra vegetation types, Kuparuk watershed, Alaska, USA. *Arct. Antarct. Alpine Res.* 37:61–67.

Franzmann, P. D., N. Springer, W. Ludwig, E. Conway de Macario, and M. Rohde. 1992. A methanogenic archaeon from Ace Lake, Antarctica: *Methanococcoides burtonii* sp. nov. *Syst. Appl. Microbiol.* 15:573–581.

Franzmann, P. D., Y. Liu, D. L. Balkwill, H. C. Aldrich, E. Conway de Macario, and D. R. Boone. 1997. *Methanogenium frigidium* sp. nov., a psychrophilic, H_2-using methanogen from Ace Lake, Antarctica. *Int. J. Sys. Bacteriol.* 47:1068–1072.

Ganzert, L., G. Jurgens, U. Münster, and D. Wagner. 2007. Methanogenic communities in permafrost-affected soils of the Laptev Sea coast, Siberian Arctic, characterized by 16S rRNA gene fingerprints. *FEMS Microbiol. Ecol.* 59:476–488.

Gilichinsky, D. A. 2002. Permafrost, p. 2367–2385. *In* G. Bitton (ed.) *Encyclopedia of Environmental Microbiology*. John Wiley and Sons, New York, NY.

Gilichinsky, D. A., V. S. Soina, and M. A. Petrova. 1993. Cryoprotective properties of water in the Earth cryollitosphere and its role in exobiology. *Orig. Life Evol. Biosph.* 23:65–75.

Gilichinsky, D., T. Vishnivetskaya, M. Petrova, E. Spirina, V. Mamykin, and E. Rivkina. 2008. Bacteria in permafrost, p. 83–102. *In* R. Margesin, F. Schinner, J.-C. Marx, and C. Gerday (ed.), *Psychrophiles: from Biodiversity to Biotechnology*. Springer, Berlin, Germany.

Gorham, E. 1991. Northern peatlands: role in the carbon cycle and probable responses to climatic warming. *Ecol. Appl.* 1:182–195.

Gough, L., G. R. Shaver, J. Carroll, D. L. Royer, and J. A. Laundre. 2000. Vascular plant species richness in Alaskan arctic tundra: the importance of soil pH. *J. Ecol.* 88:54–66.

Graef, C., A. G. Hestnes, M. M. Svenning, and P. Frenzel. 2011. The active methanotrophic community in a wetland from the High Arctic. *Environ. Microbiol. Rep.* **3:**466–472.

Grosse, G., L. Schirrmeister, and T. J. Malthus. 2006. Application of Landsat-7 satellite data and a DEM for the quantification of thermokarst-affected terrain types in the periglacial Lena-Anabar coastal lowland. *Polar Res.* **25:**51–67.

Grosskopf, R., S. Stubner, and W. Liesack. 1998. Novel euryarchaeotal lineages detected on rice roots and in the anoxic bulk soil of flooded rice microcosms. *Appl. Environ. Microbiol.* **64:**4983–4989.

Guggenberger, G., W. Zech, and H.-R. Schulten. 1994. Formation and mobilization pathways of dissolved organic matter: evidence from chemical structural studies of organic matter fractions in acid forest floor solutions. *Org. Geochem.* **21:**51–66.

Hales, B. A., C. Edwards, D. A. Ritchie, G. Hall, R. W. Pickup, and J. R. Saunders. 1996. Isolation and identification of methanogen-specific DNA from blanket bog peat by PCR amplification and sequence analysis. *Appl. Environ. Microbiol.* **62:**668–675.

Hansen, A. A., R. A. Herbert, K. Mikkelsen, L. L. Jensen, T. Kristoffersen, J. M. Tiedje, B. A. Lomstein, and K. W. Finster. 2007. Viability, diversity and composition of the bacterial community in a high Arctic permafrost soil from Spitsbergen, Northern Norway. *Environ. Microbiol.* **9:**2870–2884.

Hedderich, R., and W. Whitman. 2006. Physiology and biochemistry of the methane-producing archaea, p. 1050–1079. *In* M. Dworkin, S. Falkow, E. Rosenberg, K.-H. Schleifer, and E. Stackebrandt (ed.), *The Prokaryotes: a Handbook on the Biology of Bacteria*, 3rd ed., vol. 2. Springer, New York, NY.

Hobbie, S. E., and L. Gough. 2004. Litter decomposition in moist acidic and non-acidic tundra with different glacial histories. *Oecologia* **140:**113–124.

Hobbie, J. E., E. A. Hobbie, H. Drossman, M. Conte, J. C. Weber, J. Shamhart, and M. Weinrobe. 2009. Mycorrhizal fungi supply nitrogen to host plants in Arctic tundra and boreal forests: ¹⁵N is the key signal. *Can. J. Microbiol.* **55:**84–94.

Høj, L., R. A. Olsen, and V. L. Torsvik. 2005. Archaeal communities in High Arctic wetlands at Spitsbergen, Norway (78°N) as characterised by 16S rRNA gene fingerprinting. *FEMS Microbiol. Ecol.* **53:**89–101.

Høj, L., M. Rusten, L. E. Haugen, R. A. Olsen, and V. L. Torsvik. 2006. Effects of water regime on archaeal community composition in Arctic soils. *Environ. Microbiol.* **8:**984–996.

IPCC. 2007. *Climate Change 2007: the Physical Science Basis. Contribution of Working Group I to the Fourth Assessment Report of the Intergovernmental Panel on Climate Change*. Cambridge University Press, Cambridge, United Kingdom, and New York, NY.

Jeanthon, C., S. L'Haridon, N. Pradel, and D. Prieur. 1999. Rapid identification of hyperthermophilic methanococci isolated from deep-sea hydrothermal vents. *Int. J. Syst. Bacteriol.* **49:**591–594.

Jonasson, S., J. Castro, and A. Michelsen. 2004. Litter, warming and plants affect respiration and allocation of soil microbial and plant C, N and P in arctic mesocosms. *Soil Biol. Biochem.* **36:**1129–1139.

Jones, A., V. Stolbovoy, C. Tarnocai, G. Broll, O. Spaargaren, and L. Montanarella (ed.). 2010. *Soil Atlas of the Northern Circumpolar Region*. European Commission, Publications Office of the European Communities, Luxembourg.

Jurgens, G., F. O. Glöckner, R. Amann, A. Saano, L. Montonen, M. Likolammi, and U. Münster. 2000. Identification of novel Archaea in bacterioplankton of a boreal forest lake by phylogenetic analysis and fluorescent in situ hybridization. *FEMS Microbiol. Ecol.* **34:**45–56.

Kaiser, K., G. Guggenberger, L. Haumeier, and W. Zech. 2001. Seasonal variations in the chemical composition of dissolved organic matter in organic forest floor layer leachates of old-growth Scots pine (*Pinus sylvestris* L.) and European beech (*Fagus sylvatica* L.) stand in northeastern Bavaria, Germany. *Biogeochemistry* **55:**103–143.

Kaluzhnaya, M. G., V. A. Makutina, T. G. Rusakova, D. V. Nikitin, V. N. Khmelenina, V. V. Dmitriev, and Y. A. Trotsenko. 2002. Methanotrophic communities in the soils of the Russian northern taiga and subarctic tundra. *Microbiology* **71:**223–227.

Keough, B. P., T. M. Schmidt, and R. E. Hicks. 2003. Archaeal nucleic acids in picoplankton from great lakes on three continents. *Microb. Ecol.* **46:**238–248.

Khmelenina, V. N., V. A. Makutina, M. G. Kaluzhnaya, E. M. Rivkina, D. A. Gilichinsky, and Y. A. Trotsenko. 2002. Discovery of viable methanotrophic bacteria in permafrost sediments of northeast Siberia. *Dokl. Biol. Sci.* **384:**235–237.

Knorr, W., I. C. Prentice, J. I. House, and E. A. Holland. 2005. Long-term sensitivity of soil carbon turnover to warming. Nature 433:298–301.

Kobabe, S., D. Wagner, and E. M. Pfeiffer. 2004. Characterisation of microbial community composition of a Siberian tundra soil by fluorescence in situ hybridisation. *FEMS Microbiol. Ecol.* **50:**13–23.

Koch, K., C. Knoblauch, and D. Wagner. 2009. Methanogenic community composition and anaerobic carbon turnover in submarine permafrost sediments of the Siberian Laptev Sea. *Environ. Microbiol.* **11**:657–668.

Kotelnikova, S., A. J. L. Macario, and K. Pedersen. 1998. *Methanobacterium subterraneum* sp. nov., a new alkaliphilic, eurythermic and halotolerant methanogen isolated from deep granitic groundwater. *Int. J. Syst. Bacteriol.* **48**:357–367.

Kreyling, J., C. Beierkuhnlein, and A. Jentsch. 2010. Effects of soil freeze-thaw cycles differ between experimental plant communities. *Basic Appl. Ecol.* **11**:65–75.

Krivushin, K. V., V. A. Shcherbakova, L. E. Petrovskaya, and E. M. Rivkina. 2010. *Methanobacterium veterum* sp. nov., from ancient Siberian permafrost. *Int. J. Sys. Evol. Microbiol.* **60**:455–459.

Lee, H., E. A. G. Schuur, and J. G. Vogel. 2010. Soil CO_2 production in upland tundra where permafrost is thawing. *J. Geophys. Res.* **115**:G01009.

Liebner, S., K. Rublack, T. Stuehrmann, and D. Wagner. 2008. Diversity of aerobic methanotrophic bacteria in a permafrost soil of the Lena Delta, Siberia. *Microb. Ecol.* **57**:25–35.

Liebner, S., and D. Wagner. 2007. Abundance, distribution and potential activity of methane oxidizing bacteria in permafrost soils from the Lena Delta, Siberia. *Environ. Microbiol.* **9**:107–117.

Liebner, S., J. Zeyer, D. Wagner, C. Schubert, E.-M. Pfeiffer, and C. Knoblauch. 2011. Methane oxidation associated with submerged brown mosses reduces methane emissions from Siberian polygonal tundra. *J. Ecol.* **99**:914–922.

Lin, C., L. Raskin, and D. A. Stahl. 1997. Microbial community structure in gastrointestinal tracts of domestic animals: comparative analyses using rRNA-targeted oligonucleotide probes. *FEMS Microbiol. Ecol.* **22**:281–294.

Lomans, B. P., R. Maas, R. Luderer, H. J. M. Op den Camp, A. Pol, C. van der Drift, and G. D. Vogels. 1999. Isolation and characterization of *Methanomethylovorans hollandica* gen. nov., sp. nov., isolated from freshwater sediment, a methylotrophic methanogen able to grow on dimethyl sulfide and methanethiol. *Appl. Environ. Microbiol.* **65**:3641–3650.

Mangelsdorf, K., E. Finsel, S. Liebner, and D. Wagner. 2009. Temperature adaptation of microbial communities in different horizons of Siberian permafrost-affected soils from the Lena Delta. *Chem. Erde* **69**:169–182.

Männistö, M. K., and M. M. Häggblom. 2006. Characterization of psychrotolerant heterotrophic bacteria from Finnish Lapland. *Syst. Appl. Microbiol.* **29**:229–243.

Männistö, M. K., M. Tiirola, and M. M. Häggblom. 2009. Effect of freeze-thaw cycles on bacterial communities of Arctic tundra soil. *Microb. Ecol.* **58**:621–631.

Martineau, C., L. G. Whyte, and C. W. Greer. 2010. Stable isotope analysis of the diversity and activity of methanotrophic bacteria in soils from the Canadian high Arctic. *Appl. Environ. Microbiol.* **76**:5773–5784.

Mathrani, I. M., and D. R. Boone. 1985. Isolation and characterization of a moderately halophilic methanogen from a solar saltern. *Appl. Environ. Microbiol.* **50**:140–143.

Metje, M., and P. Frenzel. 2007. Methanogenesis and methanogenic pathways in a peat from subarctic permafrost. *Environ. Microbiol.* **9**:954–964.

Morozova, D., D. Möhlmann, and D. Wagner. 2007. Survival of methanogenic archaea from Siberian permafrost under simulated Martian thermal conditions. *Orig. Life Evol. Biosph.* **37**:189–200.

Morozova, D., and D. Wagner. 2007. Stress response of methanogenic archaea from Siberian permafrost compared to methanogens from nonpermafrost habitats. *FEMS Microbiol. Ecol.* **61**:16–25.

Muhr, J., W. Borken, and E. Matzner. 2009. Effects of soil frost on soil respiration and its radiocarbon signature in a Norway spruce forest soil. *Glob. Change Biol.* **15**:782–793.

Nelson, D. M., A. J. Glawe, D. P. Labeda, I. K. Cann, and R. I. Mackie. 2009. *Paenibacillus tundrae* sp. nov. and *Paenibacillus xylanexedens* sp. nov., psychrotolerant, xylan-degrading bacteria from Alaskan tundra. *Int. J. Syst. Evol. Microbiol.* **59**:1708–1714.

Neufeld, J. D., and W. W. Mohn. 2005. Unexpectedly high bacterial diversity in Arctic tundra relative to boreal forest soils, revealed by serial analysis of ribosomal sequence tags. *Appl. Environ. Microbiol.* **71**:5710–5718.

Ochsenreiter, T., D. Selezi, A. Quaiser, L. Bonch-Osmolovskaya, and C. Schleper. 2003. Diversity and abundance of *Crenarchaeota* in terrestrial habitats studied by 16S RNA surveys and real time PCR. *Environ. Microbiol.* **5**:787–797.

Oechel, W. C., S. J. Hastings, M. Jenkins, G. Riechers, N. E. Grulke, and G. L. Vourlitis. 1993. Recent change of arctic tundra ecosystems from a net carbon sink to a source. *Nature* **361**:520–526.

Oechel, W. C., and G. L. Vourlitis. 1994. The effects of climate charge on land-atmosphere feedbacks in arctic tundra regions. *Trends Ecol. Evol.* **9**:324–329.

Oechel, W. C., G. L. Vourlitis, S. J. Hastings, R. C. Zulueta, L. Hinzman, and D. Kane. 2000. Acclimation of ecosystem CO_2 exchange in the Alaskan Arctic in response to decadal climate warming. *Nature* **406**:978–981.

Oelbermann, M., M. English, and S. L. Schiff. 2008. Evaluating carbon dynamics and microbial activity in arctic soils under warmer temperatures. *Can. J. Soil Sci.* **88**:31–44.

Omelchenko, M. B., L. V. Vasieleva, G. A. Zavarzin, N. D. Savelieva, A. M. Lysenko, L. L. Mityushina, V. N. Khmelenina, and Y. A. Trotsenko. 1996. A novel psychrophilic methanotroph of the genus *Methylobacter. Microbiology* **65**:339–343.

Osterkamp, T. E. 2001. Subsea permafrost, p. 2902–2912. *In* J. H. Steele, S. A. Thorpe, and K. K. Turekian (ed.), *Encyclopedia of Ocean Sciences.* Academic Press, San Diego, CA.

Parinkina, O. M. 1989. Microflora of tundra soils: ecological geographical features and productivity. Nauka, Leningrad, Russia.

Pol, A., K. Heijmans, H. R. Harhangi, D. Tedesco, M. S. M. Jetten, and H. J. M. Op den Camp. 2007. Methanotrophy below pH 1 by a new *Verrucomicrobia* species. *Nature* **450**:874–878.

Post, W. M., W. R. Emanuel, P. J. Zinke, and A. G. Stangenberger. 1982. Soil carbon pools and world life zones. *Nature* **298**:156–159.

Raghoebarsing, A. A., A. Pol, K. T. van de Pas-Schoonen, A. J. P. Smolders, K. F. Ettwig, W. I. C. Rijpstra, S. Schouten, J. S. Sinninghe Damsté, H. J. M. Op den Camp, M. S. M. Jetten, and M. Strous. 2006. A microbial consortium couples anaerobic methane oxidation to dentrification. *Nature* **440**:918–921.

Ramakrishnan, B., T. Lueders, P. F. Dunfield, R. Conrad, and M. W. Friedrich. 2001. Archaeal community structures in rice soils from different geographical regions before and after initiation of methane production. *FEMS Microbiol. Ecol.* **37**:175–186.

Rinnan, R., A. Michelsen, E. Bååth, and S. Jonasson. 2007. Fifteen years of climate change manipulations alter soil microbial communities in a subarctic heath ecosystem. *Glob. Change Biol.* **13**:28–39.

Rinnan, R., S. Stark, and A. Tolvanen. 2009. Response of vegetation and soil microbial communities to warming and simulated herbivory in a subarctic heath. *J. Ecol.* **97**:788–800.

Rivkina, E. M., D. Gilichinsky, S. Wagener, J. Tiedje, and J. McGrath. 1998. Biochemical activity of anaerobic microorganisms from buried permafrost sediments. *Geomicrobiology* **15**:187–193.

Russell, N. J. 2000. Cold shock and cold acclimation in cold-adapted bacteria. *Mol. Integr. Physiol.* **126**:130–134.

Sawicka, J. E., A. Robador, C. Hubert, B. B. Jorgensen, and V. Bruchert. 2010. Effects of freeze-thaw cycles on anaerobic microbial processes in an Arctic intertidal mud flat. *ISME J.* **4**:585–594.

Schink, B., and A. J. M. Stams. 2006. Syntrophism among prokaryotes, p. 309–335. *In* M. Dworkin, S. Falkow, E. Rosenberg, K.-H. Schleifer, and E. Stackebrandt (ed.), *The Prokaryotes: a Handbook on the Biology of Bacteria,* 3rd ed., vol. 2. Springer, New York, NY.

Schuur, E. A. G., J. Bockheim, J. C. Canadell, E. Euskirchen, C. B. Field, S. V. Goryachkin, S. Hagemann, P. Kuhry, P. M. Lafleur, H. Lee, G. Mazhitova, F. E. Nelson, A. Rinke, V. E. Romanovsky, N. Shiklomanov, C. Tarnocai, S. Venevsky, J. G. Vogel, and S. A. Zimov. 2008. Vulnerability of permafrost carbon to climate change: implications for the global carbon cycle. *BioScience* **58**:701–714.

Shlimon, A. G., M. W. Friedrich, H. Niemann, N. B. Ramsing, and K. Finster. 2004. *Methanobacterium aarhusense* sp. nov., a novel methanogen isolated from a marine sediment (Aarhus Bay, Denmark). *Int. J. Syst. Evol. Microbiol.* **54**:759–763.

Sjögersten, S., and P. A. Wookey. 2009. The impact of climate change on ecosystem carbon dynamics at the Scandinavian mountain birch forest-tundra heath ecotone. *Ambio* **38**:2–10.

Steven, B., G. Briggs, C. P. McKay, W. H. Pollard, C. W. Greer, and L. G. Whyte. 2007. Characterization of the microbial diversity in a permafrost sample from the Canadian high Arctic using culture-dependent and culture-independent methods. *FEMS Microbiol. Ecol.* **59**:513–523.

Steven, B., R. Léveillé, W. H. Pollard, and L. G. Whyte. 2006. Microbial ecology and biodiversity in permafrost. *Extremophiles* **10**:259–267.

Steven, B., W. H. Pollard, C. W. Greer, and L. G. Whyte. 2008. Microbial diversity and activity through a permafrost/ground ice core profile from the Canadian high Arctic. *Environ. Microbiol.* **10**:3388–3403.

Ström, L., and T. R. Christensen. 2007. Below ground carbon turnover and greenhouse gas exchanges in a sub-arctic wetland. *Soil Biol. Biochem.* **39**:1689–1698.

Tarnocai, C. 2006. The effect of climate change on carbon in Canadian peatlands. *Glob. Planet. Change* **53**:222–232.

Tarnocai, C., J. G. Canadell, E. A. G. Schuur, P. Kuhry, G. Mazhitova, and S. Zimov. 2009. Soil organic carbon pools in the northern circumpolar permafrost region. *Glob. Biogeochem. Cycles* **23**:GB2023.

Trotsenko, Y. A., and V. N. Khmelenina. 2005. Aerobic methanotrophic bacteria of cold ecosystems. *FEMS Microbiol. Ecol.* **53**:15–26.

Trumbore, S. E., O. A. Chadwick, and R. Amundson. 1996. Rapid exchange between soil carbon and atmospheric carbon dioxide driven by temperature change. *Science* **272**:393–396.

Vasilyeva, L. V., M. V. Omelchenko, Y. Y. Berestovskaya, A. M. Lysenko, W. R. Abraham, S. N. Dedysh, and G. A. Zavarzin. 2006. *Asticcacaulis benevestitus* sp. nov., a psychrotolerant, dimorphic, prosthecate bacterium from tundra wetland soil. *Int. J. Syst. Evol. Microbiol.* **56**:2083–2088.

Vishnivetskaya, T., S. Kathariou, J. McGrath, D. Gilichinsky, and J. M. Tiedje. 2000. Low-temperature recovery strategies for the isolation of bacteria from ancient permafrost sediments. *Extremophiles* **4**:165–173.

Vorobyova, E., V. Soina, M. Gorlenko, N. Minkovskaya, N. Zalinova, A. Mamukelashvili, D. Gilichinsky, E. Rivkina, and T. Vishnivetskaya. 1997. The deep cold biosphere: facts and hypothesis. *FEMS Microbiol. Rev.* **20**:277–290.

Wagner, D. 2008. Microbial communities and processes in Arctic permafrost environments, p. 133–154. *In* P. Dion and C. S. Nautiyal (ed.) *Microbiology of Extreme Soils.* Springer, Berlin, Germany.

Wagner, D., A. Gattinger, A. Embacher, E. M. Pfeiffer, M. Schloter, and A. Lipski. 2007. Methanogenic activity and biomass in Holocene permafrost deposits of the Lena Delta, Siberian Arctic and its implication for the global methane budget. *Glob. Change Biol.* **13**:1089–1099.

Wagner, D., S. Kobabe, and S. Liebner. 2009. Bacterial community structure and carbon turnover in permafrost-affected soils of the Lena Delta, northeastern Siberia. *Can. J. Microbiol.* **55**:73–83.

Wagner, D., S. Kobabe, E. M. Pfeiffer, and H.-W. Hubberten. 2003. Microbial controls on methane fluxes from a polygonal tundra of the Lena Delta, Siberia. *Permafrost Periglacial Processes* **14**:173–185.

Wagner, D., and S. Liebner. 2009. Global warming and carbon dynamics in permafrost soils: methane production and oxidation, p. 219–236. *In* R. Margesin (ed.), *Permafrost Soils.* Springer, Berlin, Germany.

Wagner, D., A. Lipski, A. Embacher, and A. Gattinger. 2005. Methane fluxes in extreme permafrost habitats of the Lena Delta: effects of microbial community structure and organic matter quality. *Environ. Microbiol.* **7**:1582–1592.

Walker, M. D., C. H. Wahren, R. D. Hollister, G. H. Henry, L. E. Ahlquist, J. M. Alatalo, M. S. Bret-Harte, M. P. Calef, T. V. Callaghan, A. B. Carroll, H. E. Epstein, I. S. Jonsdottir, J. A. Klein, B. Magnusson, U. Molau, S. F. Oberbauer, S. P. Rewa, C. H.

Robinson, G. R. Shaver, K. N. Suding, C. C. Thompson, A. Tolvanen, O. Totland, P. L. Turner, C. E. Tweedie, P. J. Webber, and P. A. Wookey. 2006a. Plant community responses to experimental warming across the tundra biome. *Proc. Natl. Acad. Sci. USA* **103**:1342–1346.

Walker, V. K., G. R. Palmer, and G. Voordouw. 2006b. Freeze-thaw tolerance and clues to the winter survival of a soil community. *Appl. Environ. Microbiol.* **72**:1784–1792.

Wartiainen, I., A. G. Hestnes, I. R. McDonald, and M. M. Svenning. 2006a. *Methylobacter tundripaludum* sp. nov., a novel methanotrophic bacterium from Arctic wetland soil, Svalbard, Norway (78° N). *Int. J. Syst. Evol. Microbiol.* **56**:109–113.

Wartiainen, I., A. G. Hestnes, I. R. McDonald, and M. M. Svenning. 2006b. *Methylocystis rosea* sp. nov., a novel methanotrophic bacterium from Arctic wetland soil, Svalbard, Norway (78° N). *Int. J. Syst. Evol. Microbiol.* **56**:541–547.

Wartiainen, I., A. G. Hestnes, and M. M. Svenning. 2003. Methanotrophic diversity in high arctic wetlands on the islands of Svalbard (Norway)—denaturing gradient gel electrophoresis analysis of soil DNA and enrichment cultures. *Can. J. Microbiol.* **49**:602–612.

Whitman, W. B., D. C. Coleman, and W. J. Wiebe. 1998. Prokaryotes: the unseen majority. *Proc. Natl. Acad. Sci. USA* **95**:6578–6583.

Wuebbles, J., and K. Hayhoe. 2002. Atmospheric methane and global change. *Earth Sci. Rev.* **57**:177–210.

Yergeau, E., H. Hogues, L. G. Whyte, and C. W. Greer. 2010. The functional potential of high Arctic permafrost revealed by metagenomic sequencing, qPCR, and microarray analyses. *ISME J.* **4**:1206–1214.

Zak, D. R., and G. W. Kling. 2006. Microbial community composition and function across an arctic tundra landscape. *Ecology* **87**:1659–1670.

Zhang, T., R. G. Barry, K. Knowles, J. A. Heginbotton, and J. Brown. 1999. Statistics and characteristics of permafrost and ground-ice distribution in the Northern Hemisphere. *Polar Geogr.* **23**:132–154.

Zhou, J., M. E. Davey, J. B. Figueras, E. Rivkina, D. Gilichinsky, and J. M. Tiedje. 1997. Phylogenetic diversity of a bacterial community determined from Siberian tundra soil DNA. *Microbiology* **143**:3913–3919.

Zimov, S. A., E. A. G. Schuur, and F. S. Chapin III. 2006. Permafrost and the global carbon budget. *Science* **312**:1612–1613.

POLAR MARINE MICROBIOLOGY

10

INTRODUCTION

At the highest taxonomic levels, microbial communities in the polar oceans are similar to those in temperate oceans, and contain diverse representatives from the three domains of life: *Eukarya* (protists), *Bacteria*, and *Archaea*. One exception is that *Cyanobacteria*, which dominate photosynthetic productivity within the small size class (cells 3.0 to 0.2 μm) in temperate and tropical oceans, are largely absent from polar seas (Vincent, 2000), and in the Arctic are replaced by small photosynthetic protists (eukaryotic phytoplankton). The perennially cold surface waters and extreme seasonality provide conditions for maintaining distinctly polar microbial flora, and at lower taxonomic levels there are clear differences between temperate and polar oceans. Polar oceans also provide an opportunity to explore questions about microbial biogeography and latitudinal diversity gradients. For example, a number of diatom species are characteristic of polar marine waters, and recent surveys of the small-subunit rRNA gene suggest that other microbial groups also have polar species or ecotypes (Lovejoy and Potvin, 2011) Seasonality is pronounced at high latitudes, and eukaryotic phytoplankton follow seasonal trends in the upper mixed layer, with light availability the most important factor determining the onset of spring productivity. Diatoms are a major component of the phytoplankton and dominate photosynthetic production in many regions where they follow the seasonal ice edge. Diatoms are also are frequently associated with blooms following upwelling events, and polar seas have been thought to have short food chains compared to other oceans. Small, functionally and phylogenetically diverse flagellates occur throughout the year at all depths and persist over winter. Ciliates, dinoflagellates, and other protozooplankton are also common throughout the year and prey on the small flagellates as well as diatoms. Little is known about the seasonality of *Bacteria* and *Archaea*; however, as with small flagellates, they persist throughout the year throughout the water column.

Physical oceanic processes over horizontal (geographic) and vertical distances (depth) set up very different ecological conditions for microbial communities, and the various polar seas differ in their seasonal dynamics, annual productivity, and microbial community structure. What is clear is that microbes persist within water masses and are advected into and out of different geographical regions. Where

Connie Lovejoy, Département de Biologie, Institut de biologie intégrative et des systèmes (IBIS)/Québec-Océan, Université Laval, Québec City, QC G1V 0A6, Canada.

Polar Microbiology: Life in a Deep Freeze
Edited by Robert V. Miller and Lyle G. Whyte © 2012 ASM Press, Washington, DC

conditions change slowly, microbial communities may change little over vast geographical areas within horizontally flowing currents. On the other hand, if water masses are subjected to physical forces such as upwelling, downwelling, and latitudinal flow, environmental conditions may change dramatically and result in community changes.

Because of logistical constraints, seasonal dynamics of microbial communities are better known inshore compared to offshore for both the Antarctic and Arctic. While the bias from nearshore data presents a challenge, the short summer growing season, the lack of photosynthetically available radiation for much of the year, and the formation and melting of sea ice are defining characteristics of polar marine systems. Sea ice communities themselves are dealt with in Chapter 11. This chapter presents a brief overview of pelagic microbes and their diversity, vertical distribution, and influences on biogeochemistry and upper food webs. The focus is on recent advances following the application of molecular biological techniques to polar marine systems.

POLAR MICROBIOLOGY

The types of microbes found in polar oceans can be explored at various taxonomic levels, ranging from domain and phylum level down to species, phylotype, or even ecotype. Historically, the identity and diversity of bacteria in the sea were considered largely unknowable, and hence of little interest. Bacteria were classified by metabolic characteristics, which meant that a pure culture was required, and most marine microbes remain largely uncultivated (Giovannoni et al., 2007). A revolution has occurred with the application of molecular biological techniques, especially small-subunit rRNA gene surveys, and identification of *Bacteria*, *Archaea*, and picoeukaryotes (ca. <3 μm) is now possible. Such surveys first highlighted the diversity of microbes in the sea (Fuhrman and Campbell, 1998), and questions of whether there is functional redundancy in marine microbial systems or whether rare organisms carry out key processes remain hotly debated (Sogin

et al., 2006; Galand et al., 2009a; Yooseph et al., 2010). A major discovery was that nonextremophile *Archaea* were ubiquitous in the marine waters (Delong, 1992; Fuhrman et al., 1992) and that they were found at significantly high concentrations throughout the water column, including waters surrounding Antarctica (Massana et al., 1998) and the Arctic (Bano et al., 2004). Other advances in molecular marine microbiology have highlighted the previously unappreciated metabolic complexity and diversity of marine microbial communities (Yooseph et al., 2010). The Global Ocean Sampling Expedition in particular brought attention to how little was known about the metabolic and taxonomic diversity of even just surface-ocean microbes (Rusch et al., 2007). In particular, newly discovered carbon and nitrogen pathways (Francis et al., 2007) challenge our understanding of these cycles (Ward et al., 2007). Identifying the *Bacteria*, *Archaea*, and small eukaryotes in the microbial black boxes also provides a means of comparing communities from different oceans and depths, opening the possibility of discovering recurrent patterns of association among microbes. Such associations may indicate tight interdependence of communities in resource-limited environments (Farnelid et al., 2010) or other mutualistic associations such as those reported in more extreme environments (Muller et al., 2010). Among the most exciting challenges facing microbial ecologists today is understanding how the diversity of and interactions within microbial food webs affect carbon and energy cycling.

Whether the same or closely related phylotypes occur at both poles most likely depends on the survivability and life history of different groups. For example, the dinoflagellate *Polarella glacialis*, which appears to be eurytolerant and can form cysts, has been isolated from both polar regions (Montresor et al., 2003). But the Arctic ecotype of a naked prasinophyte, a *Micromonas* sp. (Lovejoy et al., 2007), with no known cyst stage, has never been recovered from Antarctic seas. Other marine microbes are likely to have bipolar distributions, as has been recently found for terrestrial cyanobacteria (Jungblut et al., 2010). At the functional level of the gene, bipolar distribution seems clear; for example, ammonia monooxygenase sequences >99% similar were recovered from both poles (Kalanetra et al., 2009). However, whether this reflects species and implied genetic exchange or the conserved nature of the genes being investigated will require single-cell sequencing and further cultivation of isolates from both poles. Here I provide a brief survey of polar microbes, mostly at higher taxonomic levels. Comparative studies at lower taxonomic levels will require agreement on the definition of a species or ecotype and well-thought-out global surveys.

Bacteria

The diversity and composition of polar bacteria have a direct effect on carbon and nitrogen cycles, with ramifications for greenhouse gas dynamics and other components of the microbial food web (Sarmento et al., 2010). Bacteria are also important for global sulfur cycling, with important feedbacks on the Arctic climate system (Vila-Costa et al., 2008). Heterotrophic bacteria primarily recycle carbon, and some data suggest that in the polar regions they process less organic carbon than in low-latitude oceans (Kirchman et al., 2009). Their activities are likely to be stimulated, however, in the vicinity of the large rivers that discharge into the Arctic Ocean, which deliver large quantities of dissolved and particulate organic carbon (Garneau et al., 2008).

Prior to the application of molecular techniques, all that was known about the identity of Bacteria in polar seas was from culture studies. Although sea ice bacteria are well represented in culture collections, few unequivocally planktonic polar bacteria have ever been cultured. Bacterial 16S rRNA gene surveys provide some information into geographic, seasonal, and depth distributions of Bacteria (Hollibaugh et al., 2007; Lovejoy et al., 2010). Most studies of bacterial diversity have relied on cloning and sequencing of the 16S rRNA gene, but the application of tag pyro-

sequencing without cloning has been applied to the Arctic (Kirchman et al., 2010; Galand et al., 2010) and the Antarctic (http://icomm. mbl.edu/). Those results indicate, as with most open ocean systems, that bacterial diversity is underestimated using more classical cloning and sequencing approaches. But whether or not diversity decreases with latitude has not been resolved, as more comparative studies are needed across all latitudes.

Fluorescent in situ hybridization (FISH) using taxon-specific probes and enumerating cells by microscopy is another method to explore diversity and distribution of specific taxonomic groups of microbes and has been applied in both the Arctic and Antarctic (Hollibaugh et al., 2007). A third approach is collecting DNA and sequencing the environmental metagenome (Rusch et al., 2007). This approach is expensive and computationally challenging and no polar marine studies have been published to date, although this is likely to change with the advent of new sequencing technologies.

The major groups of Bacteria found in low-latitude oceans are also seen in polar oceans. α-Proteobacteria, especially the SAR11 clade, generally dominate upper-ocean bacterial communities (Fuhrman and Hagström, 2008) but are relatively less common in polar waters. γ-Proteobacteria and Bacteroidetes are relatively more abundant in polar surface waters, often accounting for >20% of the total community (Lovejoy et al., 2010). In contrast, those two groups usually account for <10% of total bacterial diversity in other open-ocean surface waters (Rusch et al., 2007). Interestingly, ice-based bacterial communities are also relatively rich in γ-Proteobacteria and Bacteroidetes (Collins et al., 2010). These groups may be important for the degradation of high-molecular-weight exopolymers (Kirchman et al., 2009) associated with sea ice and sinking particles. However, the association with sea ice may also indicate that the high proportion of γ-Proteobacteria and Bacteroidetes is due to geographical and seasonal sampling biases. Indeed, recent studies have found α-Proteobacteria to be more abundant in oligotrophic Arctic basins, and communities were more similar to open-ocean low-latitude communities than to bacterial communities in more productive coastal Arctic waters (Lovejoy et al., 2010; Galand et al., 2010).

Although the largest role of marine bacteria is the degradation of organic matter, the widespread distribution of proteorhodopsin in the sea among diverse bacterial phyla (Beja et al., 2002) suggests that many marine bacteria supplement their energy requirements using light (Gómez-Consarnau et al., 2010). Proteorhodopsin genes have been reported from nearshore Arctic waters (Cottrell and Kirchman, 2009) and Antarctic sea ice (Koh et al., 2010), but the relative importance of proteorhodopsin in polar regions remains to be determined.

Polar seas are well oxygenated, and bacterial chemosynthetic primary production has not been widely studied. Methane production has been reported in ice-covered waters (Damm et al., 2010), suggesting seasonal patterns of microbial activity apart from heterotrophic pathways. Extensive exopolymer production and sinking particles during productive periods (Juul-Pedersen et al., 2010) may provide an anaerobic refuge for chemosynthetic microbes. 16S rRNA gene sequences associated with aerobic methanotrophs, though rare, have also been recovered from polar seas (Galand et al., 2010; Kirchman et al., 2010). Polar seas are likely to be similar to other marine systems where aerobic methanotrophs are credited with preventing the buildup of methane in surface waters (Tavormina et al., 2010). Cold methane and sulfide seeps have been reported on polar shelves (O'Regan and Moran, 2010; Romanovskii et al., 2005; Lichtschlag et al., 2010; Niemann et al., 2009), and chemosynthetic carbon fixation is probably locally significant, as it is elsewhere. Benthic and digenetic processes may also contribute to the suite of rare taxa found in water columns but are outside the scope of this chapter and will not be dealt with further.

Archaea

Archaea may be more important in polar seas compared to other systems. There is some evidence of relatively more Archaea (percentage of total prokaryote abundance) in Arctic surface waters than are found in low-latitude oceans, as determined by FISH-based studies (Wells et al., 2006; Kirchman et al., 2007; Garneau et al., 2006). The relative importance of Archaea is correlated to particle abundance in Arctic waters (Wells et al., 2006; Garneau et al., 2006). Archaea were reported to be more abundant in winter compared to summer in both Arctic (Alonso-Sáez et al., 2008) and Antarctic waters (Church et al., 2003; Murray et al., 1998). The two most abundant archaeal phyla in the ocean belong to the Euryarchaeota (Marine Groups [MG] II, III, and IV) and Thaumarchaeota (Spang et al., 2010) MGI, which were originally classified as part of another phylum, the Crenarchaeota. Although MGI are more often reported from deep waters and Euryarchaeota MGII from near the surface (Martin-Cuadrado et al., 2008), in most polar studies the opposite has been reported (Hollibaugh et al., 2007). The only free-living cultivated representative of MGI (Nitrosopumilus maritimus) is able to oxidize ammonia and fix inorganic carbon (Walker et al., 2010; Konneke et al., 2005), and the majority of Thaumarchaeota in the ocean appear to have the ammonia monooxygenase gene (amoA) involved in ammonium oxidation and nitrification (Francis et al., 2007). The archaeal amoA gene has been widely reported in the Arctic and Antarctica (Galand et al., 2009c; Kalanetra et al., 2009), suggesting the importance of nitrification and nonphotosynthetic inorganic carbon fixation in the upper water column of polar seas compared to other regions.

MGII Euryarchaeota are widespread and reported from throughout the world's oceans but have remained uncultivated. Although some are reported to take up amino acids (Ouverney and Fuhrman, 2000), their full metabolic capacity remains speculative, but they are likely chemolithotrophic as well (Martin-Cuadrado et al., 2008). Even less is known about MGIII Euryarchaeota, which are rare in the global oceans but appear to be common in the mesopelagic zone of the Arctic (Galand et al., 2009b) and have also been recovered from Antarctic waters (Martin-Cuadrado et al., 2008). As with MGII, there is no clear understanding of the functional role of these microorganisms in the sea, but if they are chemolithotrophic their sheer numbers suggest they could contribute to oceanic inorganic carbon fixation.

Marine Protists (*Eukarya*)

Marine protists in the pelagic zone include many photosynthetic and heterotrophic groups, as well as phylogenetically diverse mixotrophic groups, capable of both photosynthesis and phagotrophy. These photosynthetic protists (phytoplankton) are responsible for carbon fixation in polar seas via oxygenic photosynthesis. Heterotrophic protists graze on bacteria and other protists. All protists are food for zooplankton and hence support the higher trophic levels of polar marine ecosystems (Falk-Petersen et al., 2009).

Most of the strictest phototrophs are diatoms (*Bacillariophyta*). Because of the species-specific morphologies of their silica frustules (cell walls), diatoms have a well-defined taxonomy. However, while species identification is robust, diatom taxonomy at higher levels has been rearranged several times following technological advances, and molecular phylogenies are controversial (Medlin, 2009), as they sometimes contradict classical phylogenies. Classical texts divide diatoms into three major groups—rhaphid pennates (*Bacillariophyceae*), centrics (*Coscinodiscophyceae*), and araphid pennates (*Fragilariophyceae*)—based on valve structure (Round et al., 1990). The araphid species tend to occur in freshwaters. A generality of marine systems is that most pennate species are benthic, while centric species tend to occupy the water column. In polar regions pennates are also common in the ice (see Chapter 11); in addition, several ribbon-forming pennate species such as *Fragilariopsis kerguelensis* in Antarctica and other *Fragilariopsis* and *Navicula* spp. in the Arctic are usually found in the water column. Centric diatoms also occur in polar water columns, especially small *Chaetoceros* spp. (Lovejoy et al., 2002; Smetacek et al., 2002).

The diatoms of both polar regions have been well characterized, especially among ice algae (Medlin and Priddle, 1990; von Quillfeldt, 2000). Although many ice algae are reported from water column samples, these are likely transitory, and most photosynthetic activity is by specialist water column species (Lovejoy et al., 2002; Rozanska et al., 2008).

Most other phototrophic protists, including the *Prasinophyceae*, are likely to be mixotrophic (Bell and Laybourn-Parry, 2003; Sherr et al., 2003). Larger prasinophytes are easily identified and are common in both sea ice and the upper mixed layer of the water column (Daugbjerg, 2000; Lovejoy et al., 2002). They are also found in marine-derived Antarctic lakes (Bell and Laybourn-Parry, 2003). In the Arctic, one <2.0-μm-diameter, 18S rRNA ribotype of *Micromonas* is ubiquitous (Lovejoy et al., 2007) and may be the most abundant phototroph in the Arctic.

In many Antarctic waters *Haptophyceae* (Fig. 1) are periodically abundant, especially *Phaeocystis antarctica*, which can form blooms. *Phaeocystis* can occur as single cells or form large colonies similar to its temperate counterparts (Larsen et al., 2004). The oceanographic conditions that promote diatoms versus *Phaeocystis* blooms have been extensively studied, and the trade-off between light availability and nutrients is thought to be a major factor (Smith et al., 2010). Small *Haptophyceae* are also common in the Arctic, and in the western (Canadian) Arctic are mostly represented by single cells, which co-occur with *Micromonas*.

Other mixotrophic flagellates belong to a diverse array of higher taxa. Among the stramenopiles (*Heterokonta*) are *Dictyophyceae*, *Pelagophyceae*, *Raphidophyceae*, and *Chrysophyceae* (Scott and Marchant, 2005; Poulin et al., 2010), including large tree-shaped colonies of *Dinobryon balticum* that have been reported from the Arctic (Lovejoy et al., 2002). *Bolidophyceae* are also frequently recovered in Arctic 18S rRNA gene surveys. Recent work suggests that the *Parmales*, which have siliceous walls and are ubiquitous in electron microscopy studies of polar waters, are closely related

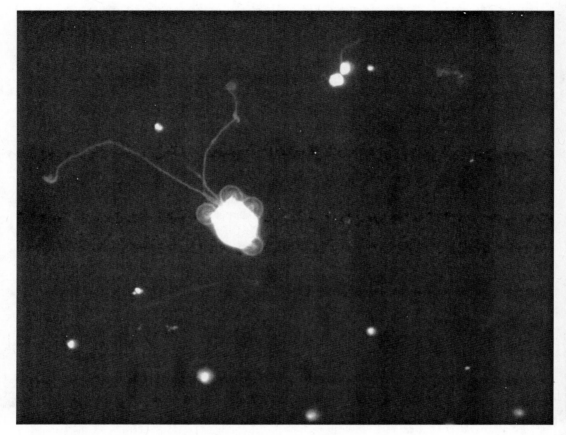

FIGURE 1 Epifluorescence micrograph of a mixotrophic haptophyte flagellate from the Beaufort Sea. 4′,6-Diamidino-2-phenylindole is used to stain nucleic acids, especially the nucleus. In this overexposed micrograph the flagella, haptonema, and organic scales can also be seen. The cell is ca. 10 × 5 microns.

to or within the flagellated *Bolidophyceae* (Ichinomiya et al., 2011).

Chlorarachniophytes, which are members of the *Cercozoa* with chlorophyll *b* (derived from a green algal secondary endosymbiosis), have been recovered from most surface 18S rRNA gene clone libraries in the Arctic (Lovejoy and Potvin, 2011). Other mixotrophs include members of the *Cryptophyceae* and *Euglenophyceae* (Lovejoy et al., 2006; Terrado et al., 2009; Lovejoy and Potvin, 2011). Also among the uncultivated flagellates are the enigmatic picobiliphytes (Not et al., 2007), which are now thought to be phagotrophic (Yoon et al., 2011). It is likely that the same major groups will be found in the Antarctic upper mixed

layer using 18S rRNA gene surveys, but none have been published to date.

Molecular 18S rRNA gene surveys have also highlighted the diversity and distribution of small heterotrophic protists (Lovejoy et al., 2010). Among the small flagellates, genetic surveys have confirmed the widespread distribution of phylogenetically diverse heterotrophic flagellates including choanoflagellates, Katablepharidophyta, *Telonemia*, cerozoans, diplomonads, and diverse marine stramenopiles (MASTs) (Massana et al., 2006). MASTs are only known from their rRNA gene sequences with the exception of one group (MAST-3). The ecology of several other MAST clades has been deduced using FISH, and they appear to be phagotrophic (Massana et al., 2004). A recent report that MAST-3 stramenophiles are related to an epibiont of algae (Gómez et al., 2011) suggests that more functional diversity may eventually be found within these clades. In the water column the diverse heterotrophic flagellates range in size from ≤2 to >20 μm, and feed on both bacteria and other protists, including phytoplankton.

Rhizaria are also frequently recovered from 18S rRNA gene libraries from the Arctic, with nearest identity to other environmental sequences from the deep ocean (Lovejoy and Potvin, 2011). These fragile cells are not well preserved from net or bottle samples, but their frequency in environmental gene surveys suggests they may be important phytoplankton predators in polar waters.

As in other oceanic regions, alveolates are common and very diverse. Dinoflagellates and ciliates have long been noted in microscopic surveys (Lovejoy et al., 2002; Okolodkov and Dodge, 1996) and are also frequently recovered in 18S rRNA gene surveys. While about half of known dinoflagellates are photosynthetic, it is likely that all prey on phytoplankton and other protists (Taylor et al., 2008). Ciliates also graze on phytoplankton, other protists, and each other (Montagnes et al., 2010). Gene surveys have also revealed several groups of uncultivated alveolates that mostly fall into two major clades (group I and group II alveolates) (López-García et al., 2001). These two groups are found in nearly all marine samples, including from the Arctic (Lovejoy et al., 2006) and Antarctic (López-García et al., 2001). These uncultivated alveolates are within, or related to, the parasitic *Syndiniales*. The most commonly recovered clade in the Arctic belongs to *Syndiniales* group II, which contains the dinoflagellate parasitoid *Amoebophyra*. All known representatives of the *Syndiniales* have complex life stages and are either parasitoid, parasitic, or commensally dependent on a host (Skovgaard et al., 2005). The diversity of these protists suggests that they are fast evolving, and many may be restricted to a single host (Guillou et al., 2008). These novel groups were first found in deep Antarctic waters (López-García et al., 2001), and in the Arctic they have been reported from all depths and seasons sampled to date when clone libraries were constructed using environmental DNA as a template.

Information on community composition of heterotrophic protists by microscopic analysis covers a much broader range of polar systems and includes data on cell abundances and biomass. However, taxonomic identification of protist species, especially the smallest flagellates and spheres, by microscopy is often less certain than identification by molecular genetics. A major challenge is to reconcile the morphological species information with environmental sequence information. Achieving this requires either extensive cultivation efforts or amplifying the appropriate marker genes from single cells, which is time-consuming and has not been carried out at this time in polar seas.

Viruses

Viruses have been retrieved from all oceanic regions and contribute to the top-down control of bacteria and protist populations (Suttle, 2007). Similarly, viruses have been reported from all polar marine systems investigated. In the Antarctic, Marchant et al. (2000) reported that viral abundance first decreased and then increased in a southward transect, with minimum concentrations in Antarctic circumpolar waters. Few process studies have been carried out, but several suggest differences between polar and lower-latitude seas (Payet and Suttle, 2008; Steward et al., 1996). The diversity and types of viruses found in the Arctic may also differ. Viral communities in the Arctic Ocean include many pan-ocean genotypes but appear to have somewhat lower diversity compared to viral communities in lower-latitude ocean regions (Angly et al., 2006). Metagenomic study of a composite Arctic virome indicated that prophage sequences are relatively more abundant in the Arctic compared to lower latitudes, and that cyanophages are rare (Angly et al., 2006), as are cyanobacteria (Waleron et al., 2007).

POLAR OCEANOGRAPHY

All oceans are density stratified, with the less dense waters in an upper mixed layer separated from deeper waters. Water is less dense when it is warm or when it is fresher. The Pacific and Atlantic Oceans are alpha oceans and stratified by temperature, with a steep temperature gradient, or thermocline, between an upper mixed layer and deeper waters (Carmack, 2007). Both the Arctic and Antarctic are beta oceans, defined as being salinity stratified with cold, fresher water in the surface mixed layer and slightly warmer, salty water at depth, separated by a halocline. Seasonal ice dynamics are responsible for this fresher surface layer, with meltwater remaining in the upper water column in summer; in winter, when ice forms, salt is rejected and brine sinks to deeper waters, leaving less-saline water nearer the surface. The Arctic Ocean also receives freshwater runoff from major rivers and is the freshest of all oceans.

General Arctic and Antarctic

The Arctic Ocean is not completely enclosed and receives inflow of Atlantic water along the deep shelves of western Spitsbergen and the Barents Sea and Pacific water via the Bering Strait, which flows across the shallow shelf of the Chukchi Sea (Carmack and Wassmann, 2006). Arctic waters must also exit the Arctic basin. Some Arctic water returns to the Atlantic as a cold current on the western side of Fram Straight. Other water exits after flowing through and over the top of the Canadian Arctic Archipelago. This Pacific-origin water finally flows into the Atlantic Ocean though Baffin Bay (Jahn et al., 2010). The complex double estuarine circulation (Carmack, 2007) combined with the effect of seasonal ice dynamics means that the Arctic Ocean and surrounding seas are physically complex, with different water masses within the upper euphotic zone (Fig. 2). This stratification in much of the Canadian Arctic results in nutrient-poor surface waters and the formation of a deep chlorophyll maximum layer above the Arctic halocline (Carmack and MacDonald, 2002). Nitrogen is the limiting nutrient for Arctic photosynthetic productivity, and recently overall nitrate concentrations have declined in the upper waters of the Canada Basin at the same time that salinity has decreased, presumably as a response to decreasing multiyear ice (Li et al., 2009).

In contrast, the Antarctic Circumpolar Current (ACC) dominates the physical structuring of Antarctic water masses and seas. The ACC is strongly influenced by interannual and subdecadal large-scale climate processes such as the El Niño Southern Oscillation and the Pacific Decadal Oscillation. The Southern Annular Mode is the dominant atmospheric mode affecting the Southern Hemisphere and drives circulation, affecting mixed-layer depth in the Southern Ocean and ACC, hence, light and nutrient availability (Sallee et al., 2010). The geographical position of the polar front is not fixed and is strongly influenced by the ACC. As in the Arctic, seasonal ice formation and melting are the major density-structuring agents of the water column, with Circumpolar Deep Water transported by the ACC. When Upper Circumpolar Deep Water intrudes onto the Antarctic shelf, such as along the western Antarctic Peninsula, it brings nutrients and increases productivity (Ducklow et al., 2007). Water column structure also influences productivity by affecting the depth of the photic zone as well as nutrient dynamics, with consequences for microbial food webs and higher trophic levels (Clarke et al., 2007). As in the Arctic, regional variations exist among different seas. Iron availability is relatively more important in the Antarctic compared to the Arctic, and in the Southern Ocean higher regions of productivity are associated with melting ice, including in the wake of icebergs (Smith et al., 2007).

Polynyas

Although seasonal and multiyear ice covers vast regions of polar seas, open-water areas called polynyas occur even in late winter and early spring in both the Arctic and Antarctic. The physical mechanism for maintaining open waters differs among polynyas, with upwelling

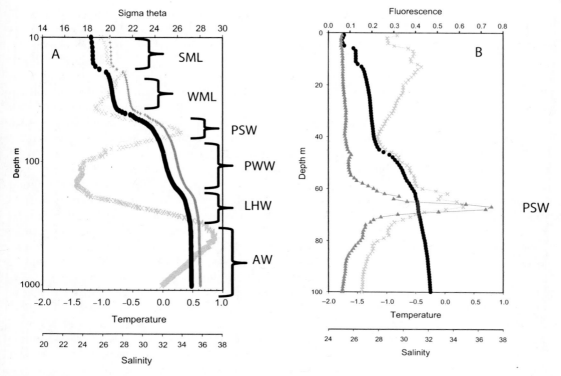

FIGURE 2 (A) Water column profile taken 15 August 2007 from the Canada Basin, showing the distinct layering of different water masses. Temperature is indicated in light gray, sigma theta in dark gray, and practical salinity units in black. SML, summer mixed layer; WML, winter mixed layer; PSW, Pacific summer water; PWW, Pacific winter water; AHW, lower halocline water; AW, deep water from the Atlantic Ocean. (B) The upper 100 m of the same water column with the addition of the fluorescence trace (dark gray triangles), indicating the chlorophyll maximum layer within the PSW.

of warmer currents, wind-driven latent heat production, or a combination of the two preventing local ice buildup (Arrigo and van Dijken, 2003; Winsor and Bjork, 2000). Ice is advected out of the immediate area as it is formed, but local brine rejection results in the formation of denser water that sinks and can carry nutrients and organisms to depth (Dethleff, 2010; Kusahara et al., 2010). Polynyas are defined mostly by local geography and tend to occur in the same region every year, but the extent and duration are influenced by climatology (Dumont et al., 2010) and iceberg drift (Martin et al., 2007). Because polynyas are ice free for longer and hence have a longer season with higher irradiance, they tend to be more productive than surrounding regions (Hollibaugh et al., 2007) and are impor-

tant for maintaining bird and marine mammal populations in the polar regions (Deming et al., 2002).

MICROBIAL FOOD WEBS AND NUTRIENT CYCLING

Microbes in the ocean carry out much of the global biogeochemical cycling and are responsible for half of global photosynthesis (Field et al., 1998). Traditionally, most trophic transfers were thought to be characterized by a linear food web beginning with nutrient input to large phytoplankton that are then eaten by zooplankton and in turn eaten by fish (the so-called NPZ model). This view remains a valuable modeling approach (Poulin and Franks, 2010), but in recent years microbial oceanography studies have focused on nutrient, carbon, and energy transfers within the microbial loop and microbial food webs (Azam and Malfatti, 2007). A microbial loop begins with organic matter in the ocean being taken up by bacteria; the bacteria are eaten by small flagellates that excrete organic matter, which is used by bacteria that are eaten by small flagellates. This respiratory spiral results in lower and lower concentrations of nutrients, especially nitrogen, in a photic zone over time and is characteristic of oligotrophic waters (Azam et al., 1994). Viruses are thought to speed up this process when they lyse microbial cells, increasing the availability of organic matter to bacteria (Suttle, 2007). A microbial food web includes the interconnections of this microbial loop to other microbes, including phytoplankton. Microbial food webs also intersect with higher food webs and allochthonous nutrient input (Fig. 3a). The timing and magnitude of nutrient input, which is linked to climatology, will thus influence biogeochemical processes, energy transfer, and the carbon cycle (Sarmento et al., 2010). For modeling purposes microbes are usually placed into biological black boxes categorized by size and assumed function, with carbon or energy transfer rates estimated among the boxes, usually based on empirical data. To increase the power of such models, a basic understanding of food web interactions is required. A functional scenario that could also operate in the polar oceans and includes *Archaea* and nitrification is shown in Fig. 3b.

Based on what was known about the function of different size classes in the early 1970s, plankton are often categorized as picoplankton (0.2 to 2 μm), nanoplankton (2 to 20 μm), and microplankton (20 to 200 μm) (Sieburth, 1978). Size was thought to reflect functionality in that most heterotrophic bacteria are picoplankton, and many flagellates, including most bacterial grazers, are of nanoplankton size. The larger, conspicuous diatoms and photosynthetic dinoflagellates are microplankton. It is now evident that there are photosynthetic picoplankton and heterotrophic microplankton and that nanoplankton are both heterotrophic

and mixotrophic. It is also evident that taxonomic composition, regardless of size, has an effect on nutrient cycling. One obvious example is that diatoms that range in size from ca. 3 to >200 μm require silica for constructing their cell walls, whereas other nonsiliceous phototrophs do not. Species are also important; with different diatom species having an influence on zooplankton reproductive success (Soreide et al., 2010).

In polar seas, microbial food webs are maintained by complex interactions among different groups of photosynthetic and heterotrophic eukaryotes as well as *Bacteria*, *Archaea*, and viruses. Dissolved organic matter (DOM) is required by heterotrophic *Bacteria*. However, DOM quality is not uniform (Obernosterer and Herndl, 1995); for example, the photosynthate released from healthy diatom communities is complex and subject to breakdown by extracellular enzymes from select bacteria (Elifantz et al., 2007). In contrast, small organic molecules are scavenged by oligotrophic bacteria and can be transported directly into cells.

Macrozooplankton in these systems graze primarily on large phytoplankton and heterotrophic protists, which include ciliates and dinoflagellates that graze on bacteria and other protists, including phytoplankton. The additional trophic steps that occur when protists graze on other protists as well as bacteria before being transferred to zooplankton mean that microbial food webs are less efficient at transferring carbon and energy to higher food webs (De Laender et al., 2010). The export of carbon out of the euphotic zone to the benthos is also linked to the community in surface waters. When nutrients begin to be depleted during a bloom, the larger, heavier diatoms sink quickly and can provide particulate organic material (POM) to the benthic community. This tight pelagic benthic coupling, especially along shallow shelves, is important for benthic feeders such as walrus (Grebmeier et al., 2006). Although much of the diatom production is associated with ice and ice edges (see Chapter 11), diatoms also persist in some open-water

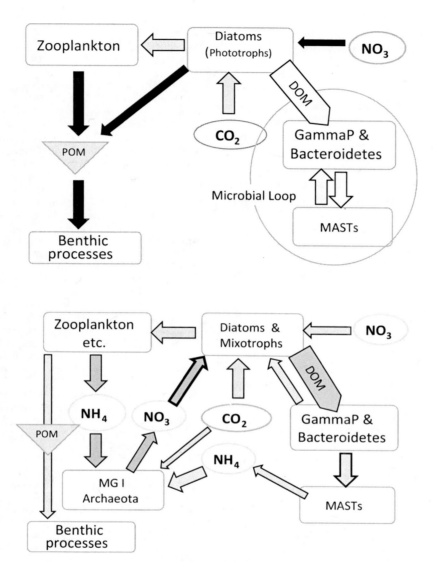

FIGURE 3 Conceptual diagram of microbial food webs from polar seas including links to macrozooplankton and advective input of nitrate and inorganic carbon fixation. Arrows indicate pathways relevant to both nitrogen and carbon cycling. The POM triangle indicates aggregation of particles and zooplankton fecal pellets, which feed the benthos. DOM is released by active phototrophs and taken up by γ-*Proteobacteria* (GammaP) and *Bacteroidetes*, the most abundant bacteria in polar surface waters. MASTs are the primary bacterial grazers, and the interactions between these two groups are a microbial loop. The upper portion shows names of organisms for a classic scenario, while the lower portion includes MGI *Archaea* and a nitrification component. In this case diatoms may be maintained for longer in the upper euphotic zone. The net effect would result in an additional pathway for inorganic carbon fixation via the *Archaea*.

regions, such as those with periodic upwelling events, cross-shelf currents, and tidal mixing (Carmack and MacDonald, 2002). However, a short seasonal production pulse followed by a community shift and lower productivity is more common. In contrast to blooms over shallow shelves, very little material reaches the sediment, with most production being recycled by microbial processes in deeper offshore regions (Manganelli et al., 2009).

CONCLUSIONS

All higher trophic levels, including whales, seals, and birds at both poles and polar bears in the Arctic, ultimately depend on microbes to convert inorganic carbon and solar energy into organic carbon, maintain it in a biologically available form, and recycle nutrients. The microbes that inhabit these moving water masses have very different environmental constraints compared to those living in soils or even in lakes and streams in the polar regions. As in all oceanic systems the main constraint on overall production is nutrient and light availability. Over much of the Arctic the strong salinity stratification of the water column means that once light is available following winter darkness, nutrients in the upper mixed layer are rapidly taken up, with renewal limited. The result is that in the euphotic zone inorganic nutrients are limiting and ongoing carbon fixation or productivity is supported by microbial communities that recycle organic material to obtain sufficient nitrogen and phosphorus for ongoing protein and nucleic acid synthesis required for growth. Such food webs are in contrast to more linear food chains where nitrate is used by larger phytoplankton, especially diatoms, which have very efficient nitrate transport systems (Song and Ward, 2007). In many parts of the Southern Ocean, including the Ross Sea, seasonal iron limitation may occur as well and ongoing production is linked to periodic Fe inputs (Coale et al., 2003), with consequences for microbial community dynamics (Rose et al., 2009).

The Arctic is already changing from historic norms because of global warming, and the Antarctic will be affected in the near future. Understanding how microbial communities interact and influence higher food webs as well as biogeochemical cycling in these systems is an urgent necessity, and with new molecular and other tools, research into these questions provides unprecedented opportunities for new discoveries.

REFERENCES

Alonso-Sáez, L., O. Sánchez, J. M. Gasol, V. Balagué, and C. Pedrós-Alio. 2008. Winter-to-summer changes in the composition and single-cell activity of near-surface Arctic prokaryotes. *Environ. Microbiol.* **10**:2444–2454.

Angly, F. E., B. Felts, M. Breitbart, P. Salamon, R. A. Edwards, C. Carlson, A. M. Chan, M. Haynes, S. Kelley, H. Liu, J. M. Mahaffy, J. E. Mueller, J. Nulton, R. Olson, R. Parsons, S. Rayhawk, C. A. Suttle, and F. Rohwer. 2006. The marine viromes of four oceanic regions. *PLoS Biol.* **4**:e368.

Arrigo, K. R., and G. L. van Dijken. 2003. Phytoplankton dynamics within 37 Antarctic coastal polynya systems. *J. Geophys. Res.* **108**:3271.

Azam, F., and F. Malfatti. 2007. Microbial structuring of marine ecosystems. *Nat. Rev. Microbiol.* **5**:782–791.

Azam, F., D. C. Smith, G. F. Steward, and Å. Hagström. 1994. Bacteria-organic matter coupling and its significance for oceanic carbon cycling. *Microb. Ecol.* **28**:167–179.

Bano, N., S. Ruffin, B. Ransom, and J. T. Hollibaugh. 2004. Phylogenetic composition of Arctic Ocean archaeal assemblages and comparison with Antarctic assemblages. *Appl. Environ. Microbiol.* **70**:781–789.

Beja, O., M. T. Suzuki, J. F. Heidelberg, W. C. Nelson, C. M. Preston, T. Hamada, J. A. Eisen, C. M. Fraser, and E. F. DeLong. 2002. Unsuspected diversity among marine aerobic anoxygenic phototrophs. *Nature* **415**:630–633.

Bell, E. M., and J. Laybourn-Parry. 2003. Mixotrophy in the Antarctic phytoflagellate, *Pyramimonas gelidicola* (Chlorophyta: Prasinophyceae). *J. Phycol.* **39**:644–649.

Carmack, E. C. 2007. The alpha/beta ocean distinction: a perspective on freshwater fluxes, convection, nutrients and productivity in high-latitude seas. *Deep Sea Res. Part 2 Top. Stud. Oceanogr.* **54**:2578–2598.

Carmack, E. C., and R. W. MacDonald. 2002. Oceanography of the Canadian shelf of the Beaufort Sea: a setting for marine life. *Arctic* **55**:29–45.

Carmack, E., and P. Wassmann. 2006. Food webs and physical-biological coupling on pan-Arctic shelves: unifying concepts and comprehensive perspectives. *Prog. Oceanogr.* **71**:446–477.

Church, M. J., E. F. DeLong, H. W. Ducklow, M. B. Karner, C. M. Preston, and D. M. Karl. 2003. Abundance and distribution of planktonic Archaea and Bacteria in the waters west of the Antarctic Peninsula. *Limnol. Oceanogr.* **48**:1893–1902.

Clarke, A., E. J. Murphy, M. P. Meredith, J. C. King, L. S. Peck, D. K. A. Barnes, and R. C.

Smith. 2007. Climate change and the marine ecosystem of the western Antarctic Peninsula. *Philos. Trans. R. Soc. Lond. B Biol. Sci.* **362:**149–166.

Coale, K. H., X. Wang, S. J. Tanner, and K. S. Johnson. 2003. Phytoplankton growth and biological response to iron and zinc addition in the Ross Sea and Antarctic Circumpolar Current along 170°W. *Deep Sea Res. Part 2 Top. Stud. Oceanogr.* **50:**635–653.

Cottrell, M. T., and D. L. Kirchman. 2009. Photoheterotrophic microbes in the Arctic Ocean in summer and winter. *Appl. Environ. Microbiol.* **75:**4958–4966.

Damm, E., E. Helmke, S. Thoms, U. Schauer, E. Nothig, K. Bakker, and R. P. Kiene. 2010. Methane production in aerobic oligotrophic surface water in the central Arctic Ocean. *Biogeosciences* **7:**1099–1108.

Daugbjerg, N. 2000. *Pyramimonas tychotreta*, sp. nov. (Prasinophyceae), a new marine species from Antarctica: light and electron microscopy of the motile stage and notes on growth rates. *J. Phycol.* **36:**160–171.

De Laender, F., D. Van Oevelen, K. Soetaert, and J. J. Middelburg. 2010. Carbon transfer in herbivore- and microbial loop-dominated pelagic food webs in the southern Barents Sea during spring and summer. *Mar. Ecol. Prog. Ser.* **398:**93–107.

DeLong, E. F. 1992. Archaea in coastal marine environments. *Proc. Natl. Acad. Sci. USA* **89:**5685–5689.

Deming, J. W., L. Fortier, and M. Fukuchi. 2002. The International North Water Polynya Study (NOW): a brief overview. *Deep Sea Res. Part 2 Top. Stud. Oceanogr.* **49:**4887–4892.

Dethleff, D. 2010. Linear model estimates of potential salt rejection and theoretical salinity increase in a standardized water column of recurrent Arctic flaw leads and polynyas. *Cold Regions Sci. Technol.* **61:**82–89.

Ducklow, H. W., K. Baker, D. G. Martinson, L. B. Quetin, R. M. Ross, R. Smith, S. E. Stammerjohn, M. Vernet, and W. Fraser. 2007. Marine pelagic ecosystems: the west Antarctic Peninsula. *Philos. Trans. R. Soc. Lond. B Biol. Sci.* **362:**67–94.

Dumont, D., Y. Gratton, and T. E. Arbetter. 2010. Modeling wind-driven circulation and landfast ice-edge processes during polynya events in northern Baffin Bay. *J. Phys. Oceanogr.* **40:**1356–1372.

Elifantz, H., A. I. Dittell, M. T. Cottrell, and D. L. Kirchman. 2007. Dissolved organic matter assimilation by heterotrophic bacterial groups in the western Arctic Ocean. *Aquat. Microb. Ecol.* **50:**39–49.

Falk-Petersen, S., T. Haug, H. Hop, K. T. Nilssen, and A. Wold. 2009. Transfer of lipids from plankton to blubber of harp and hooded seals off East Greenland. *Deep Sea Res. Part 2 Top. Stud. Oceanogr.* **56:**2080–2086.

Farnelid, H., W. Tarangkoon, G. Hansen, P. J. Hansen, and L. Riemann. 2010. Putative N_2-fixing heterotrophic bacteria associated with dinoflagellate-*Cyanobacteria* consortia in the low-nitrogen Indian Ocean. *Aquat. Microb. Ecol.* **61:**105–117.

Field, C. B., M. J. Behrenfeld, J. T. Randerson, and P. Falkowski. 1998. Primary production of the biosphere: integrating terrestrial and oceanic components. *Science* **281:**237–240.

Francis, C. A., J. M. Beman, and M. M. M. Kuypers. 2007. New processes and players in the nitrogen cycle: the microbial ecology of anaerobic and archaeal ammonia oxidation. *ISME J.* **1:**19–27.

Fuhrman, J. A., and L. Campbell. 1998. Marine ecology: microbial microdiversity. *Nature* **393:**410–411.

Fuhrman, J. A., and Å. Hagström. 2008. Bacterial and archaeal community structure and its patterns, p. 45–90. *In* D. L. Kirchman (ed.), *Microbial Ecology of the Oceans.* Wiley-Blackwell, New York, NY.

Fuhrman, J. A., K. McCallum, and A. A. Davis. 1992. Novel major archaebacterial group from marine plankton. *Nature* **356:**148–149.

Galand, P. E., E. O. Casamayor, D. L. Kirchman, and C. Lovejoy. 2009a. Ecology of the rare microbial biosphere of the Arctic Ocean. *Proc. Natl. Acad. Sci. USA* **106:**22427–22432.

Galand, P. E., E. O. Casamayor, D. L. Kirchman, M. Potvin, and C. Lovejoy. 2009b. Unique archaeal assemblages in the Arctic Ocean unveiled by massively parallel tag sequencing. *ISME J.* **3:**860–869.

Galand, P. E., C. Lovejoy, A. K. Hamilton, R. G. Ingram, E. Pedneault, and E. C. Carmack. 2009c. Archaeal diversity and a gene for ammonia oxidation are coupled to oceanic circulation. *Environ. Microbiol.* **11:**971–980.

Galand, P. E., M. Potvin, E. O. Casamayor, and C. Lovejoy. 2010. Hydrography shapes bacterial biogeography of the deep Arctic Ocean *ISME J.* **4:**564–576.

Garneau, M. E., S. Roy, C. Lovejoy, Y. Gratton, and W. F. Vincent. 2008. Seasonal dynamics of bacterial biomass and production in a coastal arctic ecosystem: Franklin Bay, western Canadian Arctic. *J. Geophys. Res.* **113:**C07S91.

Garneau, M. E., W. F. Vincent, L. Alonso-Sáez, Y. Gratton, and C. Lovejoy. 2006. Prokaryotic community structure and heterotrophic production in a river-influenced coastal arctic ecosystem. *Aquat. Microb. Ecol.* **42:**27–40.

Giovannoni, S. J., R. A. Foster, M. S. Rappe, and S. Epstein. 2007. New cultivation strategies bring more microbial plankton species into the laboratory. *Oceanography* 20:62–69.

Gómez, F., D. Moreira, K. Benzerara, and P. López-García. 2011. *Solenicola setigera* is the first characterized member of the abundant and cosmopolitan uncultured marine stramenopile group MAST-3. *Environ. Microbiol.* 13:193–202.

Gómez-Consarnau, L., N. Akram, K. Lindell, A. Pedersen, R. Neutze, D. L. Milton, J. M. González, and J. Pinhassi. 2010. Proteorhodopsin phototrophy promotes survival of marine bacteria during starvation. *PLoS Biol.* 8:e1000358.

Grebmeier, J. M., L. W. Cooper, H. M. Feder, and B. I. Sirenko. 2006. Ecosystem dynamics of the Pacific-influenced Northern Bering and Chukchi Seas in the Amerasian Arctic. *Prog. Oceanogr.* 71:331–361.

Guillou, L., M. Viprey, A. Chambouvet, R. M. Welsh, A. R. Kirkham, R. Massana, D. J. Scanlan, and A. Z. Worden. 2008. Widespread occurrence and genetic diversity of marine parasitoids belonging to *Syndiniales* (*Alveolata*). *Environ. Microbiol.* 10:3349–3365.

Hollibaugh, J. T., C. Lovejoy, and A. E. Murray. 2007. Microbiology in polar oceans. *Oceanography* 20:140–145.

Ichinomiya, M., S. Yoshikawa, M. Kamiya, K. Ohki, S. Takaichi, and A. Kuwata. 2011. Isolation and characterization of Parmales (Heterokonta/Heterokontophyta/Stramenopiles) from the Oyashio region, western North Pacific. *J. Phycol.* 47:144–151.

Jahn, A., L. B. Tremblay, R. Newton, M. M. Holland, L. A. Mysak, and I. A. Dmitrenko. 2010. A tracer study of the Arctic Ocean's liquid freshwater export variability. *J. Geophys. Res.* 115: C07015.

Jungblut, A. D., C. Lovejoy, and W. F. Vincent. 2010. Global distribution of cyanobacterial ecotypes in the cold biosphere. *ISME J.* 4:191–202.

Juul-Pedersen, T., C. Michel, and M. Gosselin. 2010. Sinking export of particulate organic material from the euphotic zone in the eastern Beaufort Sea. *Mar. Ecol. Prog. Ser.* 410:55–70.

Kalanetra, K. M., N. Bano, and J. T. Hollibaugh. 2009. Ammonia-oxidizing *Archaea* in the Arctic Ocean and Antarctic coastal waters. *Environ. Microbiol.* 11:2434–2445.

Kirchman, D. L., M. T. Cottrell, and C. Lovejoy. 2010. The structure of bacterial communities in the western Arctic Ocean as revealed by pyrosequencing of 16S rRNA genes. *Environ. Microbiol.* 12:1132–1143.

Kirchman, D. L., H. Elifantz, A. I. Dittel, R. R. Malmstrom, and M. T. Cottrell. 2007.

Standing stocks and activity of archaea and bacteria in the western Arctic Ocean. *Limnol. Oceanogr.* 52:495–507.

Kirchman, D. L., X. A. G. Morán, and H. Ducklow. 2009. Microbial growth in the polar oceans—role of temperature and potential impact of climate change. *Nat. Rev. Microbiol.* 7:451–459.

Koh, E. Y., N. Atamna-Ismaeel, A. Martin, R. O. M. Cowie, O. Beja, S. K. Davy, E. W. Maas, and K. G. Ryan. 2010. Proteorhodopsin-bearing bacteria in Antarctic sea ice. *Appl. Environ. Microbiol.* 76:5918–5925.

Konneke, M., A. E. Bernhard, J. R. de la Torre, C. B. Walker, J. B. Waterbury, and D. A. Stahl. 2005. Isolation of an autotrophic ammonia-oxidizing marine archaeon. *Nature* 437:543–546.

Kusahara, K., H. Hasumi, and T. Tamura. 2010. Modeling sea ice production and dense shelf water formation in coastal polynyas around East Antarctica. *J. Geophys. Res.* 115:C10006.

Larsen, A., G. A. F. Flaten, R. A. Sandaa, T. Castberg, R. Thyrhaug, S. R. Erga, S. Jacquet, and G. Bratbak. 2004. Spring phytoplankton bloom dynamics in Norwegian coastal waters: microbial community succession and diversity. *Limnol. Oceanogr.* 49:180–190.

Li, W. K. W., F. A. McLaughlin, C. Lovejoy, and E. C. Carmack. 2009. Smallest algae thrive as the Arctic Ocean freshens. *Science* 326:539.

Lichtschlag, A., J. Felden, V. Bruchert, A. Boetius, and D. de Beer. 2010. Geochemical processes and chemosynthetic primary production in different thiotrophic mats of the Hakon Mosby Mud Volcano (Barents Sea). *Limnol. Oceanogr.* 55:931–949.

López-García, P., F. Rodríguez-Valera, C. Pedrós-Alió, and D. Moreira. 2001. Unexpected diversity of small eukaryotes in deep-sea Antarctic plankton. *Nature* 409:603–607.

Lovejoy, C., P. E. Galand, and D. L. Kirchman. 2010. Picoplankton diversity in the Arctic Ocean and surrounding seas. *Mar. Biodiv.* 41:5–12.

Lovejoy, C., L. Legendre, M. J. Martineau, J. Bacle, and C. H. von Quillfeldt. 2002. Distribution of phytoplankton and other protists in the North Water. *Deep Sea Res. Part 2 Top. Stud. Oceanogr.* 49:5027–5047.

Lovejoy, C., R. Massana, and C. Pedrós-Alió. 2006. Diversity and distribution of marine microbial eukaryotes in the Arctic Ocean and adjacent seas. *Appl. Environ. Microbiol.* 72:3085–3095.

Lovejoy, C., and M. Potvin. 2011. Microbial eukaryotic distribution in a dynamic Beaufort Sea and the Arctic Ocean. *J. Plankton Res.* 33:431–444.

Lovejoy, C., W. F. Vincent, S. Bonilla, S. Roy, M. J. Martineau, R. Terrado, M. Potvin, R. Massana, and C. Pedrós-Alió. 2007. Distribution, phylogeny, and growth of cold-adapted pico-prasinophytes in arctic seas. *J. Phycol.* **43:**78–89.

Manganelli, M., F. Malfatti, T. J. Samo, B. G. Mitchell, H. Wang, and F. Azam. 2009. Major role of microbes in carbon fluxes during austral winter in the southern Drake Passage. *PLoS One* **4:**e6941.

Marchant, H., A. Davidson, S. Wright, and J. Glazebrook. 2000. The distribution and abundance of viruses in the Southern Ocean during spring. *Antarct. Sci.* **12:**414–417.

Martin, S., R. S. Drucker, and R. Kwok. 2007. The areas and ice production of the western and central Ross Sea polynyas, 1992-2002, and their relation to the B-15 and C-19 iceberg events of 2000 and 2002. *J. Mar. Syst.* **68:**201–214.

Martin-Cuadrado, A. B., F. Rodriguez-Valera, D. Moreira, J. C. Alba, E. Ivars-Martínez, M. R. Henn, E. Talla, and P. López-García. 2008. Hindsight in the relative abundance, metabolic potential and genome dynamics of uncultivated marine archaea from comparative metagenomic analyses of bathypelagic plankton of different oceanic regions. *ISME J.* **2:**865–886.

Massana, R., J. Castresana, V. Balague, L. Guillou, K. Romari, A. Groisillier, K. Valentin, and C. Pedrós-Alió. 2004. Phylogenetic and ecological analysis of novel marine stramenopiles. *Appl. Environ. Microbiol.* **70:**3528–3534.

Massana, R., L. J. Taylor, A. E. Murray, K. Y. Wu, W. H. Jeffrey, and E. F. DeLong. 1998. Vertical distribution and temporal variation of marine planktonic archaea in the Gerlache Strait, Antarctica, during early spring. *Limnol. Oceanogr.* **43:**607–617.

Massana, R., R. Terrado, I. Forn, C. Lovejoy, and C. Pedrós-Alió. 2006. Distribution and abundance of uncultured heterotrophic flagellates in the world oceans. *Environ. Microbiol.* **8:**1515–1522.

Medlin, L. K. 2009. The use of the terms centric and pennate. *Diatom Res.* **24:**499–501.

Medlin, L. K., and J. Priddle (ed.). 1990. *Polar Marine Diatoms.* British Antarctic Survey, Natural Environment Research Council, Cambridge, United Kingdom.

Montagnes, D. J. S., J. Allen, L. Brown, C. Bulit, R. Davidson, S. Fielding, M. Heath, N. P. Holliday, J. Rasmussen, R. Sanders, J. J. Waniek, and D. Wilson. 2010. Role of ciliates and other microzooplankton in the Irminger Sea (NW Atlantic Ocean). *Mar. Ecol. Prog. Ser.* **411:**101–115.

Montresor, M., C. Lovejoy, L. Orsini, G. Procaccini, and S. Roy. 2003. Bipolar distribution of the cyst-forming dinoflagellate *Polarella glacialis.* *Polar Biol.* **26:**186–194.

Muller, F., T. Brissac, N. Le Bris, H. Felbeck, and O. Gros. 2010. First description of giant *Archaea* (*Thaumarchaeota*) associated with putative bacterial ectosymbionts in a sulfidic marine habitat. *Environ. Microbiol.* **12:**2371–2383.

Murray, A. E., C. M. Preston, R. Massana, L. T. Taylor, A. Blakis, K. Wu, and E. F. DeLong. 1998. Seasonal and spatial variability of bacterial and archaeal assemblages in the coastal waters near Anvers Island, Antarctica. *Appl. Environ. Microbiol.* **64:**2585–2595.

Niemann, H., D. Fischer, D. Graffe, K. Knittel, A. Montiel, O. Heilmayer, K. Nothen, T. Pape, S. Kasten, G. Bohrmann, A. Boetius, and J. Gutt. 2009. Biogeochemistry of a low-activity cold seep in the Larsen B area, western Weddell Sea, Antarctica. *Biogeosciences* **6:**2383–2395.

Not, F., K. Valentin, K. Romari, C. Lovejoy, R. Massana, K. Töbe, D. Vaulot, and L. K. Medlin. 2007. Picobiliphytes: a marine picoplanktonic algal group with unknown affinities to other eukaryotes. *Science* **315:**253–255.

Obernosterer, I., and G. J. Herndl. 1995. Phytoplankton extracellular release and bacterial growth—dependence on the inorganic N-P ratio. *Mar. Ecol. Prog. Ser.* **116:**247–257.

Okolodkov, Y. B., and J. D. Dodge. 1996. Biodiversity and biogeography of planktonic dinoflagellates in the Arctic Ocean. *J. Exp. Mar. Biol. Ecol.* **202:**19–27.

O'Regan, M., and K. Moran. 2010. Deep water methane hydrates in the Arctic Ocean: reassessing the significance of a shallow BSR on the Lomonosov Ridge. *J. Geophys. Res.* **115:**B05102.

Ouverney, C. C., and J. A. Fuhrman. 2000. Marine planktonic Archaea take up amino acids. *Appl. Environ. Microbiol.* **66:**4829–4833.

Payet, J. P., and C. A. Suttle. 2008. Physical and biological correlates of virus dynamics in the southern Beaufort Sea and Amundsen Gulf. *J. Mar. Syst.* **74:**933–945.

Poulin, F. J., and P. J. S. Franks. 2010. Size-structured planktonic ecosystems: constraints, controls and assembly instructions. *J. Plankton Res.* **32:**1121–1130.

Poulin, M., N. Daugbjerg, R. Gradinger, L. Ilyash, T. Ratkova, and C. von Quillfeldt. 2010. The pan-Arctic biodiversity of marine pelagic and sea-ice unicellular eukaryotes: a first-attempt assessment. *Mar. Biodiv.* doi:10.1007/s12526-010-0058-8.

Romanovskii, N. N., H. W. Hubberten, A. V. Gavrilov, A. A. Eliseeva, and G. S. Tipenko. 2005. Offshore permafrost and gas hydrate stability zone on the shelf of East Siberian Seas. *Geo Mar. Lett.* **25:**167–182.

Rose, J. M., Y. Feng, G. R. DiTullio, R. B. Dunbar, C. E. Hare, P. A. Lee, M. Lohan, M. Long, W. O. Smith, B. Sohst, S. Tozzi, Y. Zhang, and D. A. Hutchins. 2009. Synergistic effects of iron and temperature on Antarctic phytoplankton and microzooplankton assemblages. *Biogeosciences* **6**:3131–3147.

Round, F. E., R. M. Crawford, and D. G. Mann. 1990. *The Diatoms: Biology & Morphology of the Genera.* Cambridge University Press, Cambridge, United Kingdom.

Rozanska, M., M. Poulin, and M. Gosselin. 2008. Protist entrapment in newly formed sea ice in the coastal Arctic Ocean. *J. Mar. Syst.* **74**:887–901.

Rusch, D. B., A. L. Halpern, G. Sutton, K. B. Heidelberg, S. Williamson, S. Yooseph, D. Wu, J. A. Eisen, J. M. Hoffman, K. Remington, K. Beeson, B. Tran, H. Smith, H. Baden-Tillson, C. Stewart, J. Thorpe, J. Freeman, C. Andrews-Pfannkoch, J. E. Venter, K. Li, S. Kravitz, J. F. Heidelberg, T. Utterback, Y. H. Rogers, L. I. Falcón, V. Souza, G. Bonilla-Rosso, L. E. Eguiarte, D. M. Karl, S. Sathyendranath, T. Platt, E. Bermingham, V. Gallardo, G. Tamayo-Castillo, M. R. Ferrari, R. L. Strausberg, K. Nealson, R. Friedman, M. Frazier, and J. C. Venter. 2007. The Sorcerer II Global Ocean Sampling expedition: Northwest Atlantic through eastern tropical Pacific. *Plos Biol.* **5**:e77.

Sallee, J. B., K. G. Speer, and S. R. Rintoul. 2010. Zonally asymmetric response of the Southern Ocean mixed-layer depth to the Southern Annular Mode. *Nat. Geosci.* **3**:273–279.

Sarmento, H., J. M. Montoya, E. Vazquez-Dominguez, D. Vaque, and J. M. Gasol. 2010. Warming effects on marine microbial food web processes: how far can we go when it comes to predictions? *Philos. Trans. R. Soc. Lond. B Biol. Sci* **365**:2137–2149.

Scott, F. J., and H. J. Marchant (ed.). 2005. *Antarctic Marine Protists.* Australian Biological Resources Study/Australian Antarctic Division, Canberra, Australia.

Sherr, E. B., B. F. Sherr, P. A. Wheeler, and K. Thompson. 2003. Temporal and spatial variation in stocks of autotrophic and heterotrophic microbes in the upper water column of the central Arctic Ocean. *Deep Sea Res. Part 1 Ocean Res.* **50**:557–571.

Sieburth, J. M. 1978. About bacterioplankton, p. 283–287. *In* A. Sournia (ed.), *Phytoplankton Manual.* UNESCO, Paris, France.

Skovgaard, A., R. Massana, V. Balague, and E. Saiz. 2005. Phylogenetic position of the copepod-infesting parasite *Syndinium turbo* (Dinoflagellata, Syndinea). *Protist* **156**:413–423.

Smetacek, V., C. Klaas, S. Menden-Deuer, and T. A. Rynearson. 2002. Mesoscale distribution of dominant diatom species relative to the hydrographical field along the Antarctic Polar Front. *Deep Sea Res. Part 2 Top. Stud. Oceanogr.* **49**:3835–3848.

Smith, K. L., B. H. Robison, J. J. Helly, R. S. Kaufmann, H. A. Ruhl, T. J. Shaw, B. S. Twining, and M. Vernet. 2007. Free-drifting icebergs: hot spots of chemical and biological enrichment in the Weddell Sea. *Science* **317**:478–482.

Smith, W. O., M. S. Dinniman, S. Tozzi, G. R. DiTullio, O. Mangoni, M. Modigh, and V. Saggiomo. 2010. Phytoplankton photosynthetic pigments in the Ross Sea: patterns and relationships among functional groups. *J. Mar. Syst.* **82**:177–185.

Sogin, M. L., H. G. Morrison, J. A. Huber, D. M. Welch, S. M. Huse, P. R. Neal, J. M. Arrieta, and G. J. Herndl. 2006. Microbial diversity in the deep sea and the underexplored "rare biosphere." *Proc. Natl. Acad. Sci USA* **103**:12115–12120.

Song, B., and B. B. Ward. 2007. Molecular cloning and characterization of high-affinity nitrate transporters in marine phytoplankton. *J. Phycol.* **43**:542–552.

Soreide, J. E., E. Leu, J. Berge, M. Graeve, and S. Falk-Petersen. 2010. Timing of blooms, algal food quality and *Calanus glacialis* reproduction and growth in a changing Arctic. *Glob. Change Biol.* **16**:3154–3163.

Spang, A., R. Hatzenpichler, C. Brochier-Armanet, T. Rattei, P. Tischler, E. Spieck, W. Streit, D. A. Stahl, M. Wagner, and C. Schleper. 2010. Distinct gene set in two different lineages of ammonia-oxidizing archaea supports the phylum Thaumarchaeota. *Trends Microbiol.* **18**:331–340.

Steward, G. F., D. C. Smith, and F. Azam. 1996. Abundance and production of bacteria and viruses in the Bering and Chukchi Seas. *Mar. Ecol. Prog. Ser.* **131**:287–300.

Suttle, C. A. 2007. Marine viruses—major players in the global ecosystem. *Nat. Rev. Microbiol.* **5**:801–812.

Tavormina, P. L., W. Ussler, S. B. Joye, B. K. Harrison, and V. J. Orphan. 2010. Distributions of putative aerobic methanotrophs in diverse pelagic marine environments. *ISME J.* **4**:700–710.

Taylor, F. J. R., M.Hoppenrath, and J. F. Saldarriaga. 2008. Dinoflagellate diversity and distribution. *Biodivers. Conserv.* **17**:407–418.

Terrado, R., W. F. Vincent, and C. Lovejoy. 2009. Mesopelagic protists: diversity and succession in a coastal Arctic ecosystem. *Aquat. Microb. Ecol.* **56**:25–39.

Vila-Costa, M., R. Simo, L. Alonso-Sáez, and C. Pedrós-Alió. 2008. Number and phylogenetic affiliation of bacteria assimilating dimethylsulfoniopropionate and leucine in the ice-covered coastal Arctic Ocean. *J. Mar. Syst.* **74:**957–963.

Vincent, W. F. 2000. Cyanobacterial dominance in the polar regions, p. 321–340. *In* B. Whitton and M. Potts (ed.), *Ecology of the Cyanobacteria: Their Diversity in Space and Time.* Kluwers Academic Press, Amsterdam, The Netherlands.

von Quillfeldt, C. H. 2000. Common diatom species in arctic spring blooms: their distribution and abundance. *Botanica Marina* **43:**499–516.

Waleron, M., K. Waleron, W. F. Vincent, and A. Wilmotte. 2007. Allochthonous inputs of riverine picocyanobacteria to coastal waters in the Arctic Ocean. *FEMS Microbiol. Ecol.* **59:**356–365.

Walker, C. B., J. R. de la Torre, M. G. Klotz, H. Urakawa, N. Pinel, D. J. Arp, C. Brochier-Armanet, P. S. G. Chain, P. P. Chan, A. Gollabgir, J. Hemp, M. Hugler, E. A. Karr, M. Konneke, M. Shin, T. J. Lawton, T. Lowe, W. Martens-Habbena, L. A. Sayavedra-Soto, D. Lang, S. M. Sievert, A. C. Rosenzweig, G. Manning, and D. A. Stahl. 2010. *Nitrosopumilus maritimus* genome reveals unique mechanisms for nitrification and autotrophy in globally distributed marine crenarchaea. *Proc. Natl. Acad. Sci. USA* **107:**8818–8823.

Ward, B. B., D. G. Capone, and J. P. Zehr. 2007. What's new in the nitrogen cycle? *Oceanography* **20:**101–109.

Wells, L. E., M. Cordray, S. Bowerman, L. A. Miller, W. F. Vincent, and J. W. Deming. 2006. Archaea in particle-rich waters of the Beaufort Shelf and Franklin Bay, Canadian Arctic: clues to an allochthonous origin? *Limnol. Oceanogr.* **51:**47–59.

Winsor, P., and G. Bjork. 2000. Polynya activity in the Arctic Ocean from 1958 to 1997. *J. Geophys. Res.* **105:**8789–8803.

Yoon, H. S., D. C. Price, R. Stepanauskas, V. D. Rajah, M. E. Sieracki, W. H. Wilson, E. C. Yang, S. Duffy, and D. Bhattacharya. 2011. Single-cell genomics reveals organismal interactions in uncultivated marine protists. *Science* **332:**714–717.

Yooseph, S., K. H. Nealson, D. B. Rusch, J. P. McCrow, C. L. Dupont, M. Kim, J. Johnson, R. Montgomery, S. Ferriera, K. Beeson, S. J. Williamson, A. Tovchigrechko, A. E. Allen, L. A. Zeigler, G. Sutton, E. Eisenstadt, Y. H. Rogers, R. Friedman, M. Frazier, and J. C. Venter. 2010. Genomic and functional adaptation in surface ocean planktonic prokaryotes. *Nature* **468:**60–66.

CRYOSPHERIC ENVIRONMENTS IN POLAR REGIONS (GLACIERS AND ICE SHEETS, SEA ICE, AND ICE SHELVES)

Mark Skidmore, Anne Jungblut,
Matthew Urschel, and Karen Junge

11

INTRODUCTION

The focus of this chapter is the glaciers and ice sheets, sea ice, and ice shelves of the polar regions, i.e., those latitudes above the Arctic and Antarctic Circles where glaciers and ice sheets cover a significant proportion of the land mass and where large expanses of the surface waters of the Arctic and Southern Oceans undergo an annual cycle of freezing and melting. Antarctica covers ~14 million km^2, of which 98% is glaciated, and the Greenland Ice Sheet covers 1.7 million km^2, with other smaller ice caps and glaciers in the Arctic covering an additional 0.25 million km^2 (Oerlemans, 2001; Benn and Evans, 2010). The frozen–ocean and sea-ice environments can cover an area of up to 7% of the earth's surface during winter (Dieckmann and Heller, 2010). Taken together they represent one of the largest biomes on earth and play a crucial role in structuring the whole polar ecosystem (Eicken, 1992; Priscu et al., 2008; Chapter 10, this volume). The following sections provide an overview of

the microbial ecology of and the biogeochemical cycling processes that occur in glaciers and ice sheets, sea ice, and ice shelves of the polar regions.

MICROBIOLOGY OF GLACIERS AND ICE SHEETS

Glaciers and Ice Sheets as a Microbial Habitat

Glacial systems can be characterized into three environments: supraglacial, subglacial, and englacial. Supraglacial (surficial) systems are exposed to the sun, and therefore photosynthesis is an important component of these systems. Supraglacial systems include dynamic flowing components such as surface runoff, both sheet flow and channelized, and more stable aqueous environments such as cryoconite holes and supraglacial lakes. In mountain settings these systems may also contain significant supraglacial debris. In contrast, subglacial systems are permanently dark, and thus chemolithotrophy and heterotrophy are the only viable metabolisms. However, in certain polythermal glaciers and at the margin of the Greenland Ice Sheet the subglacial system may be fed by supraglacial waters that contain photosynthetic carbon, especially later during the summer melt season (Boon et al., 2003; Bhatia et al., 2010). There

Mark Skidmore, Department of Earth Sciences, Montana State University, Bozeman, MT 59717. *Anne Jungblut,* Department of Botany, The Natural History Museum, Cromwell Road, London SW7 5BD, United Kingdom. *Matthew Urschel,* Department of Microbiology, Montana State University, Bozeman, MT 59717. *Karen Junge,* Applied Physics Laboratory, University of Washington, 1013 NE 40th Street, Seattle, WA 98105.

Polar Microbiology: Life in a Deep Freeze
Edited by Robert V. Miller and Lyle G. Whyte © 2012 ASM Press, Washington, DC

are a range of subglacial aquatic environments, including subglacial lakes and their sediments, saturated sediments in nonlake settings, and hydrologic systems, that can be distributed and channelized, connecting subglacial lakes and saturated sediments (Skidmore, 2011). In addition, there are subglacial aquatic environments in the solid-phase, accreted lake ice and basal ice as liquid water films on particles and in vein networks within the ice (Skidmore, 2011). However, most of these environments remain unexplored. To date, only three subglacial locations have been sampled for microbial analysis in the interior of ice sheets. Two of these locations are in the interior of the Antarctic Ice Sheet: Kamb Ice Stream sediments, West Antarctica, and the accreted ice of Lake Vostok, East Antarctica (Skidmore, 2011). The other location is in the interior of the Greenland Ice Sheet: debris-rich basal ice from the GISP2 ice core (Miteva et al., 2004). In the englacial environment the microbial habitat is largely confined to the liquid-vein network between ice crystals. It has received much less attention than supraglacial and subglacial habitats given its relatively low nutrient concentrations, low cell concentrations, and often low temperatures, especially in ice-sheet settings, and will not be covered in depth here (reviewed in Miteva, 2008).

Glaciers and Ice-Sheet Ecology: Diversity and Cold Adaptation

SUPRAGLACIAL ENVIRONMENTS

Cryoconite holes are small, water-filled, cylindrical holes found in the surface of glacier ice with a thin layer of sediment at the hole bottom (Fig. 1). These holes form on the ice-covered parts of polar ice masses, where there is sufficient energy for melting (Fountain and Tranter, 2008). There are two main types of cryoconite holes. The most common are as pools of surface water open to the atmosphere. The more unusual kind is found in the McMurdo Dry Valleys of Antarctica (and probably elsewhere at high latitudes), where energy-balance conditions favor frozen surfaces and internal melting. The pool of meltwater is enclosed in ice, isolating it from direct contact with the atmosphere (Fountain and Tranter, 2008).

Cryoconite holes have long been recognized as a microbial habitat in supraglacial environments. For example, Gerdel and Druet (1960) examined the biological composition of cryoconite holes on the Greenland Ice Sheet, noting the presence of filamentous algae, fungal hyphae, unicellular desmids, and rotifers. Numerous studies since have noted the biological composition of cryoconite holes in Arctic and Antarctic glacial environments via macroscopic and microscopic techniques (Wharton et al., 1981, 1985; Mueller et al., 2001, Säwström et al., 2002; Foreman et al., 2007; Anesio et al., 2009). Cell biomass is lower in the waters in cryoconite holes, ranging from 10^4 to 10^5 cells ml^{-1}, relative to the sediments in the same holes, with 10^8 to 10^9 cells g of sediment^{-1} in both Arctic and Antarctic settings (Anesio et al., 2009). Viruslike particles have been documented in Arctic cryoconites, with viruslike particle-to-bacteria ratios ranging from 0.3 to 4.5 (Säwström et al., 2002) and from 0.3 to 1.5 in sediments, relative to 7.3 to 31 in overlying waters (Anesio et al., 2007). Viruses play an important role in recycling carbon in aquatic environments, including cryoconite holes, since they cause lysis of bacterial, algal, and protozoan cells, recycling carbon to the pool before it can be passed up the food chain (Fuhrman, 1999).

There has been limited research on the microbial diversity of communities in cryoconite holes using 16S and 18S rRNA gene sequence approaches (Christner et al., 2003; Foreman et al., 2007; Edwards et al., 2011). There are some similarities in lineages represented by 16S rRNA gene sequences obtained by Christner et al. (2003) from a cryoconite hole on the Canada Glacier, Taylor Valley, Antarctica, and from numerous cryoconite holes on three Arctic glaciers (Edwards et al., 2011). Members of the bacterial lineages *Acidobacterium*, *Actinobacteria*, *Cyanobacteria*, *Planctomycetes*, and *Proteobacteria* were present in both studies, with Christner et al. (2003)

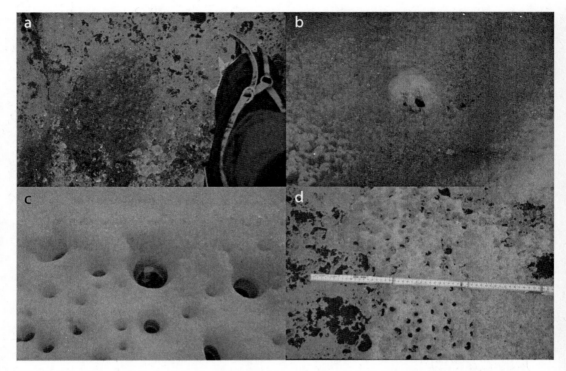

FIGURE 1 Cryoconite holes, John Evans Glacier, Nunavut, Canadian High Arctic. (a) Cryoconite with frozen lid. (b) Partially open cryoconite. (c) Open cryoconite. (d) Transition zone from individual cryoconite holes to broader zones of debris-covered surficial ice. (Photo credit: M. Skidmore.)

also finding sequences from members of the *Verrucomicrobia*, *Cytophagales*, and *Gemmimonas* and Edwards et al. (2011) finding sequences from members of *Chloroflexus* and candidate phylum OP10. Christner et al. (2003) also documented metazoan (nematode, tardigrade, and rotifer), truffle, ciliate, and green algal eukaryal sequences. No archaeal PCR products were generated in these two studies despite the use of archaeal-specific primers (Christner et al., 2003; Edwards et al., 2011). Foreman et al. (2007), in a study on three Antarctic glaciers in the McMurdo Dry Valleys, used three phylogenetic probes on DNA extracted from samples of cryoconite sediments and overlying ice. They documented a higher percentage of *Cytophaga-Flavobacterium* cells in cryoconite sediments, with β-*Proteobacteria* cells dominating the ice overlying the sediment layer.

Limited culturing work has been carried out using material from cryoconite holes, de-spite numerous studies on rates of community carbon uptake and respiration. Christner et al. (2003) note that most of the aerobic heterotrophic isolates cultured from an Antarctic cryoconite were capable of growth at 22°C; however, a few isolates would grow at 15°C but not at 22°C, with one isolate showing spontaneous cell lysis when growing cultures were moved from 15°C to 22°C. These isolates were members of the phyla *Cytophagales* and *Actinobacteria* based on 16S rRNA gene sequence analysis (Christner et al., 2003).

SUBGLACIAL ENVIRONMENTS

Bacteria have been detected in all subglacial environments sampled in polar regions, including subglacial waters, sediments, basal ice, and accreted ice (Christner et al., 2008; Skidmore, 2011) (Fig. 2 and 3). Cell concentrations as high as 10^7 cells g^{-1} have been documented in Antarctic subglacial sediments and 10^8 cells

FIGURE 2 Subglacial stream sampling, John Evans Glacier, Nunavut, Canadian High Arctic. (Photo credit: M. Skidmore.)

g^{-1} in Greenland basal ice (Lanoil et al., 2009; Miteva et al., 2009; Yde et al., 2010). Subglacial hydrology has a first-order impact on the oxygen content and therefore the redox potential, E_h, of the subglacial environment (Tranter et al., 2005). A wide range of metabolisms have been documented, primarily through enrichment culturing and inference

FIGURE 3 Taylor Glacier, Antarctica. (a) Debris-rich basal ice outcrops on northern margin. (b) Tunnel to access debris-rich ice. (c) Cutting a vertical profile into the debris-rich ice in the ice tunnel. (Photo credits: panels a and b, M. Skidmore; panel c, B. Christner.)

from a combination of 16S rRNA gene sequence data and geochemical and/or isotope
geochemical analyses (Skidmore et al., 2000,
2005, 2010; Boyd et al., 2010, 2011; Wadham
et al., 2004, 2010b; Lanoil et al., 2009; Mikucki
et al., 2009). The metabolisms include aerobic
heterotrophy, iron and sulfur oxidation, nitrate
reduction, iron reduction, sulfate reduction, and
methanogenesis and reflect a wide E_h range.

Molecular analysis of the subglacial microbial communities from a range of ice masses
using 16S rRNA gene sequences has revealed
three sequence clusters in the β-*Proteobacteria*
class that are closely related to the *Comamonas*,
Gallionella, and *Thiobacillus* groups (Lanoil et al.,
2009). Cultured isolates in these three groups
are aerobic heterotrophs and chemolithoautotrophic iron and sulfur oxidizers, respectively
(Lanoil et al., 2009). These metabolisms are
consistent with geochemical evidence that has
inferred biologically driven sulfide oxidation as
a significant source of sulfate in a number of
subglacial polar environments (Christner et al.,
2006; Skidmore et al., 2010; Wadham et al.,
2010b). Similarities in microbial community
diversity between a range of glacial environments were investigated at John Evans Glacier,
Canadian High Arctic, via terminal restriction fragment length polymorphism analysis
of bacterial 16S rRNA genes amplified from
these environments (Bhatia et al., 2006). The
analysis revealed that the subglacial water, basal
ice, and sediment communities were distinct
from those detected in supraglacial meltwater
and proglacial sediment. Thus, the subglacial
community at John Evans Glacier appears to
be predominantly autochthonous rather than
allochthonous, and it may be adapted to subglacial conditions (Bhatia et al., 2006).

Culturing work on samples from these systems has resulted in the isolation of heterotrophs
at temperatures of >0°C (Miteva et al., 2004;
Christner et al., 2006; Cheng and Foght, 2007;
Lanoil et al., 2009; Yde et al., 2010). Miteva et
al. (2004) demonstrated that only a small proportion of their isolates from Greenland basal
ice were limited to growing in the temperature
range of 2 to 18°C, with the majority able to

grow at temperatures of 25°C or higher. Cultured isolates of nitrate-, iron-, and sulfate-reducing organisms and methanogens from polar
subglacial systems have yet to be described in
the literature. To date, archaea have rarely been
found in polar subglacial systems, although Skidmore et al. (2000) demonstrated methanogenesis
at 4°C in enrichment cultures of basal ice from
John Evans Glacier, Canadian High Arctic.
Subsequent PCR amplification and sequencing of archaeal 16S rRNA gene clones from the
enrichment resulted in a single phylotype with
>98% identity to *Methanosarcina lacustris*, a psy-
chrotolerant methanogen (Boyd et al., 2010). A
few archaeal 16S rRNA gene sequences were
also amplified from Greenland silty basal ice that
were related to the psychrophilic methanogenic
species *Methanococcoides burtonii* and *Methanococcoides alaskense* (Miteva et al., 2009).

Limited research has been undertaken on
examining the physiology of organisms from
subglacial systems. However, significant numbers of ultramicrobacteria (cell volume, <0.1
μm^3) were reported from Greenland silty basal
ice (Miteva et al., 2004; Miteva and Benchley,
2005) and a proposed new species, *Chryseobacterium greenlandense* sp. nov., has been isolated from the same ice (Loveland-Curtze et
al., 2010). Interestingly, this isolate has a wide
temperature range for growth: from 1 to 37°C
(Loveland-Curtze et al., 2010). Miteva et al.
(2009) also describe significant salt tolerance
of between 12 and 20% NaCl for some of
their Greenland ice isolates, particularly from
a depth of 2,495 m. Salt and cold tolerance
of subglacial isolates have received limited
attention; however, laboratory experiments
using glacial isolates of *Sporosarcina* sp. strain
B5 (debris-rich basal ice from Taylor Glacier, Antarctica) and *Chryseobacterium* sp.
strain V3519-10 (ice from 3,519-m depth
in the Vostok Ice Core, East Antarctica) in
ice and subzero brines have been conducted
(Bakermans and Skidmore, *Environ. Microbiol.
Rep.*, in press). Bakermans and Skidmore (in
press) demonstrated respiration of acetate by
both isolates in nutrient-rich ices at temperature as low as −33°C, with 10^6 cells ml^{-1} as

the original inoculum. Further, parallel incubations using the same isolates in brine
and nutrient-rich ice (where the liquid-vein
salinity is comparable to the brine) at −5°C
show growth of both isolates in brine and evidence for growth within the ice for the isolate *Sporosarcina* sp. strain B5 (Bakermans and
Skidmore, in press). Raymond et al. (2008)
documented that *Chryseobacterium* sp. strain
V3519-10 contains an ice-binding protein
that demonstrates recrystallization-inhibition
activity at −4°C. See Chapter 5 for further
examples of physiological adaptation in cold-
adapted microorganisms.

Biogeochemical Cycling in Glacier and Ice-Sheet Communities

CARBON CYCLING

Given the significant areal coverage of ice
masses in the polar regions, carbon cycling in
both supraglacial and subglacial environments is
important to understanding the role of the polar
regions in the global carbon cycle. Numerous
studies have focused on measuring the rates of
carbon fixation, via photosynthesis, relative to
heterotrophic activity in cryoconite ecosystems,
primarily in the Arctic where open cryoconites
are common (Säwström et al., 2002; Hodson
et al., 2007; Stibal et al., 2008; Anesio et al.,
2009; Hodson et al., 2010; Telling et al., 2010;
Edwards et al., 2011). Currently there is debate
over whether these ecosystems are net primary
producers (Anesio et al., 2009) or heterotrophic
and dependent on external (aeolian) supplies of
organic matter (Stibal et al., 2008). At the present time the spatial and temporal coverage of
the studies in the literature precludes a definitive answer. However, as Hodson et al. (2010)
note, respiration and photosynthesis fluxes for
the Greenland Ice Sheet based on cryoconite
data are on the order of ~10 to 100 Gg of C a^{-1},
and thus a more accurate quantification of the
carbon balance is warranted.

Where surface waters drain to the bed, in
polythermal glaciers and at the margin of the
Greenland Ice Sheet, modern photosynthetically derived carbon can be supplied to the

subglacial environment and the microbial communities therein (Boon et al., 2003; Bhatia et al.,
2010). However, where there is no rapid route
from the surface to the bed, as in Antarctica, the
subglacial carbon cycle is controlled primarily
by the organic carbon content of the underlying sediments and microbial carbon fixation via
chemolithoautotrophic processes (Skidmore,
2011). The specific nature of carbon cycling
beneath polar ice masses, especially the large
ice sheets, remains unknown. However, estimates by Priscu et al. (2008) suggest that total
cell numbers in subglacial groundwater beneath
the Antarctic Ice Sheet are of a similar order
to those in the (global) open ocean, ~10^{29}, indicating a sizable subglacial biome. Given the
lack of connectivity between oxygenated surface waters and the subglacial environment in
Antarctica and the extended water flow paths
and water-rock residence times, anoxic conditions seem highly likely for a large proportion
of the groundwater system. Such anoxic conditions would favor anaerobic microbial metabolisms similar to those documented in other deep
terrestrial subsurface environments, such as
iron reduction, sulfate reduction, acetogenesis,
methanogenesis, fermentation, and anaerobic
methane oxidation (Skidmore, 2011).

In subglacial systems the production of car-
bonic acid can be enhanced by microbially
respired CO_2 and the release of protons from
the oxidation of sulfides accelerated by microbial activity of chemolithoautotrophs. Both of
these processes can result in enhanced weathering of both carbonate and silicate minerals
in subglacial polar environments (Skidmore et
al., 2010; Wadham et al., 2010a), thus demonstrating the key role microbial processes have
in subglacial chemical weathering processes.

NITROGEN AND PHOSPHORUS CYCLING

Bacterial activity in cryoconite holes is primarily
limited by phosphorus, based on manipulation
experiments showing that addition of P alone
in water samples from Svalbard cryoconite
holes stimulated bacterial production, while no
stimulation was detected by addition of either

nitrogen (N) or carbon (C) alone (Säwström et al., 2007). In contrast, the continuous comminution of debris in the subglacial environment should result in the replenishment of both mineralogical P and Fe sources. Further, given the energetics of diazotrophy and that it cannot be driven via photosynthetic organisms in subglacial systems, one might expect a greater degree of N limitation in subglacial environments. Limited research has been undertaken on the nitrogen cycle in subglacial environments, especially in polar systems (Wynn et al., 2007). Wynn et al. (2007) infer that microbial processes enhance nitrate production via either the oxidation of NH_4^+ or mineralization of organic nitrogen at the glacier bed from geochemical and isotopic measurements. However, no direct measurements of the microbial communities were made.

MICROBIOLOGY OF SEA ICE COMMUNITIES OF POLAR REGIONS

In-depth reviews on microorganisms in sea ice and its biogeochemistry have recently been compiled: on bacteria and viruses (Deming, 2010; Junge et al., 2011), on primary producers (Arrigo et al., 2010), on heterotrophic protists (Caron and Gast, 2010), and on sea ice biogeochemistry (Thomas et al., 2010). In the following sections, we introduce sea ice as a microbial habitat and summarize from some of the aforementioned reviews what is known to date about the abundance, activity, diversity, and ecology of prokaryotic sea-ice microorganisms. We also briefly outline the role of microorganisms in biogeochemical cycling of elements in sea ice.

Sea Ice as a Microbial Habitat

Ecology and adaptation mechanisms of microorganisms that live within sea ice are linked to its chemical and physical characteristics. When ice is formed from seawater (below $-1.86°C$), dissolved salts, air, and other "impurities" in the water (including microorganisms and inorganic and organic dissolved and particulate matter) are concentrated into a salty brine that persists as inclusions of pockets and channels within the ice (Mock and Junge, 2007, and references therein; also see Petrich and Eicken, 2010, for an in-depth review on ice development and its micro- and macrostructure). These brine channels of varying size (from micrometers to several millimeters in diameter) represent the main habitat for sea ice microorganisms (Fig. 4) (Mock and Junge 2007).

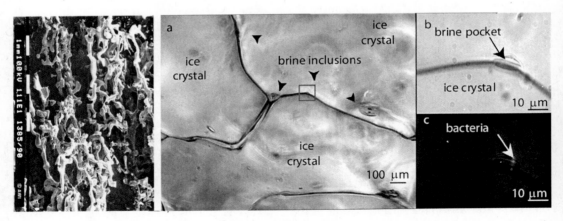

FIGURE 4 (Left) Scanning electron microscopy image of the brine channels system in columnar sea ice made visible by filling the system with epoxy resin under a vacuum. (Right) In situ microscopic images of (a) ice crystals and brine pockets and (b) detail of a brine pocket in panel a that harbors bacteria stained with the blue DNA stain 4′,6-diamidino-2-phenylindole in panel c. (Left image from Alfred-Wegener Institute for Polar and Marine Research, Bremerhaven, Germany; reprinted from Mock and Junge [2007] with permission of the publisher. Right images adapted from Junge et al. [2001] with permission of the publisher.)

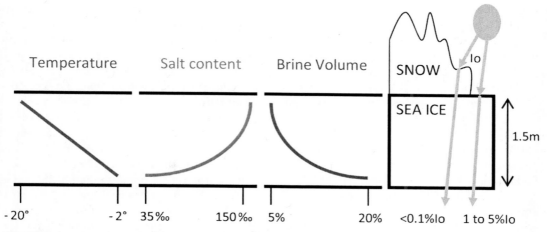

FIGURE 5 Vertical gradients of temperature, salt content, brine volume, and irradiance (Io) through sea ice. These general patterns may vary due to changes in temperature. (Adapted from Mock and Junge [2007] with permission of the publisher.)

Temperature determines the salt content (salinity) of the brine and the volume of brine channels. When temperature decreases, brine volume decreases and salinity increases. Ice at the sea ice-air interface is usually colder than ice in contact with the underlying water, resulting in vertical temperature (salinity and brine volume) gradients throughout the ice (Fig. 5). At temperatures above −5°C (and bulk ice melt salinities of 5‰) and brine volume fractions above 5%, the ice is permeable (Golden et al., 1998). Below −5°C, sea ice is considered impermeable, with no convective fluid flow occurring (Golden et al., 1998).

During the polar summer most of the ice melts and the community is released into the underlying water, providing seed populations for algae blooms in the water in addition to nutrients for bottom communities. In the Arctic, melting ponds form on the surface of the ice, containing unique ice-algae assemblages and microbial communities (Brinkmeyer et al., 2004). With continued increasing water temperatures and solar irradiance on top of the ice, the ice gets thinner and more porous, and eventually melts completely. Some ice survives the summer, and refreezing during the following winter results in thicker ice (termed multiyear ice). Sea ice of northern Greenland

and the Canadian Arctic Archipelago can be several years old, with an average thickness of 6 to 8 m. The ice in the Southern Ocean is considerably thinner, with an average thickness of only 1 m. Such differences in physical properties of the ice also result in differences in the abundance, activity, and composition of the microbial communities within (Mock and Junge, 2007).

Sea Ice Ecology: Diversity and Cold Adaptations

Annually during spring and summer, extensive microbial communities develop within sea ice (for reviews, see Mock and Thomas, 2005; Mock and Junge, 2007; Deming, 2010; Junge et al., 2011). These communities are usually dominated by ice-algae assemblages composed primarily of pennate diatoms (reviewed by Mock and Junge, 2007; Arrigo et al., 2010)—often populations are so rich that the ice appears brown-green to the naked eye. Heterotrophic bacteria represent another major group within these communities (as do viruses; recently reviewed by Deming, 2010), as evidenced by measures of high bacterial abundance, activity, and production and phylogenetically diverse bacterial assemblages. Heterotrophic protozoa, amphipods, invertebrate larvae, copepods, euphausids such as krill, nematodes, turbellarians, and some fishes are the larger protozoan and metazoan consumers found in sea ice (for in-depth reviews on these larger consumers, see various chapters in Thomas and Dieckmann, 2010).

The phylogenetic diversity of sea ice bacteria is well studied, especially during spring and summer (Gosink et al., 1993; Bowman et al., 1997; Brown and Bowman, 2001; Junge et al., 2002; Brinkmeyer et al., 2003) and more recently during winter as well (Brinkmeyer et al., 2003; Junge et al., 2004; Collins et al., 2010). Both from the Antarctic and Arctic new genera and species have been described, mostly within the divisions of the phylum *Proteobacteria*, including the classes α-*Proteobacteria*, β-*Proteobacteria*, and γ-*Proteobacteria*; the phylum *Bacteriodetes* (formerly called the *Cytophaga-Flavobacterium-Bacteroides* or CFB phylum); and the phylum *Actinobacteria* (see Table 1 in Junge et al., 2011, for a complete list of psychrophilic sea-ice bacterial isolates and genera). Many of these sea ice bacteria are stenopsychrophiles (exhibiting growth optima or maxima below 15°C). The majority of bacteria isolated from sea ice are pigmented and highly cold adapted, with some able to form gas vesicles (Sullivan and Palmisano, 1984; Grossi et al., 1984; Staley et al., 1989; Gosink and Staley, 1995).

Up to 60% of the total sea-ice bacterial population has been found to be amenable to cultivation (Junge et al., 2002; Brinkmeyer et al., 2003). This stands in marked contrast to the culturability of most seawater bacteria, at only ~0.01% of the total cell count (Amann et al., 1995), and is possibly due to the exceptionally high concentrations of labile dissolved organic matter (DOM) released by ice algae. DOM concentrations in sea-ice brines in the Arctic and Antarctic exceed surface water concentrations by factors of up to 500 (Mock and Thomas, 2005, and references therein). Thus, groups commonly associated with marine algae are repeatedly found in sea-ice bacterial culture collections (Deming, 2010).

Small-subunit rRNA gene sequence analyses of whole communities have also demonstrated a strong congruence with diversity data derived from cultivation studies. Mostly, the same phyla and genera were observed to be present (Brown and Bowman, 2001; Brinkmeyer et al., 2003), except for some clones belonging to the *Verrucomicrobia*, closest to the prosthecate aerobic genera *Verrucomicrobium* and *Prosthecobacter* (Brown and Bowman, 2001). Bacterial clone groups such as SAR11, SAR86, or archaeal groups that commonly occur in oceanic nonpolar and polar seawater samples have not been found in the Arctic and Antarctic spring/summer ice or melt ponds studied (Brinkmeyer et al., 2004). Arctic ice-melt ponds have been found to harbor β-proteobacterial genera that are not found in the interior of the ice. These genera, known only from freshwater habitats, were found to dominate in the mostly freshwater ponds (along with gram-positive species, α- and γ-proteobacterial genera occurring in more saline ponds, and members of the *Bacteriodetes* occurring in sediment-containing ponds) (Brinkmeyer et al., 2004).

The occurrence of anaerobic bacterial denitrification with high numbers of anaerobic nitrate-reducing bacteria and anaerobic ammonium oxidation in zones with high levels of nitrate, ammonium ions, and DOM indicate the presence of oxygen-deficient and anoxic zones in Arctic sea ice (Rysgaard and Glud, 2004; Rysgaard et al., 2008). Mucopolysaccharide gels and exopolymeric substances could provide such oxygen-depleted microhabitats for these species within the sea ice habitat (Mock and Thomas, 2005).

Archaea were found in small numbers (up to 4% of the population) in Arctic wintertime sea-ice samples only (Junge et al., 2004; Collins et al., 2010). During the winter, the communities of *Bacteria* and *Archaea* in the ice furthermore resembled those in the underlying water and consisted primarily of SAR11 clade α-*Proteobacteria* and Marine Group I *Crenarchaeota*, neither of which is known from spring- and summertime sea ice (Collins et al., 2010). Selection during ice formation and mortality during winter might thus play minor roles in the process of microbial succession that leads to distinctive spring and summer sea-ice bacterial communities (Collins et al., 2010). Though seasonal studies that explore the winter/spring/summer transitions are needed to explore how distinctive summer- and springtime communities develop, factors such as extensive algal growth resulting in increased highly labile DOM levels are likely important.

The enzymes and membranes that enable marine psychrophiles to live at low temperatures and high salinities are of considerable interest for biotechnological and industrial applications (see Chapter 5, this volume). Among the most-studied marine psychrophiles is *Colwellia psychrerythraea* strain 34H, an Arctic marine sediment isolate (Huston et al., 2000). It is also a common sea ice resident, as other strains of this species were first isolated from Antarctic sea ice (Bowman et al., 1998). Its cardinal growth temperatures in heterotrophic media are among the lowest for all characterized bacteria (optimum of 8°C, maximum of 19°C, and extrapolated minimum of −14.5°C) (Huston, 2003). 34H is known for its remarkable adaptations to low temperature: e.g., swimming speeds at −10°C comparable to mesophilic bacteria at 37°C (Junge et al., 2003); the production of cold-active enzymes (Huston et al., 2004); susceptibility to cold-active viruses (Wells and Deming, 2006); and the incorporation of tritiated leucine into newly made proteins to remarkably low temperatures (−196°C) (Junge et al., 2006). Possible cold-adaptation strategies revealed by whole-genome sequence analysis (Methe et al., 2005) also include the production of cryoprotective osmolytes (Collins, 2009) and exopolymers (Krembs et al., 2011). These traits are likely to make *Colwellia* species important to carbon and nutrient cycling in the cold marine environment.

Biogeochemical Cycling in Sea Ice Communities

Marine psychrophiles have been shown to play a globally significant role in biogeochemical cycling (Helmke and Weyland, 2004). In the polar regions, they are the generators and processors of polar marine primary production (Legendre et al., 1992; Chapter 10, this volume), serving as the base for the entire polar food web, ultimately feeding krill, fish, whales, penguins, and seabirds. The rate of primary production varies greatly throughout Arctic and Antarctic marine waters and ice (Mock and Thomas, 2005; Garneau et al., 2009; Kellogg and Deming, 2009; Arrigo et al., 2010; see also Chapter 10).

CARBON CYCLING

The metabolic balance between autotrophy (carbon fixation) and heterotrophy (carbon respiration) in sea ice is influenced primarily by seasonal variations in irradiance levels. During spring and summer in polar regions, when incident light levels are highest and snow cover is thinnest, higher irradiance levels lead to large blooms of photosynthesizing algae and diatoms in sea-ice habitats (Horner and Schrader, 1982; Legendre, 1990; Legendre et al., 1992; von Quillfeldt et al., 2003; Riedel et al., 2008). Algal cell concentrations in sea ice can vary by up to 6 orders of magnitude from $<10^4$ to $>10^9$ cells liter^{-1} (Arrigo et al., 2010). In Arctic sea ice, average dissolved organic carbon concentrations increase significantly as a result of the bloom, from 30 mg C m^{-2} in prebloom months to 333 mg C m^{-2} during bloom months, with postbloom values of 509 mg C m^{-2} (Riedel et al., 2008). These large algal blooms also significantly alter the gaseous composition and chemistry of sea-ice brines (Gleitz et al., 1995; Günther and Dieckmann, 1999; Thomas et al., 2001; Papadimitriou et al., 2007). CO_2 exhaustion by the photosynthesizing organisms produces increasingly alkaline conditions. Gleitz et al. (1995) reported that the pH of brines taken from first-year sea ice in the Weddell Sea, Antarctica, ranged from 8.2 to 9.9 when measured in the summer months, but dropped to 7.8 to 8.9 during the winter. At the same time, the photosynthetic activity produces large amounts of O_2, leading to supersaturated solutions with O_2 concentrations as high as 932 μmol kg^{-1} in the summer months, compared to a maximum concentration of 650 μmol kg^{-1} during low-light, winter conditions.

The high O_2 and high organic carbon levels present in sea ice during algal blooms also support larger heterotrophic assemblages during the summer months; however, the autotrophic population is still dominant (Riedel et al., 2008). In contrast, heterotrophic assemblages dominate sea-ice habitats during the low-irradiance conditions (low incident light and thicker snow cover) of winter. Riedel et al. (2008) reported that heterotrophic bacteria contributed up to 60% of total carbon in first-year, landfast Arctic sea ice with high snow cover during the late winter, compared to about 8% under low snow cover in the late summer.

NITROGEN, PHOSPHORUS, SILICA, AND IRON CYCLING

Cycling of the elements N, P, and silicon (Si) in sea ice is primarily driven by biological activity. Numerous studies have demonstrated microbial uptake of N, P, and Si, by noting depletion in the aqueous concentrations of these species relative to their theoretical dilution lines, as predicted from change in sea ice bulk salinity during melting (Meese, 1989; Dieckmann et al., 1991; Gleitz and Thomas, 1993; Gleitz et al., 1995; Kennedy et al., 2002; Arrigo et al., 2003; Papadimitriou et al., 2007). Ammonia oxidation has been demonstrated as an important component of the nitrogen cycle in Antarctic sea ice (Priscu et al., 1990). Suboxic and anoxic conditions have also been documented in brines in Arctic sea ice and dissimilatory nitrate reduction and anaerobic ammonia oxidation measured under these conditions (Rysgaard and Glud, 2004). Nitrogen has also been demonstrated to be a limiting nutrient in some sea ice systems. For example, Smith et al. (1997) reported a significantly higher growth rate and final cell concentration in nitrogen-enriched cultures of sea ice algae from the Canadian High Arctic, compared to those enriched with only phosphorus or silicate.

Silicic acid [$Si(OH)_4$] is a key element in the formation of diatomic frustules, and $Si(OH)_4$ limitation has been demonstrated in Arctic sea ice systems (Cota et al., 1990; Gosselin et al., 1990; Smith et al., 1990). For example, Cota et al. (1990) demonstrated a strong correlation between $Si(OH)_4$ and chlorophyll concentrations in bottom-ice samples collected in the Canadian High Arctic. $Si(OH)_4$ concentration in sea ice is controlled primarily by exchange with the surrounding water column as a result of advective, diffusive, and turbulent flow, depending on the physical setting (Cota et al, 1987; Vancoppenolle et al., 2010). Fe concentrations in sea ice in the Antarctic (Lannuzel et al., 2010) and Arctic (Measures, 1999) are relatively high and thus unlikely to be significantly limiting in these systems.

MICROBIOLOGY OF ICE-SHELF COMMUNITIES

Polar ice shelves are thick masses of ice floating on the ocean. They are formed through glacial ice and ice sheets pushing onto the sea or long-term accumulations of sea ice (Hawes et al., 2008; Mueller and Vincent, 2006). Until recently, polar ice shelves were seen as mostly abiotic glaciological features. However, it has become clear that their aquatic ecosystems are oases for life. The majority of the research to date on the microbiology of polar ice shelves has been carried out on the McMurdo Ice Shelf region in Antarctica and the Ellesmere Island ice shelves in the Canadian High Arctic.

Ice Shelves as a Microbial Habitat

Polar ice-shelf ecosystems are characterized by supraglacial meltwater lakes that provide habitats for rich microbial diversity (Fig. 6). These freshwater cryoecosystems contain liquid water during the summer months, but freeze over completely in the winter (Hawes et al., 2008). Microbes also likely inhabit the ice itself as well as the seasonal accumulations of snow, depending on the availability of space, nutrients, and liquid water; however, these habitats have not been explored yet.

Ice-Shelf Ecology: Diversity and Environmental Adaptation

Overall, biomass accumulations in the supraglacial lakes are dominated by benthic assemblages, while planktonic assemblages remain at

FIGURE 6 Network of supraglacial lakes and ponds on the McMurdo Ice Shelf near Bratina Island, Antarctica. (Photo credit: A. Jungblut.)

low concentrations. Cyanobacteria contribute the dominant autotrophic biomass in supraglacial ice-shelf ecosystems, and comprise at least 88% of total algal cell counts in Arctic ice-shelf mats such as the Markham Ice Shelf (Vincent et al., 2004). Filamentous, exopolymer-producing oscillatorian cyanobacteria produce complex biofilms that in turn provide refugia for diverse heterotrophic bacteria and eukaryote communities, which survive and proliferate within the boundaries of the cyanobacterial mats (Fig. 7). Microbial eukaryotes might have an overall low biomass in these cryoecosystems, yet they still display diverse communities, including various taxa within protist clades: *Chlorophyta, Ulvophyceae, Prasinophyceae, Chrysophyceae, Bacillariophyceae, Amoebazoa, Ciliophora, Dinophyceae, Fungi, Chytridiomycota,* euglenoids, and *Cercozoa,* as well as microfauna such as *Rotaria, Tardigrada,* and *Nematoda* (Howard-Williams et al., 1989; Vincent et al., 2000; Varin et al., 2010; Jungblut et al., in press).

Microbial mats on the McMurdo Ice Shelf can be stratified. Cyanobacteria are usually found in the oxic top layers, with mats often becoming anoxic toward the bottom of the mats due to heterotrophic decomposition processes (Vincent et al., 1993). A macroscopical lamination, however, is not visible in Arctic ice-shelf mats. Heterotrophic bacteria are well represented in polar benthic ice-shelf communities, and *Bacteroidetes, Proteobacteria* (α-, β-, δ-, and γ-), and *Actinobacteria* were identified from both Arctic and Antarctic ice-shelf benthic microbial communities (Sjöling and Cowan, 2003; Bottos et al., 2008; Varin et al., 2010). Recent metagenomic profiling even suggests that *Proteobacteria* contribute most of the genetic material in Arctic mats (Varin et al., 2010). Bottos et al. (2008) also described *Firmicutes, Verrucomicrobia,* and *Gemmatimonadetes* on the Markham Ice Shelf and *Fibrobacteres* on Ward Hunt Ice Shelf, albeit in lower abundances. Our understanding of archaeal communities from Arctic and Antarctic ice-shelf habitats is limited. Molecular surveys retrieved genotypes grouping within *Euryarchaeota* and

Crenarchaeota; however, their total contribution of genetic material to the Arctic ice-shelf microbial mat metagenome was estimated to be less than 1% of total identified genes based on metagenomic analyses (Sjöling and Cowan, 2003; Bottos et al., 2008; Varin et al., 2010).

Although viral assemblages have also only been studied to a limited extent, they are recognized as an integral part of microbial mat communities, with, for example, a concentration of 2.3×10^7 to 16.5×10^7 viruses cm^{-2} on the Ward Hunt Ice Shelf, Arctic (Vincent et al., 2000). Proteobacterial (α-, β-, and γ-) phages were retrieved as the dominant group within the viral assemblages from two Arctic ice-shelf microbial mat communities. Cyanobacterial phages were also abundant, and some viruses of eukaryotes were detected, which reflects the taxonomic composition of heterotrophic and phototrophic hosts in the mats (Varin et al., 2010).

Polar ice-shelf cryoecosystems with analogous habitats represent ideal sites to study biogeography. Analysis of ice-shelf heterotrophic bacteria and microbial eukaryotes suggests phylogenetic affiliation with taxa from diverse environments and climatic zones ranging from Antarctica and other cryosphere habitats to temperate ecozones. Antarctic 16S rRNA gene surveys of cyanobacteria from the McMurdo Ice Shelf suggest the presence of lineages with broad geographic distributions, but also include cyanobacteria that have thus far only been identified in Antarctica (Jungblut et al., 2005; Nadeau et al., 2001). Interestingly, molecular analyses of cyanobacterial mat assemblages of Ellesmere Island ice shelves indicate the presence of cyanobacterial ecotypes that are 99% similar based on 16S rRNA gene analyses to taxa previously thought to be endemic to Antarctica, implying the global distribution of low-temperature cyanobacterial ecotypes throughout the cold terrestrial biosphere (Jungblut et al., 2010).

The polar ice shelves are characterized by perennial cold temperatures, highly variable conductivity (100 to 50,000 μS cm^{-1}), frequent freeze-thaw cycles, short phototrophic

FIGURE 7 Highly pigmented cyanobacteria-dominated microbial mats from a supraglacial pond on the McMurdo Ice Shelf, Antarctica. (Photo credit: A. Jungblut.)

growth periods, and high UV radiation (Jungblut et al., in press). Cyanobacteria dominate these benthic communities due to their ability to tolerate these extreme conditions. High-latitude cyanobacteria tend to be psychrotrophs, as a strategy of general tolerance to a broader range of temperatures allows a faster acclimatization to rapid changes (Tang et al., 1997). Cyanobacteria have also developed an array of mechanisms to preserve cell integrity during freezing (Vincent, 1988) and overcome high UV radiation via the synthesis of a variety of pigments such as carotenoids, scytonemin, and mycosporinelike amino acids (Ross and Vincent, 1998; Hodgson et al., 2004).

In contrast, physiological studies on heterotrophic bacteria from Ellesmere ice shelves showed that many were cold adapted, with growth at temperatures as low as $-10°C$ (Bottos et al., 2008). Similarly, Mountfort and coworkers isolated two novel psychrophilic bacteria, *Clostridium vincentii* and *Psychromonas antarcticus*, from sediment of supraglacial meltwater ponds on the McMurdo Ice Shelf, Antarctica (Mountfort et al., 1997, 1998), suggesting that specifically cold-adapted physiologies may be more common among heterotrophic bacteria within these cryoecosystems.

Biogeochemical Cycling in Ice-Shelf Communities

CARBON CYCLING

Polar ice shelves have extensive accumulation of organic matter despite low temperatures and an autotrophic growing season of only approximately 70 days. A total of 34 Gg of organic matter was estimated for the Ellesmere ice shelves, with an average productivity per unit area of 129 mg of C m^{-2} day^{-1}, which is above the values obtained for the Central Arctic pack ice (Mueller et al., 2006). The measured chlorophyll-specific primary productivity and heterotrophic bacterial production for Ellesmere ice shelves are in the same range as estimates for other polar microbial ice mat systems, including the McMurdo Ice Shelf, Antarctica (Vincent and Howard-

Williams, 1989). Primary productivity, however, on a per unit area basis is higher in the Arctic than Antarctic mats, because of the large volume of standing stocks of photosynthetic pigments. Heterotrophic recycling processes are estimated to be 0.037 to 0.21 mg of C m^{-2} h^{-1} on the Ward Hunt Ice Shelf (Arctic), and minor relative to primary productivity; therefore, the net effect processes lead to is an overall gain in organic standing stocks in these ice-shelf cryoecosystems (Mueller et al., 2005).

Primary productivity by the autotrophic community in benthic ice-shelf assemblages on Markham Ice Shelf (Arctic) and McMurdo Ice Shelf (Antarctica) showed broad tolerances to salinity, irradiance, and temperatures. In contrast, heterotrophic bacterial productivity was less tolerant to increased salinity and temperature on the Ward Hunt Ice Shelf (Arctic) (Hawes et al., 1999; Mueller et al., 2005). Microbial processes such as methanogenesis, denitrification, and sulfate reduction have been detected in the anoxic sediment layers beneath the mats of the McMurdo Ice Shelf (Mountfort et al., 1999) and ice-shelf microbial mats (Jungblut et al., 2009). Biomarker analysis also detected signatures of iron- and sulfur-oxidizing bacteria on the McMurdo Ice Shelf (Jungblut et al., 2009).

NITROGEN AND PHOSPHORUS CYCLING

Cryoecosystems of polar ice shelves such as supraglacial lakes are usually ultra-oligotrophic and have low ratios of dissolved inorganic nitrogen to dissolved inorganic phosphorus. Meltwater ponds of the McMurdo Ice Shelf have low nitrogen concentrations due to the marine origin of the sediments and to fixed dinitrogen input being limited to recycled ammonium-N from sediments and snow and ice melt (Hawes et al., 1993). In particular, phytoplankton community biomass in polar freshwater aquatic ecosystems is severely constrained by limited nutrients, based on nutrient enrichment studies in a High Arctic lake (Bonilla and Vincent, 2005). However, benthic microbial assemblages seem to overcome these constraints by forming nutrient-rich microenvironments, likely via nutrient trapping within the diffusion-limited mat matrix (Vincent et al., 1993; Bonilla and Vincent, 2005). Studies of Ward Hunt and Markham Ice Shelf mats on northern Ellesmere Island showed that they have 2 to 5 orders of magnitude higher ammonium-N than is found in the overlying water column. Other nutrients, such as total dissolved nitrogen, nitrate, nitrite, total dissolved phosphorus, and soluble reactive phosphorus, are 2 orders of magnitude more concentrated in the microbial mat than in the overlying water (Mueller and Vincent, 2006). Recent metagenomic profiling by Varin et al. (2010) identified diverse nutrient scavenging and recycling systems from High Arctic ice-shelf microbial mats, including genes for transport proteins and enzymes for converting larger molecules into more readily assimilated inorganic forms using allantoin degradation, cyanate hydrolysis, and phosphonatases. Nitrogen-related genes were dominated by ammonium assimilation systems, implying that the microbial mats are sites of intense mineralization, but not nitrogen oxidation, since nitrification genes were absent.

On the McMurdo Ice Shelf (Antarctica), most active dinitrogen fixation in meltwater ponds in the Antarctic to date has been attributed to the cyanobacterial genus *Nostoc*. Using environmental *nifH* gene surveys and acetylene reduction assays, Fernández-Valiente et al. (2001) estimated that biological N_2 fixation contributes 1 g m^{-2} per year from the microbial mats out of a total dinitrogen requirement of approximately 3 g m^{-2} per year for these ice-based freshwater systems, in the same range as shown for temperate rice fields.

CONCLUSIONS

Microbial investigations on polar glaciers, ice sheets, and ice shelves are still largely in their infancy, with sea-ice research being somewhat more established. Many basic questions remain unanswered; for example, how is nitrogen cycled in the subglacial environment and what is the role of microbes in this process? Are surficial glacial environments carbon sinks?

None of these environments is static, and recent observations indicate that climate change is most rapid in high-latitude regions (IPCC, 2007; Vincent, 2010). Arctic sea ice has shown a dramatic decrease in thickness and extent due to increasing overall temperatures in Arctic regions (Stroeve et al., 2007, Serreze et al., 2007). How sea ice microorganisms with critical ecological roles might respond to these changes is an important and urgent question that warrants attention (Kirchman et al., 2009; Krembs et al., 2011). Similarly, ice-shelf cryoecosystems depend on ice integrity and thus will likely be affected by more rapid temperature changes. Changes in melt dynamics for the Greenland Ice Sheet, including the increase in supraglacial lakes in the ablation zone, also modify surficial and in certain locations subglacial hydrology, which in turn affects the microbial communities and their biogeochemical cycles.

ACKNOWLEDGMENTS
M.S. was supported by National Science Foundation (NSF) grants EAR 0525567 and OPP 0636770. M.U. was supported by an NSF-IGERT fellowship from NSF grant DGE 0654336 to Montana State University. K.J. was supported by NSF-OPP grants 0739783 and 1023462.

REFERENCES
Amann, R. I., W. Ludwig, and K. H. Schleifer. 1995. Phylogenetic identification and in situ detection of individual microbial cells without cultivation. *Microbiol. Rev.* **59**:143–169.
Anesio, A. M., A. J. Hodson, A. Fritz, R. Psenner, and B. Sattler. 2009. High microbial activity on glaciers: importance to the global carbon cycle. *Glob. Change Biol.* **15**:955–960.
Anesio, A. M., B. Mindl, J. Laybourn-Parry, A. J. Hodson, and B. Sattler. 2007. Viral dynamics in cryoconite holes on a high Arctic glacier (Svalbard). *J. Geophys. Res.* **112**:G04S31.
Arrigo, K. R., T. Mock, and M. P. Lizotte. 2010. Primary producers and sea ice, p. 283–375. *In* D. N. Thomas and G. S. Dieckmann (ed.), *Sea Ice: an Introduction to Its Physics, Chemistry, Biology, and Geology.* Blackwell Science, Ltd., Oxford, United Kingdom.
Arrigo, K. R., D. H. Robinson, R. B. Dunbar, A. R. Leventer, and M. P. Lizotte. 2003. Physical control of chlorophyll *a*, POC, and PON

distributions in the pack ice of the Ross Sea, Antarctica. *J. Geophys. Res.* **108:**3316.

Bakermans, C., and M. L. Skidmore. 2011. Microbial metabolism in ice and brine at −5°C. *Environ. Microbiol.* **13:**2269–2278.

Benn, D. I., and D. J. A. Evans. 2010. *Glaciers and Glaciation.* Hodder Education, London, United Kingdom.

Bhatia, M. P., S. B. Das, K. Longnecker, M. A. Charette, and E. B. Kujawinski. 2010. Molecular characterization of dissolved organic matter associated with the Greenland ice sheet. *Geochim. Cosmochim. Acta* **74:**3768–3784.

Bhatia, M., M. Sharp, and J. Foght. 2006. Distinct bacterial communities exist beneath a high Arctic polythermal glacier. *Appl. Environ. Microbiol.* **72:**5838–5845.

Bonilla, S., and W. F. Vincent. 2005. Benthic and planktonic algal communities in a high Arctic lake: pigment structure and contrasting responses to nutrient enrichment. *J. Phycol.* **41:**1120–1130.

Boon, S., M. Sharp, and P. Nienow. 2003. Impact of an extreme melt event on the runoff and hydrology of a high Arctic glacier. *Hydrol. Processes* **17:**1051–1072.

Bottos, E. M., W. F. Vincent, C. W. Greer, and L. G. Whyte. 2008. Prokaryotic diversity of arctic ice shelf microbial mats. *Environ. Microbiol.* **10:**950–966.

Bowman, J. P., J. J. Gosink, S. A. McCammon, T. E. Lewis, D. S. Nichols, P. D. Nichols, J. H. Skerratt, J. T. Staley, and T. A. McMeekin. 1998. *Colwellia demingiae* sp. nov., *Colwellia hornerae* sp. nov., *Colwellia rossensis* sp. nov., and *Colwellia psychrotropica* sp. nov.: psychrophilic Antarctic species with the ability to synthesize docosahexaenoic acid (22:6ω3). *Int. J. Syst. Bacteriol.* **48:**1171–1180.

Bowman, J. P., S. A. McCammon, M. V. Brown, D. S. Nichols, and T. A. McMeekin. 1997. Diversity and association of psychrophilic bacteria in Antarctic sea ice. *Appl. Environ. Microbiol.* **63:**3068–3078.

Boyd, E. S., R. K. Lange, A. Mitchell, J. R. Havig, T. L. Hamilton, M. J. Lafrenière, E. L. Shock, J. W. Peters, and M. Skidmore. 2011. Diversity, abundance, and potential activity of nitrifying and nitrate reducing microbial assemblages in a subglacial ecosystem. *Appl. Environ. Microbiol.* **77:**4778–4787.

Boyd, E. S., M. Skidmore, C. Bakermans, A. Mitchell, and J. W. Peters. 2010. Methanogenesis in subglacial sediments. *Environ. Microbiol. Rep.* **2:**685–692.

Brinkmeyer, R., F. O. Glöckner, E. Helmke, and R. Amann. 2004. Predominance of β-

Proteobacteria in summer melt pools on Arctic pack ice. *Limnol. Oceanogr.* **49:**1013–1021.

Brinkmeyer, R., K. Knittel, J. Jürgens, H. Weyland, R. Amann, and E. Helmke. 2003. Diversity and structure of bacterial communities in Arctic versus Antarctic pack ice. *Appl. Environ. Microbiol.* **69:**6610–6619.

Brown, M. V., and J. P. Bowman. 2001. A molecular phylogenetic survey of sea-ice microbial communities (SIMCO). *FEMS Microbiol. Ecol.* **35:**267–275.

Caron, D. A., and R. J. Gast. 2010. Heterotrophic protists associated with sea ice, p. 327–356. *In* D. N. Thomas and G. S. Dieckmann (ed.), *Sea Ice: an Introduction to Its Physics, Chemistry, Biology, and Geology.* Blackwell Science, Ltd., Oxford, United Kingdom.

Cheng, S. M., and J. M. Foght. 2007. Cultivation-independent and -dependent characterization of Bacteria resident beneath John Evans Glacier. *FEMS Microbiol. Ecol.* **59:**318–330.

Christner, B. C., B. H. Kvitko II, and J. N. Reeve. 2003. Molecular identification of Bacteria and Eukarya inhabiting an Antarctic cryoconite hole. *Extremophiles* **7:**177–183.

Christner, B. C., G. Royston-Bishop, C. M. Foreman, B. R. Arnold, M. Tranter, K. A. Welch, W. B. Lyons, A. I. Tsapin, M. Studinger, and J. C. Priscu. 2006. Limnological conditions in subglacial Lake Vostok, Antarctica. *Limnol. Oceanogr.* **51:**2485–2501.

Christner, B. C., M. L. Skidmore, J. C. Priscu, M. Tranter, and C. M. Foreman. 2008. Bacteria in subglacial environments, p. 51–71. *In* R. Margesin, F. Schinner, J.-C. Marx, and C. Gerday (ed.), *Psychrophiles: from Biodiversity to Biotechnology.* Springer, Berlin, Germany.

Collins, R. E. 2009. Microbial evolution in sea ice: communities to genes. Ph.D. thesis. University of Washington, Seattle, WA.

Collins, R. E., G. Rocap, and J. W. Deming. 2010. Persistence of bacterial and archaeal communities in sea ice through an Arctic winter. *Environ. Microbiol.* **12:**1828–1841.

Cota, G. F., S. T. Kottmeier, D. H. Robinson, W. O. Smith Jr., and C. W. Sullivan. 1990. Bacterioplankton in the marginal ice zone of the Weddell Sea: biomass, production and metabolic activities during austral autumn. *Deep Sea Res.* **37:**1145–1167.

Deming, J. W. 2010. Sea ice bacteria and viruses, p. 247–282. *In* D. N. Thomas and G. S. Dieckmann (ed.), *Sea Ice: an Introduction to Its Physics, Chemistry, Biology, and Geology.* Blackwell Science, Ltd., Oxford, United Kingdom.

Dieckmann, G. S., and H. H. Heller. 2010. Importance of sea ice: an overview, p. 1–22. *In* D. N. Thomas and G. S. Dieckmann (ed.), *Sea Ice:*

an Introduction to Its Physics, Chemistry, Biology, and Geology. Blackwell Science, Ltd., Oxford, United Kingdom.

Dieckmann, G. S., M. Spindler, M. A. Lange, S. F. Ackley, and H. Eicken. 1991. Antarctic sea ice: a habitat for the foraminifer *Neogloboquadrina pachyderma*. *J. Foramin. Res.* **21**:184–191.

Edwards, A., A. M. Anesio, S. M. Rassner, B. Sattler, B. Hubbard, W. T. Perkins, M. Young, and G. W. Griffith. 2011. Possible interactions between bacterial diversity, microbial activity and supraglacial hydrology of cryoconite holes in Svalbard. *ISME J.* **5**:150–160.

Eicken, H. 1992. The role of sea ice in structuring Antarctic ecosystems. *Polar Biol.* **12**:3–13.

Fernández-Valiente, E., A. Quesada, C. Howard-Williams, and I. Hawes. 2001. N₂-fixation in cyanobacterial mats from ponds on the McMurdo Ice Shelf, Antarctica. *Microb. Ecol.* **42**:338–349.

Foreman, C. M., B. Sattler, J. A. Mikucki, D. L. Porazinska, and J. C. Priscu. 2007. Metabolic activity and diversity of cryoconites in the Taylor Valley, Antarctica. *J. Geophys. Res.* **112**: G04S32.

Fountain, A. G., and M. Tranter. 2008. Introduction to special section on Microcosms in Ice: the Biogeochemistry of Cryoconite Holes. *J. Geophys. Res.* **113**:G02S91.

Fuhrman, J. A. 1999. Marine viruses and their biogeochemical and ecological effects. *Nature* **399**:541–548.

Garneau, M. E., W. F. Vincent, R. Terrado, and C. Lovejoy. 2009. Importance of particle-associated bacterial heterotrophy in a coastal Arctic ecosystem. *J. Mar. Sys.* **75**:185–197.

Gerdel, R. W., and F. Drouet. 1960. The cryoconite of the Thule area, Greenland. *Trans. Am. Microsc. Soc.* **79**:256–272.

Gleitz, M., M. R. Loeff, D. N. Thomas, G. S. Dieckmann, and F. J. Miller. 1995. Comparison of summer and winter inorganic carbon, oxygen and nutrient concentrations in Antarctic sea ice brine. *Mar. Chem.* **51**:81–91.

Gleitz, M., and D. N. Thomas. 1993. Variation in phytoplankton standing stock, chemical composition and physiology during sea-ice formation in the southeastern Weddell Sea, Antarctica. *J. Exp. Mar. Biol. Ecol.* **173**:211–230.

Golden, K. M., S. F. Ackley, and V. I. Lytle. 1998. The percolation phase transition in sea ice. *Science* **282**:2238–2241.

Gosink, J. J., R. L. Irgens, and J. T. Staley. 1993. Vertical distribution of bacteria in arctic sea ice. *FEMS Microbiol. Lett.* **102**:85–90.

Gosink, J. J., and J. T. Staley. 1995. Biodiversity of gas vacuolate bacteria from Antarctic sea ice and water. *Appl. Environ. Microbiol.* **61**:3486–3489.

Gosselin, M., L. Legendre, J.-C. Therriault, and S. Demers. 1990. Light and nutrient limitation of sea-ice microalgae (Hudson Bay, Canadian Arctic). *J. Phycol.* **26**:220–232.

Grossi, S. M., S. T. Kottmeier, and C. W. Sullivan. 1984. Sea ice microbial communities. III. Seasonal abundance of microalgae and associated bacteria. *Microb. Ecol.* **10**:231–242.

Günther, S., and G. S. Dieckmann. 1999. Seasonal development of algal biomass in snow-covered fast ice and the underlying platelet layer in the Weddell Sea, Antarctica. *Antarct. Sci.* **11**:305–315.

Hawes, I., C. Howard-Williams, and A. G. Fountain. 2008. Ice-based freshwater ecosystems, p. 103–115. *In* W. F. Vincent and J. Laybourn-Perry (ed.), *Polar Lakes and Rivers*. Oxford University Press, Oxford, United Kingdom.

Hawes, I., C. Howard-Williams, and R. D. Pridmore. 1993. Environmental control of microbial biomass in the ponds of the McMurdo Ice Shelf, Antarctica. *Arch. Hydrobiol.* **127**:271–287.

Hawes, I., R. Smith, C. Howard-Williams, and A.-M. Schwarz. 1999. Environmental conditions during freezing, and response of microbial mats in ponds of the McMurdo Ice Shelf, Antarctica. *Antarct. Sci.* **11**:198–208.

Helmke, E., and H. Weyland. 2004. Psychrophilic versus psychrotolerant bacteria—occurrence and significance in polar and temperate marine habitats. *Cell. Mol. Biol.* **50**:553–561.

Hodgson, D. A., W. Vyverman, E. Verleyen, K. Sabbe, P. Leavitt, A. Taton, A. Squier, and B. Keely. 2004. Environmental factors influencing the pigment composition of *in situ* benthic microbial communities in east Antarctic lakes. *Aquat. Microb. Ecol.* **37**:247–263.

Hodson, A., A. M. Anesio, F. Ng, R. Watson, J. Quirk, T. Irvine-Fynn, A. Dye, C. Clark, P. McCloy, J. Kohler, and B. Sattler. 2007. A glacier respires: quantifying the distribution and respiration CO₂ flux of cryoconite across an entire Arctic supraglacial ecosystem. *J. Geophys. Res.* **112**: G04S36.

Hodson, A., C. Boggild, E. Hanna, P. Huybrechts, H. Langford, K. Cameron, and A. Houldsworth. 2010. The cryoconite ecosystem on the Greenland ice sheet. *Ann. Glaciol.* **51**: 123–129.

Horner, R., and G. C. Schrader. 1982. Relative contributions of ice algae, phytoplankton, and benthic microalgae to primary production in nearshore regions of the Beaufort Sea. *Arctic* **35**:485–503.

Howard-Williams, C., R. Pridmore, M. Downes, and W. Vincent. 1989. Microbial biomass, photosynthesis and chlorophyll *a* related pigments in the ponds of the McMurdo Ice Shelf, Antarctica. *Antarct. Sci.* **1**:125–131.

Huston, A. L. 2003. Bacterial adaptation to the cold: in situ activities of extracellular enzymes in the North Water polynya and characterization of a cold-active aminopeptidase from *Colwellia psychrerythraea* strain 34H. Ph.D. thesis. University of Washington, Seattle, WA.

Huston, A. L., B. B. Krieger-Brockett, and J. W. Deming. 2000. Remarkably low temperature optima for extracellular enzyme activity from Arctic bacteria and sea ice. *Environ. Microbiol.* **2:**383–388.

Huston, A. L., B. Methe, and J. W. Deming. 2004. Purification, characterization and sequencing of an extracellular cold-active aminopeptidase produced by marine psychrophile *Colwellia psychrerythraea* strain 34H. *Appl. Environ. Microbiol.* **70:**3321–3328.

IPCC. 2007. *Climate Change 2007: the Physical Science Basis. Contribution of Working Group I to the Fourth Assessment Report of the Intergovernmental Panel on Climate Change.* Cambridge University Press, Cambridge, United Kingdom and New York, NY.

Jungblut, A. D., M. A. Allen, B. P. Burns, and B. A. Neilan. 2009. Lipid biomarker analysis of cyanobacterial dominated microbial mats in meltwater ponds on the McMurdo Ice Shelf, Antarctica. *Organ. Geochem.* **40:**258–269.

Jungblut, A. D., I. Hawes, D. Mountfort, B. Hitzfeld, D. R. Dietrich, B. P. Burns, and B. A. Neilan. 2005. Diversity within cyanobacterial mat communities in variable salinity meltwater ponds of McMurdo Ice Shelf, Antarctica. *Environ. Microbiol.* **7:**519–529.

Jungblut, A. D., C. Lovejoy, and W. F. Vincent. 2010. Global distribution of cyanobacterial ecotypes in the cold biosphere. *ISME J.* **4:**191–202.

Jungblut, A. D., D. R. Mueller, and W. F. Vincent. Arctic ice shelf ecosystems. *In* L. Copland and D. R. Mueller (ed.), *Arctic Ice Shelves and Ice Islands*, in press. Springer, Berlin, Germany.

Junge, K., B. C. Christner, and J. T. Staley. 2011. Diversity of psychrophilic bacteria from sea ice—and glacial ice communities, p. 793–815. *In* K. Horikoshi, G. Antranikian, A. T. Bull, F. T. Robb, and K. O. Stetter (ed.), *Extremophiles Handbook.* Springer, Berlin, Germany.

Junge, K., H. Eicken, and J. W. Deming. 2003. Motility of *Colwellia psychrerythraea* strain 34H at subzero temperatures. *Appl. Environ. Microbiol.* **69:**4282–4284.

Junge, K., H. Eicken, and J. W. Deming. 2004. Bacterial activity at −2 to −20°C in Arctic wintertime sea ice. *Appl. Environ. Microbiol.* **70:** 550–557.

Junge, K., H. Eicken, B. D. Swanson, and J. W. Deming. 2006. Bacterial incorporation of leucine into protein down to −20°C with evidence for potential activity in sub-eutectic saline ice formations. *Cryobiology* **52:**417–429.

Junge, K., F. Imhoff, T. Staley, and J. W. Deming. 2002. Phylogenetic diversity of numerically important Arctic sea-ice bacteria cultured at subzero temperature. *Microb. Ecol.* **43:**315–328.

Junge, K., C. Krembs, J. Deming, A. Stierle, and H. Eicken. 2001. A microscopic approach to investigate bacteria under in situ conditions in sea-ice samples. *Ann. Glaciol.* **33:**304–310.

Kellogg, C. T. E., and J. W. Deming. 2009. Comparison of free-living, suspended particle, and aggregate-associated bacterial and archaeal communities in the Laptev Sea. *Aquat. Microb. Ecol.* **57:**1–18.

Kennedy, H., D. N. Thomas, G. Kattner, C. Haas, and G. S. Dieckmann. 2002. Particulate organic matter in Antarctic summer sea ice: concentration and stable isotopic composition. *Mar. Ecol. Prog. Ser.* **238:**1–13.

Kirchman, D. L., X. A. G. Morán, and H. Ducklow. 2009. Microbial growth in the polar oceans—role of temperature and potential impact of climate change. *Nat. Rev. Microbiol.* **7:**451–459.

Krembs, C., J. W. Deming, and H. Eicken. 2011. Exopolymer alteration of physical properties of sea ice and implications for ice habitability and biogeochemistry in a warmer Arctic. *Proc. Natl. Acad. Sci. USA* **108:**3653–3658.

Lannuzel, D., V. Schoemann, J. de Jong, B. Pasquer, P. van der Merwe, F. Masson, J.-L. Tison, and A. Bowie. 2010. Distribution of dissolved iron in Antarctic sea ice: spatial, seasonal, and inter-annual variability. *J. Geophys. Res.* **115:** G03022.

Lanoil, B., M. Skidmore, J. C. Priscu, S. Han, W. Foo, S. W. Vogel, S. Tulaczyk, and H. Engelhardt. 2009. Bacteria beneath the West Antarctic Ice Sheet. *Environ. Microbiol.* **11:**609–615.

Legendre, L. 1990. The significance of microalgal blooms for fisheries and for the export of particulate organic carbon in oceans. *J. Plankton Res.* **12:**681–699.

Legendre, L., S. F. Ackley, G. S. Dieckmann, B. Gulliksen, R. Horner, T. Hoshiai, I. A. Melnikov, W. S. Reeburgh, M. Spindler, and C. W. Sullivan. 1992. Ecology of sea ice biota. *Polar Biol.* **12:**429–444.

Loveland-Curtze, J., V. Miteva, and J. Brenchley. 2010. Novel ultramicrobacterial isolates from a deep Greenland ice core represent a proposed new species, *Chryseobacterium greenlandense* sp. nov. *Extremophiles* **14:**61–69.

Measures, C. I. 1999. The role of entrained sediments in sea ice in the distribution of aluminum and iron in the surface waters of the Arctic Ocean. *Mar. Chem.* **68:**59–70.

Meese, D. A. 1989. *The Chemical and Structural Properties of Sea Ice in the Southern Beaufort Sea.* CRREL Report 89-25. Cold Regions Research and Engineering Lab, Hanover, NH.

Methe, B. A., K. E. Nelson, J. W. Deming, B. Momen, E. Melamud, X. Zhang, J. Moult, R. Madupu, W. C. Nelson, R. J. Dodson, L. M. Brinkac, S. C. Daugherty, A. S. Durkin, R. T. DeBoy, J. F. Kolonay, S. A. Sullivan, L. Zhou, T. M. Davidsen, M. Wu, A. L. Huston, M. Lewis, B. Weaver, J. F. Weidman, H. Khouri, T. R. Utterback, T. V. Feldblyum, and C. M. Fraser. 2005. The psychrophilic lifestyle as revealed by the genome sequence of *Colwellia psychrerythraea* 34H through genomic and proteomic analyses. *Proc. Natl. Acad. Sci. USA* **102:**10913–10918.

Mikucki, J. A., A. Pearson, D. T. Johnston, A. V. Turchyn, J. Farquhar, D. P. Schrag, A. D. Anbar, J. C. Priscu, and P. A. Lee. 2009. A contemporary microbially maintained subglacial ferrous "ocean." *Science* **324:**397–400.

Miteva, V. 2008. Bacteria in snow and glacier ice, p. 31–50. *In* R. Margesin, F. Schinner, J.-C. Marx, and C. Gerday (ed.), *Psychrophiles: from Biodiversity to Biotechnology.* Springer, Berlin, Germany.

Miteva, V. I., and J. E. Brenchley. 2005. Detection and isolation of ultrasmall microorganisms from a 120,000-year-old Greenland glacier ice core. *Appl. Environ. Microbiol.* **71:**7806–7818.

Miteva, V. I., P. P. Sheridan, and J. E. Brenchley. 2004. Phylogenetic and physiological diversity of microorganisms isolated from a deep Greenland glacier ice core. *Appl. Environ. Microbiol.* **70:**202–213.

Miteva, V., C. Teacher, T. Sowers, and J. Brenchley. 2009. Comparison of the microbial diversity at different depths of the GISP2 Greenland ice core in relationship to deposition climates. *Environ. Microbiol.* **11:**640–656.

Mock, T., and K. Junge. 2007. Psychrophilic diatoms: mechanisms for survival in freeze-thaw cycles, p. 343–364. *In* J. Seckbach (ed.), *Algae and Cyanobacteria in Extreme Environments.* Springer, Dordrecht, The Netherlands.

Mock, T., and D. N. Thomas. 2005. Recent advances in sea-ice microbiology. *Environ. Microbiol.* **7:**605–619.

Mountfort, D. O., H. F. Kaspar, M. Downes, and R. A. Asher. 1999. Partitioning effects during terminal carbon and electron flow in sediments of a low-salinity meltwater pond near Bratina Island, McMurdo Ice Shelf, Antarctica. *Appl. Environ. Microbiol.* **65:**5493–5499.

Mountfort, D. O., F. A. Rainey, J. Burghardt, H. F. Kaspar, and E. Stackebrandt. 1997. *Clostridium vincentii* sp. nov., a new obligately anaerobic, saccharolytic, psychrophilic bacterium isolated form low-salinity pond sediment of the McMurdo Ice Shelf, Antarctica. *Arch. Microbiol.* **167:**54–60.

Mountfort, D. O., F. A. Rainey, J. Burghardt, H. F. Kaspar, and E. Stackebrandt. 1998. *Psychromonas antarcticus* gen. nov., sp. nov., a new aerotolerant anaerobic, halophilic psychrophile isolated from pond sediment of the McMurdo Ice Shelf, Antarctica. *Arch. Microbiol.* **169:**231–238.

Mueller, D. R., and W. F. Vincent. 2006. Microbial habitat dynamics and ablation control on the Ward Hunt Ice Shelf. *Hydrol. Processes* **20:**857–876.

Mueller, D. R., W. F. Vincent, S. Bonilla, and I. Laurion. 2005. Extremotrophs, extremophiles and broadband pigmentation strategies in a high arctic ice shelf ecosystem. *FEMS Microbiol. Ecol.* **53:**73–87.

Mueller, D. R., W. F. Vincent, and M. O. Jeffries. 2006. Environmental gradients, fragmented habitats, and microbiota of a northern ice shelf cryoecosystem, Ellesmere Island, Canada. *Arctic Antarctic Alpine Res.* **38:**593–607.

Mueller, D. R., W. F. Vincent, W. H. Pollard, and C. H. Fritsen. 2001. Glacial cryoconite ecosystems: a bipolar comparison of algal communities and habitats. *Nova Hedwigia* **123:**171–195.

Nadeau, T. L., E. C. Milbrandt, and R. W. Castenholz. 2001. Evolutionary relationships of cultivated Antarctic oscillatorians (cyanobacteria). *J. Phycol.* **37:**650–654.

Oerlemans, J. 2001. *Glaciers and Climate Change.* Swets & Zeitlinger, Lisse, The Netherlands.

Papadimitriou, S., D. N. Thomas, H. Kennedy, C. Haas, H. Kuosa, A. Krell, and G. S. Dieckmann. 2007. Biogeochemical composition of natural sea ice brines from the Weddell Sea during early austral summer. *Limnol. Oceanogr.* **52:**1809–1823.

Petrich, C., and H. Eicken. 2010. Growth, structure and properties of sea ice, p. 23–78. *In* D. N. Thomas and G. S. Dieckmann (ed.), *Sea Ice: an Introduction to Its Physics, Chemistry, Biology, and Geology.* Blackwell Science, Ltd., Oxford, United Kingdom.

Priscu, J. C., M. T. Downes, L. R. Priscu, A. C. Palmisano, and C. W. Sullivan. 1990. Dynamics of ammonium oxidizer activity and nitrous-oxide (N_2O) within and beneath Antarctic sea ice. *Mar. Ecol. Prog. Ser.* **62:**37–46.

Priscu, J., S. Tulaczyk, M. Studinger, M. C. Kennicutt II, B. Christner, and C. M. Foreman. 2008. Antarctic subglacial water: origin, evolution, and ecology, p. 119–135. *In* W. F. Vincent and J. Laybourn-Perry (ed.), *Polar Lakes and Rivers.* Oxford University Press, Oxford, United Kingdom.

Raymond, J. A., B. C. Christner, and S. C. Schuster. 2008. A bacterial ice-binding protein from the Vostok ice core. *Extremophiles* **12**:713–717.

Riedel, A., C. Michel, M. Gosselin, and B. LeBlanc. 2008. Winter-spring dynamics in sea-ice carbon cycling in the coastal Arctic Ocean. *J. Mar. Syst.* **74**:918–932.

Ross, J. C., and W. F. Vincent. 1998. Temperature dependence of UV radiation effects on Antarctic cyanobacteria. *J. Phycol.* **34**:118–125.

Rysgaard, S., and R. N. Glud. 2004. Anaerobic N_2 production in Arctic sea ice. *Limnol. Oceanogr.* **49**:86–94.

Rysgaard, S., R. N. Glud, M. K. Sejr, M. E. Blicher, and H. J. Stahl. 2008. Denitrification activity and oxygen dynamics in Arctic sea ice. *Polar Biol.* **31**:527–537.

Säwström, C., J. Laybourn-Parry, W. Granéli, and A. M. Anesio. 2007. Heterotrophic bacterial and viral dynamics in Arctic freshwaters: results from a field study and nutrient-temperature manipulation experiments. *Polar Biol.* **30**:1407–1415.

Säwström, C., P. Mumford, W. Marshall, A. Hodson, and J. Laybourn-Parry. 2002. The microbial communities and primary productivity of cryoconite holes in an Arctic glacier (Svalbard 79°N). *Polar Biol.* **25**:591–596.

Serreze, M. C., M. M. Holland, and J. Stroeve. 2007. Perspectives on the Arctic's shrinking sea-ice cover. *Science* **315**:1533–1536.

Sjöling, S., and D. A. Cowan. 2003. High 16S rDNA bacterial diversity in glacial meltwater lake sediment, Bratina Island, Antarctica. *Extremophiles* **7**:275–282.

Skidmore, M. 2011. Microbial communities in Antarctic subglacial aquatic environments, p. 61–81. *In* M. J. Siegert, M. C. Kennicutt II, and R. A. Bindschadler (ed.), *Antarctic Subglacial Aquatic Environments*. AGU Press, Washington, DC.

Skidmore, M., S. P. Anderson, M. Sharp, J. Foght, and B. D. Lanoil. 2005. Comparison of microbial community compositions of two subglacial environments reveals a possible role for microbes in chemical weathering processes. *Appl. Environ. Microbiol.* **71**:6986–6997.

Skidmore, M., C. Bakermans, T. Brox, B. Christner, and S. Montross. 2009. Microbial respiration at sub-zero temperatures in laboratory ices. *Geochim. Cosmochim. Acta* **73**:A1234.

Skidmore, M. L., J. M. Foght, and M. J. Sharp. 2000. Microbial life beneath a high Arctic glacier. *Appl. Environ. Microbiol.* **66**:3214–3220.

Skidmore, M., M. Tranter, S. Tulaczyk, and B. Lanoil. 2010. Hydrochemistry of ice stream beds—evaporitic or microbial effects? *Hydrol. Processes* **24**:517–523.

Smith, R. E. H., M. Gosselin, S. Kudoh, B. Robineau, and S. Taguchi. 1997. DOC and its relation to algae in bottom ice communities. *J. Mar. Syst.* **11**:71–80.

Smith, R. E. H., W. G. Harrison, L. R. Harris, and A. W. Herman. 1990. Vertical fine structure of particulate matter and nutrients in sea ice of the High Arctic. *Can. J. Fish. Aquat. Sci.* **47**:1348–1355.

Staley, J. T., R. L. Irgens, and R. P. Herwig. 1989. Gas vacuolate bacteria found in Antarctic sea ice with ice algae. *Appl. Environ. Microbiol.* **55**:1033–1036.

Stibal, M., M. Tranter, L. G. Benning, and J. Rehak. 2008. Microbial primary production on an Arctic glacier is insignificant in comparison with allochthonous organic carbon input. *Environ. Microbiol.* **10**:2172–2178.

Stroeve, J., M. M. Holland, W. Meier, T. Scambos, and M. Serreze. 2007. Arctic sea ice decline: faster than forecast. *Geophys. Res. Lett.* **34**:L09501.

Sullivan, C. W., and A. C. Palmisano. 1984. Sea ice microbial communities: distribution, abundance, and diversity of ice bacteria in McMurdo Sound, Antarctica, in 1980. *Appl. Environ. Microbiol.* **47**:788–795.

Tang. E. P. Y., R. F. Tremblay, and W. F. Vincent. 1997. Cyanobacterial dominance of polar freshwater ecosystems: are high-latitude matformers adapted to low temperatures? *J. Phycol.* **33**:171–181.

Telling, J., A. Anesio, J. Hawkings, M. Tranter, J. L. Wadham, A. Hodson, T. Irvine-Fynn, and M. L. Yallop. 2010. Measuring rates of gross photosynthesis and net community production in cryoconite holes: a comparison of field methods. *Ann. Glaciol.* **51**:153–162.

Thomas, D. N., and G. S. Dieckmann (ed.). *Sea Ice: an Introduction to Its Physics, Chemistry, Biology, and Geology.* Blackwell Science, Ltd., Oxford, United Kingdom.

Thomas, D. N., G. Kattner, R. Engbrodt, V. Gianelli, H. Kennedy, C. Haas, and G. S. Dieckmann. 2001. Dissolved organic matter in Antarctic sea ice. *Ann. Geol.* **33**:297–303.

Thomas, D. N., S. Papadimitriou, and C. Michel. 2010. Biogeochemistry of sea ice, p. 425–467. *In* D. N. Thomas and G. S. Dieckmann (ed.), *Sea Ice: an Introduction to Its Physics, Chemistry, Biology, and Geology.* Blackwell Science, Ltd., Oxford, United Kingdom.

Tranter, M., M. Skidmore, and J. Wadham. 2005. Hydrological controls on microbial communities in subglacial environments. *Hydrol. Processes* **19**:995–998.

Vancoppenolle, M., H. Goosse, A. de Montety, T. Fichefet, B. Tremblay, and J.-L. Tison.

2010. Modeling brine and nutrient dynamics in Antarctic sea ice: the case of dissolved silica. *J. Geophys. Res.* **115:**C02005.

Varin, T., C. Lovejoy, A. D. Jungblut, W. F. Vincent, and J. Corbeil. 2010. Metagenomic profiling of Arctic microbial mat communities as nutrient scavenging and recycling systems. *Limnol. Oceanogr.* **55:**1901–1911.

Vincent, W. F. 1988. *Microbial Ecosystems of Antarctica.* Cambridge University Press, Cambridge, United Kingdom.

Vincent, W. F. 2010. Microbial ecosystem responses to rapid climate change in the Arctic. *ISME J.* **4:**1087–1090.

Vincent, W. F., R. W. Castenholz, M. T. Downes, and C. Howard-Williams. 1993. Antarctic cyanobacteria: light, nutrients, and photosynthesis on the microbial mat environments. *J. Phycol.* **29:**745–755.

Vincent, W. F., J. A. E. Gibson, R. Pienitz, V. Villeneuve, P. A. Broady, P. B. Hamilton, and C. Howard-Williams. 2000. Ice shelf microbial ecosystems in the High Arctic and implications for life on Snowball Earth. *Naturwissenschaften* **87:**137–141.

Vincent, W. F., and C. Howard-Williams. 1989. Microbial communities in southern Victoria Land streams (Antarctica) II. The effects of low temperature. *Hydrobiology* **172:**39–49.

Vincent, W. F., D. R. Mueller, and S. Bonilla. 2004. Ecosystems on ice: the microbial ecology of Markham Ice Shelf in the high Arctic. *Cryobiology* **48:**103–112.

von Quillfeldt, C. H., W. G. Ambrose Jr., and L. M. Clough. 2003. High number of diatoms species in first-year ice from the Chukchi Sea. *Polar Biol.* **26:**806–818.

Wadham, J. L., S. Bottrell, M. Tranter, and R. Raiswell. 2004. Stable isotope evidence for microbial sulphate reduction at the bed of a polythermal high Arctic glacier. *Earth Planet. Sci. Lett.* **219:**341–355.

Wadham, J. L., M. Tranter, A. J. Hodson, R. Hodgkins, S. Bottrell, R. Cooper, and R. Raiswell. 2010a. Hydro-biogeochemical coupling beneath a large polythermal Arctic glacier: implications for subice sheet biogeochemistry. *J. Geophys. Res.* **115:**F04017.

Wadham, J. L., M. Tranter, M. Skidmore, A. J. Hodson, J. Priscu, W. B. Lyons, M. Sharp, P. Wynn, and M. Jackson. 2010b. Biogeochemical weathering under ice: size matters. *Glob. Biogeochem. Cycles* **24:**GB3025.

Wells, L. E., and J. W. Deming. 2006. Characterization of a cold-active bacteriophage on two psychrophilic marine hosts. *Aquat. Microb. Ecol.* **45:**15–29.

Wharton, R. A., C. P. McKay, G. M. Simmons, and B. C. Parker. 1985. Cryoconite holes on glaciers. *Bioscience* **35:**499–503.

Wharton, R. A., W. C. Vinyard, B. C. Parker, G. M. Simmons, and K. G. Seaburg. 1981. Algae in cryoconite holes on Canada Glacier in southern Victoria Land, Antarctica. *Phycologia* **20:**208–211.

Wynn, P. M., A. J. Hodson, T. H. E. Heaton, and S. R. Chenery. 2007. Nitrate production beneath a High Arctic glacier, Svalbard. *Chem. Geol.* **244:**88–102.

Yde, J. C., K. W. Finster, R. Raiswell, J. P. Steffensen, J. Heinemeier, J. Olsen, H. P. Gunnlaugsson, and O. B. Nielsen. 2010. Basal ice microbiology at the margin of the Greenland ice sheet. *Ann. Glaciol.* **51:**71–79.

CHALLENGES TO LIVING IN POLAR AND SUBPOLAR ENVIRONMENTS

IV

LOW-TEMPERATURE LIMITS OF MICROBIAL GROWTH AND METABOLISM

P. Buford Price

12

INTRODUCTION

Life on Mars before Life on Earth?

Arguments have been made that life could have originated on Mars as readily as on Earth after the cessation of a period of heavy bombardment around 3.8 billion years (3.8 Ga) ago while both planets still contained extensive icy regions.

Perron et al. (2007) recently made a convincing case that a global ocean covered much of Mars during its first few hundred million years. The Sun was only ~70% as luminous then as it is today. This led to freezing of most of the Mars ocean and to the possibility of a cold origin of life, in which oligonucleotides of adenine (A), guanine (G), cytosine (C), and uracil (U) were synthesized in veins in solid ice that remain liquid at temperatures well below the melting point of pure H_2O. The next section gives a brief explanation of binary eutectic compounds, characterized by having an intermediate composition with a melting temperature in thermal equilibrium that is lower than that of either pure compound.

To synthesize a strand of RNA requires bonding of ribose and phosphate to each

nucleotide base in the strand. Phosphate is available in Martian and terrestrial rocks and oceans. Until recently, bonding of ribose has presented a barrier. In a scientific breakthrough, Ricardo et al. (2004) discovered an efficient inorganic pathway for the synthesis of ribose from chemically available boron ions, thus avoiding the usual condensation of the ribose into tars. Kirschvink et al. (2006) argued that high concentrations of boron ions are more likely to have been present on the early Martian surface than on Earth. Evaporitic environments capable of concentrating borate minerals were likely present on ancient Mars. This borate-dependent ribose synthesis pathway on Mars removed the barrier to synthesis of DNA and RNA.

The heavy meteorite bombardment before 3.8 Ga ago could have ejected matter into orbit from Mars. Had cellular life then existed on Mars, these ejecta might have transported cells within rock crevices, protected from solar UV, to Earth. The presence of organics in the Martian meteorite ALH84001 found in Antarctica demonstrates that some rocks from Mars reached Earth relatively undamaged. That meteorite was briefly the topic of great excitement due to the presence of a nanometer-scale object with a bacillus-like shape that some thought was evidence for a primitive life

P. Buford Price, Physics Department, University of California, Berkeley, CA 94720.

Polar Microbiology: Life in a Deep Freeze
Edited by Robert V. Miller and Lyle G. Whyte © 2012 ASM Press, Washington, DC

form. Despite the doubts regarding a life form in that meteorite, it is now known that carbonaceous chondrites such as Murchison have deposited as many as 80 different amino acids and several nucleotide bases onto Earth (Martins et al., 2008).

Sleep and Zahnle (1998) pointed out that deep subsurface refuges from impacts were safer on Mars than on Earth. Its habitable zone extended deeper. Due to its greater age and relatively placid history, life could have originated on Mars before the Earth-Moon system even formed.

Lazcano and Miller (1994) estimated the time durations for life to arise on Earth and to evolve to cyanobacteria. Starting with the paleontological evidence that late accretion impacts may have killed off putative life on Earth as recently as 3.8 Ga ago and that stromatolite-building phototactic bacteria were already in existence by 3.5 Ga ago, this leaves 300 Ma to go from the prebiotic soup to the RNA world and thence to cyanobacteria, which they argued is more than sufficient for life to have arisen. Taking into account the time duration of various steps, including loss of intermediate prebiotic compounds due to passage of the entire ocean through deep-sea vents in less than 10 Ma, and assuming that the rate of gene duplication of ancient bacteria was comparable to the present values, they concluded that the time interval from the beginning of the primitive soup to cyanobacteria was no more than 10^7 years.

Evidence Favoring an Icy Origin of Life on Earth or Mars

The origin of life on Earth or Mars is generally thought to have required the availability of organic compounds synthesized from simpler precursor molecules. The hypothesis of an RNA world depends on the self-assembly of a monomer unit consisting of one of four bases—the pyrimidines C and U and the purines A and G—bonded together into a phosphate-sugar-base unit called a nucleotide. The concentration of prebiotic compounds such as HCN in the early terrestrial ocean

at 100°C and pH 7 has been estimated to be only $\sim 4 \times 10^{-12}$ M, which is far too weak for synthesis of nucleotide bases and amino acids to compete with their decomposition by hydrolysis (Miyakawa et al., 2002). In a series of experiments, Miller, Orgel, and their collaborators (Miller and Orgel, 1974) exploited the fact that freezing concentrates impurities into eutectic phases of liquid in veins within ice in oceans or lakes. As an ice crystal forms, essentially only the water molecules join the growing crystal, while impurities like salt or cyanide are excluded into the liquid phase. In polycrystalline ice the impurities become concentrated into quasi-one-dimensional liquid veins at triple junctions of individual crystals. In thermodynamic equilibrium the veins remain in the liquid phase by having their diameters shrink with decreasing temperature, and the concentration of prebiotic molecules in the liquid veins becomes orders of magnitude higher than in a nonfrozen ocean or lake. The water activity in the veins becomes so low that polymerization of the prebiotic molecules is favored over hydrolysis. In this way Sanchez et al. (1966) synthesized liquid purines in concentrated HCN at -18°C, and Cleaves et al. (2006) synthesized liquid pyrimidines from cyanoacetaldehyde and urea at -20°C.

Two recent experiments illustrate the success of eutectic freezing in leading to polymerization of nucleobases. Miyakawa et al. (2002) formed a eutectic solution of NH_4CN at -78°C and left it in a sealed vessel. After 27 years at -78°C they detected monomers of 7 different amino acids and 11 types of nucleobases in the eutectic phase, from which they concluded that a cold Earth is more favorable for chemical evolution than a warm Earth. (An examination of the substructure of that ice in a stereomicroscope would have shown that the veins were filled with a dark-brown liquid at the triple junctions of clear ice crystals.) Later Trinks et al. (2005) showed that starting with NH_4CN in veins in sterile sea ice, chains of RNA up to 420 bases long were synthesized when the ice was cycled between -7 and -24°C for 1 year.

An argument against a hot origin of life is the short lifetime of nucleobases (Levy and Miller, 1998) and of ribose (Larralde et al., 1995) against decomposition at high temperatures. For example, in order for life to have arisen at 100°C or higher, the nucleobases and ribose could not have accumulated unless their syntheses occurred on time scales of only a few days and a few hours, respectively. Furthermore, the ester linkage of ribose and phosphoric acid in RNA is unstable against hydrolysis, a problem that is avoided if the compounds are trapped in veins in ice.

MICROBIAL SIZES IN GLACIAL ICE

Chapter 11 deals with sea ice, which contains brine pockets and large channels that allow it to be in intimate contact with a rich variety of ocean life: eukarya, bacteria, archaea, and viruses. Our discussion excludes sea ice, subglacial lakes, as well as permafrost, which contains a high percentage of rock particles. Microbes in glacial ice on Greenland and Antarctica were transported by wind from oceans and land surfaces and have undergone sorting by sizes and shapes during transit to polar ice surfaces. In his discussion of sorting of airborne particles, Jaenicke (1980) showed that their size distribution peaks between 0.1 and 3 μm, with maximum residence times of 2 to 5 days at heights of ~1 km to the middle of the troposphere. Particles smaller than 0.1 μm have

large mechanical mobility and quickly attach to larger aerosols, and particles larger than ~3 μm have a large sedimentation rate determined by Stokes' law. Unbiased measurements show that the great majority of microbes in glacial ice are between 0.1 and 3 μm in size. For example, in contrast to the 10-fold-higher concentration of viruses than microbes in the ocean, the concentration of viruses in ice is quite small, except perhaps for those incorporated into cells. Using flow cytometry to study sizes in the West Antarctic Ice Sheet (WAIS) Divide ice core, Priscu et al. (J. Priscu, C. Foreman, and J. McConnell, presented at WAIS Divide Science Meeting, Kings Beach, CA, 4 to 5 October 2007) derived the size distribution of bacteria and mineral dust particles (Fig. 1, left and middle). Miteva and Brenchley (2005) noted that many of the so-called dwarf microbes in the silty ice at the bottom of the Greenland Ice Sheet Project (GISP2) ice core were able to pass through a 0.2-μm Nuclepore filter (Fig. 1, right). Cultivations for several months at 2 to 5°C gave colonies of cells down to 0.2 μm or smaller. They raised the question of whether they are a distinct class or are starved, dormant, dwarf forms of normal-sized microbes. The present author (unpublished data) has found that the ratio of the number of microbes in melted samples from the WAIS Divide ice core that passed through a 0.2-μm Nuclepore filter and were caught on

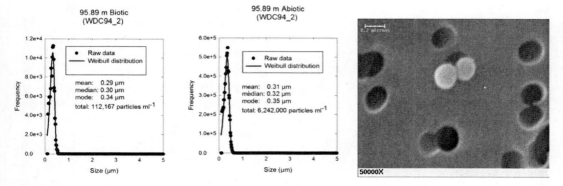

FIGURE 1 Size distributions of biotic (left) and abiotic (middle) particles in WAIS Divide ice. (Right) Dwarf cells on 0.2-μm filter. (Left and middle panels courtesy of John Priscu, reproduced with permission. Right image reprinted from Miteva and Brenchley [2005], with permission of the publisher.)

a 0.1-μm filter ranged from ~0.1 to ~1 times the number caught on a 0.2-μm filter. Christner et al. (2008) used flow cytometry to study cell sizes at depths of 1,686, 2,334, and 3,612 m in Vostok ice. They found modal sizes of 0.7, 3.5, and 0.7 μm, respectively. D'Elia et al. (2008, 2009) centrifuged melted ice from the Vostok accretion ice in order to concentrate the small number of bacteria and eukaryotic cells (fungi and algae). They did not measure the size distribution.

THREE ANSWERS TO THE QUESTION OF HABITATS FOR MICROBIAL SURVIVAL AND METABOLISM IN ICE

In any terrestrial ice (or Martian ice), a tiny fraction of the volume is interlaced with liquid veins within which microbes adapted to harsh conditions can survive. Price (2000) was the first to propose that microbes smaller than a few microns in size (thus, all bacteria and archaea) can obtain water, energy, carbon, and other essential elements if they are segregated into liquid veins that are present at triple junctions of the ice grains in all polycrystalline ice below the firn layer. Antarctic ice is acidic and its major impurity is SO_4^{2-}, with a eutectic temperature of −73°C. Since this temperature is lower than that of the coldest ice on Earth (−56°C), all Antarctic ice that is deep enough to have a well-developed polycrystalline structure will contain liquid veins. They arise as follows. Micron-sized droplets of acids deposited as aerosols are essentially insoluble in ice crystals. Coarsening of deep ice takes place by migration of grain boundaries, which sweep through and scavenge the droplets. The acid ultimately ends up concentrated along veins kept liquid by virtue of its low eutectic temperature. The concentration of acid in veins, C_{vein}, is governed by the free-energy requirement that in equilibrium the liquid be on the freezing line. The colder the ice, the more concentrated must be the acid solution in order to keep the liquid from freezing. The ratio of vein diameter to average grain diameter is proportional to $\sqrt{(C_{bulk}/C_{vein})}$, where C_{bulk} is the concentration of acid in meltwater, typically ~1 μM for Antarctic ice. Price (2000) derived the dependence of acid molarity and of vein diameter on temperature for ice with the impurity composition that is typical of Antarctica. For example, for an average crystal size of 2 cm and an ice temperature of −10°C, corresponding to a location 500 m above bedrock at Vostok Station, the vein diameter would be ~7 μm and the acid molarity would be ~2 M. The drawing in Fig. 2 shows two bacteria in the vein structure (Price, 2000).

Several species of archaea and bacteria are now known to live in acidic mines with pH as low as 0 (~2.5 M if H_2SO_4). One species of a *Thiobacillus*-like bacterium lives in a Greenland mine (83°N) at a pH of 0 and a temperature of ≈−30°C 6 months of the year. The Antarctic glacial environment is even harsher. There is high pressure (up to ~400 bars), low temperature, highly acidic medium in the veins, low nutrient level, and complete darkness. The following equation gives an example of a reaction of methanosulfonate, an aerosol constituent, which provides energy and a source of carbon and sulfur for biosynthesis when catalyzed by microbes in a liquid vein.

$$CH_3SO_3^- \rightarrow HS^- + HCO_3^- + H^+$$

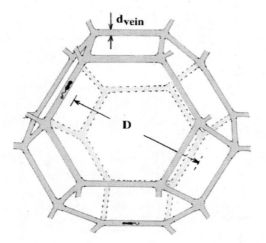

FIGURE 2 Sketch of microbes confined to liquid veins in glacial ice. (Reprinted from Price [2000] with permission.) (Copyright 2000, National Academy of Sciences, U.S.A.)

This equation represents the decomposition of methanosulfonate, which would yield 55 kcal/mol at −10°C.

Other elements such as nitrogen, phosphorus, and iron necessary for life are also present in the veins. Using a scanning spectrofluorometer to scan ice cores at −25°C in the National Ice Core Laboratory, Rohde and Price (2007) mapped the concentration of tryptophan (Trp), a proxy for microbial cells, and F420, a coenzyme found in detectable concentrations only in methanogens. In their study of Greenland ice they found that the fluorescence intensity occasionally shot up a factor of nearly 10^2 as the fluorimeter crossed localized linear regions within the ice. They interpreted these spikes as evidence for the beam crossing a liquid vein containing a high concentration of microbes.

A second microbial habitat in ice is surfaces of mineral grains, especially those with layer structures, which are separated from solid ice by a thin layer of what has come to be called "unfrozen water." Microbes attached to such surfaces are also coated with unfrozen water within which nutrients can be transported. Wettlaufer (1999) developed the theory, which accounts for a host of environmental phenomena such as the ability of ice skaters to glide at temperatures far below freezing. His equation for equilibrium undercooling as a function of thickness of the unfrozen water contains four terms: an expression for depression of the freezing point of an ionic solution; the short-range attractive van der Waals contribution within a few tenths of a nanometer of a surface; the Coulombic contribution due to charges on the mineral surface; and the Gibbs-Thomson contribution, which depends on curvature of the walls of any small pores present. On clay grains the effect is quite large: the thickness of the unfrozen layer decreases slowly from several tens of nanometers just below 0°C to ~0.3 nm at ~−80°C. Experiments have shown that the unfrozen water still has considerable mobility at such a low temperature.

Sheridan et al. (2003) and Tung et al. (2006) discovered that the silty ice in the bottom 13 m of the GISP2 ice core contains 10^8 to 10^{10} cells/cm^3, which had preexisted in the underlying clay-rich wetland before Greenland acquired its present glacier cover. The high concentration correlated with huge excesses of CO_2 and CH_4 (Tison et al., 1998) that Tung et al. attributed to the in situ metabolism of microbes such as iron-reducing bacteria and methanogenic archaea over the last 10^5 to 10^6 years. They found that more than 90% of the cells in the silty ice are attached to Fe-rich clay grains up to several microns in diameter. Their measurements of the distribution of microbes on grain surfaces (Fig. 3, left) showed that the number of attached cells was proportional to grain diameter (or perimeter) instead of grain area, which suggested that the microbes accessed nutrients at grain edges. This led them to propose a mechanism (Fig. 3, right) by which a shuttle molecule such as phenol in a thin unfrozen liquid film on a clay grain transports an electron from an attached microbial cell to an edge of one of the clay layers that contains ferric iron. The shuttle mechanism is analogous to the way in which micalike minerals conduct direct current along layers (but not perpendicular to layers) when a battery is applied. Instead of a battery, diffusion along a concentration gradient drives electrons from an edge into the interior surfaces, where eventually almost all of the ferric ions are converted into ferrous ions, and the clay eventually collapses after providing energy to Fe-reducing bacteria on its external surfaces (Tung et al., 2006).

The third habitat is the most general one. The known existence in ice of microbes unable to adapt to the hostile chemical environment in a vein, of eukarya much too large to fit into a vein, and of aerobes coexisting with strict anaerobes at the same depth and within ~1 cm of each other in an ice core implies the need for an additional habitat. Rhode and Price (2007) proposed a habitat that accommodates those requirements. They showed that bacteria, archaea, and even eukarya as large as tens of microns in size, at all depths, independent of oxygen content in the ice, can metabolize by redox reactions with dissolved small molecules

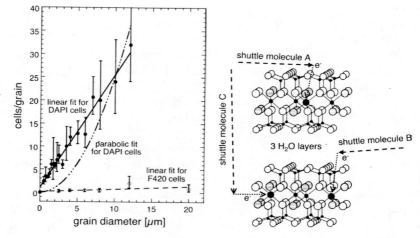

FIGURE 3 (Left) Linear fits to diameter (or perimeter) instead of area imply that cells on clay grains get access to their food at edges. (Right) Fe(III) reduction via a shuttle molecule that transports an electron to one of three locations with access to an Fe(III) ion (large black discs in octahedral planes). Dashed lines show examples of paths of shuttle molecules; dotted lines show paths of electrons. Several cells might be attached to outer surfaces of clay grains within a coating of unfrozen water. (Reprinted from *Astrobiology*, Tung et al. [2006], with permission of the publisher.)

diffusing through the crystal structure, without the need for a vein habitat. They calculated that the metabolic rate is limited only by the requirement that the diffusion rate (D) of the molecular participants in the redox reactions be at least as great as $\approx 10^{-15}$ m^2 s^{-1} at the temperature of the ice. For most molecules with up to five atoms (e.g., CO_2, H_2, O_2, N_2, CO, and CH_4), D is sufficiently large to maintain metabolism. Both nutrient and waste molecules move by interstitial diffusion at a rate that allows only ~1 redox reaction/h at $-10°C$ or 1 reaction/week at $-32°C$. Thus, in a single ice grain several millimeters in size, microbes behave as a noninteracting community. An interesting consequence is that methanogens can exist in the same crystal grain within a few millimeters of aerobes, which reduce the concentration of O_2 by virtue of their metabolism. Many strict anaerobes, including methanogens, express the enzymes catalase and superoxide dismutase to eliminate the toxic by-products of O_2 inside their cells. A case in point is that Abyzov et al. (1998a) found that eukarya up to several tens of microns in size and much too large to fit into veins can survive in ice. When isolated in ice grains, microbes that would normally metabolize via large molecules such as acetate or ethanol may be able to metabolize by expressing genes that allow them to use smaller molecules such as CO or H_2.

ADAPTATION OF NONPSYCHROPHILES TO LIFE AT LOW TEMPERATURE

Depending on wind speed and direction, microbes are swept up from diverse terrestrial and oceanic environments and blown onto glacial ice. Ocean surfaces are particularly interesting, as we will see toward the end of this chapter. The surface contains marine gels that are three-dimensional networks of biopolymers. They may serve as nutrients and attachment surfaces for microbes that are ejected from the ocean by bursting bubbles (Leck and Bigg, 2005). As soon as they are deposited onto snow surfaces of glacial ice sheets, they are confronted with temperatures ranging from $-31°C$ in central Greenland ice to $-56°C$ in East Antarctic ice. Most inblown microbes were not adapted to such extremely low temperatures and might not survive sudden immersion in surface snow. The decrease in microbial concentration in the first tens of meters, as seen later in Fig. 6, may result in part from cell death. It is interesting to note that *Escherichia coli* and several other mesophiles were found to undergo minimal cell lysis during exposure for 2 months in an Antarctic Ocean environment at −1.8°C (Smith et al., 1994). Subsequent incubation at various temperatures showed that the prolonged polar marine exposure had lowered optimal growth temperatures enough that they were able to form colonies in Antarctic seawater at −1.8°C, which indicated that physiological adaptation had occurred during the 2-month-long exposure. Furthermore, the cold-adapted microbes were no longer able to form colonies at 37°C. The conclusion is that the changes in membrane lipids, expression of cold shock proteins, and mutational events during exposure shifted their optimal growth temperatures downward.

Systematic live/dead staining tests as a function of depth might provide information on the rate of survival of microbes incorporated into glacial ice. Two different stains—usually propidium iodide and SYTO 9 mixed together—are applied to cells. The SYTO 9 responds to nucleic acids throughout the cell. A cell with an intact membrane appears yellowish, whereas a dead cell has a damaged membrane that is penetrated by the propidium iodide and appears reddish. The method is not completely reliable, as it depends on the ratio of concentrations of the two stains and perhaps also on the temperature of the cell at the time of staining. Miteva et al. (2006) reported that only a few percent of the cells at various depths in the GISP2 ice were judged alive on the basis of their tests. D'Elia et al. (2008, 2009) carried out live/dead tests on bacteria and fungi from Lake Vostok accretion ice and found ratios ranging from 0 to 100%.

Not enough measurements have yet been made to determine whether there is a systematic dependence of live/dead ratios of cells on depth and thus on exposure time in glacial ice.

At the end of this chapter it is suggested that flow cytometry might be used to evaluate for each cell whether it is alive or is dormant or dead.

METABOLIC RATES OF MICROBES FOR GROWTH, MAINTENANCE, AND SURVIVAL IN ICE

Figure 4 shows that metabolism of microbial communities operates at three distinct levels, depending on availability of nutrients and space for microbes to move. With an unlimited supply of nutrients in a volume sufficiently large that there is no competition for the nutrients, the growth rate shown by the line labeled "growth" that passes through the cluster of inverted triangles obeys an Arrhenius equation with an exponential factor $\exp(-U/RT)$. The activation energy U is approximately 110 kJ, and the average metabolic rate appears as a straight line on a semilog plot when expressed in grams of carbon per gram of cell carbon taken up per unit time as a function of $1/T$, with T in kelvins. The empirical rule of thumb that rates double over a given change of Celsius temperature is valid only over a limited range of temperature. The Arrhenius relation represents the correct physicochemical behavior for a process with a single activation energy, whereas an expression in terms of doubling time for a given increment of Celsius temperature could, at best, provide an approximate behavior over a narrow range of temperatures.

The line labeled "maintenance" that passes through the solid squares applies when a microbial community has reached a concentration when competition for available nutrients is inadequate to maintain growth, and a steady state of constant concentration at a given temperature is maintained. Such limitation is seen in the late stage of growth in a petri dish and limits the extent of microbial "bloom" in the ocean. Maintenance metabolism is the energy required for osmotic regulation, maintenance of intracellular pH, futile cycles, turnover of macromolecules, motility, and energy dissipation by proton leak and ATP hydrolysis,

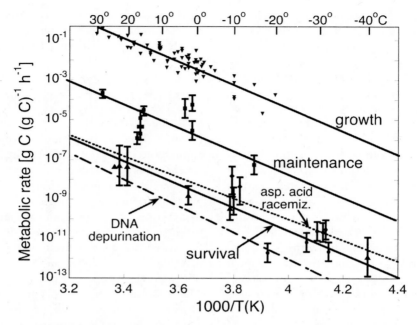

FIGURE 4 Arrhenius plot showing data on microbial metabolism. See text for explanation. (Reprinted from Price and Sowers [2004] and augmented by points from Tung et al. [2005], both with permission of the publisher.) (Copyright 2004, 2005, National Academy of Sciences, U.S.A.)

exclusive of biomass production. The activation energy for maintenance is approximately the same, ~110 kJ, as that for growth, but with a preexponential factor that is a thousand times smaller.

The line labeled "survival" that passes through the triangular points applies when the microbes are immobilized in a closed community in which there is no room for growth or even motion and the food supply is so limited that it can reach the microbes only by diffusion through ice, shale, or compact sediment or soil. The activation energy for survival metabolism turns out also to be ~110 kJ, but with an ~10^6-fold smaller preexponential factor than for unlimited growth. This survival level would apply to putative terrestrial or Martian microbes frozen into veins in ice or onto mineral grain surfaces in ice. The metabolic rate at −40°C was derived by assuming that the excess of N_2O gas found by Sowers (2001) at about the same depth in Vostok ice where Abyzov et al. (1998a) had found a large increase in cell

concentration was due to production of N_2O by nitrifying bacteria. Miteva et al. (2007) obtained additional data confirming the excess of N_2O at that depth. If the correlation of excess N_2O to excess microbes is accepted as causal, it implies that microbes can metabolize at a temperature as low as −40°C.

Some of the data with triangular symbols in Fig. 4 were obtained by Tung et al. (2005), who noticed that Brook et al. (1996) had found methane concentrations up to ~10^3-fold higher at depths of 2,954, 3,018, and 3,036 m, nearly 100 m above bedrock, in the GISP2 ice core. At those same intervals, Tung et al. found excess cell concentrations and F420 autofluorescence produced in situ by methanogens that accounted quantitatively for the excess methane. At depths just a few tens of centimeters above and below the depths with high methanogen concentrations, the F420 concentration had dropped to normal levels, which supported the interpretation that the excess methanogens at discrete locations in the ice had been making methane in situ for hundreds of thousands of years. Later, using scanning spectrofluorimetry, Rohde et al. (2008) found eight depths in GISP2 ice where large excesses of nitrifying bacteria coincided in depth with locations where Sowers et al. (2003) had found narrow spikes of excess N_2O. The metabolic rates derived from those data are included in Fig. 4.

Correlations such as these of methane/methanogen spikes and N_2O/nitrifier spikes show that a sufficiently high concentration of microbes in ice can sometimes distort the climate record and, moreover, can be used to calculate the incredibly low rate of survival metabolism. For a gaseous metabolic product in a closed system, Price and Sowers (2004) inferred the rate by measuring the concentration, n_j, of microbes and average carbon mass per microbe, m_j, both applied to microbes of type j that have been metabolizing for a time t in a sealed system in which they are immobilized and unable to grow. Then Y_j, the measured concentration of biogenic gas of type j generated at temperature T, is inferred from the following expression:

$$R(T) = [(g\ carbon)\ (g\ carbon)^{-1}\ year^{-1}]$$
$$= 2.33 Y_j(T) g_j / f_j n_j m_j t$$

where the coefficient 2.33 is a conversion factor that applies if the gas yield Y_j of type j is expressed in ppbv relative to air in the ice, g_j is the fraction of metabolic products leading to carbon, n_j (in cm^{-3}) is the concentration of living, metabolizing microbes producing gas of type j, m_j (in fg) is the average mass per cell of type j, t (in years) is the retention time of the gas in the ice, and the air content in the ice core at standard temperature and pressure is ~0.09 cm^3 (g ice)$^{-1}$ for GISP2 ice.

Two comments are in order. First, it should be pointed out that no correction was made for the fraction of microbes that were dead or dormant and no longer metabolizing. The large spread in the data for survival metabolism may be due in part to the varying live/dead fraction. Second, it should be noted that a single rate with an Arrhenius temperature dependence on $\exp(-U/kT)$

applies to a community of microorganisms, not to a single species. By contrast, a single species of microbes in a culture would have followed not an Arrhenius curve but a bell-shaped curve with a peak temperature that depends on whether they are psychrophiles, mesophiles, or thermophiles.

A few data have been obtained on production of metabolic products at low temperature on a laboratory time scale. For example, Miteva et al. (2007) have reported the production of N_2O by the nitrifier *Nitrosomonas cryotolerans* at −12 and −32°.

EVIDENCE THAT THE MAIN ROLE OF SURVIVAL METABOLISM IS TO REPAIR MACROMOLECULAR DAMAGE

In Fig. 4 the approximate agreement of the data on survival rates of an immobilized microbial community with rates of racemization and depurination shown in dotted and dashed lines (extrapolated from laboratory measurements at higher temperatures) suggests that microbial communities imprisoned in environments such as veins or clay grains in ice use their limited source of nutrients almost entirely to repair molecular damage, enabling them to stay alive for millions of years, even if their habitat is too confining for them to grow or move toward sources of nutrients. The dotted line is an extrapolation of data on spontaneous amino acid racemization measured at higher temperature (Brinton et al., 2002), and the dashed line is an extrapolation of data on the rate of spontaneous DNA depurination (Lindahl and Nyberg, 1972). Despite being immobilized, the microbes survive because gaseous nutrients diffuse toward them and gaseous waste products diffuse away. Clarke (2003) has shown that almost all cells contain an enzyme called methyltransferase that can repair aging proteins, and he suggests that additional repair enzymes may also exist.

Death ultimately results from exhaustion of a limited supply of nutrients or from the accumulation of incompletely repaired double-strand breaks.

LIMITS TO GROWTH AND SURVIVAL OF MICROBES IN ICE

The question of how long a microbe metabolizing in ice can live without reproducing is fascinating. First is the problem of confinement in narrow veins, compaction against clay grains, confinement within an ice grain, and dependence on diffusion of reactants and products of redox reactions. Veins a few microns in diameter or even narrower are spacious enough to allow growth of ultrasmall cells (<0.2 μm in size) that have been discovered in Greenland ice (Miteva and Brenchley, 2005) and in Antarctic ice (Priscu et al., presented).

Even if available volume is not a barrier to growth of such small cells, there remains the far larger metabolic energy that would be required for growth than for survival (Fig. 4). Price (2009) has recently discussed the ultimate constraints on lifetime of a nongrowing cell confined in ice. Cells will probably not be killed by bacteriophages that were present before they were transported onto ice, because the phage lytic cycle is typically about 90% of the host division cycle, independent of the rate of division of the host cell (C. Suttle, personal communication). Cells that cannot divide will not undergo lysis by phages that also cannot divide. On Earth, shielding by overlying ice quickly attenuates solar UV, protons, neutrons, and electrons in the cosmic rays, leaving only muons, which fall off slowly with depth. At depths greater than ~100 m, even the flux of muons is low enough not to be a threat. However, organic compounds and microbial life near the surface of Mars may not survive solar UV, since its atmosphere is too thin to filter out UV. Estimates from lunar sample studies suggest that micrometeorite impacts on Mars result in vertical mixing of the regolith to a depth of the order of 1 to several meters, ensuring that potential life forms will from time to time be directly exposed to solar UV, solar flares, and strong oxidants. Microbes in the purest polar ice, far above sources on nutrients in underlying silt, are able to survive for a few 10^5 years, depending on the concentration of nutrients deposited as aerosols relative to the concentration of microbes. A fraction of sporeformers transported as aerosols may be able to complete the process of sporulation before reaching ice, but the time to complete the process ranges from ~8 h in air at 20°C to several months in air at 0°C. Using a sensitive technique for detecting dipicolinic acid in spore coats, Yung et al. (2007) detected ~370 spores/ml at a depth of 94 m in GISP2 ice and showed that ~80% of the spores were germinable.

It would be interesting to collect spores and cells from ice near the bottom of Dome C, where the age has been shown to be 740,000 years, and measure the live/dead ratio. It had generally been thought that spores would survive far longer than vegetative cells, because they shut down all metabolic activity and require no nutrients until an improvement in external conditions triggers their return to the vegetative state. However, a study by Johnson et al. (2007) has suggested that vegetative cells survive longer than spores. They detected survival of some non-spore-forming bacteria in permafrost samples of ages up to 0.5 Ma and showed that the DNA in nonsporeformers degrades far more slowly than DNA in endospore formers. The reason may be that spores may return to the vegetative state too late to succeed in repairing the accumulated molecular damage. Johnson et al. concluded that this long-term survival is closely tied to cellular metabolic activity and ongoing DNA repair that over a long time is superior to sporulation as a mechanism in sustaining bacterial viability.

Price (2009) calculated that, even for the oldest glacial ice, microbial lifetime is not shortened by radiation damage from uranium-238, thorium-232, or potassium-40 in mineral dust in ice. Instead, death of those cells adapted to the hostile conditions in glacial ice is probably due to exhaustion of available nutrients. By contrast, in permafrost microbial death is more likely due to alpha-particle radiation damage from U and Th in the soil and rocks intermixed with ice. The alpha particles cause double-chain breaks at distances within the 10- to 20-μm range of the alpha particles.

WIND TRANSPORT AND DEPOSITION OF MICROBES AND DUST INTO ICE

Spectrofluorimeters are proving to be very powerful tools for measuring concentrations of autofluorescent fluorophores as a function of depth in lakes, boreholes in glacial ice, and ice cores. Using an early version powered by 375-nm excitation with light-emitting diodes, Bramall (2007) mapped chlorophyll (Chl) and NADH (a coenzyme found in all living cells) as a function of depth in Lake Tahoe. Later he used a version powered by a 224-nm HeAg laser to excite protein fluorescence (Trp) as a function of depth in the South Pole Remote Earth Science and Seismological Observatory (SPRESSO) borehole near the South Pole. Finally, he developed a miniaturized biospectrallogger (mini-BSL) with a 224-nm laser in a cylinder only 5 cm in diameter that operated in a borehole in the ice over Lake Vida, Antarctica. He showed that the mini-BSL was able to detect a single *Bacillus megaterium* cell on a background of iron-rich biotite with a single 100-μs laser pulse. Because of the high background of fluorescence induced in a borehole drilling fluid, spectrofluorometers can be used only in water, in air-filled boreholes, and to examine ice cores in air.

Figure 5 shows the Berkeley Fluorescence Spectrometer (BFS), designed and built by Rohde (2010), with its 224-nm laser surrounded by seven photon counters. It automatically sends pulses down into a moving 1-m-long ice core in a dark room at −20°C at NICL (U.S. National Ice Core Laboratory). It maps autofluorescence of Trp, Chl, and volcanic ash. When it is used to study short-term variations in microbial deposition rate, the beam is focused into a conical region with a diameter of ~0.5 mm, and the laser pulses provide spectra every 700 μm along a 1-m core in 2 min, giving 1,400 readings per m. It takes about 2 min to change from one ice core to another and to adjust the vertical position of the BFS just above a flat surface of the core. Typically ~400 m of ice can be examined per 5-day week. Rohde et al. (2008) found highly variable microbial deposition rates onto the West Antarctic and GISP2 ice, recorded with the BFS set with its 224-nm laser excitation source collimated to a narrow beam optimized to detect fluctuations in arrival times. The smaller the laser spot, the smaller the volume sampled and the greater the fluctuations in time. The arrival rates of bacteria and nonmicrobial dust blown from African desert sources to an air collector on Barbados showed similar patterns of seasonal and daily

FIGURE 5 Berkeley Fluorescence Spectrometer showing vertical laser surrounded by seven photon counters just above an ice core on the moving translation stage in a dark lab at −20°C at NICL. The student in the middle is removing an ice core that was previously scanned.

fluctuations (Prospero et al., 2005). Data on fluctuations in arrival times as a function of depth in ice cores provide useful meteorological information on wind transport speeds as a function of time over many thousands of years.

CONCENTRATIONS OF CHL AND TRP AS A FUNCTION OF DEPTH IN ICE CORES

Contamination of ice surfaces is always a problem that must be eliminated in studies of microbes in ice. To evaluate contamination, Rohde (2010) made raster scans of organic fluorescence in cross sections of ice. He found that dry-drilled WAIS Divide ice has no significant organic contamination at its surface, whereas fluid-drilled WAIS Divide ice is contaminated by drill fluid that had penetrated along microcracks to depths of 1 to 3 mm. In

Vostok and Siple Dome ice the contamination is worse: it extended as deep as 1 cm into the core. It is now routine in ground-truth studies of microbes in ice for biologists to discard the outer several millimeters to 1 cm from the ice surfaces in order to avoid contamination due to drill fluids and handling.

To smooth over the fluctuations discussed in the previous section, the BFS beam was set to a diameter of ~2 mm and weakly focused over a depth of ~2 cm. Figure 6 (Rohde, 2010) shows the results of measurements by Rohde, Bay, and Price of fluorescence intensity of Chl and Trp from six polar regions: Greenland (GISP2), WAIS Divide, Siple Dome, Ross Ice Drainage System (RIDS, in West Antarctica), South Pole, and Vostok Station (East Antarctica). To reduce clutter, each data point was averaged over all readings in an entire section

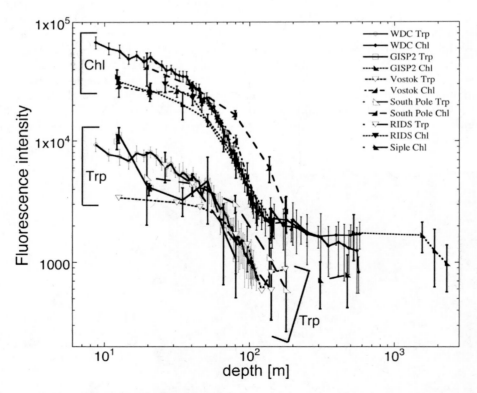

FIGURE 6 Measurements with the BFS of Chl and Trp fluorescence intensity versus depth in ice cores from five sites in Antarctica and one site in Greenland. See text for explanation. (Data from R. A. Rohde, R. Bay, P. B. Price, and D. Tosi [unpublished].)

of ice core of length between 0.5 and 1 m. The full dataset is an appendix to Rohde's thesis. The photon counters above the ice-core surface were set to accept emitted fluorescence in seven wavelength bins centered on 300, 320, 340, 360, 380, 670, and 710 nm. The shortest five passbands were used to select Trp fluorescence from cells inside the ice, and the 670- and 710-nm passbands were used to distinguish Chl, with a sharp peak at 670 nm and no emission at 710 nm, from those mineral grains that emit at both 670 and 710 nm. Grains within a calibrated ratio of intensities at both 710 and 670 nm have been found to be due to volcanic ash (Rohde, 2010).

The data in Fig. 6 provide by far the largest collection on proxies for the microbial content of glacial ice ever assembled. The microbial content of ice is seen to be fairly homogeneous on a decimeter scale, in contrast to the heterogeneity on a millimeter scale. Two features should be noted. (i) The intensity of Chl fluorescence differs by no more than a factor of ~3 from site to site at a given depth, and the same is true of the variation of Trp fluorescence from site to site. (ii) The intensities of both Chl and Trp fluorescence at each of the six locations decrease with depth down to ~120 m and then level off at greater depths.

In seeking to interpret the rapid decrease in fluorescence intensities of Trp in the top 120 m of ice and the flattening of intensity values at greater depth, Price and his student Tong Liu carried out ground-truth measurements of cell concentrations in ice from several sites in Antarctica and Greenland. Some of their results are shown in Fig. 7 (T. Liu and P. B. Price, unpublished data). The solid circles and triangles show results of epifluorescence microscopy using a filter set with excitation at 480 ± 15 nm and emission at 535 ± 20 nm to measure concentrations of cells that were filtered onto a 0.2-μm polycarbonate membrane and then stained with SYBR Gold to reveal DNA-containing cells. The Trp concentration measured with the BFS is a good proxy for concentration of cells of roughly the same carbon content, since all cells have about the same concentration of Trp per unit mass. The points labeled with open circles and triangles in Fig. 7

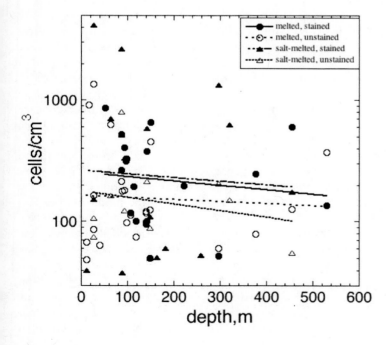

FIGURE 7 Cell concentrations in glacial ice from Greenland and Antarctica that was melted under various conditions and measured with epifluorescence microscopy via their NADH autofluorescence (unstained) and with SYBR Gold staining. See text for explanation. (Data from Liu and Price [unpublished].)

show results of epifluorescence microscopy of unstained ice, using a filter set with excitation at 350 ± 25 nm and emission at 460 ± 25 nm to map the blue autofluorescence of cells containing NADH. Cells with NADH in the oxidized form, NAD^+, do not fluoresce. The similarity in magnitudes of the points with and without staining in Fig. 7 shows that NADH in the cells from the polar ice is mostly in the reduced form. Triangles indicate cell concentrations in ice that melted in a salt solution chosen to reduce the osmotic stress on cell membranes during the transition from solid ice to liquid water. The straight lines are exponential fits to the data on the semilog plot. Although the spread in data is very large, it appears that in all four cases indicated by the four symbols, there is a far weaker decrease in cell concentrations than of the Chl and Trp fluorescence measured with the BFS in solid ice. The concentration of stained cells is roughly 50% greater than that of unstained cells, which is probably due in part to their greater visibility to the human observer.

The main conclusion is that the depth dependence of cell concentration seen with epifluorescence microscopy is far weaker than the ~20-fold decrease with depth of the Chl and Trp fluorescence shown in Fig. 6. It is likely that most of the decrease in Fig. 6 is a geometric effect: the laser beam diffuses markedly in passing through the highly granular firn ice at shallow depths but remains well collimated in passing through fully dense ice at depths greater than ~100 m. The weak decrease in microbial concentration with depth seen in Fig. 7 suggests that both psychrophiles and nonpsychrophiles are equally able to adapt to the lower temperatures, lower nutrient availability, and immobility in ice than in oceans and soil.

Figure 8 shows measurements of cell concentrations in Vostok and GISP2 by various biologists, some made with epifluorescence microscopy and others made with flow cytometry. Both the magnitudes and fluctuations in concentrations from one dataset to another in Fig. 8 are larger than in Fig. 7. In general, biologists have been more interested in tax-

onomy and physiology of microorganisms in ice than in systematic measurements of concentration versus depth.

SUBMICROMETER-SIZE CYANOBACTERIA FROM OCEAN TO ICE

The Mystery of the Missing Phototrophs

Unlike the straightforward procedure just discussed in which epifluorescence microscopy and flow cytometry have been used to calibrate the Trp fluorescence measurements that relate to cell concentrations, ground-truth searches by the author's group using epifluorescence microscopy have failed to find any Chl autofluorescence of microbial cells in glacial ice. After discussing the discovery of an annual modulation of the Chl fluorescence, I will give what seems like the best explanation of the fact that no Chl autofluorescence has yet been seen in melted ice samples viewed with epifluorescence microscopy.

Annual Modulation of Concentration of Chl in Glacial Ice

With the BFS, Price and Bay made an exciting discovery: that Chl fluorescence intensity in the WAIS Divide core ice shows an annual modulation with deposition time. Figure 9 (left) shows four peaks in a 1-m length of WAIS Divide ice core, which correspond to ~25 cm per year of accumulation. They also found annual Chl layers in all of the GISP2 ice they examined (12 to 81 m) and in the three depths in RIDS ice (26, 40, and 141 m) they examined. As seen in Fig. 9 (top), they found that Trp fluorescence in WAIS Divide core ice also shows an annual modulation, usually at a weaker level than the modulation of Chl fluorescence.

They interpreted the modulation as evidence that photosynthesizing microbes show an annual maximum in their rate of transport from ocean surfaces and their subsequent deposition into ice. By correlating depths of Chl maxima with data on various ions in the same core provided by Cole-Dai and Ferris (J. Cole-Dai

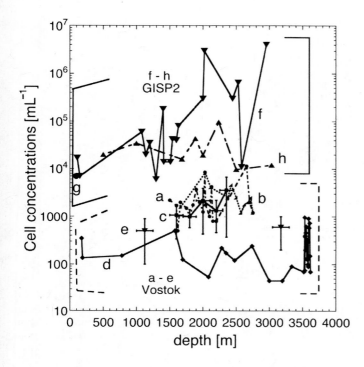

FIGURE 8 Cell concentrations in GISP2 and Vostok ice. Some of the large scatter may be due to different techniques and criteria used in cell identification. (Some data points have been taken from the following references: a, Abyzov et al. [1998b]; b, Abyzov et al. [2007]; c, Abyzov [2004]; d, Christner et al. [2006]; e, Abyzov et al. [2004]; f, Miteva et al. [2009]; g, Yung et al. [2007]; h, Tung et al. [2005].)

and D. Ferris, presented at WAIS Divide Science Meeting, Kings Beach, CA, 4 to 5 October 2007), Price and Bay concluded that the concentration of phototrophs in the ice correlates in time and phase with the concentration of non–sea-salt (nss) SO_4^{2-} ions, with a maximum in January for Antarctic ice. The correlation with nss-SO_4 is to be expected, for the following reason: dimethyl sulfate (DMS) is produced by its precursor compound dimethyl sulfoniopropionate, which is released by marine phototrophs in the upper ocean; in the atmosphere DMS is oxidized to form nss-SO_4^{2-} and methane sulfonate aerosols. DMS production is highest when the amount of Chl is highest, which is when the sea surface temperature is highest. Chl in ocean surface waters has been mapped by SeaWiFs satellite ocean color data (Gregg, 2008; Abbott et al., 2000). It reaches maxima in Antarctic summer (January).

For WAIS Divide, Price and Bay analyzed exactly the same WAIS Divide ice cores as did Cole-Dai and Ferris, who showed that the peak concentrations of nss-SO_4^{2-} occurred in January. They found that the phases are the same for Chl and nss-SO_4^{2-}, as exemplified in Fig. 9 (bottom). They estimate that the depths of maximum amplitudes coincide to within 2 to 3 cm, which corresponds to maxima occurring within a few days to weeks of each other. The amplitude of the Chl intensity modulation is typically ~30% for depths up to ~100 m, gradually weakening and blurring with increasing depth. The peak in arrivals of Chl-bearing microbes in January indicates that Chl reaches a peak in the Antarctic summer. This is to be expected, as increased sunlight in summer leads to ocean warming and thinning, melting, and breakup of sea ice and increased light input to phototrophic organisms living on the bottom of the sea ice.

The "Invisible" Cyanobacteria

To try to understand the nature of the putative phototrophic cells responsible for the Chl fluorescence as a function of depth in ice cores shown in Fig. 6, Liu and Price searched for autofluorescing cyanobacteria from WAIS Divide core and RIDS ice that was melted, then

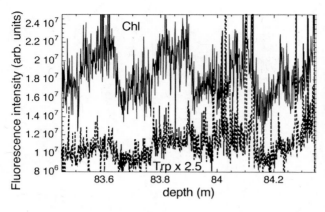

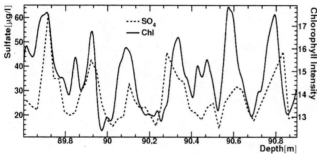

FIGURE 9 (Top) Annual modulation of both Chl and Trp from BFS data in WAIS Divide core ice over 4 years. (Bottom) Agreement of phases of annual modulation of Chl (BFS data) and SO_4^{2-}. (BFS data from Price and Bay [unpublished]; SO_4^{2-} data provided by Cole-Dai and Ferris at WAIS Divide Science Meeting, 2007.)

either stained or unstained, then filtered onto 0.2-μm polycarbonate filters, and finally examined with epifluorescence microscopy. In preparation for that study, they first examined laboratory samples of unstained cyanobacteria ~2 μm in size and found that the cells showed intense red autofluorescence when viewed with a Chl filter set (455 ± 30 nm excitation and 660 ± 50 nm emission). With the same illumination they were unable to detect autofluorescing particles from melted samples of glacial ice, even though the Chl fluorescence recorded by the BFS (Fig. 6) in scans of solid ice was easily detectable at all depths. Marine cyanobacteria of the genus *Synechococcus*, 0.8 to 1.5 μm in size, are known to be autofluorescent at wavelengths in the orange (545 ± 30 nm excitation, 590 ± 50 nm emission) due to the phycoerythrin (PE) pigment they contain. However, this emission is at too short a wavelength to be responsible for the fluorescence recorded in the BFS scans in the emission band at 670 ± 20 nm in Fig. 6. The autofluorescence

of cells deposited in the ice and measured with the BFS is simply too faint to be seen when individual cells are removed from the ice and examined with epifluorescence microscopy.

How can the two conflicting observations be rationalized? In epifluorescence microscopy a Chl-containing cell with a diameter less than or about 1 μm would have a volume less than 10^{-6} as large as that of the region of an ice core hit by a single pulse of the BFS laser beam. One possible explanation of the missing Chl autofluorescence searched for with epifluorescence microscopy is that a single cell may fluoresce too weakly to be visible even with a 100× oil-immersion objective, whereas a large population of such cells might produce enough fluorescence for the BFS to detect. A second consideration has to do with the photophysics of bleaching. During the scans of ice cores at NICL with the BFS that led to the data in Fig. 6, Rohde (2010) sent a huge number of pulses of laser light into ice in order to test for the onset of photobleaching. Even after having injected 20,000 pulses into a spot during 1 hour, he found no evidence of photobleaching, which suggested that ice provides an environment that protects microbes from photobleaching.

The author conjectured that the culprit is *Prochlorococcus marinus*, a species of marine cyanobacteria now well known to biological oceanographers but still unknown to most biologists who specialize in studies of microbes in glacial ice. *P. marinus* is the smallest (~0.5 µm) and most abundant photosynthetic cell on Earth. According to Penny Chisholm of Massachusetts Institute of Technology, who discovered the species in ocean samples using a flow cytometer on a ship (Chisholm et al., 1988), shallow seawater at mid-latitudes contains concentrations as high as several times 10^5 cells/ml. Using fluorescence microscopy with an image intensifier attached to a CCD camera, Albertano et al. (1997) detected the faint, rapidly bleached red autofluorescence of individual submicron *Prochlorococcus* cells with epifluorescence microscopy and reported that one commonly encountered group had a diameter of 0.5 to 0.6 µm. Moore and Abbott (2002) determined their physiology and molecular phylogeny and grouped them into two categories: strains that exist in shallow ocean waters and are adapted to bright sunlight, and strains at depths of 80 to 200 m that are adapted to light intensities as much as 10^4-fold weaker. Cells of the near-surface strains autofluoresce very faintly and photobleach rapidly with epifluorescence microscopy. Both strains contain unique light-harvesting pigments consisting predominantly of divinyl derivatives of Chl *a* and Chl *b* that are unique to the genus *Prochlorococcus*. The genomes of both strains have been sequenced (Rocap et al., 2002; Dufresne et al., 2003). The *Prochlorococcus* spp. are of great ecological importance: they are some of the main primary producers of the phytoplankton, which are responsible for half of the photosynthesis on Earth; and they are prominent actors in global oceanic function, in global biogeochemical cycles, and in the evolution of climate.

In order to confirm the author's conjecture about the faintly autofluorescing cyanobacteria, he and his students now use flow cytometry in the way that Chisholm did in 1988. Berkeley's Cytopeia Influx Cytometer can excite each cell falling through phosphate-buffered solution simultaneously with up to five lasers at wavelengths of 366, 405, 488, 567, and 632 nm, and can detect autofluorescence of NADH (blue emission), flavin adenine dinucleotide (FAD^+) (green), Chl (red), and PE (orange), as well as fluorescence of stained samples. PE is a pigment in many cyanobacteria and is particularly bright in the two closely related genera *Prochlorococcus* and *Synechococcus*. Side scatter and forward scatter of cells by a laser beam provide a useful measure of size. Figure 10 shows examples of autofluorescence of unstained cells at several depths in glacial ice from various locations. In order to view only cyanobacteria, the plots are generated by triggering on cells with greater than some threshold concentration of chlorophyll.

In contrast to the findings of microbial oceanographers, who have reported that *Prochlorococcus* inhabits only latitudes between ~45°N and ~45°S, the author and his colleagues have found that every glacial ice sample examined by them contains detectable concentrations of *Prochlorococcus* and *Synechococcus* cells and little else. The first vertical column in Fig. 10 shows flow cytograms of cultures of the two well-known strains of *Prochlorococcus* (MED04, found mainly in the top few meters of seawater; and MIT9313, found mainly at depths as great as ~100 m) and *Synechococcus*, more widely distributed in latitude than *Prochlorococcus*. By comparing the patterns of dots in the 12 cytograms, one can see several interesting features: (i) the cultures are not axenic (pure), in that both the shallow-water culture and the deep-water culture have admixtures from other depths, and the *Synechococcus* culture contains an admixture of *Prochlorococcus* cells; (ii) the data in the 9 cytograms to the right of the cultures are dominated by *Prochlorococcus* and *Synechococcus*, almost to the exclusion of all other taxa that contain both red and orange fluorophores; (iii) in WAIS

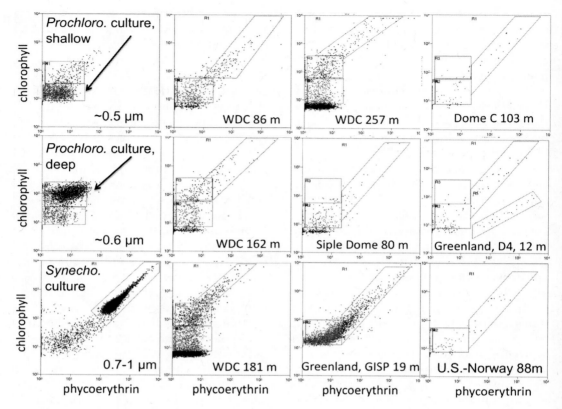

FIGURE 10   Flow cytograms of autofluorescence of Chl and PE in melted ice from Greenland GISP (on the divide) and D4 (west of the divide), West Antarctica (WAIS Divide [WDC] and Siple Dome), and East Antarctica (Dome C and along a U.S.-Norway traverse). Rectangular boxes denote *Prochlorococcus* (high-Chl and low-Chl); inclined boxes denote *Synechococcus* (at Greenland D4, two likely strains are found).

Divide core ice at depths of 162, 181, and 257 m, a comparable concentration of both strains of *Prochlorococcus* is present; and (iv) in Greenland ice from both D4 and GISP, two strains of *Synechococcus* are present. The concentrations of the two taxa of cyanobacteria in various polar locations have been determined by Price and his colleagues. They cannot be inferred directly by counting the dots in the boxes in Fig. 10, because of differences in the volumes of ice sampled and by the necessity to use a centrifuge to increase cell concentrations from locations such as East Antarctica that have very low cell counts. Based on data acquired by the author in 2010, the median ratio of concentrations of cells of *Prochlorococcus* to those of *Synechococcus*

for all sites is ~7 to 1. Due to the constraint that the flow cytometer was set to trigger on Chl, non–chlorophyll-bearing cells did not show up in Fig. 10. Comparison of concentrations of cells recorded by flow cytometry and by epifluorescence microscopy has shown that the concentration of cells of the cyanobacteria transported onto glacial ice and detected by flow cytometry without staining is about an order of magnitude lower than the total concentration of unstained bacterial cells imaged by epifluorescence microscopy at various wavelengths. We next discuss how heterotrophic bacteria without Chl can be imaged without staining by using filters that select the autofluorescence of the coenzymes NADH and FAD^+.

LIVE/DORMANT TESTS ON INDIVIDUAL CELLS

Redox fluorimetry, pioneered by Britton Chance (1991), enables one to study metabolism and respiration of cells containing mitochondria or a dehydrogenase. The idea is to measure the relative intensities of both reduced NADH and oxidized flavoprotein, FAD^+, using excitation wavelengths of ~360 and ~460 nm to observe their blue and green autofluorescence, respectively. The intensity of fluorescence is high for the reduced form of NADH and almost zero for the oxidized form. The reverse is true for the flavoproteins: the intensity is high for the oxidized form and almost zero for the reduced form. Both NADH and FAD are bound to proteins. The fluorescence intensity of the two biomolecules in the same cell can be used to trace its cellular metabolic activity: The cellular autofluorescence changes according to the redox conditions obtained within the cell.

Price (2000) and independently Linda Powers (Estes et al., 2003) proposed that autofluorescence could be used to distinguish living from dormant or dead cells by comparing the relative intensity of fluorescence from both NADH and FAD^+. One could, for example, use flow cytometry or epifluorescence microscopy to measure NADH fluorescence with excitation at 366 nm and emission at ~466 nm and FAD fluorescence with excitation at 488 nm and emission at ~550 nm. The great majority of published papers on this topic deal with eukaryotic cells for which the fluorophores are usually located in the mitochondria. Whether signals indicative of cellular viability can be detected due to fluorescence of NADH and FAD^+ in dehydrogenase in bacteria and archaea should be explored. As a first step, Liu and Price subjected cells in several samples of West Antarctic ice to "cold shock" by first removing them from their normal repository at −20°C and leaving them for a week at −50°C. Following this shock, they melted the samples and then froze them at −10°C. Examination of unstained samples of that ice with epifluorescence microscopy showed that less than 10% of the cells were detectable by NADH auto-fluorescence, indicating that most of the cells had oxidized to NAD^+, the nonfluorescent form. Experiments to evaluate whether they were actually dead or could slowly recover after prolonged storage at −20°C have not been done. Addition of a new filter set to look for FAD fluorescence together with NADH fluorescence may provide a rigorous test of the live/dormant method without the need for a live/dead stain.

A NEW TOOL FOR THE STUDY OF MICROBIAL EVOLUTION

Prochlorococcus and *Synechococcus* cells have the optimal size for transport from near-surface ocean waters onto the high plateaus of Antarctica and Greenland. They comprise the dominant majority of submicron Chl-bearing cells at all depths (at least in temperate oceans). This leads to the possibility of studying their evolution over as many as 10^7 generations by collecting them at depths down to several thousand meters, corresponding to more than 10^5 years. The idea is that cells transported onto glacial ice preserve an unaltered record of their genomes since they were deposited as aerosols from nearby oceans. As long as they remain in the ocean their communities evolve with time, but not once they become trapped in a growing ice sheet. By using new techniques of single-cell genomics, it should be possible to track changes in their genome as a function of depth in the ice and thus to infer their mutation rates in the ocean before they reached the ice. There are two reasons why the genomes do not continue to evolve for microbes in ice: one is that they are trapped and cannot pass on their mutations to progeny; the other is that their mutation rates decrease with temperature along an Arrhenius line similar to those in Fig. 4 but translated downward by a factor of at least 10^3. Thus, cells in ice not only mutate at extremely low rates, but they do not divide and thus do not pass on the changes in their genomes.

ACKNOWLEDGMENTS

I am supported in part by National Science Foundation Antarctic Glaciology grant ANT-0738658. I have enjoyed fruitful collaborations with former students Camas Tung, Nathan Bramall, Robert Rohde, Tong Liu, Joyce Lee, and Elaine Lee; with senior researchers Ryan Bay, Lisa Moore, and Delia Tosi; and with technical specialists Gordon Vrdoljak, Steve Ruzin, Denise Schichsnes, and Hector Nolla. Cultures were provided by Lisa Moore.

REFERENCES

Abbott, M. R., J. R. Richman, R. M. Letelier, and J. S. Bartlett. 2000. The spring bloom in the Antarctic Polar Frontal Zone as observed from a mesoscale array of bio-optical sensors. *Deep Sea Res. Part 2 Top. Stud. Oceanogr.* 47:3285–3314.

Abyzov, S. S. 2004. Use of different methods for discovery of ice-entrapped microorganisms in ancient layers of the Antarctic glacier. *Adv. Space Res.* 33:1222–1230.

Abyzov, S., M. Fukuchi, S. Imura, H. Kanda, I. Mitskevich, T. Naganuma, M. Poglazova, L. Savatyugin, and M. Ivanov. 2004. Biological investigations of the Antarctic ice sheet: review, problems and prospects. *Polar Biosci.* 17:106–116.

Abyzov, S. S., I. N. Mitskevich, and M. Ivanov. 2007. Microbiology of the Antarctic glacier above the Lake Vostok, p. 11–39. *In* S. Abyzov and D. Perovich (ed.), *Climate Change and Polar Research.* Luso-American Development Foundation, Lisbon, Portugal.

Abyzov, S. S., I. N. Mitskevich, and M. N. Poglazova. 1998a. Microflora of the deep glacier horizons of central Antarctica. *Microbiology* 67:547–555.

Abyzov, S. S., I. N. Mitskevich, M. N. Poglazova, N. I. Barkov, V. Y. Lipenkov, N. E. Bobin, B. B. Koudryashov, and V. M. Pashkevich. 1998b. Antarctic ice sheet as a model in search of life on other planets. *Adv. Space Res.* 22:363–368.

Albertano, P., D. DiSomma, and E. Capucci. 1997. Cyanobacterial picoplankton from the Central Baltic Sea: cell size classification by image-analyzed fluorescence microscopy. *J. Plankton Res.* 19:1405–1416.

Benner, S. A., K. G. Devine, L. N. Matveeva, and D. H. Powell. 2000. The missing organic molecules on Mars. *Proc. Natl. Acad. Sci. USA* 97:2425–2430.

Bidle, K. D., S. H. Lee, D. R. Marchant, and P. G. Falkowski. 2007. Fossil genes and microbes in the oldest ice on Earth. *Proc. Natl. Acad. Sci. USA* 104:13455–13460.

Bramall, N. E. 2007. The remote sensing of microorganisms. Ph.D. thesis. University of California—Berkeley, Berkeley, CA.

Brinton, K. K. F., A. I. Tsapin, D. Gilichinsky, and G. D. McDonald. 2002. Aspartic acid racemization and age-depth relationships for organic carbon in Siberian permafrost. *Astrobiology* 2:77–82.

Brook, E. J., T. Sowers, and J. Orchardo. 1996. Rapid variations in atmospheric methane concentrations during the past 110,000 years. *Science* 273:1087–1091.

Chance, B. 1991. Optical method. *Annu. Rev. Biophys. Biophys. Chem.* 20:1–28.

Chisholm, S. W., R. J. Olson, E. R. Zettler, R. Goericke, J. B. Waterbury, and N. A. Welschmeyer. 1988. A novel free-living prochlorophyte occurs at high cell concentrations in the oceanic euphotic zone. *Nature* 334:340–343.

Christner, B. C., G. Royston-Bishop, C. M. Foreman, B. R. Arnold, M. Tranter, K. A. Welch, W. B. Lyons, A. I. Tsapin, M. Studinger, and J. C. Priscu. 2006. Limnological conditions in subglacial Lake Vostok, Antarctica. *Limnol. Oceanogr.* 51:2485–2501.

Christner, B. C., M. L. Skidmore, J. C. Priscu, M. Tranter, and C. M. Foreman. 2008. Bacteria in subglacial environments, p. 51–71. *In* R. Margesin, F. Schinner, J.-C. Marx, and C. Gerday (ed.), *Psychrophiles: from Biodiversity to Biotechnology.* Springer, Berlin, Germany.

Clarke, S. 2003. Aging as war between chemical and biochemical processes: protein methylation and the recognition of age-damaged proteins for repair. *Ageing Res. Rev.* 2:263–285.

Cleaves, H. J., K. E. Nelson, and S. L. Miller. 2006. The prebiotic synthesis of pyrimidines in frozen solution. *Naturwissenschaften* 93:228–231.

D'Elia, T., R. Veerapaneni, and S. O. Rogers. 2008. Isolation of microbes from Lake Vostok accretion ice. *Appl. Environ. Microbiol.* 74:4962–4965.

D'Elia, T., R. Veerapaneni, V. Theraisnathan, and S. O. Rogers. 2009. Isolation of fungi from Lake Vostok accretion ice. *Mycologia* 101:751–763.

Dufresne, A., M. Salanoubat, F. Partensky, F. Artiguenave, I. M. Axmann, V. Barbe, S. Duprat, M. Y. Galperin, E. V. Koonin, F. Le Gall, K. S. Makarova, M. Ostrowski, S. Oztas, C. Robert, I. B. Rogozin, D. J. Scanlan, N. Tandeau de Marsac, J. Weissenbach, P. Wincker, Y. I. Wolf, and W. R. Hess. 2003. Genome sequence of the cyanobacterium *Prochlorococcus marinus* SS120, a nearly minimal oxyphototrophic genome. *Proc. Natl. Acad. Sci. USA* 100:10020–10025.

Estes, C., A. Duncan, B. Wade, C. Lloyd, W. Ellis, and L. Powers. 2003. Reagentless detection of microorganisms by intrinsic fluorescence. *Biosens. Bioelectron.* 18:511–519.

Gregg, W. W. 2008. Assimilation of SeaWiFS ocean chlorophyll data into a three-dimensional global ocean model. *J. Mar. Syst.* 69:205–225.

Jaenicke, R. 1980. Atmospheric aerosols and global climate. *J. Aerosol Sci.* **11:**577–588.

Johnson, S. S., M. B. Hebsgaard, T. R. Christensen, M. Mastepanov, R. Nielsen, K. Munch, T. Brand, M. T. P. Gilbert, M. T. Zuber, M. Bunce, R. Rønn, D. Gilichinsky, D. Froese, and E. Willerslev. 2007. Ancient bacteria show evidence of DNA repair. *Proc. Natl. Acad. Sci. USA* **104:**14401–14405.

Kirschvink, J. L., B. P. Weiss, and N. J. Beukes. 2006. Boron, ribose, and a Martian origin for terrestrial life. *Geochim. Cosmochim. Acta* **70:**S320.

Larralde, R., M. P. Robertson, and S. L. Miller. 1995. Rates of decomposition of ribose and other sugars: implications for chemical evolution. *Proc. Natl. Acad. Sci. USA* **92:**8158–8160.

Lazcano, A., and S. L. Miller. 1994. How long did it take for life to begin and evolve to cyanobacteria? *J. Mol. Evol.* **39:**546–554.

Leck, C., and E. K. Bigg. 2005. Biogenic particles in the surface microlayer and overlaying atmosphere in the central Arctic Ocean during summer. *Tellus B Chem. Phys. Meteorol.* **57:**305–316.

Levy, M., and S. L. Miller. 1998. The stability of the RNA bases: implications for the origin of life. *Proc. Natl. Acad. Sci. USA* **95:**7933–7938.

Lindahl, T., and N. Nyberg. 1972. Rate of depurination of native deoxyribonucleic acid. *Biochemistry* **11:**3610–3618.

Martins, Z., O. Botta, M. L. Fogel, M. A. Sephton, D. P. Glavin, J. S. Watson, J. P. Dworkin, A. W. Schwartz, and P. Ehrenfreund. 2008. Extraterrestrial nucleobases in the Murchison meteorite. *Earth Planet. Sci. Lett.* **270:**130–136.

Miller, S. L., and L. E. Orgel. 1974. *The Origins of Life on the Earth.* Prentice-Hall, Princeton, NJ.

Miteva, V. I., and J. E. Brenchley. 2005. Detection and isolation of ultrasmall microorganisms from a 120,000-year-old Greenland glacier ice core. *Appl. Environ. Microbiol.* **71:**7806–7818.

Miteva, V., T. Sowers, and J. Brenchley. 2007. Production of N₂O by ammonia-oxidizing bacteria at subfreezing temperatures as a model for assessing the N₂O anomalies in the Vostok ice core. *Geomicrobiol. J.* **24:**451–459.

Miteva, V., T. Sowers, C. Olsen, and J. E. Brenchley. 2006. Geochemical and molecular data support a biogenic origin of methane in the basal Greenland ice, p. 371. In *Proceedings of the 11th International Symposium on Microbial Ecology (ISME 11), Vienna, Austria, 20-25 August 2006.* ISME Society, Wageningen, The Netherlands.

Miteva, V., C. Teacher, T. Sowers, and J. Brenchley. 2009. Comparison of the microbial diversity at different depths of the GISP2 Greenland ice core in relationship to deposition climates. *Environ. Microbiol.* **11:**640–656.

Miyakawa, S., H. I. Cleaves, and S. L. Miller. 2002. Cold origin of life: B. Implications based on pyrimidines and purines produced from frozen ammonium cyanide solutions. *Orig. Life Evol. Biosph.* **32:**209–218.

Moore, J. K., and M. R. Abbott. 2002. Surface chorophyll concentrations in relation to the Antarctic Polar Front: seasonal and spatial patterns from satellite observations. *J. Mar. Syst.* **37:**69–86.

Perron, J. T., J. X. Mitrovica, M. Manga, I. Matsuyama, and M. A. Richards. 2007. Evidence for an ancient martian ocean in the topography of deformed shorelines. *Nature* **447:**840–843.

Price, P. B. 2000. A habitat for psychrophiles in deep Antarctic ice. *Proc. Natl. Acad. Sci. USA* **97:**1247–1251.

Price, P. B. 2009. Microbial genesis, life and death in glacial ice. *Can. J. Microbiol.* **55:**1–11.

Price, P. B., and T. Sowers. 2004. Temperature dependence of metabolic rates for microbial growth, maintenance, and survival. *Proc. Natl. Acad. Sci. USA* **101:**4631–4636.

Prospero, J. M., E. Blades, G. Mathison, and R. Naidu. 2005. Interhemispheric transport of viable fungi and bacteria from Africa to the Caribbean with soil dust. *Aerobiologia* **12:**1–19.

Ricardo, A., M. A. Carrigan, A. N. Olcott, and S. A. Benner. 2004. Borate minerals stabilize ribose. *Science* **303:**196.

Rocap, G., D. L. Distel, J. B. Waterbury, and S. W. Chisholm. 2002. Resolution of *Prochlorococcus* ecotypes by using 16S-23S ribosomal DNA internal transcribed spacer sequences. *Appl. Environ. Microbiol.* **68:**1180–1191.

Rohde, R. A. 2010. The development and use of the Berkeley Fluorescence Spectrometer to characterize microbial content and detect volcanic ash in glacial ice. Ph.D. thesis. University of California—Berkeley, Berkeley, CA.

Rohde, R. A., and P. B. Price. 2007. Diffusion-controlled metabolism for long-term survival of single isolated microorganisms trapped within ice crystals. *Proc. Natl. Acad. Sci. USA* **104:**16592–16597.

Rohde, R. A., P. B. Price, R. C. Bay, and N. E. Bramall. 2008. In situ microbial metabolism as a cause of gas artifacts in ice. *Proc. Natl. Acad. Sci. USA* **105:**8667–8672.

Sanchez. R., J. Ferris, and L. Orgel. 1966. Conditions for purine synthesis: did prebiotic synthesis occur at low temperatures? *Science* **153:**72–73.

Sheridan, P. P., V. I. Miteva, and J. E. Brenchley. 2003. Phylogenetic analysis of anaerobic psychrophilic enrichment cultures obtained from a Greenland glacier ice core. *Appl. Environ. Microbiol.* **69:**2153–2160.

Sleep, N. H., and K. Zahnle. 1998. Refugia from asteroid impacts on early Mars and the early Earth. *J. Geophys. Res.* **103:**28529–28544.

Smith, J. J., J. P. Howington, and G. A. McFeters. 1994. Survival, physiological response, and recovery of enteric bacteria exposed to a polar marine environment. *Appl. Environ. Microbiol.* **60:**2977–2984.

Sowers, T. 2001. N$_2$O record spanning the penultimate deglaciation from the Vostok ice core. *J. Geophys. Res.* **106:**31903–31914.

Sowers, T., R. B. Alley, and J. Jubenville. 2003. Ice core records of atmospheric N$_2$O covering the last 106,000 years. *Science* **301:**945–948.

Tison, J.-L., R. Souchez, E. W. Wolff, J. C. Moore, M. R. Legrand, and J. de Angelis. 1998. Is a periglacial biota responsible for enhanced dielectric response in basal ice from the Greenland Ice Core Project ice core? *J. Geophys. Res.* **103:**18885–18894.

Trinks, H., W. Schröder, and C. K. Biebricher. 2005. Ice and the origin of life. *Orig. Life Evol. Biosph.* **35:**429–445.

Tung, C., N. E. Bramall, and P. B. Price. 2005. Microbial origin of excess methane in glacial ice and implications for life on Mars. *Proc. Natl. Acad. Sci. USA* **102:**18292–18296.

Tung, C., P. B. Price, N. E. Bramall, and G. Vrdoljak. 2006. Microorganisms metabolizing on clay grains in 3-km-deep Greenland basal ice. *Astrobiology* **6:**69–86.

Wettlaufer, J. S. 1999. Impurity effects in the premelting of ice. *Phys. Rev. Lett.* **82:**2516–2519.

Yung, P. T., H. S. Shafaat, S. A. Connon, and A. Ponce. 2007. Quantification of viable spores from a Greenland ice core. *FEMS Microbiol. Ecol.* **59:**300–306.

CLIMATE CHANGE, OZONE
DEPLETION, AND LIFE AT THE POLES

Helen A. Vrionis, Karen Warner,
Lyle G. Whyte, and Robert V. Miller

13

INTRODUCTION

As observed in the work highlighted in this text, the poles do not represent vast, frozen, uninhabitable wastelands but are heterogeneous environments rich in microbial life capable of surviving via a variety of different lifestyles. Previous chapters have presented information on the diversity of both the habitats that exist at the poles and the organisms that have been shown to dwell in these unique, often stark environments. The reciprocal interaction of microbes with their environment remains an area of interest for which much information is still required. In light of increasing interest in the functioning of extremophilic environments and growing concerns over global warming, this chapter attempts to address some of the factors that influence life in cold polar habitats (Amoroso et al., 2010; Vincent et al., 2009; Wagner et al., 2009).

Often viewed as geographically separated but equivalent domains, the Arctic and Ant-

arctic exhibit many similarities in subhabitats and climatic extremes. In reality, however, the two represent uniquely distinct environments, with the Arctic being composed primarily of water enclosed by various large and small land and ice masses, while Antarctica is largely terrestrial, comprising a large ice mass (in some cases several kilometers deep) surrounded by water (see http://nsidc.org/seaice/characteristics/difference.html at the National Snow and Ice Data Center for more information). Conditions in polar habitats are not static but are highly influenced by diurnal and seasonal changes in a variety of parameters, including nutrient concentration, gas flux, temperature, solar radiation, and humidity (Doran et al., 2002; Illeris et al., 2003). The interactions and feedback in physical and biochemical parameters at each pole are unique. Often functioning at thermodynamic limits, the cryobiosphere is especially sensitive to changes in climate and itself plays an important role in gas fluxes and environmental shifts at the poles and, by extension, on the earth as a whole (Meltofte et al., 2008). This sensitivity and its consequent global impact make an understanding of polar bioclimatic interactions critical to predicting future climate change trends.

Helen A. Vrionis and Lyle G. Whyte, Department of Natural Resource Sciences, McGill University, Macdonald Campus, 21,111 Lakeshore Road, Ste.-Anne-de-Bellevue, QC H9X 3V9, Canada. *Karen Warner and Robert V. Miller,* Department of Microbiology and Molecular Genetics, Oklahoma State University, 307 Life Sciences East, Stillwater, OK 74078.

Polar Microbiology: Life in a Deep Freeze
Edited by Robert V. Miller and Lyle G. Whyte © 2012 ASM Press, Washington, DC

FACTORS AFFECTING LIFE IN POLAR (CRYOSPHERIC) ENVIRONMENTS

Basic Elements: the Importance of Water

The poles are not immune to the "water equals life" dogma that results in large habitat variations for all respiring organisms. The fundamental importance of water manifests itself in vast socioeconomic differences in human and all life on Earth, and forms the foundation of the search for life on other planets (Diaz and Schulze-Makuch, 2006). The geology and climatic conditions at the poles result in a variety of unique water-based or water-permeated habitats, which include permafrost soils, saline cold-water springs, supraglacial lakes on ice shelves, epishelf lakes in fjords, deep meromictic lakes, and shallow lakes, ponds, and streams (Vincent et al., 2009). Water considerations in polar environments include habitats that involve circulating water (groundwater, ponds, springs, lakes, and oceans), niches dominated by frozen water (snow cover, permafrost, and various ice forms), and the influence of precipitation (rain, snow, and hail) and are greatly influenced by changes in water nature or quality.

The Arctic, which is largely dominated by water (Arctic Ocean, Bering Strait, Beaufort Sea, Fram Strait, Chukchi Sea, Baffin Bay, etc.) interspersed with numerous archipelagic and individual islands, and home to various ice bodies, is relatively well precipitated. Precipitation in the High Arctic occurs mostly as snow and ranges annually between 60 and 160 mm (Meltofte et al., 2008). Water in Arctic soils tends to pool just above the permafrost table, and this layer often proves to be a strong supporter of microbial growth. Despite the limited precipitation, ground environments in the Arctic are often wet since moisture evaporates slowly and drainage conditions are poor. The large land-ice mass of Antarctica (14,000,000 km²), on the other hand, is primarily considered a cold desert, receiving very little precipitation and having most of its lakes (at least 145) below the sur-face of the Antarctic Ice Sheet (Christner et al., 2001; Amoroso et al., 2010). The dryness of Antarctic air prevents reabsorption of radiated heat, which would help release moisture from frozen structures, hence perpetuating the dry, cold nature of the southern polar region. Furthermore, the high average elevation of the southern continent creates regions of low pressure where water availability is especially low.

In addition to the oceans, both the Arctic and Antarctica contain a range of unfrozen aqueous environments (marine ocean waters and·salt springs, interstitial water occurring in fine veins within ice bodies) that persist even at temperatures below $-20°C$. Production of ice or permafrost leads to exclusion of nutrients and salts, concentrating them in aqueous phases whose freezing point is then depressed, providing a habitat for cold-tolerant, often halophilic or haloterant, microbes (Vincent et al., 2009). These relatively "rich" pockets have been shown to support active microbial populations (Vincent et al., 2004; Steven et al., 2008). Additional factors not associated with nutrient concentration but enabling aqueous freshwater environments (lakes and streams) to persist below $0°C$ include geothermal heating, high pressure arising from the weight of overlying ice, and the insulating effect of ice bodies and active layers (Christner et al., 2001; Amoroso et al., 2010).

Hydrogeology plays a critical role in defining microbial habitats. The degree of waterlogging in the active layer above permafrost soils directly influences the diversity and activity of microbial communities by controlling the diffusion of nutrients, O_2, and other electron acceptors within the system. Whereas interstitial water flows may promote oxygen transport into ice, saturation of soils during thaws and increased nutrient flow in permafrost soils may promote anaerobicity and growth of microbes utilizing alternate electron acceptors such as sorbed $Fe(III)$ and $Mn(IV)$ hydroxides (Lovley and Phillips, 1986; Myers and Nealson, 1990; Mehta et al., 2005) or dissolved SO_4 (Walker et al., 2009).

The Carbon Factor

CARBON CONCENTRATION AND QUALITY

In polar terrestrial environments the greatest concentrations of organic matter are associated with the upper surface (active layer) overlying permafrost and as dissolved organic matter (DOM) within the oceans. A large portion of the carbon input results from the activity of primary producers. In aquatic environments, the organic composition is highly influenced by bacterioplankton populations, which contribute to DOM through the release of cellular components, but also take up and utilize more labile organic substrates (Dyda et al., 2009). Significant sources of organic matter in polar waters include sea-ice-associated diatoms and ice algae, ice-rafted debris (including heterogeneous ice-entrained biomass and sediment), peat (picked up by water flow), and aeolian deposition (Stein and Macdonald, 2004).

Variations in wind temperature and precipitation influence the state and mobility of carbon compounds into and through polar ecosystems. As a consequence of the environmental influence on microbial and other floral and faunal activities, and the climatic impact on water phase and erosion, the concentration of inputs and nutrient quality of organic matter that occurs in polar environments is highly seasonally variable (Montserrat Sala et al., 2008). Coastal erosion is hypothesized to play a large role in ocean material budgets, as does riverine input. The Arctic Ocean's material budget is highly influenced by organic influx from rivers, receiving about 11% of the global runoff volume (Shiklomanov, 1998; Stein and Macdonald, 2004). In general, the terrestrial input on the Arctic is high as a result of its enclosure by landmasses (Brinkmeyer et al., 2003). By contrast, Antarctic ocean bodies like the Ross Sea experience much less inflow from rivers and streams, and by extension contain lower concentrations of organic matter.

An additional consideration is that, as in temperate environments, the presence of organics does not guarantee their bioavailability (Wagner et al., 2005). Entrapment of microbes and/or carbon compounds within ice layers/structures may inhibit the contact necessary for enzymatic transformation. Alternatively, cells may be sequestered from additional nutrients or cofactors required for active metabolism. Adaptations in the cell membrane to prevent loss of cell integrity as a result of freeze-thaw may prevent/alter transport of carbon substrates or other components across cell membranes. Depending on the combination of compounds, the low levels of organics present may be insufficient to overcome cell catabolite repression mechanisms and activate degradative pathways. Additionally, the carbon compounds that exist at or reach the poles may not stimulate the necessary cometabolic pathways. Furthermore, for complex molecules, insufficient energy to overcome enzyme activation barriers may inhibit metabolic usage of specific organics.

CO_2 AND CH_4 FLUXES—SINKS AND SOURCES

The cold nature of the poles favors slow rates of chemical and biological transformations, and as a result approximately one-third of global soil carbon is preserved in northern latitudes, primarily trapped in frozen ground areas (Wagner et al., 2009). A range of geological, environmental, and microbiological conditions and activities influence how this carbon is sequestered, modified, released, and cycled.

The state/form of carbon reservoirs and how they are handled biogeochemically is largely dependent on the redox conditions in the associated reservoir area. Organics present in aerobic regions such as thawed active layers more effectively undergo mineralization and conversion to CO_2. Carbon turnover occurs in low-redox niches, such as those in water layers overlying permafrost and within brine-saturated ice pockets, but the rates are usually considerably lower and are often exceeded by deposition rates (Cowan and Ah Tow, 2004). Biological anaerobic pathways often involve contributions of numerous community members to turnover of complex organics in a stepwise fashion, with differing final products that may or may not lead to CO_2 production. In many cases with anaerobic respiratory microbes, the ability to make contact with solid inorganic minerals (such as Fe and Mn oxides) determines whether transformation of carbon compounds will occur, and hence the nature of the water phase (liquid, slurry, or ice) and degree of water saturation at any given time has a profound influence on these activities. If conditions are sufficiently anaerobic, low-molecular-weight organics like CO_2 and acetic acid can themselves be used as terminal electron acceptors for methanogenesis (Wagner et al., 2009). A large part of the polar climate question lies in determining if increased melting and hence greater saturation and anaerobic conditions will result in increased greenhouse gas (e.g., methane) emissions to the atmosphere, further contributing to the global warming phenomenon.

Seasonal variations in microbial gas (CH_4 and CO_2) production have been documented at both poles (Adushkin and Kudryavstev, 2010). The community structure at a given site and the sum response to minute-scale changes in oxygen, temperature, and water control the balance, conversion, and movement of carbon compounds into various phases. Both methanogens and methanotrophs have been shown to be able to maintain their populations under suboptimal conditions, i.e., survive through dry and saturated periods, respectively (Le Mer and Roger, 2001). The balance in activity between these two groups of organisms and their interaction in the complex communities in which they exist are the determining factors of whether a particular environment will be a source or sink for carbon compounds.

Methane release to the atmosphere from Arctic tundras is estimated to be about 25% of the release from natural sources, corresponding to between 17 and 42 Tg/year (Christensen et al., 1996; Fung et al., 1991; Le Mer and Roger, 2001). This release is not uniform, with different CH_4 emission and oxidation rates being observed, for instance, at the rim versus the center of ice-wedge polygons or in highly saturated versus dry soils. Substrate limitation due to decreased organic matter quality and bioavailability may limit gas production (for example, methane) despite conditions of high carbon concentration (Wagner et al., 2005). In many cases CH_4 produced by microorganisms at depth is oxidized by methanotrophs in the upper oxidative layers before it reaches the atmosphere.

Significance of Salts and Noncarbon Nutrients

Of the inorganic molecules important to life, salts (NaCl, Mg and Ca carbonates) play a particularly important role in life in polar environments. Their role is important not only in contributing to cellular osmotic balance but, as mentioned above, as solutes that (along with other solids) depress the freezing point of water, allowing it to remain in aqueous form at subzero temperatures. In the case of the formation of sea ice, gravity pushes brine-filled pockets downward so that freezing to solid ice occurs from the top layer down (http://www.eoearth.org/article/Arctic_Ocean). Both heat-driven evaporation and cold-driven cryodesiccation decrease water content, in the first case concentrating salts and sorbed nutrients in the upper layers of exposed soils, while in the latter case generating ice that results in increased solute and nutrient concentrations within the excluded water that occurs in liquid veins and unfrozen liquid films within and on the surface of the ice. The different salt concentrations in these brine waters create distinct niches for halotolerant and halophilic microbes.

The increased nutrient concentration in ice veins may promote the growth of increased biomass and lead to shifts in microbial diversity from more oligotrophic organisms to communities with higher trophic complexity. Salt- and nutrient-concentrated waters above permafrost layers and various cryopegs represent some of the most diversity-rich niches in polar environments (Gilichinsky et al., 2005). It cannot be ignored, however, that the inherently oligotrophic nature of many polar habitats presents a selective pressure for organisms able to survive under limited-nutrient concentrations. Conditions at the poles (as in many extreme environments) involve microorganisms operating at the thermodynamic limits of life (Knoblauch et al., 1999), often highly dependent on syntrophic interactions to create the conditions enabling their survival (Stams and Plugge, 2009). Rather than being beneficial, sudden influxes of high concentrations of nutrients in certain environments may actually "shock" communities, disturbing the internal balance and decreasing overall microbial activity.

Oxygen limitation is a key characteristic of the saturated layer above permafrost, the water-permeated cracks that occur in ice and firnified snow, the deeper, saturated portion of active layers, and various other polar environments. Under anaerobic conditions, iron and other transition metals, including manganese and selenium, play a direct role in microbial growth, themselves acting as terminal electron acceptors. The redox levels in any particular niche influence the dominant terminal electron-accepting process, the type of syntrophic interactions, and the nature of the carbon conversion reactions (fermentation versus respiration) that occur (Cord-Ruwisch et al., 1988).

Sulfur cycling largely contributes to carbon turnover in marine environments, with sulfate-reducing bacteria being responsible for up to 50% of organic carbon mineralization in sediments (Jørgensen, 1982). Psychrophilic sulfate-reducing bacteria isolated from permanently cold Arctic marine sediments were the first known sulfate reducers to grow below 0°C (Knoblauch et al., 1999). Examination of Arctic cold springs has revealed chemolithoautotrophic bacteria capable of growing by both sulfide and thiosulfate oxidation (Niederberger et al., 2009). Preliminary studies at Blood Falls in Antarctica (Ovadnevaite et al., 2009) have provided evidence of an interesting lifestyle whereby bacteria obtain energy from the reduction of sulfate to sulfite, which then reacts with iron, replenishing the sulfate concentration (Mikucki et al., 2009).

One of the most important factors in the contribution of sulfur-containing compounds in polar environments is their occurrence in the atmosphere as aerosols or as part of sand and snow sulfate deposits arising from volcanic eruptions (e.g., Mount Agung, Mount Pinatubo, Mount Erebus) (Legrand and Wagenbach, 1999; Mazzera et al., 2001; Ovadnevaite et al., 2009). Sulfuric acid and sulfate aerosols can have numerous direct and indirect impacts on polar environments, including influencing droplet number in cloud formation, increasing cloud albedo (indirect effect), and possibly contributing to changes in precipitation (Kiehl and Briegleb, 1993; Ovadnevaite et al., 2009). The impact of sulfur aerosols on macro- and microclimate and habitats has been associated with their role in reflecting solar radiation back to space (Kiehl and Briegleb, 1993). A further influence of sulfates on microbial niches is through the acidity they contribute to the environment. This can be neutralized by ammonium, but sulfate concentrations at the poles tend to exceed those of ammonium. Acidity affects numerous microbial activities, including carbon turnover rates, nitrogen fixation, and nitrogen mineralization (Mancinelli, 1986).

Mineral composition and the forms that are released have profound influences on polar communities. P-, N-, and S-containing nutrients may help promote carbon turnover and growth, and some metals (Mg, Fe, and Ni) may act as beneficial cofactors for enzyme activity. Accumulation of manganese complexes has been indicated to provide a first level of protection against ionizing radiation (Daly, 2009). Alternatively, release of mineral-bound metals such as As and Hg, or other compounds, e.g., CN, within seasonally confined habitats, may greatly contribute to toxicity and inhibition of microbial activities. This is also the case for initially bacterial-promoting activities such as sulfate reduction where confined pockets within permafrost prevent escape of toxic H_2S.

Extensive studies in more temperate environments have clearly shown that microbial utilization of organic matter is largely influenced by the ratio of C to N to P (Lein et al., 2010). Though the ratios for polar organisms may differ due to slower metabolism and protein turnover rates, a requirement for N and P remains critical to microbial growth and influences the nature of the metabolic pathways that are activated (Mindl et al., 2007; Thingstad et al., 2008; Chang et al., 2010). The main inputs of phosphorus (and other organic and inorganic matter) to continental polar environments come from aeolian deposits and sediment release as a result of glaciation activities. Coastal regions and areas along animal migration routes also receive faunal inputs, largely in the form of animal fecal matter (Hodson et al., 2004; Mindl et al., 2007). A broad combination of physical, chemical, and biological activities contribute to the liberation, transformation, and movement of minerals throughout polar environments. The frozen nature of many habitats at the poles may, as with carbon, partition microbes away from essential compounds so that the presence of particular nutrients in a proximal area does not necessarily guarantee their biological accessibility.

Latitude, Altitude, and Solar Irradiance

THE IMPORTANCE OF LIGHT
Most of the solar radiance that the earth receives is from the visible and near-infrared spectrum, about 70% of which reaches the earth (the other 30% being deflected back) (http://www.gcrio.org/CONSEQUENCES/winter96/sunclimate.html). Solar energy drives the water cycle, impacting on water movement and availability, cloud formation, and subsequent precipitation. Solar irradiance, in fact, is the prime source of energy driving biological and climatic processes on earth. Differential heating of air masses largely powers the movement of winds, with associated influences on erosion and weather. Vertical changes in energy drive local geochemical and niche modifications, while on the global scale these transformations are driven by horizontal energy influences on

water and air (Kuhn, 2009). At the poles the strength of light energy has a profound impact on determining both the nature (e.g., ice versus aqueous) and stability of microbial niches. The degree of heating (see "Temperature—Diurnal and Seasonal Variations," below) and permeation at the earth's surface influences the depth of the thawed active layer in terrestrial ecosystems and the photic zone in deeper water bodies. Temperature-driven variations in water flow and terrestrial permeation throughout the year greatly influence nutrient transport and microbial activity.

Although most bacteria are not driven by the type of circadian rhythms that influence the lives of humans and most other living organisms, they are far from immune to the influences of light. This influence is especially critical in polar environments, where limited plant life exists. The light and heat energy provided by the sun drives the microbial photosynthesis and primary production that form the basis of polar biological loops (Laybourn-Parry, 2009). An additional effect of solar radiance on polar environments is introduction of organic compounds as a result of the "grasshopper effect" (global distillation). Introduction of these chemicals has detrimental effects (see "Stratospheric Ozone Depletion," below), but their photodegradation also helps enrich certain niches with easily utilizable substrates, with associated impacts on bacterial growth rates, bacterial size, and secondary production (Chrost and Faust, 1999).

A key issue for microbial (and all) life at the poles is UV radiation. As a result of decreased ozone (see "Stratospheric Ozone Depletion," below) and other factors, microbes at the poles experience higher levels of UV radiation. Much like a day at the beach perfecting a tan, microbial growth is influenced by a balance between beneficial energy that supports life and excessive irradiation that leads to desiccation and decreased cell viability. Radiation in the UV range is highly toxic to cells and is believed to be a large contributing factor in limiting microbial numbers and diversity in various polar environments. Although both poles are subject to winter periods of 24-hour darkness and summer days of almost 24-hour light, the earth's increased proximity to the sun during the Southern Hemisphere's summer results in Antarctica experiencing a considerably higher degree of irradiance than its northern counterpart (McMinn et al., 2000). This effect is further exaggerated by the higher average elevation of the southern continent. At a fundamental level the quality of light and wavelengths received affects microbial pigment recruitment and various physiological responses (Orbaek et al., 2002).

TEMPERATURE—DIURNAL AND SEASONAL VARIATIONS

Polar climates are based on short, cold summers and long, extremely cold winters. During the summer months there is up to 24 hours of daylight, while during much of their respective winters the Arctic and Antarctic are in complete darkness. Portions of polar landscapes exist in a permafrost state, with a thin upper layer that undergoes alternative freeze-thaw cycles based on variations in air temperature. Lower Arctic soils experience steep thermal gradients, while a decrease in thermal gradient is observed in the soils of the High Arctic, attributed to the presence of a thin, compacted snow cover that prevents excessive heating and promotes rapid cooling (Tarnocai, 1980).

The shallow angle at which the sun's rays hit the poles results in the energy being spread over a larger area, greatly diminishing the heating power relative to solar energy at more equatorial latitudes. Any heat input at polar latitudes, however, can have a profound impact on permafrost, snow, glaciers, cryoconite, and various other polar environments. Freeze-thaw activities within rock surfaces greatly contribute to erosion, niche transformation, and nutrient release. Partial snowmelt followed by rapid freeze-thaw cycling (particularly on a diurnal scale) can lead to compaction of snow into an ice layer. Such layers subsequently act to limit water and heat permeation into subsurface environments, leading to decreases in active-layer thickness. Pooling of precipita-

tion in valley areas followed by cooling winds contributes to formation of large ice bodies, as seen in the Antarctic Dry Valleys (Doran et al., 2002).

On the Antarctic continent, ice formation at the coasts is largely determined by interior landmass temperatures, which affect both melt conditions and nutrient release. At both poles, depth increases associated with glacial melt (and to a lesser extent snowmelt) not only alter the salinity and nutrient nature in respective water bodies but can alter the transport and exchange with adjacent/surrounding sediments and waters. Increased depth, particularly in ponds and streams, changes the level of light intensity and heat, as well as the nutrient and oxygen concentrations that reach underlying sediments, with critical impacts on associated biogeochemical cycles. Changes in water chemistry, temperature, and depth have profound influences on the halocline, the important layer of cold water that protects sea ice from warmer water inflows that occur at depth (Sigman et al., 2007). Losses in this insulation layer promote ice melt, with associated changes in albedo, heat absorption, and carbon release from oceans. Even more critically, changes in water stratification and heat distribution impact on the ocean conveyor flows that play a large role in defining climate on the global scale (http://www.whoi.edu/oceanus/viewArticle.do?id=9206). For instance, it is the relative warmth of the oceans that provides the moderating effect on the northern landscape and which is the reason that the Arctic does not experience the extreme temperatures seen on the Antarctic continent (Doran et al., 2002).

In addition to water chemistry, the frequency, length, and timing of freeze-thaw cycles determine the extent of the Arctic Ocean that is covered by the Arctic ice pack and the degree of glacial and permafrost thawing at both poles (Doran et al., 2002; Vincent et al., 2004). Furthermore, within the poles themselves, changes are not homogeneous. Differences in wind temperatures and conditions at different areas and altitudes of Antarctica give rise to variable landscape responses; cooling on one coast occurs simultaneously with warming, permafrost melt, and increases in active-layer depth on the other.

Seasonal temperature conditions select for activity of certain community members, while others become dormant only to be "reawakened" by subsequent cyclic changes in ambient conditions. Thawing in near-surface sediments may promote aerobic heterotrophic activity, whereas permeation of meltwater into soils may lead to stimulation of anaerobic metabolisms. Seasonal variations in swamping/desiccation may create niches that undergo periodic aerobic/anaerobic activity cycles. Increased water exposure in the Arctic Ocean or at Antarctic coasts may stimulate growth of primary producers. Temperature, irradiance, and other influences on productivity and organic transformations during warm summer months largely determine winter viability through their impact on the concentration and complexity of nutrient compounds, particularly carbon.

The number of "ice-free" days and the timing of their occurrence have profound influences on microbial activity. Like the flower garden that blooms in a stretch of warm winter days only to be frostbitten some days later by the return of cold conditions, an early melt may stimulate bacterial activity, but the air temperature, precipitation, and nutrient conditions may be insufficient to sustain growth and a subsequent stress may result in cell death. Similarly, an extended growth period may result in microbes' becoming more nutrient or desiccation stressed and/or potentially lead to depletion of nutrient and carbon resources that would normally have supported a winter microbial community (Meltofte et al., 2008).

Niches on mountain slopes vary largely based on whether they occur on exposed or protected slopes. Protected slopes not only likely experience milder temperature extremes, but are less prone to cryodesiccation and nutrient stripping. In addition to contributing to erosion, the circulation created by wind and water movements influences the mixing of gases and organic matter. Together with redox variations (primarily vertical), this mixing

plays a key role in driving microbial activities, promoting fermentative and slower anaerobic activities in deeper, darker niches, and heterotrophic and occasional phototrophic activities in shallower layers (Cota et al., 1996; Tarnocai, 2009).

STRATOSPHERIC OZONE DEPLETION

As previously mentioned, the UV fraction of solar light is damaging to cells of all types. Microorganisms are particularly vulnerable. Exposure to UV radiation can cause mutation or can be lethal due to the damage it causes to DNA and other cellular components. As such, solar UV may be an important regulator of the sizes of bacterial populations in many environments, especially in the polar regions (Jeffrey et al., 1997; Miller et al., 1999; De Fabo, 2005). Solar UV can be divided into three components, UVA, UVB, and UVC, based on wavelength. UVC includes the shortest, most energetic wavelengths (approximate 190 to 280 nm), while UVA covers the longest, least-damaging wavelengths (315 to 400 nm). Because of its relative higher energy, exposure to UVC produces more damage than an equal amount of exposure to UVB or UVA. Likewise, UVB is more damaging than UVA (Miller et al., 1999). Photodegradation can also impact on carbon compounds, influencing the type of energy molecules available, a particularly important factor in nutrient-limited environments such as the poles.

Much of the UV radiation from the sun does not reach the earth's surface due to absorption by ozone in the stratosphere (Jeffrey et al., 1997; Miller et al., 1999; De Fabo, 2005). This ozone layer absorbs all of the UVC produced by the sun as well as a large portion of the UVB radiation (Miller et al., 1999). Ironically, formation of ozone itself occurs largely as a result of impact between UV and atmospheric oxygen. As a result of decreased sunlight, the poles experience a much thinner protective ozone layer (Jokela et al., 2008).

Microorganisms and other life forms on Earth have evolved systems to repair damage to DNA caused by UV radiation and other damaging environmental agents such as desiccation (Walker, 1984; Battista, 1997; Miller et al., 1999). These repair systems often cause mistakes to be made in reproducing the DNA sequences containing the damage, ultimately leading to mutation. Because of this and other negative consequences to the cell, living organisms have evolved repair systems that are capable of eliminating the damage to cellular DNA and other components that they encounter in their environments. Mutations may lead to phenotypes that offer competitive advantages, or exposure to increased levels of solar UV radiation could have disastrous effects (De Fabo, 2005). Living organisms do not have excess repair capacity and are living on the brink of disaster (Jagger, 1985).

Destruction of the stratospheric ozone layer caused by pollutants has been observed since the 1980s (Miller et al., 1999; De Fabo, 2005) and is leading to just such an increase in exposure to solar UV radiation, particularly to UVB wavelengths. While this reduction is most apparent and recognized in the creation of the Antarctic ozone hole during the spring and summer months (Madronich et al., 1991; Schoeberl and Hartmann, 1991), a similar, albeit smaller, ozone depletion occurs over the Arctic (Brune et al., 1991; Ausatin et al., 1992). Thinning has been observed throughout the mid-latitudes of the Northern Hemisphere as well (De Fabo, 2005). Data on ozone levels from 1996 to 2005 are available from NASA's Total Ozone Mapping Spectrometer (TOMS) (http://jwocky.gsfc.nasa.gov/eptoms/ep.html) and demonstrate the increasing problem over time. Greenhouse gases exacerbate the effects of the depletion of ozone. When these gases, including ozone, become trapped in the troposphere, they act to reduce the outpouring of heat from the earth's surface. This cools the stratosphere, resulting in stratospheric clouds that further destroy ozone and increase the levels of harmful UVB transmitted to the earth's surface (Miller et al., 1999; De Fabo, 2005; Vincent et al., 2007).

Many factors affect the levels of solar UV irradiation encountered by microorganisms. DOM, particularly colored DOM (CDOM), is an important factor in protecting aquatic microorganisms from solar radiation. UVB accelerates the decomposition of CDOM entering aquatic systems (Zepp et al., 2003). This both reduces protection from UV irradiation and has important effects on carbon-cycling dynamics, especially in the ocean. Different pigments absorb different wavelengths of light and by extension support different hierarchies of organisms and associated metabolic transformations.

Leavitt et al. (1997, 2003a) studied changes in climate that altered the amounts of DOM in alpine and subalpine lakes. They quantified UV exposure in Crowfoot Lake in Alberta, Canada over the past 12,000 years (Leavitt et al., 2003a). When external DOM inputs were the lowest and UV exposure highest (during the past 4,000 years), algal abundance was 10- to 25-fold lower than during times when UV exposure was reduced due to higher levels of DOM in the lake. In a similar study, Pienitz and Vincent (2000) concluded that the changes in CDOM over the Holocene in a subarctic lake produced an inferred change in biological exposure to UV radiation that was at least 2 orders of magnitude greater than changes due to moderate ozone depletion.

Scully et al. (1997) compared photochemical effects of increased UV radiation due to ozone depletion in several lakes including Lake Erie, a lake clear of CDOM; and Brookes Bay, a lake colored by CDOM. They discovered that hydrogen peroxide and singlet oxygen production was increased by 70% with increased exposure to UVB in Lake Erie, while it was only increased by 25% in the colored lake by exposure to higher levels of UVB. These data highlight the importance of CDOM in protecting aquatic microbes as well as the effects of increased solar UVB on the photochemistry of these systems. Under depleted ozone conditions, carbon monoxide also increased, suggesting that increased UV exposure led to the destruction of DOM, with the released carbon being lost to the overlying atmosphere.

Antarctic lakes are especially susceptible to UV exposure since they have very limited sources of UV-absorbing DOM (Pienitz and Vincent, 2000). Changes in UV exposure and its consequences are controlled almost exclusively by changes in solar production, atmospheric transmission, lake depth, and ice and snow cover (Moorhead et al., 1997; Doran et al., 2002; Vincent et al., 2007). These changes have profound impacts on microbial communities and by extension their carbon-cycling activities.

Using an approach similar to their studies at Crowfoot Lake in Canada, Leavitt et al. (2003b) studied several Antarctic lakes and found that shallow lakes on this continent have been exposed to fluctuating amounts of solar UV radiation during the Holocene. Higher levels of UV exposure occurred in the preceding interglacial period (Hodgson et al., 2004). These investigators assessed the abundance of photoprotective pigments in lake sediments and found that they were relatively more abundant in samples older than 4,000 years. During the past 13,000 years benthic cyanobacterial communities have been exposed to levels of UV radiation that varied more than fourfold. Today photoprotective pigments are common in algal communities in these shallow lakes of Antarctica, and algal abundance varies directly with the levels of UV to which they are exposed (Hodgson et al., 2004). Similar results were obtained in a study of 17 shallow Arctic lakes in Canada and Alaska (Bonilla et al., 2009). These authors observed that planktonic photoprotective pigments in cyanobacteria and algae were directly correlated with the penetration of UV light into the lakes. They were inversely correlated with temperature and the amount of CDOM.

The effect of mild artificial UV conditions on the photosynthetic potential of two red algal species isolated from Arctic waters was studied by Holzinger et al. (2004). They found that UV exposure reduced photochemical efficiencies of photosystem II. While the photosynthetic apparatus was severely influenced in both species, only *Palmaria palmate* showed photoinhibition under the experimental conditions; *Odonthalia dentate* was not photoinhibited. Many of the membrane structures of both species were damaged by exposure to UV, including their mitochondria. Such physiological impacts on individual microbes have profound effects on the functioning of communities as a whole.

Callaghan et al. (2004) studied the past changes in Arctic terrestrial ecosystems due to changes in climate and UV radiation exposure. These investigators studied the period from the last glacial maximum to the present day, focusing primarily on the past 21,000 years. Mass extinctions occurred during the most recent large-magnitude global warming, and the Arctic biota have been at their minimum for the past 10,000 years. Evidence from the early Holocene suggests that as the earth warms due to greenhouse gas emissions the tree line will advance into the tundra, increasing the risk of species extinction.

Unlike these past episodes of global warming, the amount of current co-occurring environmental changes, including enhanced levels of UVB exposure, deposition of nitrogen compounds from the atmosphere, pollution by metals and acids, and habitat fragmentation, will put these ecosystems under unprecedented environmental stress. Likewise, temperature changes associated with increases in atmospheric greenhouse gases have reduced the ice cover of many Arctic and Antarctic lakes. These changes are exposing underwater microbial communities to increases in solar UV greater than the increase in exposure due to the reduction in stratospheric ozone (Wrona et al., 2006). These synergistic effects of reduced ozone, CDOM, and ice are acting to dramatically increase DNA-damaging stress on these communities, perhaps to levels exceeding their repair capacity.

Molecular Studies on the Response of Polar Communities to Enhanced Solar UV Exposure. Chapter 6 and many insightful reviews (Bakermans, 2009) attempt to address many of the adaptive mechanisms that enable microbes to grow/survive in the extreme conditions they encounter at the poles. Here we will address the mechanisms by which microbes repair UV damage in their most fundamental molecule, the DNA that encodes the proteins and molecules that mediate all their life functions. The product of the recA gene, the RecA protein, is a regulator of light-independent DNA repair in bacteria (Walker, 1984; Miller, 2000). While photoreactivation will repair DNA damage, it requires visible light (Walker, 1984; Jagger, 1985; Miller, 2000). When visible light is not available to the cell or when the damage exceeds the capacity of the photoreactivation system to repair it, light-independent repair mechanisms (commonly known as the SOS repair regulon) are induced (Walker, 1984).

In the presence of pyrimidine dimers and other photoproducts that produce DNA damage, DNA synthesis in bacteria is inhibited, leading to single-stranded regions (ssDNA) in the chromosome. These ssDNAs activate RecA to relieve repression by the LexA protein exerted on the SOS regulon (Walker, 1984). Expression of the recA gene is part of this regulatory network and the concentration of RecA increases in the cell, sometimes becoming the most abundant protein. This release from repression is only transitory, however. Following repair of the ssDNA damage, RecA is no longer activated and LexA reestablishes repression (Elasri and Miller, 1998, 1999; Elasri et al., 2000). With the reestablishment of repression, the RecA concentration in bacterial cells returns to noninduced levels (Elasri and Miller, 1998). Thus, RecA concentration in the bacterial cell is directly proportional to the concentration of pyrimidine dimers in the cell's DNA (Miller et al., 1999; Booth et al., 2001a, 2001b), and the level of RecA in a cell is therefore an indicator of both the level of DNA damage and the cell's DNA repair capacity. This marker of repair activity may potentially help identify which community members may be best able to adapt to a particular set of conditions and, hence,

help in prediction of the direction of carbon flux.

This marker possibility is strengthened by the fact that the recA gene is found in all bacteria (Miller, 2000) but is not present in archaea or in eukaryotes, although they contain similar genes (Miller, 2000). RecA is highly conserved, and antisera derived from Escherichia coli will cross-react with virtually all RecA proteins from other bacterial species (Kokjohn and Miller, 1988; Miller and Kokjohn, 1990).

By investigating levels of recA mRNA in marine bacterial populations Lyons et al. (1998) estimated the concentrations of RecA in bacteria in the Gulf of Mexico. Their study demonstrated that light-independent repair was essential for the repair of DNA damage caused by daily exposure of bacterioplankton to solar UVB radiation. The primary form of DNA damage was shown to be formation of pyrimidine dimers and other photoproducts, and this occurred on a diel cycle (Jeffrey et al., 1996a, 1996b).

Booth et al. (2001b) used the RecA protein as a dosimeter for DNA damage and repair capacity to measure the relative levels of ozone depletion-induced DNA damage in bacteria in the Gulf of Mexico. Like Lyons et al., these investigators also found that the concentration of RecA antigen in marine bacteria in the Gulf followed a diel cycle, with a minor peak at noon and a substantial peak during the early hours following sunset (Fig. 1 shows an example). These authors suggested that the noonday peak indicated that the capacity of the photoreactivation repair system (light-dependent repair) was exceeded, while the induction during the dark hours was usual and necessary to completely repair the damage remaining in the cell when visible light was no longer available. These same investigators obtained similar results when investigating the bacterioplankton of the ocean waters of the Gerlache Strait near Palmer Station, Antarctica (Booth et al., 2001a). They analyzed both recA mRNA and RecA antigen in this community. Similar patterns of expression were also observed in a γ-proteobacterium isolate from the strait. In each case there was a diel induction at midday and a second at dusk.

During the 2001 season Warner and Miller collected data on RecA antigen concentrations in marine bacterioplankton near Palmer Station (unpublished observations). While they also observed peaks in RecA concentration at midday and early evening, the midday peaks were found to exceed those in the evening (Fig. 1), unlike those observed in the Gulf of Mexico. Moreover, although the data were scattered, a direct correlation between an increased ratio of midday-to-evening induction and the extent of the stratospheric ozone depletion was observed. During the period from October to December 2001, the ozone hole over Antarctica closed from a covering of more that 22 million km^2 in early October to less than 6 million km^2 in early December (TOMS project data; see http://jwocky.gsfc.nasa.gov/eptoms/ep.html). The stratospheric ozone layer that is usually thinnest in September thickened greatly during this period (Color Plate 10). As it thickened, the level of RecA induction at midday was reduced, indicating that DNA damage and therefore the necessity to induce

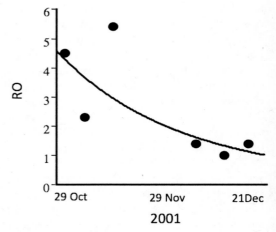

FIGURE 2 Ratio (RO) of noonday-to-evening RecA induction peaks (see Fig. 1) as a function of date. There is a direct correlation of this ratio with the extent of stratospheric ozone (see Color Plate 10). As the hole becomes smaller, this ratio becomes smaller (Warner and Miller, unpublished data).

light-independent repair at this time of day was reduced (Fig. 2). These data indicate that cellular repair capacity is taxed to the maximum by the increased levels of UV irradiation during ozone thinning. Indeed, Booth et al. (2001a) found that increased levels of solar UV exposure negatively affected the number of viable bacterioplankton.

The stratosphere's influence on bacterioplankton is complicated. Influence from greenhouse gases and many environmental factors work synergistically to tax the survival strategies of polar bacterioplankton. While additional studies will help us to understand the important ways in which the upper layers of the earth's atmosphere affect biological organisms, it is clear that the consequences of ozone thinning are dramatic to microbial life. Even though some attempts have been made to reduce and even reverse the destruction of the ozone layer, it will be many years before positive effects are realized. The fabulous ability of microbes to evolve, repair damage, and adapt and the rate at which they can do so will determine the fate of polar microbial life and its impact on global carbon and climate.

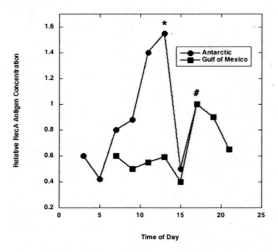

FIGURE 1 Relative concentrations of RecA antigen in marine bacterioplankton in the Gulf of Mexico and Antarctica. The midday (*) and evening (#) peaks are indicated. The 17:00 (evening) reading was normalized to one for comparative purposes (Warner and Miller, unpublished data).

NITROGEN SPECIES

A number of other compounds have an impact on ozone and climate; this is particularly true for nitrogen, which is transformed in snow from nitrates (the form in which it is most commonly deposited) to molecules that are reactive in the atmosphere (NO_x and HONO). Microbial nitrous oxide production appears to be enhanced at higher water–activity concentrations (Stres et al., 2008), and atmospheric release of nitrogen species with a biological origin from snow has been shown to occur even under conditions of low or absent light without ice-/snowmelt (Amoroso et al., 2010). Cycling of nitrogen has important climatic consequences, but also its presence in snow and other polar niches strongly determines microbial activity and dominance as oftentimes it is a limiting nutrient for organisms unable to fix N_2. Both the concentration and the form in which the nitrogen exists are critical factors. Nitrous oxide has a global warming potential 300 times greater than that of CO_2 and its aerosol impact on acidity can have additional impacts on microbial life.

Geological Transformations

The geological record reveals the dramatic impact that climate-driven alterations in water-phase chemistry and wind velocity play in material deposition, landscape shaping, and destruction/accumulation of ice and glaciers. Increases and decreases in freeze length time and frequency associated with climate warming or cooling contribute extensively to upheaval and rearrangement of soil horizons, shearing of particulates, and translocation of organics and solutes, and hence change the overall physicochemical nature and physical structure of permafrost habitats. Mixing of layers has a profound impact on microbes, affecting not only the degree of light and moisture that they receive but also the quality and concentration of bioavailable nutrients.

Erosion of surface sediments can decrease both the depth of the active layer as well as the level of nutrients available for transport to deeper layers. Wind and water circulation also provide transport routes for movement of microbes into new niches. Transfer of nutrients to different aqueous and ice layers and the size and degree of incorporation of particles into permafrost and other ice structures influence albedo. Associated changes in heat maintenance and transfer have influences on microbial activities that themselves act to influence particle breakdown and niche reorganization. Changes in soil and sediment grain size and geometry have strong effects on the surface area available for microbial colonization and the colocalization of microbes with nutrients and gases. The temperature at the time of these various erosive activities influences whether transport and nutrient enrichment will be more localized (as in the case of aqueous erosion followed by rapid cooling) or longer range (in cases where extended thaws contribute to increased water flow). Features such as the depth of ice layers and the size and degree of water veins in ice are largely influenced by climatic and hydrogeological factors at the time of formation.

Snow is well ventilated and can act as a photochemical reactor, heavily influencing the chemistry of the overlying atmosphere (Amoroso et al., 2010). Gas pockets, sediment pore volumes, and linkage between these via subsurface water flow result in landscape-level transport and exchange of gaseous and dissolved forms of carbon. In ice bodies at least three different habitat types can be predicted: gas-filled veins, liquid-saturated veins, and thin liquid films surrounding mineral grains that may be entrained in the frozen structure itself. Movement of liquid and gas transports carbon through these numerous, functionally distinct polar ecosystems, where it is biochemically transformed via methanogenic, methanotrophic, and a variety of organic matter degradative pathways (both mineralizing and fermentative) (Judd and Kling, 2002). The interaction of terrestrial and aquatic mechanisms that drive physical partitioning of carbon molecules between aqueous, gas, and solid phases and the balance of the above microbial carbon transformation mechanisms will have the determinative effect on climate (Yergeau et al., 2010).

METABOLIC TRANSITIONS

A key question that arises when considering extreme environments such as those that exist at the poles is in regards to the true metabolic state of the microbial communities: are microbes actively growing and multiplying; maintaining basal metabolism and perhaps evolving with time; or in a dormant state? Furthermore, what are the factors that determine the switch between these various modes? Temperature obviously plays an important role; however, most bacteria exhibit a broad growth temperature profile, so this is not the only factor influencing microbial production. This point is supported by data that show that although microbial production in polar systems is generally lower than in more temperate environments, production during peak annual seasons often overlaps with the minimal rates observed at lower latitudes (Kirchman et al., 2009b).

As in all environments, microbial survival at the poles is based on community interactions, and biological responses of individual members to environmental changes are nonuniform. What may be an advantageous shift for one member of the community may not be so for other members, resulting in a decrease in growth, metabolic quiescence, or perhaps even death for another. Any attempt to understand microbial transformations must take into account that three different types of alterations determine observed changes in microbial activities. (i) The first of these is shifts in rates and degrees of activity of active community members as conditions fluctuate. (ii) The second is changes in the specific pathways recruited by microbes in response to varying ambient conditions. (iii) Finally, there are changes in the fundamental composition of the active fraction of communities. These key considerations lead to yet other questions, as previously discussed relating to microbial mutation, repair, and adaptation. The sum parcel of observed activity will reflect the influence of a large range of environmental factors at both the macro and micro levels. How these factors interact to impact on overall function is a complex but essential consideration for understanding activity in current polar environments and, even more so, for efforts to comprehend and predict the responses to and impact on global warming.

Since many bacteria rely on other organisms to ferment or break down complex nutrients to simpler carbon compounds that they can then use, loss of one group of organisms (or at least their metabolic pathways) may result in breakdown of communities, turnover rate downshifts, and/or moves to completely different carbon-transforming pathways. Shifts may involve any variety of blocks at particular intermediates in mineralization processes, resulting in lower CO_2 release. Channeling into previously secondary fermentative or anaerobic pathways may result in increased production of short organic acids or methane.

Phototrophy has a large influence on the turnover of carbon in polar environments. Annual temperature variations and gradients within different niches affect the interaction between bacteria and primary producers, with an associated impact on carbon fluxes (Kirchman et al., 2009b; Berggren et al., 2010). Microbial pigment production has been shown to be related to the concentration of particulate matter in the growth environment. Distinct differences in phototrophic communities between oxic and anoxic layers of polar water columns have been observed (Antoniades et al., 2009). Not surprisingly, euphotic depth has been shown to greatly influence the ratio of bacterial production to primary production in polar environments (Rich et al., 1997), but as with temperature the relationship is complex. As pointed out above, degree of ice cover (also snow cover in terrestrial ecosystems) influences penetration of solar radiation, mixing regimes, and maintenance of salt and temperature stratification. Together these factors have an impact on the nature of microbial metabolism and carbon assimilation and dissimilation. Alterations that increase light penetration will enable microbes to increase their production per unit biomass.

The balance of electron shunting through various carbon turnover pathways versus biomass production influences both instantaneous carbon fluxes and the shifts in the nature and concentration of microbes and carbon compounds driving community interaction and carbon cycling in subsequent seasons. Microbial characteristics that are considered are bulk estimates of biomass (mass of cellular carbon per m^2), production (mass of cellular carbon produced per day per m^2), and growth rates (per capita daily change in abundance) (Kirchman et al., 2009b). High biomass production may result in biofilm communities and microniches that provide increased protection from harsh conditions. Changes in biomass ratios can fundamentally alter the redox state and function of a particular niche. For instance, high oxygen consumption can subsequently protect oxygen-sensitive enzymes, promoting anaerobic carbon flux pathways. Imbalances can disrupt intricate syntrophic relationships and the thermodynamic possibilities they enable (McInerney et al., 2009). The complex interaction of various physicochemical parameters (e.g., nutrients, light penetration, redox structure, presence of potential electron acceptors) and microbial concentrations and lifestyle are major determinants of microbial activities on the seasonal scale, and in longer-term considerations of global warming impacts.

A number of organisms are capable of mixotrophy, an ability to switch from phototrophy to heterotrophic growth when light levels become low. Mixotrophs have been shown to be able to survive 6 months (and perhaps longer periods) of darkness and to be able to rapidly resume photosynthesis when light returns. Mixotrophy allows organisms to not only survive, but to grow and contribute to active food chains and the survival of other community members (Jones et al., 2009). In polar habitats where extended periods of darkness are the norm, be it as a result of polar winters or shielding beneath frozen layers, mixotrophs can have a profound influence

on survival and carbon turnover. How mixotrophic populations adapt to future changes in polar environments will likely play a large role in determining the role of polar environments as either sources or sinks of atmospheric greenhouse gases.

Bacteria often exhibit differential responses to organic matter substrates, and this has led to the classification of two functionally divergent groups: generalists that exhibit high metabolic flexibility and quick adaptation to new substrates, and specialists that display more limited niche adaptation and highly defined nutritional requirements (Dyda et al., 2009). The versatility of metabolic generalists and the independence of autotrophic microbes enable colonization of new environments, which are then slowly modified to conditions that support heterotrophs and more specialized microbes. Interactions between these lifestyles contribute to niche stabilization and functioning of cohesive microbial ecosystems. However, due to large seasonal variations in growth conditions, bacterial periodicity is a reality in many of the polar niches and disruption of the delicate balance can result in intertrophic mismatches (Meltofte et al., 2008). Such breakdowns in community functioning are difficult to predict but will be key determinants of the nature and direction of carbon cycling in polar environments.

As highlighted in Chapter 6, cold-adapted microorganisms have developed a variety of resistance mechanisms, enabling them to adapt and survive at decreased temperatures. In some cases adaptation to one set of stresses strengthens an organism's ability to deal with another, as was observed for the survival of both *E. coli* and *Deinococcus* strains exposed to both decreased air pressure and subzero temperature (Diaz and Schulze-Makuch, 2006). Though rates of enzyme activity at cooler temperatures are largely diminished, preliminary evidence suggests that turnover rates at cooler temperatures may be equally limited, such that organisms need not synthesize proteins as frequently in order to maintain a target level of activity. Though a discussion of adaptive evolution is beyond the scope of this chapter, as revealed in the discussion of DNA repair systems in response to UV damage, microbes never cease to amaze with their ability to adapt to the most stark and extreme ambient conditions, and the story of polar community response to global warming is certain to be full of many wonderful surprises.

CLIMATE CHANGE—INTEGRATING RECIPROCAL INFLUENCES OF BIOTIC AND ABIOTIC FACTORS ON CLIMATE AND HABITAT CHANGE

Environmental and (Micro)biological Shifts Contributing to Global Warming

Changes in climate and the subsequent influences on ecosystems involve a variety of interacting dynamic and multidimensional factors. In addition to the variety of variables outlined above, singular events (such as the volcanic eruption of Mount Pinatubo) can also have far-reaching effects that extend beyond the local environment and limited time/season in which they occur. As pointed out above, ice melt in one area resulting in increased freshwater flow into streams may have a strong impact on the ocean halocline miles away, with subsequent impacts on ocean stratification and global wind movements. These alterations redefine the nature and even existence of numerous niches, with profound influences on biological functions and by extension climate. It is these impacts on already fragile environments that have led to the belief that if the poles and their ice caps go, so does the planet (http://www.whoi.edu/oceanus/viewArticle.do?id=9206).

Observed temperature increases over the past century and considerable decreases in polar glaciers have already raised much concern regarding the degree of dangerous global warming and associated consequences for life as we know it. Niche alterations have profound influences on microbes, and since microbial activities play a central role in global cycling of key greenhouse gases like CO_2 and CH_4, feedbacks will themselves have profound impacts on climate. The primary factor that determines CO_2 concentrations in the atmosphere is the balance between photosynthesis and respiration, activities that are dominated by completely different mechanisms in aquatic versus terrestrial environments (Singh et al., 2010).

A wealth of scientific effort has been (and continues to be) committed to the development of models aimed at predicting long-term climate effects, and the outcome is hotly debated. Not all models have attempted to address the microbial variable in efforts to predict future climate, but it is clear that the impact of microbial activities (aerobic and anaerobic respiration, fixation, and transformation) will be considerable. Models attempt to define algorithms that promote fit to observed data. Climate prediction efforts are based on combining information from activities occurring on vastly different time scales: the hourly to monthly life cycles of various r- and k-strategists; seasonal shifts in temperature, illumination, and humidity; the stochastic advances of human technology; the erosive activities of blowing winds and flowing waters; the millennial scale of various geological transformations, to name but a few. In some cases combinations of environmental conditions can mask microbial activities and vice versa (microbial activity may complicate abiotic transformations), so that observed measures do not reflect the true involvement of various factors (biotic or abiotic), hence skewing predictive efforts. Furthermore, the degree of heterogeneity, the number of interacting factors, the natural variation of biological activity, and the impact of various feedbacks greatly muddy the waters of extrapolative efforts. It is impossible to present all the available theories and possibilities, but the following sections will attempt to summarize some of the key information, place it in a polar context, and guide the reader to appropriate sources for further reading.

One of the most common predictions is that climate warming will stimulate microbial decomposition of soil carbon, producing a positive feedback that will further increase greenhouse gas release to the atmosphere and contribute to rising global temperatures (Friedlingstein et al., 2006). Terrestrial ecosystems release approximately 115 billion tons of carbon to the atmosphere (60 billion from autotrophic respiration and 55 billion from heterotrophic respiration). The amount for oceans is about 92 billion tons (58 billion and 34 billion for autotrophic and heterotrophic respiration, respectively) (http://www.eoearth.org/article/Carbon_dioxide). With the terrestrial environment and particularly the Arctic believed to sequester between approximately 25 and 40% of global carbon concentrations (International Arctic Science Committee, 2010), changes in sequestering activities in these environments can have profound consequences for atmospheric greenhouse gas concentrations and subsequent climate effects. Increased runoff may contribute to erosion of the top layers of soil and ice, leading to loss of the richer, primary-producing communities that are associated with these aerated, more seasonally warmed segments of polar ecosystems. Adaptation of the deeper, more anaerobic communities now exposed to the potentially drier, warmer surface is uncertain. Increased heat transfer to depths, however, may promote activity of anaerobic communities, which may include increased activity of methanogenic microbes, fueling hypotheses of microbiological feedback to global warming.

Additional hypotheses related to positive warming feedback suggest that increases in biomass will contribute to larger concentrations of substrates (C but also N, P, etc.) as a result of cell turnover. The increased microbial activity may contribute to enhanced levels of organic degradation and increased fluxes of CO_2 to the atmosphere. The physiology of the most competitive species will determine the degree of carbon efficiency and the amount and nature (e.g., redox) of carbon compounds that are released.

Two main hypotheses that have been promoted to explain increases in atmospheric methane concentrations during the last deglaciation are releases from clathrate methane gas hydrates and northern peatlands (Walter et al.,

2007). Potential atmospheric release from these sites, as well as evidence that at least some polar waters that are increasingly being exposed by melting ice exhibit decreased microbial ability to sequester CO_2 (Cai et al., 2010), are some of the strongest evidence supporting concerns regarding polar environments, greenhouse gases, and climate change.

The reality of efforts at climate modeling is that numerous variables will control the direction of biological transformations of greenhouse gas and the final nature of sites as sources or sinks. A long-term picture of microbial activities requires predicting not only what changes will occur but the ambient conditions when they occur (timing). For instance, a snowmelt-associated influx of nutrients into a site during a polar summer will have distinctly different influences than a similar influx during winter (Meltofte et al., 2008). Even if winter temperatures are elevated as a result of global warming, the decreased illumination, solar warmth, etc., at this time will impose different limitations on microbial abilities to access and utilize nutrients. In addition to timing, the length of thawed warm seasons and the adaptive ability of the microbial members in various ecosystems (some of which may benefit from habitat changes, while conditions for others become suboptimal or even lethal) will determine the long-term direction of carbon flow (source or sink) of any particular niche.

Adaptations Mitigating Large-Scale Variations in Polar Climate and Subsequent Global Impact

As pointed out above, despite considerable effort, absolute knowledge of how microbial populations and their trophic interactions will be affected by global warming remains ambiguous. The geological record and the more defined nature of chemical reactions allow some prediction of environmental responses to changes in temperature, moisture, radiation, and other abiotic parameters. In this section, a variety of hypotheses regarding potential long-term shifts that may help counteract global warming effects are presented.

Global warming-induced drying in the Southern Hemisphere has decreased net primary production, counteracting net primary production increases in the north (Zhao and Running, 2010). Though an in-depth numerical evaluation of global CO_2 balances is beyond the scope of this work (excellent coverage of this subject can be found in Luyssaert et al., 2007), it is clear that there will be as many different influences on climate as there are different niches.

The possibility that warming in the Arctic and ice melt will contribute to the release of large stores of CO_2 and CH_4 cannot be unequivocally disputed at present. Such a shift in temperature at the poles, however, will likely also promote increased metabolic rates of certain portions of the microbial communities present in these environments. Turnover of the associated increased biomass could further promote growth of microbial communities and tundra-adapted plant species. Enhanced plant cover in previously barren landscapes could provide increased nutrients and protection to rhizospheric microbial communities (Luyssaert et al., 2007) and promote carbon sequestration. Increased radiation may also preferentially support growth of bacterial phototrophic primary producers, which may help counterbalance increased CO_2 emissions.

As previously mentioned, one of the most common hypotheses that has arisen out of global warming theory is the notion that ice melt and associated erosion will release or promote increased concentrations of microbes, gases, and nutrients to polar environments, and that metabolism by these bacteria will contribute to greater greenhouse gas release to the atmosphere, with a feedback contribution to further global warming. The question arises, however, as to whether microbes that have been shielded from solar radiation in oligotrophic interfacial waters and seasonally frozen permafrost soils will be able to adapt and grow at high rates. A microbial-enzyme soil carbon response model has suggested that in fact an attenuation response may be triggered that in the long term will maintain carbon level

loss at approximately current levels (Allison et al., 2010). This attenuation is predicted to be achieved through a decline in the fraction of assimilated carbon that is allocated to growth, also known as microbial carbon use efficiency (Allison et al., 2010). The information gained from current efforts to identify the genetic potential present in polar environments (outlined in Chapters 7 and 8), including the presence of metabolic capabilities that though at the moment dormant may be triggered by changes in ambient conditions, are critical aspects for predicting the fate of carbon cycling and subsequent global climate consequences.

Nutrient dilution as a result of increased ice melt may result in further oligotrophy in polar environments. Decreased availability of nutrient substrates may promote growth of autotrophic microorganisms that will capture CO_2 and utilize it to synthesize organic compounds for growth and survival. This capture would as a consequence limit the amount of CO_2 escaping to the atmosphere. Increases in the abundance of water-saturated soils resulting from permafrost melting may push many environments into low-redox conditions where anaerobic and fermentative metabolisms will dominate. Though such activity may contribute to increased methane release, these processes are considerably slower, resulting in both qualitative and quantitative changes in microbial carbon turnover processes (Wagner et al., 2009). A subtlety that cannot be overlooked is that many changes happen in degrees. Perhaps part of an ice pack melts, or the thinner portion of a stream thaws while its source remains frozen. These fragmented changes result in new subbiomes with unique responses and distinct relationships with both their "parent" niche and other surrounding environments. Efforts to predict microbial-climatic interactions must consider balances at many scales, from the micro to the global, and the heterogeneity that exists even within similar parameters when applied to different sites. For instance, efforts to predict degree of permafrost loss and changes in gas exchanges must consider that in many of the polar regions of North America, permafrost thickness is between 50 and 100 m, whereas in Siberia it is >500 m (Meltofte et al., 2008).

As mentioned previously, freezing point depression arising from high salt concentrations enables aqueous environments to persist at the poles at temperatures well below 0°C. Dilution resulting from melting ice bodies and snow may counteract this phenomenon, leading to new ice formation with resequestration of released microbes and carbon compounds. This fresh ice will have higher albedo than dense, multiyear blue ice. In this scenario, the temperature depression associated with the increased reflection of radiant heat may provide a counterbalance to the warming effect, with a subsequent net-zero change in overall polar temperatures on a long-term scale. The effects of these changes from a microbial perspective still await answers from culturing aimed at unequivocally demonstrating bacterial multiplication at temperatures below −12°C (Breezee et al., 2004).

With loss of ice cover and increases in polynya areas, the euphotic zone in newly exposed waters will likely result in increased primary production and utilization of CO_2 (Shiah and Ducklow, 1995; Friedlingstein et al., 2006). Additionally, observations of enhanced fluxes of organic carbon to the seafloor have been attributed to these recent losses of ice cover (Gobell et al., 2001). Whether these fluxes will help diminish the pool of carbon available to contribute to greenhouse gas emissions remains to be seen. Another interesting consideration for the fate of greenhouse gases in polar waters is the bioavailability of introduced carbon sources. For instance, although the Arctic Ocean has a greater input of dissolved organic carbon relative to the Ross Sea (Antarctica) as a result of greater riverine input, bacteria do not appear to exhibit a higher rate of growth. This is because terrestrial organic carbon is not believed to be a strong supporter of life in polar oceans, even if temperatures increase (Kirchman et al., 2009a).

Many of the hypotheses regarding the effect of global warming at the poles have focused on the high concentration of carbon trapped in the permafrost and the concept that its release will result in massive increases in microbial activity that will produce positive feedback, further exaggerating climate change. Respiration of old carbon in thawed sites, however, may be balanced by increased absorption by plant biomass. Evidence and extrapolation from a study at an Alaskan tundra site showed that the key factor determining if a site would act as a net source or net sink of carbon is highly related to the rate of permafrost loss and release of trapped (old) carbon (Schuur et al., 2009). Indications from multiyear tracking and simulation have shown that extensively thawed sites result in increased carbon release to the environment, primarily in winter. In contrast, moderately thawed sites appear to act as net sinks for atmospheric carbon. In the end, the influence of permafrost carbon on ecosystem balances will depend on actual carbon emission levels, thaw rates, the form of carbon gases released, nutrient and water availability, the degree of plant life stimulated and supported, and the nature/physiology of the dominant microbial communities (Dutta et al., 2006; Canadell et al., 2007).

As mentioned previously, it is well-known from years of studies in temperate soils and bioremediation efforts that carbon concentrations are only part of the story influencing microbial growth and proliferation. The concentrations of other nutrients, and very importantly the ratio of C to N to P, are fundamental to the dynamics of communities. Even in the tropics, uncontrolled growth of bacteria does not occur. Beyond a certain carrying capacity, nutrient ratios, predation control (bacteriophages, protists), and internal regulation/communication (metabolic control and signal transduction, i.e., biofilms) impose a balance on microbial activity (Clarke et al., 2010). Limitations in knowledge regarding the interaction of these factors and abiotic parameters and balances in polar bacterial metabolic activities is a large stumbling block in efforts to close the CO_2 and CH_4 loops and predict climate influences in these environments.

CONCLUSION

The critical role that microbial physiology plays in carbon cycling (and by extension global climate) highlights the importance of continuing research into the diversity, structure, and nature of life in polar environments. This is further indicated by the fact that although the microbial properties in polar systems seem to vary more than among their low-latitude counterparts, a large part of our current understanding of microbial physiology is based on extrapolations made from studies of mesophilic microbes.

The question remains: from a geomicrobiological standpoint, what does a 1, 5, or even 10°C shift in temperature mean? For years microbiologists have used the classical microbial temperature profile presented in any microbiology textbook as a guide to identify and define the optimal temperature values for microbial activity. For polar environments, however, it is likely that many bacteria are functioning under suboptimal conditions. What their levels of activity, response, and range of operation will be is very much uncertain and cannot necessarily be extrapolated from observations made in laboratory microcosms or limited in situ examinations. Certainly, increased temperatures may stimulate degradative activity and release of increased concentrations of greenhouse gases to the environment. The stimulatory effect, however, may also benefit autotrophic carbon-fixing organisms as well as methanotrophic bacteria that consume greenhouse gases.

Almost as important as identifying the low-temperature limit of life is identifying the upper temperature limits of polar psychrophilic communities. Do psychrophilic communities achieve mesophilic rates at higher temperatures or does their activity just break down? The question of diversity loss looms large. Will true psychrophiles downshift to a slower metabolism, slip into quiescence, or be irreversibly eliminated? Once quiesced, which members can be stimulated to return to an active state? What conditions will promote the return? Will the same pathways be active and will they function within the same ranges? These are some of the key considerations that culture-dependent and -independent (molecular) studies of the poles are attempting to address. The surface has barely been scratched in terms of identifying (let alone understanding) the biological and genomic potential of the poles. It is this precise information that would help us understand the biogeochemical changes that influenced the large-scale climatic changes that have contributed to the creation of current Earth.

The poles are changing—this fact is unequivocal. Polar communities exist at the edge of thermodynamic limits in a fine balance we do not yet fully understand. The consequences of disruption of these polar equilibriums remain to be seen. The microbial response, diversity shifts, and sum total of microbial and environmental interactions are very difficult to predict. Many studies (Dyda et al., 2009) have attempted to simulate changes that may occur with climate change to help predict some of the phylogenetic changes. However, due to large spatial variability and the immense heterogeneity (in nutrients, both carbon based and inorganic; light; temperature; geology; and humidity) throughout the poles, only hypotheses can be made. It is clear, however, that it will be the microbes—single-celled, nonnucleated, submicromolar-sized living organisms—that will determine if the poles will become a source or a sink for atmospheric carbon release. In the pre-antibiotic era it was pathogens that were considered the scourge of human existence; when it comes to climate change, microbes may prove to be our collective saviors, or contribute to a global catastrophe on a scale for which there is no precedent in human history.

REFERENCES

Adushkin, V. V., and V. P. Kudryavstev. 2010. Global methane flux into the atmosphere and its seasonal variations. *Earth Environ. Sci.* **46**:350–357.

Allison, S. D., M. D. Wallenstein, and M. A. Bradford. 2010. Soil-carbon response to warming

dependent on microbial physiology. *Nat. Geosci.*
3:336–340.

Amoroso, A., F. Domine, G. Esposito, S. Mo-
rin, J. Savarino, M. Nardino, M. Montagnoli,
J. M. Bonneville, J. C. Clement, A. Ianniello,
and H. J. Beine. 2010. Microorganisms in dry
polar snow are involved in the exchanges of reactive nitrogen species with the atmosphere. *Environ.
Sci. Technol.* 44:714–719.

Antoniades, D., J. Veillette, M. J. Martineau, C.
Belzile, J. Tomkins, R. Pienitz, S. Lamoreux,
and W. F. Vincent. 2009. Bacterial dominance
of phototrophic communities in a High Arctic lake
and its implications for paleoclimate analysis. *Polar
Sci.* 3:147–161.

Ausatin, J., N. Bulchart, and K. Shine. 1992.
Possibility of an Arctic ozone hole in a doubled
CO_2 climate. *Nature* 360:221–225.

Bakermans, C. 2009. Limits for microbial life at subzero temperatures, p. 17–28. *In* C. Tarnocai (ed.),
Arctic Permafrost Soils. Springer, Berlin, Germany.

Battista, J. 1997. Against all odds: the survival strategies of *Deinococcus*. *Ann. Rev. Microbiol.* 51:203–
224.

Berggren, M., H. Laudon, A. Jonsson, and M.
Jansson. 2010. Nutrient constraints on metabolism
affect the temperature regulation of aquatic bacterial growth efficiency. *Microb. Ecol.* 4:894–903.

Bonilla, S., M. Rautio, and W. F. Vincent. 2009.
Phytoplankton and phytobenthos pigment strate-
gies: implications for algal survival in the changing
Arctic. *Polar Biol.* 32:1293–1303.

Booth, M. G., L. Hutchinson, M. Brumsted, P.
Aas, R. B. Coffin, R. C. Downer Jr., C. A.
Kelley, M. M. Lyons, J. D. Pakulski, S. L.
Holder Sandvik, W. H. Jeffrey, and R. V.
Miller. 2001a. Quantification of *recA* gene ex-
pression as an indicator of repair potential in ma-
rine bacterioplankton communities of Antarctica.
Aquat. Microb. Ecol. 24:51–59.

Booth, M. G., W. H. Jeffrey, and R. V. Miller.
2001b. RecA expression in response to solar UVR
in marine bacterium *Vibrio natriegens*. *Microb. Ecol.*
42:531–539.

Breezee, J., N. Cady, and J. T. Staley. 2004.
Subfreezing growth of the sea ice bacterium "*Psychromonas ingrahamii*." *Microb. Ecol.* 47:300–304.

Brinkmeyer, R., K. Knittel, J. Jürgens, H.
Weyland, R. Amann, and E. Helmke. 2003.
Diversity and structure of bacterial communities
in Arctic versus Antarctic pack ice. *Appl. Environ.
Microbiol.* 69:6610–6619.

Brune, W. H., J. G. Anderson, D. W. Toohey,
D. W. Fahey, S. R. Kawa, R. L. Jones, D. S.
McKenna, and L. R. Poole. 1991. The potential for ozone depletion in the Arctic polar stratosphere. *Science* 252:1260–1266.

Cai, W. J., L. Chen, B. Chen, Z. Gao, S. H. Lee,
J. Chen, D. Pierrot, K. Sullivan, Y. Wang,
X. Hu, W. J. Huang, Y. Zhang, S. Xu, A.
Murata, J. M. Grebmeier, E. P. Jones, and
H. Zhang. 2010. Decrease in the carbon dioxide
uptake capacity in ice-free Arctic Ocean Basin. *Sci-
ence* 329:556–559.

Callaghan, T. V., L. O. Björn, Y. Chernov,
T. Chapin, T. R. Christensen, B. Huntley,
R. A. Ims, M. Johansson, D. Jolly, S. Jonasson,
N. Matveyeva, N. Panikov, W. Oechel, and
G. Shaver. 2004. Past changes in Arctic terres-
trial ecosystems, climate and UV radiation. *Ambio*
33:398–403.

Canadell, J. G., C. Le Quere, M. R. Raupach,
C. B. Field, E. T. Buitenhuis, P. Ciais, T. J.
Conway, N. P. Gillett, R. A. Houghton, and
G. Marland. 2007. Contributions to accelerating
atmospheric CO_2 growth from economic activity,
carbon intensity, and efficiency of natural sinks.
Proc. Natl. Acad. Sci. USA 104:18866–18870.

Chang, W., M. Dyen, L. Spagnuolo, P. Simon,
L. G. Whyte, and S. Ghoshal. 2010. Biodegradation of semi- and non-volatile hydrocarbons
in aged, contaminated soils from a sub-Arctic site:
laboratory pilot-scale experiments at site temperatures. *Chemosphere* 80:319–326.

Christensen, T. R., I. C. Prentice, J. Kaplan,
A. Haxeltine, and S. Sitch. 1996. Methane flux
from northern wetlands and tundra—an ecosystem
source modelling approach. *Tellus B Chem. Phys.
Meteorol.* 48:652–661.

Christner, B. C., E. Mosley-Thompson, L. G.
Thompson, and J. N. Reeve. 2001. Isolation of
bacteria and 16S rDNAs from Lake Vostok accretion ice. *Environ. Microbiol.* 3:570–577.

Chrost, R. J., and M. A. Faust. 1999. Consequences of solar radiation on bacterial secondary
production and growth rates in subtropical coastal
water (Atlantic Coral Reef off Belize, Central
America). *Aquat. Microb. Ecol.* 20:39–48.

Clarke, S., R. E. Mielke, A. Neal, P. Holden,
and J. L. Nadeau. 2010. Bacterial and mineral
elements in Arctic biofilm: a correlative study using fluorescence and electron microscopy. *Microsc.
Microanal.* 16:153–165.

Cord-Ruwisch, R., H.-J. Seitz, and R. Conrad.
1988. The capacty of hydrogenotrophic anaerobic
bacteria to compete for traces of hydrogen depends
on the redox potential of the termnial electron acceptor. *Arch. Microbiol.* 149:350–357.

Cota, G. F., L. R. Pomeroy, W. G. Harrison,
E. P. Jones, F. Peters, W. M. J. Sheldon, and
T. R. Weingartner. 1996. Nutrients, primary production and microbial heterotrophy in the southeastern Chukchi Sea: Arctic summer nutrient depletion
and heterotrophy. *Mar. Ecol. Prog. Ser.* 135:247–258.

Cowan, D. A., and L. Ah Tow. 2004. Endangered Antarctic environments. *Annu. Rev. Microbiol.* **58**:649–690.

Daly, M. J. 2009. A new perspective on radiation resistance based on *Deinococcus radiodurans*. *Nat. Rev. Microbiol.* **7**:237–245.

De Fabo, E. 2005. Arctic stratospheric ozone depletion and increased UVB radiation: potential impacts to human health. *Int. J. Circumpolar Health* **64**:509–522.

Diaz, B., and D. Schulze-Makuch. 2006. Microbial survival rates of *Escherichia coli* and *Deinococcus radiodurans* under low temperature, low pressure, and UV-irradiation conditions, and their relevance to possible martian life. *Astrobiology* **6**:332–347.

Doran, P. T., J. C. Priscu, W. B. Lyons, J. E. Walsh, A. G. Fountain, D. M. McKnight, D. L. Moorhead, R. A. Virginia, D. H. Wall, G. D. Clow, C. H. Fritsen, C. P. McKay, and A. N. Parsons. 2002. Antarctic climate cooling and terrestrial ecosystem response. *Nature* **415**:517–520.

Dutta, K., E. A. G. Schuur, J. C. Neff, and S. A. Zimov. 2006. Potential carbon release from permafrost soils of northeastern Siberia. *Glob. Change Biol.* **12**:2336–2351.

Dyda, R. Y., M. T. Suzuki, M. Y. Yoshinga, and H. R. Harvey. 2009. The response of microbial communities to diverse organic matter sources in the Arctic Ocean. *Deep Sea Res. Part 2 Top. Stud. Oceanogr.* **56**:1249–1263.

Elasri, M. O., and R. V. Miller. 1998. A *Pseudomonas aeruginosa* biosensor responds to exposure to ultraviolet radiation. *Appl. Microbiol. Biotechnol.* **50**:455–458.

Elasri, M. O., and R. V. Miller. 1999. Study of the response of a biofilm bacterial community to UV radiation. *Appl. Environ. Microbiol.* **65**:2025–2031.

Elasri, M. O., T. Reid, S. Hutchins, and R. V. Miller. 2000. Response of a *Pseudomonas aeruginosa* biofilm community to DNA damagng chemotherapeutic agents. *FEMS Microbiol. Ecol.* **33**:21–25.

Friedlingstein, P., P. Cox, R. Betts, L. Bopp, W. von Bloh, V. Brovkin, P. Cadule, S. Doney, M. Eby, I. Fung, G. Bala, J. John, C. Jones, F. Joos, T. Kato, M. Kawamiya, W. Knorr, K. Lindsay, H. D. Matthews, T. Raddatzh, P. Ranyer, C. Reick, E. Roeckner, K. G. Schnitzler, R. Schnur, K. Strassman, A. J. Wearver, C. Yoshikawa, and N. Zeng. 2006. Climate carbon cycle feedback analysis: results from the C4MIP Model Intercomparison. *J. Climate* **19**:3337–3353.

Fung, I., J. John, J. Lerner, E. Matthews, M. Prather, L. P. Steele, and P. J. Fraser. 1991. Three-dimensional model synthesis of the global methane cycle. *J. Geophys. Res.* **96**:13033–13065.

Gilichinsky, D., E. Rivkina, C. Bakermans, V. Shcherbakova, L. Petrovskaya, S. Ozerskaya, N. Ivanushkina, G. Kochkina, K. Laurinavichuis, S. Pecheritsina, R. Fattakhova, and J. M. Tiedje. 2005. Biodiversity of cryopegs in permafrost. *FEMS Microbiol. Ecol.* **53**:117–128.

Gobell, C., B. Sundby, R. W. Macdonald, and J. N. Smith. 2001. Recent changes in organic carbon flux to Arctic Ocean seep basins: evidence from acid volatile sulfide, manganese and rhenium discord in sediments. *Geophys. Res. Lett.* **28**:1743–1746.

Hodgson, D. A., W. Vyverman, E. Verleyen, K. Sabbe, P. Leavitt, A. Taton, A. Squier, and B. Keely. 2004. Environmental factors influencing the pigment composition of *in situ* benthic microbial communities in east Antarctic lakes. *Aquat. Microb. Ecol.* **37**:247–263.

Hodson, A. J., P. Mumford, and D. Lister. 2004. Suspended sediment and phosphorus in proglacial rivers: bioavailability and potential impacts upon the P status of ice-marginal receiving waters. *Hydrol. Processes* **18**:2409–2422.

Holzinger, A., C. Lütz, U. Karsten, and C. Wiencke. 2004. The effect of ultraviolet radiation on ultrastructure and photosynthesis in the red microalgae *Palmaria palmate* and *Odonthalia dentata* from Arctic waters. *Plant Biol.* **6**:568–577.

Illeris, L., A. Michelsen, and S. Jonasson. 2003. Soil plus root respiration and microbial biomass following water, nitrogen, and phosphorus application at a High Arctic semi desert. *Biogeochemistry* **65**:15–29.

International Arctic Science Committee. 2010. Effects of climate change on landscape and regional processes and feedbacks to the climate system in the Arctic. *In* C. J. Cleveland (ed.), *Encyclopedia of Earth*. Environmental Information Coalition, National Council for Science and the Environment, Washington, DC.

Jagger, J. 1985. *Solar-UV Actions in Living Cells.* Praeger Publishers, New York, NY.

Jeffrey, W. H., P. Aas, M. M. Lyons, R. B. Coffin, R. J. Pledger, and D. L. Mitchell. 1996a. Ambient solar radiation-induced photodamage in marine bacterioplankton. *Photochem. Photobiol.* **64**:419–427.

Jeffrey, W. H., R. V. Miller, and D. L. Mitchell. 1997. Detection of ultraviolet radiation induced DNA damage in microbial communities of the Gerlache Strait. *Antarct. J.* **32**:85–87.

Jeffrey, W. H., R. J. Pledger, P. Aas, S. Hager, R. B. Coffin, R. Von Haven, and D. L. Mitchell. 1996b. Diel and depth profiles of DNA photodamage in bacterioplankton exposed to ambient solar ultraviolet radiation. *Mar. Ecol. Prog. Ser.* **137**:283–291.

Jokela, K., K. Leszczynski, and R. Visuri. 2008. Effects of Arctic ozone depletion and snow on UV exposure in Finland. *Photochem. Photobiol.* **58**:559–566.

Jones, H., C. S. Cockell, C. Goodson, N. Price, A. Simpson, and B. Thomas. 2009. Experiments on mixotrophic protists and catastrophic darkness. *Astrobiology* **9**:563–571.

Jørgensen, B. B. 1982. Mineralization of organic-matter in the sea bed—the role of sulfate reduction. *Nature* **296**:643–645.

Judd, K. E., and G. W. Kling. 2002. Production and export of dissolved C in arctic tundra mesocosms: the roles of vegetation and water flow. *Biogeochemistry* **60**:213–234.

Kiehl, J. T., and B. P. Briegleb. 1993. The relative roles of sulphate aerosols and greenhouse gases in climate forcing. *Science* **260**:311–314.

Kirchman, D. L., V. Hill, M. T. Cottrell, R. Gradinger, R. R. Malmstrom, and A. Parker. 2009a. Standing stocks, production and respiration of phytoplankton and heterotrophic bacteria in the western Arctic Ocean. *Deep Sea Res. Part 2 Top. Stud. Oceanogr.* **56**:1237–1248.

Kirchman, D. L., X. A. G. Morán, and H. Ducklow. 2009b. Microbial growth in the polar oceans—role of temperature and potential impact of climate change. *Nat. Rev. Microbiol.* **7**:451–459.

Knoblauch, C., K. Sahm, and B. B. Jørgensen. 1999. Psychrophilic sulfate-reducing bacteria isolated from permanently cold arctic marine sediments: description of *Desulfofrigus oceanense* gen. nov., sp. nov., *Desulfofrigus fragile* sp. nov., *Desulfofaba gelida* gen. nov., sp. nov., *Desulfotalea psychrophila* gen. nov., sp. nov. and *Desulfotalea arctica* sp. nov. *Int. J. Syst. Bacteriol.* **49**:1631–1643.

Kokjohn, T. A., and R. V. Miller. 1988. Characterization of the *Pseudomonas aeruginosa recA* gene: the Les⁻ phenotypye. *J. Bacteriol.* **170**:578–582.

Kuhn, M. 2009. The climate of snow and ice as boundary condition for microbial life, p. 3–15. *In* C. Tarnocai (ed.), *Arctic Permafrost Soils*. Springer, Berlin, Germany.

Laybourn-Parry, J. 2009. No place too cold. *Science* **324**:1521–1522.

Leavitt, P. R., B. F. Cumming, J. P. Smol, M. Reasoner, R. Pienitz, and D. A. Hodgson. 2003a. Climate control of UV radiation impacts on lakes. *Limnol. Oceanogr.* **48**:2062–2069.

Leavitt, P. R., D. A. Hodgson, and R. Pienitz. 2003b. Past UVR environments and impacts in lakes, p. 509–545. *In* E. W. Helbling and H. Zagarese (ed.), *UV Effects in Aquatic Organisms and Ecosystems*. Royal Society of Chemistry Publishers, Cambridge, United Kingdom.

Leavitt, P. R., R. D. Vinebrooke, D. B. Donald, J. P. Smol, and D. W. Schinder. 1997. Past ultraviolet environments in lakes derived from fossil pigments. *Nature* **388**:457–459.

Legrand, M., and D. Wagenbach. 1999. Impact of Cerro Hudson and Pinatubo volcanic eruptions on the Antarctic air and snow chemistry. *J. Geophys. Res.* **104**:1581–1596.

Lein, A. Y., A. S. Savvichev, I. I. Rusanov, G. A. Pavlova, N. A. Belyaev, K. Craine, N. V. Pimenov, and M. V. Ivanov. 2010. Biogeochemical processes in the Chukchi Sea. *Earth Environ. Sci.* **42**:221–239.

Le Mer, J., and P. Roger. 2001. Production, oxidation, emission and consumption of methane by soils: a review. *Eur. J. Soil Sci.* **37**:25–50.

Lovley, D., and E. Phillips. 1986. Organic matter mineralization with reduction of ferric iron in anaerobic sediments. *Appl. Environ. Microbiol.* **51**:683–689.

Luyssaert, S., I. Inglima, M. Jung, A. D. Richardson, M. Reichstein, D. Papale, S. L. Piao, E. D. Schulze, L. Wingate, G. Matteucci, L. Aragao, M. Aubinet, C. Beers, C. Bernhofer, K. G. Black, D. Bonal, J. M. Bonnefond, J. Chambers, P. Ciais, B. Cook, K. J. Davis, A. H. Dolma, B. Gielen, M. Goulden, J. Grace, A. Granier, A. Grelle, T. Griffis, T. Grunwald, G. Guidolotti, P. Hanson, R. Harding, D. Y. Hollinger, L. R. Hutyra, P. Kolar, B. Kruijt, W. Kutsch, F. Lagergren, T. Laurila, J. Mateus, M. Migliavacca, L. Misson, L. Montagnani, J. Moncrieff, E. Moors, J. W. Munger, E. Nikinmaa, S. V. Ollinger, G. Pita, C. Rebmann, O. Roupsard, N. Saigusa, M. J. Sanz, G. Seufert, C. Sierra, M. L. Smith, J. Tang, R. Valentini, T. Vesala, and I. A. Janssens. 2007. Carbon dioxide balance of boreal, temperate, and tropical forests derived from a global database. *Glob. Change Biol.* **13**:2509–2537.

Lyons, M. M., P. Aas, J. D. Pakulski, L. Van Waasbergen, R. V. Miller, D. L. Mitchell, and W. H. Jeffrey. 1998. DNA damage induced by ultraviolet radiation in coral-reef communities. *Mar. Biol.* **130**:537–543.

Madronich, S., L. O. Björn, M. Ilyas, and M. M. Caldwell. 1991. Changes in biologically active ultraviolet radiation reaching the earth's surface, p. 1–13. *In* J. C. van der Leun, M. Tevini, and R. C. Worrest (ed.), *UNEP Environmental Effects Panel Report—1991 Update*. United Nations Environmental Programme, Nairobi, Kenya.

Mancinelli, R. 1986. Alpine tundra soil bacterial responses to increased soil loading rates of acid precipitation, nitrate, and sulphate, Front Range, Colorado, U.S.A. *Arct. Alp. Res.* **18**:269–275.

Mazzera, D. M., D. H. Lowenthal, J. C. Chow, and J. G. Watson. 2001. Sources of PM10 and sulphate aerosol at McMurdo station, Antarctica. *Chemosphere* **45**:347–356.

McInerney, M. J., J. R. Sieber, and R. P. Gunsalus. 2009. Syntrophy in anaerobic global carbon cycles. Curr. Opin. Biotechnol. 20:623–632.

McMinn, A., C. Ashworth, and K. G. Ryan. 2000. In situ net primary productivity of an Antarctic fast ice bottom algal community. Aquat. Microb. Ecol. 2:177–185.

Mehta, T., M. V. Coppi, S. E. Childers, and D. R. Lovley. 2005. Outer membrane c-type cytochromes required for Fe(III) and Mn(IV) oxide reduction in Geobacter sulfurreducens. Appl. Environ. Microbiol. 71:8634–8641.

Meltofte, H., T. R. Christensen, B. Elberling, M. C. Forchhammer, and M. Rasch (ed.). 2008. High-Arctic Ecosystem Dynamics in a Changing Climate. Elsevier, New York, NY.

Mikucki, J. A., A. Pearson, D. T. Johnston, A. V. Turchyn, J. Farquhar, D. P. Schrag, A. D. Anbar, J. C. Priscu, and P. A. Lee. 2009. A contemporary microbially maintained subglacial ferrous "ocean." Science 324:397–400.

Miller, R. V. 2000. recA: the gene and its protein product, p. 43–54. In S. Luria (ed.), Encyclopedia of Microbiology, 2nd ed., vol. 4. Academic Press, San Diego, CA.

Miller, R. V., W. Jeffrey, D. Mitchell, and M. Elasri. 1999. Bacterial responses to ultraviolet light. ASM News 65:535–541.

Miller, R. V., and T. A. Kokjohn. 1990. General microbiology of recA: environmental and evolutionary significance. Ann. Rev. Microbiol. 44:365–394.

Mindl, B., A. M. Anesio, K. Meirer, A. J. Hodson, J. Laybourn-Parry, R. Sommaruga, and B. Sattler. 2007. Factors influencing bacterial dynamics along a transect from supraglacial runoff to proglacial lakes of a high Arctic glacier. FEMS Microbiol. Ecol. 59:307–317.

Montserrat Sala, M., R. Terrado, C. Lovejoy, F. Unrein, and C. Pedrós-Alió. 2008. Metabolic diversity of heterotrophic bacterioplankton over winter and spring in the coastal Arctic Ocean. Environ. Microbiol. 10:942–949.

Moorhead, D. L., C. F. Wolf, and R. A. Wharton Jr. 1997. Impact of light regimes on productivity patterns of benthic microbial mats in an Antarctic lake: a modeling study. Limnol. Oceanogr. 42:1561–1569.

Myers, C. R., and K. H. Nealson. 1990. Respiration-linked proton translocation coupled to anaerobic reduction of manganese(IV) and iron(III) in Shewanella putrefaciens MR-1. J. Bacteriol. 172:6232–6238.

Niederberger, T. D., N. N. Perreault, J. R. Lawrence, J. L. Nadeau, R. E. Mielke, C. W. Greer, D. T. Andersen, and L. G. Whyte. 2009. Novel sulfur-oxidizing streamers thriving in perennial cold saline springs of the Canadian high Arctic. Environ. Microbiol. 11:616–629.

Ørbaek, J. B., I. Svenoe, and D. O. Hessen. 2002. Spectral properties and UV attenuation in Arctic freshwater systems, p. 57–72. In D. O. Hessen (ed.), UV Radiation and Arctic Ecosystems. Springer, Berlin, Germany.

Ovadnevaite, J., D. Ceburnis, K. Plauskaite-Sukiene, R. Modini, R. Dupuy, I. Rimselyte, M. Ramonet, K. Kvietkus, Z. Ristovski, H. Berresheim, and C. D. O'Dowd. 2009. Volcanic sulphate and Arctic dust plumes over the North Atlantic Ocean. Atmos. Environ. 43:4968–4974.

Pienitz, R., and W. F. Vincent. 2000. Effect of climate change relative to ozone depletion on UV exposure in subarctic lakes. Nature 404:484–487.

Rich, J., M. Gosselin, E. Sherr, E. Sherr, and D. L. Kirchman. 1997. High bacterial production, uptake and concentrations of dissolved organic matter in the central Arctic Ocean. Deep Sea Res. Part 2 Top. Stud. Oceanogr. 44:1645–1662.

Schoeberl, M. R., and D. L. Hartmann. 1991. The dynamics of the stratospheric polar vortex and its relation to springtime ozone depletions. Science 251:46–52.

Schuur, E. A. G., J. G. Vogel, K. G. Crummer, H. Lee, J. O. Sickman, and T. E. Osterkamp. 2009. The effect of permafrost thaw on old carbon release and net carbon exchange from tundra. Nature 459:556–559.

Scully, N. M., W. F. Vincent, D. R. S. Lean, and W. J. Cooper. 1997. Implications of ozone depletion for surface-water photochemistry: sensitivity of clear lakes. Aquat. Sci. 59:260–274.

Shiah, F. K., and H. W. Ducklow. 1995. Multiscale variability in bacterioplankton abundance, production, and specific growth rate in a temperate salt-marsh tidal creek. Limnol. Oceanogr. 40:55–66.

Shiklomanov, I. A. 1998. World Water Resources: a New Appraisal and Assessment for the 21st Century. UNESCO, Paris, France.

Sigman, D. M., A. M. de Boer, and G. H. Haug. 2007. Antarctic stratification, atmospheric water vapor, and Heinrich events: a hypothesis for Late Pleistocene deglaciations, p. 335–349. In A. Schmittner, J. Chiang, and S. Hemmings (ed.), Ocean Circulation: Mechanisms and Impacts. AGU Press, Washington, DC.

Singh, B. K., R. D. Bardgett, P. Smith, and D. S. Reay. 2010. Microorganisms and climate change: terrestrial feedbacks and mitigation options. Nat. Rev. Microbiol. 8:779–790.

Stams, A. J., and C. M. Plugge. 2009. Electron transfer in syntrophic communities of anaerobic bacteria and archaea. Nat. Rev. Microbiol. 7:568–577.

Stein, R., and R. W. Macdonald. 2004. Organic carbon budget: Arctic Ocean vs. Global Ocean,

p. 315–322. *In* R. Stein and R. W. Macdonald (ed.), *The Organic Carbon Cycle in the Arctic Ocean.* Springer, Berlin, Germany.

Steven, B., W. H. Pollard, C. W. Greer, and L. G. Whyte. 2008. Microbial diversity and activity through a permafrost/ground ice core profile from the Canadian high Arctic. *Environ. Microbiol.* **10:**3388–3403.

Stres, B., T. Danevcic, L. Pal, M. M. Fuka, L. Resman, S. Leskovec, J. Hacin, D. Stopar, I. Mahne, and I. Mandic-Mulec. 2008. Influence of temperature and soil water content on bacterial, archaeal and denitrifying microbial communities in drained fen grassland soil microcosms. *FEMS Microbiol. Ecol.* **66:**110–122.

Tarnocai, C. 1980. Summer temperatures of cryosolic soils in the North-Central Keewatin, NWT. *Can. J. Soil Sci.* **60:**311–327.

Tarnocai, C. 2009. Arctic permafrost soils, p. 3–16. *In* R. Margesin (ed.), *Permafrost Soils.* Springer, Berlin, Germany.

Thingstad, T. F., R. G. J. Bellerby, G. Bratbak, K. Y. Borsheim, J. K. Egge, M. Heldal, A. Larsen, C. Neill, J. Nejstgaard, S. Norland, R. A. Sandaa, E. F. Skjoldal, T. Tanaka, R. Thyrhaug, and B. Topper. 2008. Counterintuitive carbon-to-nutrient coupling in an Arctic pelagic ecosystem. *Nature* **455:**387–391.

Vincent, W. F., D. R. Mueller, and S. Bonilla. 2004. Ecosystems on ice: the microbial ecology of Markham Ice Shelf in the high Arctic. *Cryobiology* **48:**103–112.

Vincent, W. F., M. Rautio, and R. Pienitz. 2007. Climate control of biological UV exposure in polar and alpine aquatic ecosystems, p. 227–250. *In* J. B. Ørbæk, R. Kallenborn, I. Tombre, E. N. Hegseth, S. Falk-Petersen, and A. H. Hoel (ed.), *Arctic Alpine Ecosystems and People in a Changing Environment.* Springer, Berlin, Germany.

Vincent, W. F., L. G. Whyte, C. Lovejoy, C. W. Greer, I. Laurion, C. A. Suttle, J. Corbeil, and D. R. Mueller. 2009. Arctic microbial ecosystems and impacts of extreme warming during the International Polar Year. *Polar Sci.* **3:**171–180.

Wagner, D., S. Kobabe, and S. Leibner. 2009. Bacterial community structure and carbon turnover in permafrost-affected soils of the Lena Delta, northeastern Siberia. *Canadian Journal of Microbiology* **55:**73–83.

Wagner, D., A. Lipski, A. Embacher, and A. Gattinger. 2005. Methane fluxes in permafrost habitats of the Lena Delta: effects of microbial community structure and organic matter quality. *Environ. Microbiol.* **7:**1582–1592.

Walker, C., Z. He, Z. K. Yang, J. J. Ringbauer, Q. He, J. Zhou, G. Voordouw, J. D. Wall, A. P. Arkin, T. C. Hazen, S. Stolyar, and D. A. Stahl. 2009. The electron transfer system of syntrophically grown *Desulfovibrio vulgaris. J. Bacteriol.* **191:**5793–57801.

Walker, G. 1984. Mutagenesis and induced responses in deoxyribonucleic acid damage in *Escherichia coli. Microbiol. Rev.* **48:**60–93.

Walter, K. M., M. E. Edwards, G. Grosse, S. A. Zimov, and F. S. Chapin III. 2007. Thermokarst lakes as a source of atmospheric CH_4 during the last deglaciation. *Science* **318:**633–635.

Wrona, F. J., T. D. Prowse, J. D. Reist, J. E. Hobbie, L. M. Lévesque, R. W. Macdonald, and W. F. Vincent. 2006. Effects of ultraviolet radiation and contaminant-related stressors on Arctic freshwater ecosystems. *Ambio* **35:**388–401.

Yergeau, E., H. Hogues, L. G. Whyte, and C. W. Greer. 2010. The functional potential of high Arctic permafrost revealed by metagenomic sequencing, qPCR and microarray analyses. *ISME J.* **4:**1–9.

Zepp, R. G., T. V. Callaghan, and D. J. Erickson III. 2003. Interactive effects of ozone depletion and climate change on biogeochemical cycles. *Photochem. Photobiol. Sci.* **2:**51–61.

Zhao, M., and S. W. Running. 2010. Drought-induced reduction in global terrestrial net primary production from 2000 through 2009. *Science* **329:**940–943.

LIFE IN ICE ON OTHER WORLDS

Christopher P. McKay,
Nadia C. S. Mykytczuk, and
Lyle G. Whyte

14

INTRODUCTION

A key goal for astrobiology is the discovery of a second genesis of life: life forms representing an independent origin of life distinct from life on Earth. There are many reasons to be optimistic that life will be found beyond Earth. First, life is composed of elements such as C, H, N, O, P, and S that are common in the universe. Second, all life on Earth requires liquid water to grow or reproduce, and we have clear evidence of past liquid water on Mars and present liquid water below the ice on Europa and Enceladus. Third, the organic molecules of life, such as amino acids, can be easily produced in nonbiological processes, and organics are found in many places in the outer solar system. Fourth, life appeared quite early in Earth's history, soon after the surface cooled enough for liquid water to be present. Finally, we know that meteorites from Mars have landed on Earth and that these rocks could have transported life (Weiss et al., 2000). Thus, life may spread from planet to planet. All this bodes well for the search for life beyond Earth. And so we search.

Christopher P. McKay, NASA Ames Research Center, Moffett Field, CA 94035. *Nadia C. S. Mykytczuk and Lyle G. Whyte*, Department of Natural Resource Sciences, McGill University, Macdonald Campus, 21,111 Lakeshore Road, Ste.-Anne-de-Bellevue, QC H9X 3V9, Canada.

What we search for is not just life, but convincing evidence of a second genesis of life (McKay, 1998, 2001). All life on Earth represents a common genetic and biochemical system descendent from a common ancestor. If we find evidence of life on another world, we will want to compare its genetic and biochemical details to those of life on Earth and determine if that life is sufficiently different to imply an independent and separate origin. The alternative is that a single form of life simply spread between the planets. To test for a second genesis we need biologically intact material—either dead or alive. Fossils are not enough for this test.

The other worlds in our solar system that are the most promising targets in the search for life are Mars, Europa (a moon of Jupiter), and Enceladus (a moon of Saturn). On Mars we have direct evidence of past liquid water and current evidence of ice-rich ground. The data indicate that in the past Mars had a range of environments comparable to the range of environments for life on Earth (Davis and McKay, 1991). On Europa we have definite evidence that below the ice surface there is an ocean kept warm by tidal forces (Pappalardo et al., 1999; Kivelson et al., 2000). Enceladus, a small moon (~500 km in diameter), has a plume of ice and organics emanating from its

Polar Microbiology: Life in a Deep Freeze
Edited by Robert V. Miller and Lyle G. Whyte © 2012 ASM Press, Washington, DC

south pole region (Porco et al., 2006; Waite et al., 2006, 2009).

Titan represents a special case in the search for life because it is the only other world besides Earth with a liquid present on its surface. That liquid is a mixture of methane and ethane (Stofan et al., 2007). Life based on liquid methane instead of liquid water remains at the outer reaches of speculation (Benner et al., 2004; Smith and McKay, 2005; Schulze-Makuch and Grinspoon, 2005).

As it happens, all the worlds we search for life are cold. The average surface temperature of Mars is −60°C, with lows of −130°C and highs (noontime at summer at the ground level) of 20°C. Europa has an average surface temperature of −160°C and Enceladus of about −190°C. However, as is often the case in the polar regions of Earth, liquid water can often be found below cold ice or in association with ice and dirt. Thus, on these cold, ice-rich worlds there are potential habitats for life. In addition, cold conditions are excellent for preservation of evidence of life—evidence suitable for determination of the genetic and biochemical nature of that life as part of the search for a second genesis.

For the foreseeable future we can probably only access the surface and near subsurface of Mars, the very surface of Europa, and the plume of Enceladus. Thus, we are unlikely to be able to access any subsurface or subice liquid-water habitats that might contain living organisms. However, the surface and plume samples we could obtain with near-term missions might yield the remains of such organisms. Thus, what we're likely to find is organic material that may be the remains of microorganisms that may, or may not, be a separate genesis of life. The challenge will then be to distinguish between abiotic organic material—which is prevalent in the solar system—and organic material that was produced by biological systems.

Given the cold targets in our search for life, we turn to the polar environments on Earth to address several relevant questions. These include: how does microbial life survive in cold,

icy conditions? How long can microorganisms remain cold and dormant? What are the low-temperature limits of microbial growth and metabolic activity, and can we detect activity in situ given the extremely low rates expected? How is evidence of past microbial life preserved in frozen materials? And can we construct robotic instruments for future missions sensitive but specific enough to detect evidence of microbial activity in other solar systems? The answers to these questions derived from studies on Earth will shape the search for evidence of life on other worlds.

MARS

Mars: the Primary Target in the Search for Life

Mars is the world that is the main focus of astrobiology interest. There are four reasons for this, two fundamental and two practical. The first and most important reason for interest in Mars is the direct evidence of liquid water on its surface in the past. Mars clearly had rivers, lakes, and possibly even oceans. The remains of the dry riverbeds and canyons can be seen today on the surface. Much of the water that flowed in these fluvial features is still present on Mars, frozen into the subsurface and polar ice caps. The second fundamental reason for interest in Mars is the presence of CO_2 and N_2 in atmosphere and the presence of P and S in the soil. All the essential elements needed for life are present on Mars. There are also two practical reasons: the environment on Mars is cold, dry, and with low pressure. These are ideal conditions for preservation of organic remnants of life. And finally, Mars is close to Earth. Spacecraft flight times to Mars average 6 months, in contrast to flight times of 6 years or more for the Saturn system. Furthermore, solar power is a practical option for Mars missions.

Surface Environment

The surface of Mars today is a not a promising site for possible life. The strong UV radiation (wavelengths down to 190 nm) and the low water activity militate against life at the surface.

However, there is the possibility that in the recent past Mars had conditions that could have supported life. In particular, it has long been appreciated that the ice-cemented areas in the northern plains of Mars are possibly the best location on Mars for recent habitability. The presence of ice just below the surface provides a source of H_2O. The atmospheric surface pressure over the northern plains is well above the triple point of water, so the liquid phase even of pure water would be stable against boiling—this is in contrast with the ice-rich southern polar regions, which are at high elevation. (Note that the pressure at the Viking 2 landing site located at 48°N never fell below 750 Pa; the triple point of water is 610 Pa.) Thus, all that would be needed to provide liquid-water activity capable of supporting life is

sufficient energy to warm the subsurface ice. As shown in Fig. 1, this is thought to have occurred as recently as 5 Ma ago, when calculations indicate that Mars had an orbital tilt of 45°, compared to the present value of 25° (Laskar et al., 2002). The summer insolation in the polar regions of Mars at summer solstice for an obliquity of 45° is about twice that for an obliquity of 25°. Thus, when Mars had an obliquity of 45°, the polar regions received roughly the same level of summer sunlight as Earth's polar regions do at the present time.

Models suggest that the high insolation levels at the polar regions on Mars 5 Ma ago could have produced surface melting of the ground ice. Costard et al. (2002) computed peak temperatures for different obliquities for varying surface properties and slopes. They

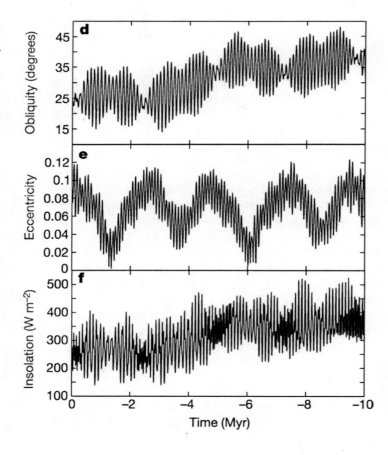

FIGURE 1 Orbital variations and north polar insolation on Mars over the past 20 million years (Myr, million years). (Reprinted from Macmillan Publishers Ltd. [*Nature*] [Laskar et al., 2002], copyright 2002.)

found that, compared with the present north polar cap temperature high of ~−60°C, peak temperatures are >0°C at the highest obliquities, and temperatures above −20°C occur for an obliquity as low as 35° (Costard et al., 2002). They suggested this as a possible cause of the gullies observed by Malin and Edgett (2000). Richardson and Michna (2005) show that when obliquity is 45°, melting can occur 50 days per year in the high northern latitudes. The higher temperatures also enhance the hydrological cycle, evaporating water ice from the northern cap, which can then precipitate as snowfall in the northern plains.

The sunlight levels of 200 to 500 W m^{-2} in Fig. 1 can be compared to the top-of-the-atmosphere flux at the equator on Earth, 436 W m^{-2}, and to the average daily summer flux in the Dry Valleys of Antarctica, 300 W m^{-2} (Clow et al., 1998).

Subsurface Environments: Ice-Cemented Ground, Perchlorate, and Eutectic Brines

Phoenix, the most recent mission to Mars, was the first—and to date only—successful polar lander to reach Mars. Phoenix landed at 68°N in a location known from the Odyssey results to contain subsurface ice. Ice was indeed found as expected, just a few centimeters below the surface (Smith et al., 2009). Like the Viking landers of the 1970s, Phoenix scooped up dirt and placed it into its instrument suite—the only mission other than Viking to do so. Like Viking, Phoenix added water to the soil and searched for organic compounds by thermal release. Low levels of carbonate in the soil were detected in confirmation of previous results. While the ice was expected, and the carbonates as well, there were two major unexpected—and to this day still unexplained—discoveries by Phoenix. The first was the presence of segregated ("light-colored") ice in the surface materials, and the second was the presence of perchlorate, presumably as magnesium perchlorate, at levels of ~0.5%. These discoveries have important implications for the search for evidence of life on Mars.

It had been known for a long time that there is Cl in the soil of Mars. Direct determination of Cl by Viking, Pathfinder, and the Mars Explorer Rovers indicated about 0.5% by weight (Clark et al., 1976; Reider et al., 2004). In addition, orbital determination of Cl abundance by the gamma ray spectrometer on the Mars Odyssey mission indicated 0.2 to 0.8% by weight over the mid-latitudes (±50°) on Mars (Keller et al., 2007). However, these instruments detected the element Cl and gave no information about its chemical form. It was assumed to be NaCl (Baird et al., 1976). The detection that much, if not most, of the Cl at the Phoenix landing site is in the form of perchlorate was a surprise. The detection of perchlorate was specific and somewhat accidental to the sensor used on Phoenix—the Hofmeister electrode (Hecht et al., 2009). The detection is robust, and other possible explanations, such as nitrates, can be ruled out (Hecht et al., 2009).

Perchlorate has three independent and important implications in considerations of life on Mars: organic detection, freezing-point depression, and electron acceptor. Perchlorate is the most oxidized form of the element chlorine, but it is not reactive. However, if heated to above ~350°C it decomposes, releasing reactive oxygen. Thus, the Viking and Phoenix thermal processing of the soils would have destroyed the very organics they were attempting to detect (Navarro-González et al., 2010). Perchlorates are highly soluble salts with low eutectic temperatures. As an example, a saturated solution of $Mg(ClO_4)_2$ has a freezing point of −67°C—within the range of the diurnal temperature cycle of the Phoenix landing site in the summer. Thus, perchlorates may be the basis for liquid solutions even on present Mars (Rennó et al., 2009; Stoker et al., 2010; Catling et al., 2010). Perchlorates are also metabolically active. It is known that microorganisms on Earth are capable of using perchlorates as electron acceptors, allowing anaerobic microbial respiration to occur where perchlorate replaces oxygen as the terminal electron acceptor (Coates and Achenbach, 2004). In principle, perchlorates could form a viable redox couple with any organic material or iron-rich basaltic rocks on Mars. Thus, perchlorates establish the possibility of chemosynthetic autotrophy on Mars.

Numerical simulations developed based on the Viking results predicted the presence of ground ice in the polar latitudes on Mars (Mellon and Jakosky, 1995). These models were based on the assumption of ground ice filling the pore spaces in the regolith and exchanging with the atmospheric moisture only by the exchange of vapor. The observations of ground ice by Mars Odyssey conformed to the predictions of these models and appeared to confirm them. The one disagreement between the orbital observations and the models was the evidence in some locations for ice concentrations that exceed the expected pore volume (Boynton et al., 2002). Phoenix did reveal ice-cemented ground below the surface that appeared consistent with vapor-deposited ice. The mean depth was 4.6 cm (Mellon et al., 2009) and varied considerably in a way that seemed to correlate with slope and thermal inertia variations in the overlying soil (Mellon et al., 2009). However, in addition to ice-cemented soil there was relatively pure light-toned ice (Fig. 2). This ice was unexpected, and Mellon et al. (2009) suggest it appears most consistent with the formation of excess ice by soil-ice segregation, such as would occur by thin-film migration and the formation of ice lenses, needle ice, or similar ice. Many of these processes require a liquid phase—perhaps created by the presence of the strong eutectic solution of perchlorate.

The detection of perchlorate on Mars, and the strong effect of this salt on lowering the freezing point of solutions, has opened up discussion of liquids on Mars at temperatures well below freezing (Rennó et al., 2009). However, it is important to note that the liquid that forms would not be suitable for life due to the low water activity. The water activity of any saline solution in equilibrium with ice (as is the eutectic) is determined by its temperature. This is because in the equilibrium mixture the water activity of the solution must be equal to the water activity of ice (not to be confused with the ice activity of ice, which is always unity). Thus, if brines form in contact with the ground ice on Mars, their water activity is determined from their temperature. Figure 3 is a plot of the water activity as a function of temperature of solutions in contact with ice. Note that for a temperature of $-67°C$—the eutectic of magnesium perchlorate—the water activity is 0.5, well below the limit of growth for microorganisms. Perchlorate on Mars may allow liquid water at Martian temperatures, but not microbial growth in that liquid at those low subzero temperatures.

Earth Analogs and Their Microbial Communities

The high-elevation valleys in the Antarctica Dry Valleys provide an analog for what the northern polar regions of Mars might have been like during periods of high obliquity. Mean annual ground temperatures in University Valley (78°S, elevation 1,700 m) and in other high-elevation areas in the Dry Valleys of Antarctica are $-25°C$, and summer air temperatures do not rise above freezing. Ice-cemented ground is present in University Valley at depths that range from a few to over 50 cm below the surface (McKay, 2009). Located below the surface, this ice is well below the depth of the active zone (McKay et al., 1998; McKay, 2009) and the maximum summer temperatures remain below freezing. For example, ice 30 cm below the surface in University Valley had a summer temperature of $-10°C$ (McKay, 2009). Studies of the microbes in the ground ice below dry permafrost in University Valley show that there is an adapted microbial community (Tamppari et al., in press), and RNA data show that there is microbial activity (J. DiRuggiero, personal communication).

Indeed, it has recently been shown that microbial life can grow and/or be biologically active at subfreezing temperatures if films of water are present (Bakermans, in press). Investigations into the low-temperature limits for growth in microorganisms continue to push

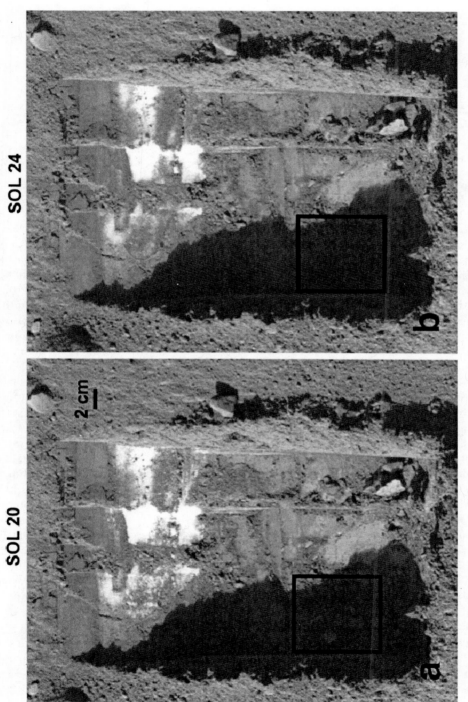

FIGURE 2 Subsurface ice at the Phoenix landing site on Mars, 68°N latitude. The flat areas visible at the bottom of the trench were hard, indicating the presence of ice-cemented ground, as expected. This material was spectrally identical to the soil. The lighter-colored, relatively softer ice seen was unexpected. Some of the light-colored ice exposed by digging has evaporated in the 4-sol (Martian day) interval, ruling out salt or carbonate as an alternative explanation to ice.

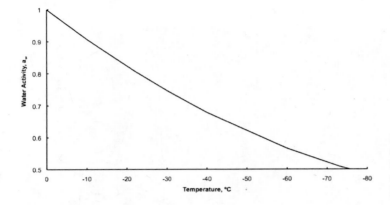

FIGURE 3 Water activity of a system in contact with ice as a function of temperature. (Based on vapor pressure data from Weast [1976].)

the known limits well below zero, with current published evidence at −12°C for microbial division (Breezee et al., 2004), −15 to −20°C for microbial respiration (Panikov et al., 2006; Steven et al., 2008), and −80°C for microbial ATP synthesis (Amato and Christner, 2009). All of the above reports were derived from species/communities from cold and dry environments from Earth's polar regions. The low-temperature limits for these cold-tolerance investigations fall within the temperature range experienced on present-day Mars and could permit survival and potentially growth, particularly in subsurface environments. Jakosky et al. (2003) discuss the potential habitability of Mars's polar regions as a function of obliquity. They conclude that temperatures of ice covered by a dust layer can become high enough (−20°C) that liquid-brine solutions form and microbial activity is possible. This suggestion is made more relevant by the discovery of perchlorate.

The availability of liquid water within the Martian subsurface (permafrost or regolith) would be concentrated into eutectic brines (described above). As such, the microorganisms that could survive and potentially remain viable under such growth conditions would most likely be halophilic cryophiles. To inform our search for possible model life forms, we look to discoveries of extremely cold-tolerant and metabolically active species at subzero temperatures. For example, we have recently isolated

and characterized such an isolate, *Planococcus halocryophilus* sp. nov., from High Arctic permafrost; it is able to grow at −15°C with a generation time of ~40 days when grown in liquid brines with 18% NaCl and 7% glycerol (N. C. S. Mykytczuk, R. R. Wilhelm, and L. G. Whyte, submitted for publication). This strain is also metabolically active at temperatures as low as −25°C (Mykytczuk et al., submitted). At subzero temperatures, this organism survives in a dense matrix of exopolysaccharide-forming aggregates of embedded cells (Fig. 4). These aggregates could be a survival strategy facilitating the exchange of limiting nutrients, cellular components, and genetic material under stressful conditions. Rivkina et al. (2000) have shown that microorganisms can function in ice-soil mixtures at temperatures as low as −20°C, living in the thin films of interfacial water. Physiological and genomic analyses of *P. halocryophilus* sp. nov. provide insights into the cellular mechanisms that allow this cryo- and halotolerant bacterium to overcome extreme temperature, water availability/osmotic pressure, and limited-nutrient constraints occurring in permafrost, not unlike those occurring in the subsurface of Mars.

Recent evidence of geomorphological features linked to water movement, including gullies (Malin et al., 2006) and spring systems (Grasby and Londry, 2007; Rossi et al., 2008), further support the presence of brines, and relic or current microbial habitats, amidst the frozen

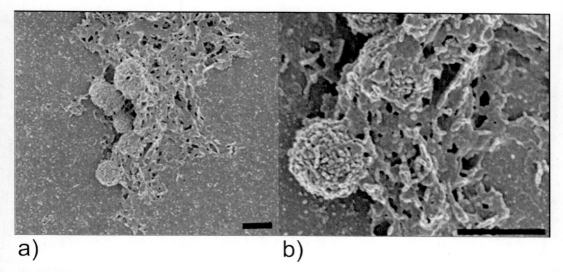

a)   b)

FIGURE 4 Scanning electron microscopy image of *Planococcus halocryophilus* sp. nov. grown at −15°C. (a) Cells (single and diplococci) clustered in dense exopolysaccharide and semicrystalline aggregates. (b) Closeup of single cell adhered to the aggregate. Bars, 1.0 μm.

Martian subsurface. For example, terrestrial cold saline spring systems in thick permafrost regions of the Canadian High Arctic support diverse microbial ecosystems. The Gypsum Hill saline springs on Axel Heiberg Island flow through a region of ~600 m of continuous permafrost and are a sulfide (25 to 100 ppm)- and sulfate (2,300 to 3,700 mg/liter)-abundant system that can serve as an analog for the potential sulfate-rich brine that could have originated in the presence of sulfur-rich groundwater on Mars (Andersen et al., 2002; Mangold et al., 2008). Sulfate is also a possible terminal electron acceptor supporting anaerobic respiration. Active biogeochemical sulfur cycling in the Gypsum Hill system can illustrate whether a self-contained community (of sulfur-reducing and sulfur-oxidizing microorganisms) can function in such a cryoenvironment and if the sulfates could preserve potential biosignatures of such microbial activity (Ono et al., 2006; Niederberger et al., 2010).

The discovery of methane in the Martian atmosphere (Formisano et al., 2004; Mumma et al., 2009) has provided a compelling line of investigation into signs of life. On Earth, >90 to 95% of methane is produced through the ac-tivity of methanogenic *Archaea* that can generate methane under anaerobic conditions using a number of different metabolic substrates, including acetate, CO_2, and H_2, and a number of different C_1 compounds. The presence and activity of methanogens in diverse cold and saline environments (reviewed in Chapter 2 of this volume) illustrate that this group of *Archaea* may be well suited to survival in extreme environments and could also be a source of methane on Mars (Onstott et al., 2006; Atreya et al., 2007). The subsequent claims that methane shows temporal and spatial variability and is perhaps emanating from hot spots or plumes from the Martian subsurface (Hand, 2008; Mumma et al., 2009; Geminale et al., 2010) could suggest a biological origin, either being slowly released from subsurface deposits or currently being regenerated by subsurface microbial methanogenesis by halophilic methanogens. The estimated 10 to 60 ppbv CH_4 in the Mars atmosphere also appears to show a 1,000-fold lower residence time (~200 days) than would be predicted from photolysis reactions (Lefèvre and Forget, 2009). Although variable methane is still largely debated due to inconsistencies with what is known about

Mars's atmospheric chemistry (Zahnle et al., in press), there still remains the possibility that a component of the methane sink could result not only through abiotic reactions with surface soils but through the metabolic activity of viable microbial methanotrophs in the Martian subsurface. Methanotrophs can oxidize methane to CO_2 via aerobic or anaerobic pathways, the latter being a viable metabolism on a planet devoid of oxygen and performed by anaerobic methane-oxidizing *Archaea* (ANME).

A unique terrestrial analog for potential methane ecosystems in cryoenvironments can be found at the Lost Hammer Spring in the Canadian High Arctic, which provides a model of how a methane seep can form in a cryoenvironment characterized by thick permafrost, as well as a mechanism that could possibly be contributing to reported methane plumes on Mars. This highly unique hypersaline (23% salinity), subzero ($-5°C$) environment supports a viable microbial community capable of activity at temperatures as low as $-10°C$ (Niederberger et al., 2010). Although isotopic analysis of the ~50% methane emitted from Lost Hammer Spring was consistent with a thermogenic origin, the methane itself can act as an energy and carbon source for sustaining anaerobic methane oxidation-based microbial metabolism, or methanotrophy, rather than methanogenesis (Niederberger et al., 2010). The presence of ANME cells in the Lost Hammer source sediments illustrates the possibility that methane emanating from the frozen subsurface can support viable microbial metabolism within these extreme environmental constraints. As an example, methane clathrates, hypothetically formed from geologic sources of methane such as serpentinization in the deep Martian subsurface, could be used as energy and carbon sources from ANME metabolism.

The upcoming Mars Science Laboratory and ExoMars missions hope to shed light on the origin of Martian methane, and may or may not point to the activity of methanogens or ANME residing in the Martian subsurface. The combined presence of methane

in the Martian atmosphere and perchlorate in the soils of the polar regions could also provide a complete, energetically plausible form of methanotrophy that could theoretically support a microbial biome through methane oxidation with perchlorate as the terminal electron acceptor (Coates, 2009).

Favorable Martian Sites for Past and Present Life

Is has been suggested that ground ice could protect organic material on Mars from destruction by oxidants, and as a result, organics from biological or meteorite sources will be detectable in polar ice-rich ground at significant concentrations (Smith and McKay, 2005). It would seem, then, that the northern polar regions of Mars, such as the Phoenix landing site, are arguably the most likely site to support recent life on Mars (Stoker et al., 2010). The near-surface ice likely provided adequate water activity during periods of high obliquity. CO_2 is present in the atmosphere. Perchlorate in the soil together with iron in basaltic rock provides a possible energy and carbon source. Furthermore, the presence of organics must once again be considered, as the results of the Viking gas chromatograph-mass spectometer are now suspect, given the discovery of the thermally reactive perchlorate.

While the northern plains represent the most likely site of recent life due to the melting of near-surface ice, the southern highlands represent the best location to find long-frozen remains of ancient life on Mars. Smith and McKay (2005) have suggested that the ancient cratered terrain in the southern highlands of Mars near 80°S, 180°W would be a suitable target for a deep drill to recover the frozen remains of life that might have been present on Mars 3 to 4 Ga ago. The map of Mars in Color Plate 11 shows crater distribution, ground ice, and crustal magnetism on Mars. The crustal magnetism is thought to have formed very early in Mars's history, more than 4.5 Ga ago. For the crustal magnetism to persist implies that these surfaces have not been severely heated or shocked. The region between 60 and 80°S at 180°W is heavily cratered, preserves crustal magnetism, and has ground ice present. This is the suggested target site for drilling to find the frozen remains of ancient Martian life (Smith and McKay, 2005).

Thus, Mars provides an opportunity to search for recent life—only 5 to 10 million years old—or ancient life 3 to 4 billion years old. In either case the challenge will be to determine if any organic materials recovered are the remains of a biological system and if that system represents an alien biochemistry.

EUROPA

In the outer solar system there are two worlds that potentially have liquid water under layers of ice: Europa and Enceladus. The icy surface of Europa has features that appear to have formed by ice floating on water, including linear cracks, icebergs, and fractured ice rafts (Pappalardo et al., 1999). The evidence that the subsurface water is still present comes from the disturbance that Europa makes as it moves through Jupiter's magnetic field. The disturbance indicates a slightly conductive global ocean (Kivelson et al., 2000). The thickness of the ice is uncertain, but most models favor an ice thickness exceeding many kilometers (Pappalardo et al., 1999). This would make the liquid-water environment out of reach of exploration missions for some time. However, there are models of Europa that argue for thinner ice with an ongoing exchange between the ocean and the surface (Greenberg, 2010).

In the water of Europa's ocean there may be life (Chyba and Phillips, 2001). There are many ecosystems on Earth that thrive and grow in water that is continuously covered by ice. These are found in both the Arctic and Antarctic regions. In addition to the polar oceans where sea ice diatoms perform photosynthesis under the ice cover, there are perennially ice-covered lakes in the Antarctic continent in which microbial mats based on photosynthesis are found in the water beneath a 4-meter ice cover. The light penetrating these thick ice covers is minimal—about 1% of the incident light. Using these Earth-based systems

as a guide, it is possible that sunlight penetrating through the cracks (the observed streaks) in the ice of Europa could support a transient photosynthetic community. Alternatively, if there are hydrothermal sites on the bottom of the Europan ocean, it may be possible that chemosynthetic life could survive there—by analogy to life at hydrothermal vent sites at the bottom of Earth's oceans. The biochemistry of hydrothermal sites on Earth does depend on O_2 produced at Earth's surface. On Europa, a chemical scheme like that suggested for subsurface life on Mars would be appropriate (H_2 + CO_2) (Boston et al., 1992; McCollom, 1999).

The surface of Europa is crisscrossed by streaks that are slightly darker than the rest of the icy surface and may represent cracks where the water has come to the surface from the ocean below. Evidence for life in the ocean may thus be deposited on the surface. Unfortunately, the harsh radiation environment would quickly kill any life forms (McKay, 2002), and on a short time scale compared to the age of the surface would likely destroy any biological information in the organic molecules (Cooper et al., 2001; Hand et al., 2010). But the implied flux of water to the surface must also represent a flux of ice back into the ocean. This flux could carry with it oxidants produced by the surface radiation. This in turn could be an energy source for life in the ocean (Cooper et al., 2001; Hand et al., 2010).

ENCELADUS

Enceladus is a small, icy moon of Saturn, with a diameter of about 500 km. Cassini data have revealed about a dozen or so jets of fine, icy particles emerging from the south polar region of Enceladus (Porco et al., 2006). The jets have also been shown to contain simple organic compounds and gases (Waite et al., 2006, 2009). These jets are evidence for activity driven by some geophysical energy source, but the nature of the energy source remains unclear, as does the source of the ice and water vapor. However, it is possible that a liquid-water environment exists beneath the south polar cap that may have been a site for the origin of life and in which plausible ecosystems might exist (McKay et al., 2008). Several theories for the origin of life on Earth would apply to Enceladus. These are (i) origin in an organic-rich mixture, (ii) origin in the redox gradient of a submarine vent, and (iii) panspermia. There are at least three microbial ecosystems on Earth that provide analogs for possible ecologies on Enceladus, because they do not rely on sunlight, oxygen, or organics produced at the surface by photosynthesis (McKay et al., 2008). Two of these ecosystems are found deep in volcanic rock, and the primary productivity is based on the consumption by methanogens of hydrogen produced by rock reactions with water. The third ecosystem is found deep below the surface in South Africa and is based on sulfur-reducing bacteria consuming hydrogen and sulfate, both ultimately produced by radioactive decay.

If there is subsurface life in the liquid-water reservoirs on Enceladus, then the geysers are carrying these organisms out into space. Here they would quickly become dormant in the cold vacuum of space and would then be killed by solar UV radiation. But these dead, frozen microbes would still retain the biochemical and genetic molecules of the living forms. As a target for future orbiter or sample return missions, Enceladus rates high because fresh samples of interest are jetting into space ready for collection. Detailed in situ analysis of the organics in the plume may indicate if biological sources are involved (McKay et al., 2008). A definitive analysis will probably require a sample return.

OTHER ICY WORLDS

In addition to Mars, Europa, and Enceladus, there are other worlds of interest to astrobiology—and they are also icy worlds. These include Titan, comets, and other small, ice-rich outer solar system objects.

Titan is unique in that it has a liquid on its surface that is not water, but liquid methane. It would appear to be a very different medium than liquid water due to both the low temperature (−180°C) and its nonpolar nature. Spec-

ulations about life in liquid methane (Benner et al., 2004; Smith and McKay, 2005; Schulze-Makuch and Grinspoon, 2005) are not well supported, and there are no terrestrial analogs that can be usefully applied. While there are organisms known on Earth that can live in pure hydrocarbon environments, they do so by producing water and sequestering it within themselves (Marcano et al., 2002). There is no relevant analog for life at temperatures as low as the surface of Titan.

Comets provide an interesting possibility for life in that the interior of large comets may heat up enough to form liquid water in the interior (Wallis, 1980; Irvine et al., 1980). The small, icy moons of the outer solar system may also have had an initial period with liquid water present. Thus, life may be present in many worlds and in many forms.

A key question that arises in the case of life in comets or in the early phases of small, icy moons is how long life can remain viable in a frozen, dormant state. The long-term viability of microorganisms in permafrost has been considered for Earth and Mars (Smith and McKay, 2005) and is limited by two processes: thermal decay and radiation. On Mars, and in the outer solar system, the time spent frozen may be as much as three billion to four billion years, much longer than the age of the oldest permafrost on Earth. But the temperatures on Mars are also much lower, <−70°C, so thermal decay would not limit the long-term survival of life in permafrost (Kanavarioti and Mancinelli, 1990; Bada and McDonald, 1995). And in the outer solar system the temperatures are lower still. Low-level radioactivity from U, Th, and K in permafrost in Siberia is equivalent to 2 mGy/year (Smith and McKay, 2005), causing a dose sufficient to sterilize bulk soils in 100 Ma. Concentrations of U, Th, and K in the Martian soil are expected to be similar to the values for Earth soils based on the Martian meteorites (Stoker et al., 1993) and Odyssey measurements. While radiation might cause sufficient damage to frozen microorganisms to kill them, it would not destroy all their biomolecules. Therefore, organisms frozen in Martian permafrost could be used for biochemical and genetic analysis. On Europa the radiation doses are much higher and would be almost instantly lethal (Cooper et al., 2001; McKay, 2002), and could be high enough to affect the preservation of the biomolecules themselves on time scales as short as 10 Ma (Cooper et al., 2001; Hand et al., 2010).

CONCLUSION

The likely targets in the search for life on other worlds in our solar system are all cold, icy worlds. Thus, it is not surprising that studies of cold region microbiology are very relevant to astrobiology. There are four ways in which polar studies inform our search: (i) showing how water, salts, and ice interact to create habitable environments when mean temperatures are below freezing; (ii) determining the rates and processes that allow microbial life to grow at temperatures below 0°C; (iii) investigating how long microorganisms can remain dormant in the frozen state and how the cumulative effects of temperature and radiation limit viability; and (iv) determining the long-term effect of temperature and radiation on frozen biological material and the total time and dose before the biological origin of the organic material is no longer preserved in the record. Addressing these questions will help prepare us to search for life on Mars, Europa, Enceladus, and beyond.

REFERENCES

Acuña, M. H., J. E. P. Connerney, N. F. Ness, R. P. Lin, D. Mitchell, C. W. Carlson, J. McFadden, K. A. Anderson, H. Rème, C. Mazelle, D. Vignes, P. Wasilewski, and P. Cloutier. 1999. Global distribution of crustal magnetism discovered by the Mars Global Surveyor MAG/ER experiment. *Science* **284**:790–793.

Amato, P., and B. C. Christner. 2009. Energy metabolism response to low-temperature and frozen conditions in *Psychrobacter cryohalolentis*. *Appl. Environ. Microbiol.* **75**:711–718.

Andersen, D. T., W. H. Pollard, C. P. McKay, and J. Heldmann. 2002. Cold springs in permafrost on Earth and Mars. *J. Geophys. Res.* **107**:1–7.

Atreya, S. K., P. R. Mahaffy, and A.-S. Wong. 2007. Methane and related trace species on Mars: origin, loss, implications for life, and habitability. *Planet Space Sci.* **55**:358–369.

Bada, J. L., and G. D. McDonald. 1995. Amino acid racemization on Mars: implications for the preservation of biomolecules from an extinct Martian biota. *Icarus* **114**:139–143.

Baird, A. K., P. Toulmin III, B. C. Clark, H. J. Rose Jr., K. Keil, R. P. Christian, and J. L. Gooding. 1976. Mineralogic and petrologic implications of Viking geochemical results from Mars: interim report. *Science* **194**:1288–1293.

Bakermans, C. Psychrophiles: life in the cold. *In* R. P. Anitori (ed.), *Extremophiles: Microbiology and Biotechnology*, in press. Horizon Scientific Press, Hethersett, United Kingdom.

Barlow, N. 1997. Mars: impact craters, p. 196–202. *In* J. H. Shirley and R. W. Fairbridge (ed.), *Encyclopedia of Planetary Sciences*. Chapman and Hall, London, United Kingdom.

Benner, S. A., A. Ricardo, and M. A. Carrigan. 2004. Is there a common chemical model for life in the universe? *Curr. Opin. Chem. Biol.* **8**:672–689.

Boston, P. J., M. V. Ivanov, and C. P. McKay. 1992. On the possibility of chemosynthetic ecosystems in subsurface habitats on Mars. *Icarus* **95**:300–308.

Boynton, W. V., W. C. Feldman, S. W. Squyres, T. H. Prettyman, J. Brückner, L. G. Evans, R. C. Reedy, R. Starr, J. R. Arnold, D. M. Drake, P. A. J. Englert, A. E. Metzger, I. Mitrofanov, J. I. Trombka, C. d'Uston, H. Wänke, O. Gasnault, D. K. Hamara, D. M. Janes, R. L. Marcialis, S. Maurice, I. Mikheeva, G. J. Taylor, R. Tokar, and C. Shinohara. 2002. Distribution of hydrogen in the near surface of Mars: evidence for subsurface ice deposits. *Science* **297**:81–85.

Breezee, J., N. Cady, and J. T. Staley. 2004. Subfreezing growth of the sea ice bacterium "*Psychromonas ingrahamii*." *Microb. Ecol.* **47**:300–304.

Catling, D. C., M. W. Claire, K. J. Zahnle, R. C. Quinn, B. C. Clark, M. H. Hecht, and S. Kounaves. 2010. Atmospheric origins of perchlorate on Mars and in the Atacama. *J. Geophys. Res.* **115**:E00E11.

Chyba, C. F., and C. B. Phillips. 2001. Possible ecosystems and the search for life on Europa. *Proc. Natl. Acad. Sci. USA* **98**:801–804.

Clark, B. E., A. K. Baird, H. J. Rose Jr., P. Toulmin III, K. Keil, A. J. Castro, W. C. Kelliher, C. D. Rowe, and P. H. Evans. 1976. Inorganic analyses of Martian surface samples at the Viking landing sites. *Science* **194**:1283–1288.

Clow, G. D., C. P. McKay, G. M. Simmons Jr., and R. A. Wharton Jr. 1988. Climatological observations and predicted sublimation rates at Lake Hoare, Antarctica. *J. Climate* **1**:715–728.

Coates, J. D. 2009. The possibility of methane oxidation coupled to microbial perchlorate metabolism. In Abstracts of Workshop on Methane on Mars: Current Observations, Interpretation and Future Plans. European Space Agency ESRIN, Frascati, Italy.

Coates, J. D., and L. A. Achenbach. 2004. Microbial perchlorate reduction: rocket-fueled metabolism. Nat. Rev. Microbiol. 27:569–580.

Cooper, J. F., R. E. Johnson, B. H. Mauk, H. B. Garrett, and N. Gehrels. 2001. Energetic ion and electron irradiation of the icy Galilean satellites. Icarus 149:133–159.

Costard, F., F. Forget, N. Mangold, and J. P. Peulvast. 2002. Formation of recent Martian debris flows by melting of near-surface ground ice at high obliquity. Science 295:110–113.

Davis, W. L., and C. P. McKay. 1996. Origins of life: a comparison of theories and application to Mars. Orig. Life Evol. Biosph. 26:61–73.

Feldman, W. C., W. V. Boynton, R. L. Tokar, T. H. Prettyman, O. Gasnault, S. W. Squyres, R. C. Elphic, D. J. Lawrence, S. L. Lawson, S. Maurice, G. W. McKinney, K. R. Moore, and R. C. Reedy. 2002. Global distribution of neutrons from Mars: results from Mars Odyssey. Science 297:75–78.

Formisano, V., S. K. Atreya, N. Ignatiev, and M. Giuranna. 2004. Detection of methane in the atmosphere of Mars. Science 306:1758–1761.

Geminale, A., V. Formisano, and G. Sindoni. 2010. Mapping methane in Martian atmosphere with PFS-MEX data. Planet. Space Sci. doi:10.1016/j.pss.2010.07.011.

Grasby, S. E., and K. L. Londry. 2007. Biogeochemistry of hypersaline springs supporting a mid-continent marine ecosystem: an analogue for Martian springs? Astrobiology 7:662–683.

Greenberg, R. 2010. Transport rates of radiolytic substances into Europa's ocean: implications for the potential origin and maintenance of life. Astrobiology 10:275–283.

Hand, E. 2008. Plumes of methane identified on Mars. Nature 455:1018.

Hand, K. P., C. P. McKay, and C. Pilcher. 2010. Spectroscopic and spectrometric differentiation between abiotic and biogenic material on icy worlds. Proc. Int. Astron. Union 6:165–176.

Hecht, M. H., S. P. Kounaves, R. C. Quinn, S. J. West, S. M. M. Young, D. W. Ming, D. C. Catling, B. C. Clark, W. V. Boynton, J. Hoffman, L. P. DeFlores, K. Gospodinova, J. Kapit, and P. H. Smith. 2009. Detection of perchlorate and the soluble chemistry of Martian soil: findings from the Phoenix Mars Lander. Science 325:64–67.

Irvine, W. M., S. B. Leschine, and F. P. Schloerb. 1980. Thermal history, chemical composition and relationship of comets to the origin of life. Nature 283:748–749.

Jakosky, B. M., K. H. Nealson, C. Bakermans, R. E. Ley, and M. T. Mellon. 2003. Subfreezing activity of microorganisms and the potential habitability of Mars' polar region. Astrobiology 3:343–350.

Kanavarioti, A., and R. L. Mancinelli. 1990. Could organic matter have been preserved on Mars for 3.5 billion years? Icarus 84:196–202.

Keller, J. M., W. V. Boynton, S. Karunatillake, V. R. Baker, J. M. Dohm, L. G. Evans, M. J. Finch, B. C. Hahn, D. K. Hamara, D. M. Janes, K. E. Kerry, H. E. Newsom, R. C. Reedy, A. L. Sprague, S. W. Squyres, R. D. Starr, G. J. Taylor, and R. M. S. Williams. 2007. Equatorial and midlatitude distribution of chlorine measured by Mars Odyssey GRS. J. Geophys. Res. 111:E03S08.

Kivelson, M. G., K. K. Khurana, C. T. Russell, M. Volwerk, R. J. Walker, and C. Zimmer. 2000. Galileo magnetometer measurements: a stronger case for a subsurface ocean at Europa. Science 289:1340–1343.

Laskar, J., B. Levrard, and J. F. Mustard. 2002. Orbital forcing of the Martian polar layered deposits. Nature 419:375–377.

Lefèvre, F., and F. Forget. 2009. Observed variations of methane on Mars unexplained by known atmospheric chemistry and physics. Nature 460:720–723.

Malin, M. C., and K. S. Edgett. 2000. Evidence for recent groundwater seepage and surface runoff on Mars. Science 288:2330–2335.

Malin, M. C., K. S. Edgett, L. V. Posiolova, S. M. McColley, and E. Z. Dobrea. 2006. Present-day impact cratering rate and contemporary gully activity on Mars. Science 314:1573–1577.

Mangold, N., A. Gendrin, B. Gondet, S. LeMouelic, C. Quantin, V. Ansan, J.-P. Bibring, Y. Langevin, P. Masson, and G. Neukum. 2008. Spectral and geological study of the sulfate-rich region of West Candor Chasma, Mars. Icarus 194:519–543.

Marcano, V., P. Benitez, and E. Palacios-Prü. 2002. Growth of a lower eukaryote in nonaromatic hydrocarbon media ≥ C_{12} and its exobiological significance. Planet. Space Sci. 50:693–709.

McCollom, T. M. 1999. Methanogenesis as a potential source of chemical energy for primary biomass production by autotrophic organisms in hydrothermal systems on Europa. J. Geophys. Res. 104:30729–30742.

McKay, C. P. 1998. The search for extraterrestrial biochemistry on Mars and Europa, p. 219–227. In J. Chela-Flores and F. Raulin (ed.), Exobiology: Matter, Energy, and Information in the Origin and

Evolution of Life in the Universe. Kluwer Academic Publishers, Dordrecht, The Netherlands.

McKay, C. P. 2001. The search for a second genesis of life in our Solar System, p. 269–277. *In* J. Chela-Flores, T. Owen, and F. Raulin (ed.), *First Steps in the Origin of Life in the Universe.* Kluwer Academic Publishers, Dordrecht, The Netherlands.

McKay, C. P. 2002. Planetary protection for a Europa surface sample return: the Ice Clipper Mission. *Adv. Space Res.* **30:**1601–1605.

McKay, C. P. 2009. Snow recurrence sets the depth of dry permafrost at high elevations in the Dry Valleys of Antarctica. *Antarct. Sci.* **21:**89–94.

McKay, C. P., M. T. Mellon, and E. I. Friedmann. 1998. Soil temperatures and stability of ice-cemented ground in the McMurdo Dry Valleys, Antarctica. *Antarct. Sci.* **10:**31–38.

McKay, C. P., C. C. Porco, T. Altheide. W. L. Davis, and T. A. Kral. 2008. The possible origin and persistence of life on Enceladus and detection of biomarkers in the plume. *Astrobiology* **8:**909–919.

Mellon, M. T., R. E. Arvidson, H. G. Sizemore, M. L. Searls, D. L. Blaney, S. Cull, M. H. Hecht, T. L. Heet, H. U. Keller, M. T. Lemmon, W. J. Markiewicz, D. W. Ming, R. V. Morris, W. T. Pike, and A. P. Zent. 2009. Ground ice at the Phoenix landing site: stability state and origin. *J. Geophys. Res.* **114:**E00E07.

Mellon, M. T., and B. M. Jakosky. 1995. The distribution and behavior of Martian ground ice during past and present epochs. *J. Geophys. Res.* **100:**11781–11799.

Mumma, M. J., G. L. Villanueva, R. E. Novak, T. Hewagama, B. P. Bonev, M. A. DiSanti, A. M. Mandell, and M. D. Smith. 2009. Strong release of methane on Mars in northern summer 2003. *Science* **323:**1041–1045.

Navarro-González, R., E. Vargas, J. de la Rosa, A. C. Raga, and C. P. McKay. 2010. Reanalysis of the Viking results suggests perchlorate and organics at mid-latitudes on Mars. *J. Geophys. Res.* **115:**E12010.

Niederberger, T. D., N. N. Perreault, S. Tille, B. S. Lollar, G. Lacrampe-Couloume, D. Andersen, C. W. Greer, W. Pollard, and L. G. Whyte. 2010. Microbial characterization of a subzero, hypersaline methane seep in the Canadian High Arctic. *ISME J.* **4:**1326–1339.

Ono, S., B. Wing, D. Johnston, D. Rumble, and J. Farquhar. 2006. Mass-dependent fractionation of quadruple stable sulfur isotope system as a new tracer of sulfur biogeochemical cycles. *Geochim. Cosmochim. Acta* **70:**2238–2252.

Onstott, T. C., D. McGown, J. Kessler, B. S. Lollar, K. K. Lehmann, and S. M. Clifford. 2006. Martian CH_4: sources, flux and detection. *Astrobiology* **6:**377–395.

Panikov, N. S., P. W. Flanagan, W. C. Oechel, M. A. Mastepanov, and T. R. Christensen. 2006. Microbial activity in soils frozen to below −39°C. *Soil Biol. Biochem.* **38:**785–794.

Pappalardo, R. T., M. J. S. Belton, H. H. Breneman, M. H. Carr, C. R. Chapman, G. C. Collins, T. Denk, S. Fagents, P. E. Geissler, B. Giese, R. Greeley, R. Greenberg, J. W. Head, P. Helfenstein, G. Hoppa, S. D. Kadel, K. P. Klaasen, J. E. Klemaszewski, K. Magee, J. M. Moore, W. B. Moore, G. Neukum, C. B. Phillips, L. M. Prockter, G. Schubert, D. A. Senske, R. J. Sullivan, B. R. Tufts, E. P. Turtle, R. Wagner, and K. K. Williams. 1999. Does Europa have a subsurface ocean? Evaluation of the geological evidence. *J. Geophys. Res.* **104:**24015–24055.

Porco, C. C., P. Helfenstein, P. C. Thomas, A. P. Ingersoll, J. Wisdom, R. West, G. Neukum, T. Denk, R. Wagner, T. Roatsch, S. Kieffer, E. Turtle, A. McEwen, T. V. Johnson, J. Rathbun, J. Veverka, D. Wilson, J. Perry, J. Spitale, A. Brahic, J. A. Burns, A. D. DelGenio, L. Dones, C. D. Murray, and S. Squyres. 2006. Cassini observes the active south pole of Enceladus. *Science* **311:**1393–1401.

Reider, R., R. Gellert, R. C. Anderson, J. Brückner, B. C. Clark, G. Dreibus, T. Economou, G. Klingelhöfer, G. W. Lugmair, D. W. Ming, S. W. Squyres, C. d'Uston, H. Wänke, A. Yen, and J. Zipfel. 2004. Chemistry of rocks and soils at Meridiani Planum from the Alpha Particle X-ray Spectrometer. *Science* **306:**1746–1749.

Rennó, N. O., B. J. Bos, D. Catling, B. C. Clark, L. Drube, D. Fisher, W. Goetz, S. F. Hviid, H. U. Keller, J. F. Kok, S. P. Kounaves, K. Leer, M. Lemmon, M. B. Madsen, W. J. Markiewicz, J. Marshall, C. McKay, M. Mehta, M. Smith, M. P. Zorzano, P. H. Smith, C. Stoker, and S. M. M. Young. 2009. Possible physical and thermodynamical evidence for liquid water at the Phoenix landing site, *J. Geophys. Res.* **114:**E00E03.

Richardson, M. I., and M. A. Mischna. 2005. Long-term evolution of transient liquid water on Mars. *J. Geophys. Res.* **110:**E03003.

Rivkina, E. M., E. I. Friedmann, C. P. McKay, and D. A. Gilichinsky. 2000. Metabolic activity of permafrost bacteria below the freezing point. *Appl. Environ. Microbiol.* **66:**3230–3233.

Rossi, A. P., G. Neukum, M. Pondrelli, S. van Gasselt, T. Zegers, E. Hauber, A. Chicarro, and B. Foing. 2008. Large-scale spring deposits on Mars? *J. Geophys. Res.* **113:**E08016.

Schulze-Makuch, D., and D. H. Grinspoon. 2005. Biologically enhanced energy and carbon cycling on Titan? *Astrobiology* **5:**560–567.

Smith, H. D., and C. P. McKay. 2005. Drilling in ancient permafrost on Mars for evidence of a second genesis of life. *Planet. Space Sci.* 53:1302–1308.

Smith, P. H., L. K. Tamppari, R. E. Arvidson, D. Bass, D. Blaney, W. V. Boynton, A. Carswell, D. C. Catling, B. C. Clark, T. Duck, E. De-Jong, D. Fisher, W. Goetz, H. P. Gunnlaugsson, M. H. Hecht, V. Hipkin, J. Hoffman, S. F. Hviid, H. U. Keller, S. P. Kounaves, C. F. Lange, M. T. Lemmon, M. B. Madsen, W. J. Markiewicz, J. Marshall, C. P. McKay, M. T. Mellon, D. W. Ming, R. V. Morris, W. T. Pike, N. Rennó, U. Staufer, C. Stoker, P. Taylor, J. A. Whiteway, and A. P. Zent. 2009. H₂O at the Phoenix landing site. *Science* 325:58–61.

Squyres, S. W., and M. H. Carr. 1986. Geomorphic evidence for the distribution of ground ice on Mars. *Science* 231:249–252.

Steven, B., W. H. Pollard, C. W. Greer, and L. G. Whyte. 2008. Microbial diversity and activity through a permafrost/ground ice core profile from the Canadian high Arctic. *Environ. Microbiol.* 10:3388–3403.

Stofan, E. R., C. Elachi, J. I. Lunine, R. D. Lorenz, B. Stiles, K. L. Mitchell, S. Ostro, L. Soderblom, C. Wood, H. Zebker, S. Wall, M. Janssen, R. Kirk, R. Lopes, F. Paganelli, J. Radebaugh, L. Wye, Y. Anderson, M. Allison, R. Boehmer, P. Callahan, P. Encrenaz, E. Flamini, G. Francescetti, Y. Gim, G. Hamilton, S. Hensley, W. T. K. Johnson, K. Kelleher, D. Muhleman, P. Paillou, G. Picardi, F. Posa, L. Roth, R. Seu, S. Shaffer, S. Vetrella, and R. West. 2007. The lakes of Titan. *Nature* 445:61–64.

Stoker, C. R., J. L. Gooding, T. Roush, A. Banin, D. Burt, B. C. Clark, G. Flynn, and O. Gwynne. 1993. The physical and chemical properties and resource potential of Martian surface soils, p. 659–707. *In* J. Lewis, M. S. Matthews, and M. L. Guerrieri (ed.), *Resources of Near-Earth Space*. University of Arizona Press, Tucson, AZ.

Stoker, C. R., A. Zent, D. C. Catling, S. Douglas, J. R. Marshall, D. Archer, B. Clark, S. P. Kounaves, M. T. Lemmon, R. Quinn, N. Rennó, P. H. Smith, and S. M. M. Young. 2010. Habitability of the Phoenix landing site. *J. Geophys. Res.* 115: E00E20.

Tamppari, L. K., R. M. Anderson, P. D. Archer Jr., S. Douglas, S. P. Kounaves, C. P. McKay, D. W. Ming, Q. Moore, J. E. Quinn, P. H. Smith, S. Stroble, and A. P. Zent. Effects of aridity on soils and habitability: McMurdo Dry Valleys as an analog for the Mars Phoenix landing site. *Antarct. Sci.*, in press.

Waite, J. H., Jr., M. R. Combi, W.-H. Ip, T. E. Cravens, R. L. McNutt Jr., W. Kasprzak, R. Yelle, J. Luhmann, H. Niemann, D. Gell, B. Magee, G. Fletcher, J. Lunine, and W.-L. Tseng. 2006. Cassini ion and neutral mass spectrometer: Enceladus plume composition and structure. *Science* 311:1419–1422.

Waite, J. H., Jr., W. S. Lewis, B. A. Magee, J. I. Lunine, W. B. McKinnon, C. R. Glein, O. Mousis, D. T. Young, T. Brockwell, J. Westlake, M.-J. Nguyen, B. D. Teolis, H. B. Niemann, R. L. McNutt Jr., M. Perry, and W.-H. Ip. 2009. Liquid water on Enceladus from observations of ammonia and ⁴⁰Ar in the plume. *Nature* 460:487–490.

Wallis, M. K. 1980. Radiogenic melting of primordial comet interiors. *Nature* 284:431–433.

Weast, R. C. (ed.). 1976. *CRC Handbook of Chemistry and Physics*, 57th ed. CRC Press, Cleveland, OH.

Weiss, B. P., J. L. Kirschvink, F. J. Baudenbacher, H. Vali, N. T. Peters, F. A. Macdonald, and J. P. Wikswo. 2000. A low temperature transfer of ALH84001 from Mars to Earth. *Science* 290:791–795.

Zahnle, K., R. S. Freedman, and D. C. Catling. 2011. Is there methane on Mars? *Icarus* 212: 493–503.

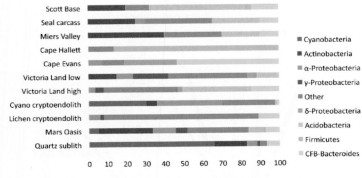

Scott Base
Seal carcass
Miers Valley
Cape Hallett
Cape Evans
Victoria Land low
Victoria Land high
Cyano cryptoendolith
Lichen cryptoendolith
Mars Oasis
Quartz sublith

0 10 20 30 40 50 60 70 80 90 100

Abundance in 16S rRNA gene clone library (%)

- Cyanobacteria
- Actinobacteria
- α-Proteobacteria
- γ-Proteobacteria
- Other
- δ-Proteobacteria
- Acidobacteria
- Firmicutes
- CFB-Bacteroides

COLOR PLATE 1 (CHAPTER 1)
Prokaryotic diversity found in Antarctic terrestrial soils. Key: CFB-Bacteroides, *Cytophaga-Flexibacter-Bacteroidetes*. "Other" taxa include *Gemmatimonas, Planctomycetales*, β-*Proteobacteria, Thermus-Deinococcus*, and *Verrucomicrobia*. (References: Scott Base and Cape Hallett, Aislabie et al. [2008]; Seal carcass and Miers Valley, Smith et al. [2006]; Cape Evans, Shravage et al. [2007]; Victoria Land high and low, Niederberg et al. [2008]; Cyano cryptoendolith and Lichen cryptoendolith, de La Torre et al. [2003]; Mars Oasis, Newsham et al. [2009]; Quartz sublith, Smith et al. [2000].)

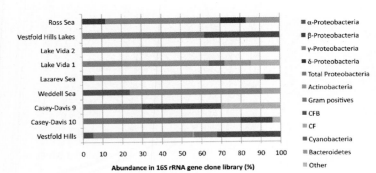

Ross Sea
Vestfold Hills Lakes
Lake Vida 2
Lake Vida 1
Lazarev Sea
Weddell Sea
Casey-Davis 9
Casey-Davis 10
Vestfold Hills

0 10 20 30 40 50 60 70 80 90 100

Abundance in 16S rRNA gene clone library (%)

- α-Proteobacteria
- β-Proteobacteria
- γ-Proteobacteria
- δ-Proteobacteria
- Total Proteobacteria
- Actinobacteria
- Gram positives
- CFB
- CF
- Cyanobacteria
- Bacteroidetes
- Other

COLOR PLATE 2 (CHAPTER 1)
Prokaryotic diversity in Antarctic aquatic habitats as assessed by metagenomic 16S rRNA gene clone libraries. Key: CFB, *Cytophaga-Flexibacter-Bacteroidetes*; CF, *Cytophaga-Flavobacterium*. (References: Ross Sea, Gentile et al. [2006]; Vestfold Hills Lakes, Van Trappen et al. [2002]; Lake Vida 2, Mondino et al. [2009]; Lake Vida 1, Mosier et al. [2006]; Lazarev Sea and Weddell Sea, Brinkmeyer et al. [2003]; Casey-Davis 9 and 10, Brown and Bowman [2001]; Vestfold Hills, Bowman et al. [1997].)

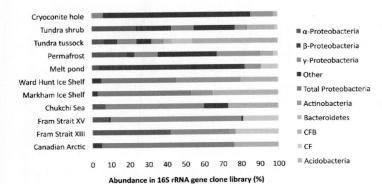

Cryoconite hole
Tundra shrub
Tundra tussock
Permafrost
Melt pond
Ward Hunt Ice Shelf
Markham Ice Shelf
Chukchi Sea
Fram Strait XV
Fram Strait XIII
Canadian Arctic

0 10 20 30 40 50 60 70 80 90 100

Abundance in 16S rRNA gene clone library (%)

- α-Proteobacteria
- β-Proteobacteria
- γ-Proteobacteria
- Other
- Total Proteobacteria
- Actinobacteria
- Bacteroidetes
- CFB
- CF
- Acidobacteria

COLOR PLATE 3 (CHAPTER 1)
Prokaryotic diversity in Arctic terrestrial and aquatic habitats as determined by culture-independent 16S rRNA gene clone libraries. Key: CFB, *Cytophaga-Flexibacter-Bacteroidetes*; CF, *Cytophaga-Flavobacterium*. "Other" taxa include *Cyanobacteria, Firmicutes, Gemmatimonas*, gram-positive bacteria, novel taxa, *Verrucomicrobia*, and unidentified ribotypes. (References: Cryoconite hole, Christner et al. [2003]; Tundra shrub and Tundra tussock, Wallenstein et al. [2007]; Permafrost, Steven et al. [2007]; Melt pond, Brinkmeyer et al. [2004]; Ward Hunt Ice Shelf and Markham Ice Shelf, Bottos et al. [2008]; Chukchi Sea, Junge et al. [2002]; Fram Strait XV and XIII, Brinkmeyer et al. [2003]; Canadian Arctic, Brown and Bowman [2001].)

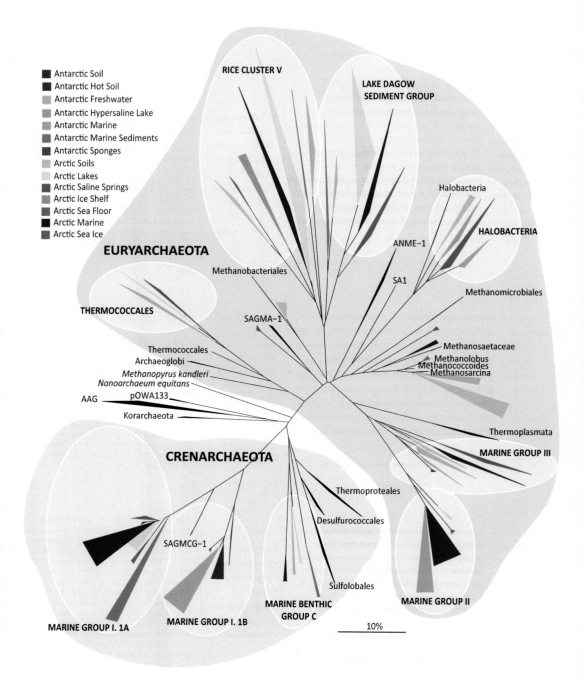

COLOR PLATE 4 (CHAPTER 2) Comparison of the phylogenetic diversity of archaeal 16S rRNA gene sequences from the Arctic and Antarctic. The sequences used for this analysis are from studies of Arctic and Antarctic terrestrial and aquatic environments (described in Table 3). A total of 1,392 sequences were included in the analysis and the tree was constructed using ARB, with DNADIST and neighbor-joining analysis. Polar archaeal sequences are found in three groups in the *Crenarchaeota* (Marine Group I [MGI] 1.1a, MGI 1.1b, and Marine Benthic Group C) and eight groups in the *Euryarchaeota* (*Thermococcales*, *Halobacteria*, MGII and MGIII, methanogens, SAGMA-1, Rice Cluster V [RC-V], and Lake Dagow Sediment [LDS] group). In several studies of *Archaea* the RC-V and LDS groups have been part of the PENDANT-33 (Schleper et al., 2005) or the "Sediment Archaea group." The largest groups of sequences in the analysis were MGI 1.1a (44%), "Sediment Archaea group" (16%), MGII (12%), and MGI 1.1b (8%), with almost 54% of sequences in the *Crenarchaeota*.

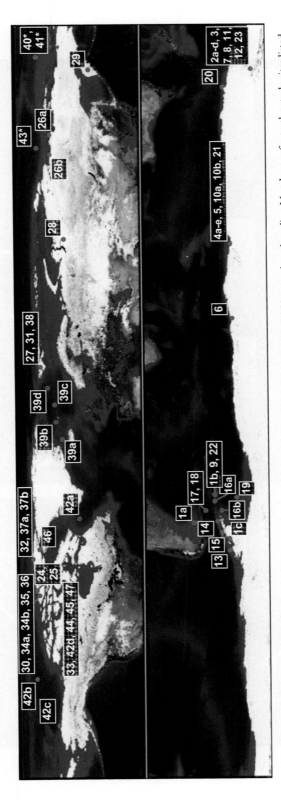

COLOR PLATE 5 (CHAPTER 2) Polar sites surveyed for the presence of *Archaea* via 16S rRNA gene-based studies. Numbers refer to the study sites listed in Table 3. *, consists of various sample sites located within the vicinity of the indicated region.

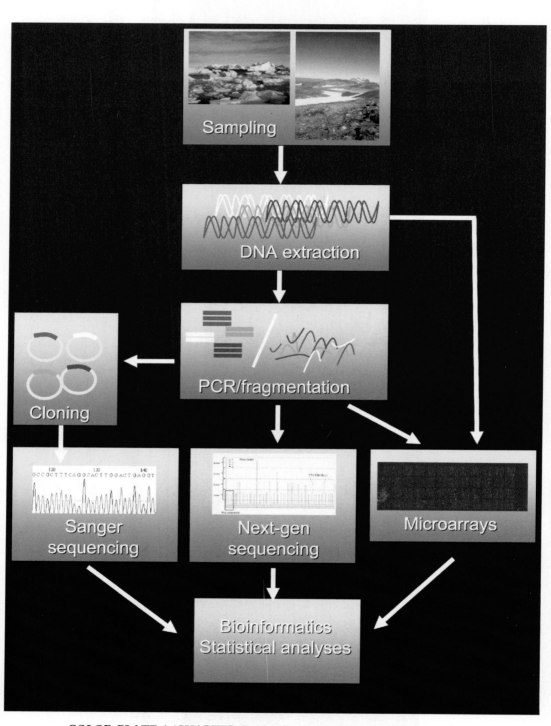

COLOR PLATE 6 (CHAPTER 7) Methods available for metagenomic analyses.

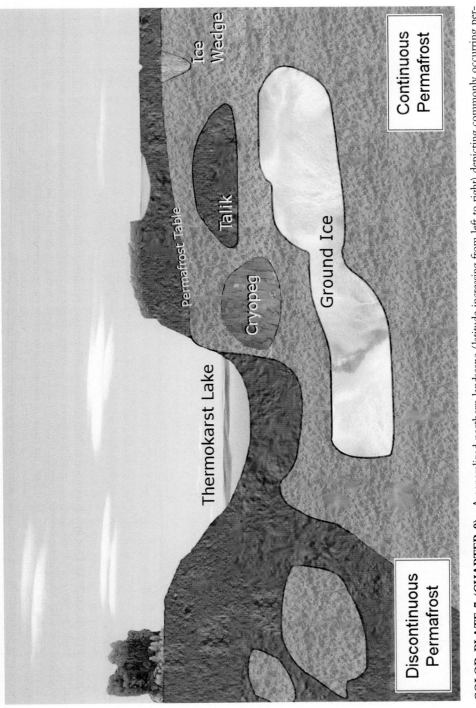

COLOR PLATE 7 (CHAPTER 9) A generalized northern landscape (latitude increasing from left to right) depicting commonly occurring permafrost features. Position and scale of features are approximated, since the size of taliks, cryopegs, and ground ice varies greatly. The general depth of the active layer varies from <0.5 m to between 2 and 3 m deep in the sub-Arctic, while permafrost appears at a depth of 0.5 m and extends upwards of 1.5 km. Thermokarst features result from thawing permafrost, where pore space once occupied by ice subsides and creates surface depressions and melt-water pools. Taliks are associated with thermokarst features and other water-dominated features.

The image contains the following labels: Thermokarst Lake, Permafrost Table, Talik, Cryopeg, Ground Ice, Ice Wedge, Discontinuous Permafrost, Continuous Permafrost.

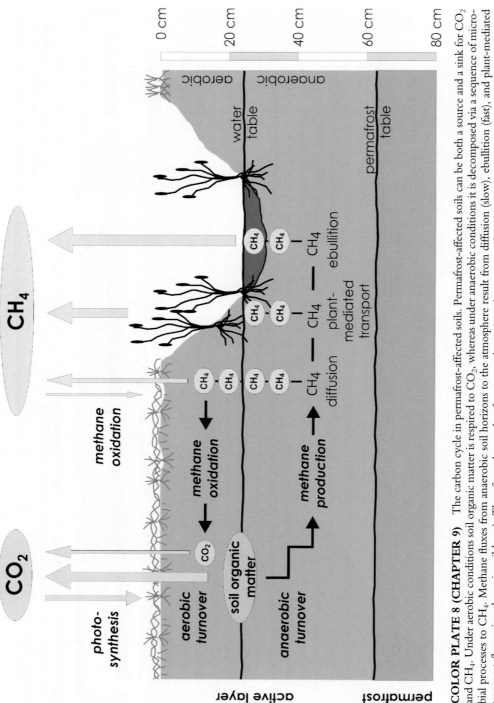

COLOR PLATE 8 (CHAPTER 9) The carbon cycle in permafrost-affected soils. Permafrost-affected soils can be both a source and a sink for CO_2 and CH_4. Under aerobic conditions soil organic matter is respired to CO_2, whereas under anaerobic conditions it is decomposed via a sequence of microbial processes to CH_4. Methane fluxes from anaerobic soil horizons to the atmosphere result from diffusion (slow), ebullition (fast), and plant-mediated transport (bypassing the oxic soil layer). Therefore, the mode of transport determines the amount of methane that is reoxidized by microorganisms in aerobic soil horizons. Photosynthesis may function as an important sink for CO_2 in permafrost environments whereby biomass is produced. In contrast, the consumption of atmospheric methane (negative methane flux) in the upper surface layer of the soils plays only a minor role for the methane budget. The thickness of the arrows reflects the importance of the above processes. (Modified according to Wagner and Liebner [2009].)

a.

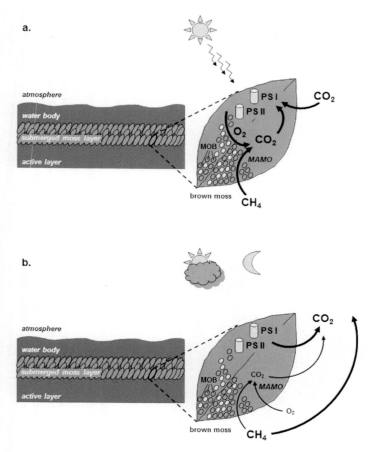

COLOR PLATE 9 (CHAPTER 9) Scheme illustrating methanotrophy associated with submerged mosses (MAMO). (a) In the presence of light, methane oxidizers (MOB) associated with the moss plant use O_2 produced via photosynthesis (PS = photosystem) to convert CH_4 into CO_2. CH_4 derives mainly from methanogenesis within the active layer, but atmospheric CH_4 is also consumed. The CO_2 is directly recycled by the moss through photosynthesis. (b) In the absence of light and when light intensity is low, respectively, MAMO is suppressed through the lack of O_2, leading to a significant increase of CH_4 emissions at the water-atmosphere boundary. (Modified according to Liebner et al. [2011].)

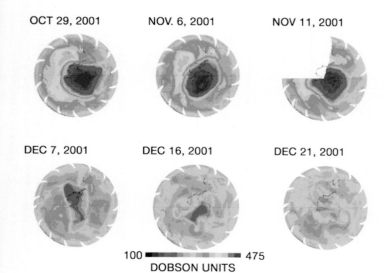

OCT 29, 2001   NOV. 6, 2001   NOV 11, 2001

DEC 7, 2001   DEC 16, 2001   DEC 21, 2001

100 ▬▬▬▬ 475
DOBSON UNITS

COLOR PLATE 10 (CHAPTER 13) TOMS project (http://jwocky.gsfc.nasa.gov/eptoms/ep.html) images of stratospheric ozone levels in Antarctica during October to December 2001 showing the reduction in the ozone hole that occurred during this period.

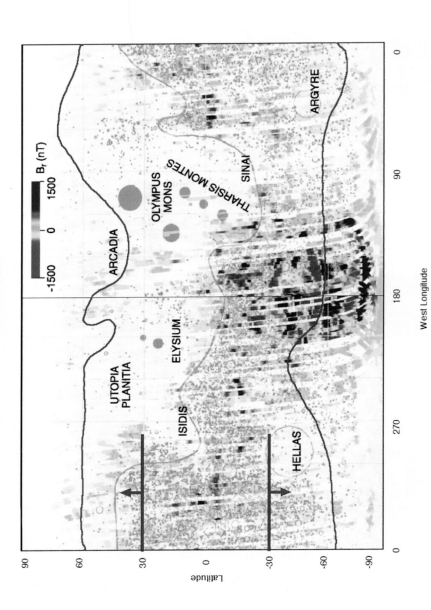

COLOR PLATE 11 (CHAPTER 14) Map showing crater distribution, ground ice, and crustal magnetism on Mars. The suggested target site for deep drilling to search for evidence of ancient life on Mars is the region between 60 and 80°S at 180°W, where the ground is heavily cratered, crustal magnetism is preserved, and ground ice is present. Each green circle represents a crater with a diameter greater than 15 km based on the crater distribution in Barlow (1997). The filled green circles are volcanic craters. The boundary between the smooth northern plains and the cratered southern highlands is shown with a green line. The southern regions of Mars are more heavily cratered and therefore considered to be older. The solid blue lines show the extent of near-surface ground ice as determined by the Odyssey mission (Feldman et al., 2002). Ground ice is present near the surface polarward of these lines. Crater morphology indicates deep ground ice poleward of 30° (Squyres and Carr, 1986), shown here by dark blue lines and arrows. Also shown in this figure is the crustal magnetism discovered by Acuña et al. (1999). The crustal magnetism is shown as red for positive and blue for negative. Full scale is 1,500 nT. The typical strength of Earth's magnetic field at the surface is 50,000 nT. (Reprinted from Smith and McKay [2005] with permission of the publisher.)

INDEX